AERÖS Conference 2017

IOP Conference Series: Materials Science and Engineering
Volume 270

Putrajaya, Malaysia
12 December 2017

ISBN: 978-1-5108-5336-2
ISSN: 1757-8981

Printed from e-media with permission by:

Curran Associates, Inc.
57 Morehouse Lane
Red Hook, NY 12571

Some format issues inherent in the e-media version may also appear in this print version.

Printed by Curran Associates, Inc. (2018)

For permission requests, please contact the Institute of Physics
at the address below.

Institute of Physics
Dirac House, Temple Back
Bristol BS1 6BE UK

Phone: 44 1 17 929 7481
Fax: 44 1 17 920 0979

techtracking@iop.org

Additional copies of this publication are available from:

Curran Associates, Inc.
57 Morehouse Lane
Red Hook, NY 12571 USA
Phone: 845-758-0400
Fax: 845-758-2633
Email: curran@proceedings.com
Web: www.proceedings.com

Table of contents

Volume 270

AEROS Conference 2017
12 December 2017, Putrajaya, Malaysia

Accepted papers received: 14 November 2017
Published online: 1 December 2017

Preface

Papers

Aerodynamics

Propulsion

Aerospace System Design

Structures and Materials

IOP Conf. Series: Materials Science and Engineering **270** (2017) 011001 doi:10.1088/1757-899X/270/1/011001

PREFACE

In conjunction with the establishment of Aerospace Society Malaysia (AEROS Malaysia) in January 2017, this inaugural edition of the biennial AEROS Conference series is organized as the key platform for local and international aerospace communities to come together and share their recent researches in development and progress of the aerospace field. Moreover, the conference also aims to become the catalyst for potential collaborations between the academia and the aerospace government agencies and industry players, which is one of the main objectives for the establishment of AEROS Malaysia.

This first AEROS Conference is held in Putrajaya, Malaysia on 12th December, 2017. After extensive peer review process, 39 paper submissions have been accepted into the conference and included in the conference proceedings. The selected papers cover a wide range of aerospace topics such as structures and materials, aerodynamics, flight dynamics and control, propulsion, aerospace application and other related areas.

I like thank the conference participants for their contributions and the conference reviewers for their expertise and valuable comments. I also like to thank the conference organizing committee for their commitment and tremendous support to the success of this inaugural AEROS Conference 2017.

Dr. Fairuz Izzuddin Romli
Chairman of AEROS Malaysia
Editor of AEROS Conference 2017 Proceedings

TOWARD GREATER HEIGHTS

IOP Conf. Series: Materials Science and Engineering **270** (2017) 011002 doi:10.1088/1757-899X/270/1/011002

Peer review statement

All papers published in this volume of *IOP Conference Series: Materials Science and Engineering* have been peer reviewed through processes administered by the proceedings Editors. Reviews were conducted by expert referees to the professional and scientific standards expected of a proceedings journal published by IOP Publishing.

IOP Conf. Series: Materials Science and Engineering **270** (2017) 012001 doi:10.1088/1757-899X/270/1/012001

An analysis on 45° sweep tail angle for blended wing body aircraft to the aerodynamics coefficients by wind tunnel experiment

M Z A Abd Latif[1,*], M A Ahmad[1], R E Mohd Nasir[1], W Wisnoe[1] and M R Saad[2]

[1]Flight Technology and Test Center (FTTC), Faculty of Mechanical Engineering, UiTM Shah Alam, Selangor, Malaysia
[2]Universiti Pertahanan Nasional Malaysia (UPNM), Faculty of Engineering, Kem Sungai Besi, Kuala Lumpur, Malaysia

*zulfazli.arief@gmail.com

Abstract. This paper presents the analysis of a model from UiTM Blended Wing Body (BWB) UAV, Baseline V that has been tested at UPNM high speed wind tunnel. Baseline V has a unique design due to different NACA sections used for its fuselage, body, wing root, midwing, wingtip, tail root, tail tip and the tail is swept 45° backward. The purpose of this experiment is to study the aerodynamic characteristics when the tail sweeps 45° backward. The experiments are conducted several times using 71.5% scaled down model at about 49.58 m/s airspeed or 25 Hz. The tail angle deflection is fixed and set at zero angle. All the data obtained is analyzed and presented in terms of coefficient of lift, coefficient of drag and also lift-to-drag ratio, and is plotted against various angles of attack. The angles of attack used for this experiments are between -10° to +30°. The blockage correction such as solid blockage, wake blockage and streamline curvature blockage are calculated in order to obtain true performance of the aircraft. From the observation, Baseline V shows that the aircraft tends to stall at around +15°. The maximum L/D ratio achieved for Baseline V is 20.8, however it decreases slightly to 20.7 after blockage corrections.

1. Introduction

The aviation industry nowadays is facing a serious problem with fuel efficiencies, noise and emission related to the aircraft operation [1]. The idea of blended wing body (BWB) aircraft is fundamentally based on flying wing designs, which are likely to become the configuration of future aircraft designs [2, 3]. BWB has smaller ratio of wetted surface area to volume and has less interference drag due to its smoother overall figure, which gives a good lift-to-drag ratio [4]. However, the flying quality of BWB is poor in some cases due to its tailless design, although BWB has better aerodynamic efficiency with higher lift-to-drag ratio [5, 6]. The horizontal and vertical tails offer stability for an aircraft. Elevator and rudder act as control surfaces and are located at the tail. Some information is needed to determine the configuration of tail surfaces, which include position of the centre of gravity (C.G.) of the aircraft, change in C.G. location throughout the flight and intended level of stability. In order to determine the C.G. of the aircraft, the weight of vertical and horizontal tails must be determined depending on their size [7].

Flight Technology and Test Centre (FTTC) in UiTM Shah Alam has started the research on BWB since 2005, focusing on the aspect of design and fundamental aerodynamics [8]. There are five designs that have already been made and studied, which are referred to as Baseline I, Baseline II, Baseline III, Baseline IV and Baseline V. Table 1 below shows the aerodynamic efficiency of each design, which is measured by its maximum lift-to-drag ratio (L/D_{max}).

Table 1: Aerodynamic efficiency of each BWB design

Year	Design	(L/D_{max})
2005	 Baseline I	14
2009	 Baseline II	24
2012	 Baseline III	13
2013	 Baseline IV	15
2014	 Baseline V	32

The first BWB Baseline-I is designed under UiTM research teams. It has a span of four meter and is equipped with large central elevator for longitudinal control and stability. Baseline-I aircraft has a poor aerodynamics performance with L/D less than 10 [9]. The second BWB aircraft, known as the Baseline-II, is designed based on lessons learned from Baseline-I. Planform shape and twisted angle airfoil along spanwise location as recommended by Bolsunovsky, and also the Inverse-Twist Method are applied in the Baseline-II aircraft design [10]. Baseline-II has a high aerodynamic efficiency with L/D of 24. Next, Baseline-III is designed to further study on aerodynamics efficiency of BWB aircraft. Baseline-III concept is inspired from the shape of flying birds. However, the results show that it has major flaws in its aerodynamics efficiency with its L/D value is only around 12 [11]. Baseline-IV is also a bird-inspired BWB aircraft designed by referring to the existing configuration of Baseline-III aircraft. For this design, the L/D is increased but the value is still not as efficient as to be expected for a BWB aircraft. Baseline V is then designed as shown in Figure 1 and it has underwent a wind tunnel experiment at UTM Skudai with its tail design blended at the side of its body. At elevator deflection angle equals to 0°, the L/D reaches its maximum of 32. The L/D is at its minimum, which is 16, when the deflection angle is at +20°.

This project is focusing on Baseline V design that is equipped with 45° sweep angle of the tail. The experiment has been conducted at Pusat Maritim Universiti Pertahanan Nasional Malaysia. The 45° of sweep tail design is chosen because hypothetically, the tail will provide more stability compared to at 0° angle. The area of the 45° tail is identical to that of the 0° tail. The purpose of the experiment is to

IOP Conf. Series: Materials Science and Engineering **270** (2017) 012001 doi:10.1088/1757-899X/270/1/012001

determine the aerodynamic characteristics for this Baseline V design such as L/D, lift coefficient and drag coefficient.

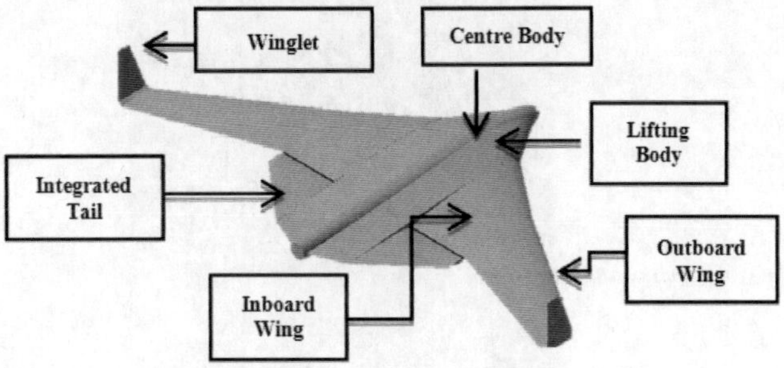

Figure 1: Baseline V configuration

2. Model Fabrication and Wind Tunnel Setup

The actual model wingspan for Baseline V is 1.4m. For this wind tunnel testing, the model has been scaled down to 71.5% from its actual size and only half part of the model is used. The structure of the model is 3-D printed and layered with fibreglass as shown in Figure 2 and Figure 3. The dimension of the model can be seen in Figure 4. The experiments are conducted at Universiti Pertahanan Nasional Malaysia (UPNM) high speed wind tunnel (Figure 5 and Figure 6). It is a suction type wind tunnel with test section area of 0.3m x 0.3m x 1.25m and it is equipped with 6-Component External Balance. The experiments in this study only uses three components, which are lift, drag and moment, and they are carried out at airspeed of 49.58 m/s or 25 Hz with Reynolds number of 7.0×105. The test is conducted at velocity 49.58 m/s to get the same Reynolds number with the actual size of the aircraft, which will be conducted at velocity 15 m/s. In order to achieve accurate results, the Reynolds number for both models must be the same by increasing the velocity for the scaled-down model. The pitching angle (angle of attack) is varied from -10° to +30°. The tail deflection is set at 0°.

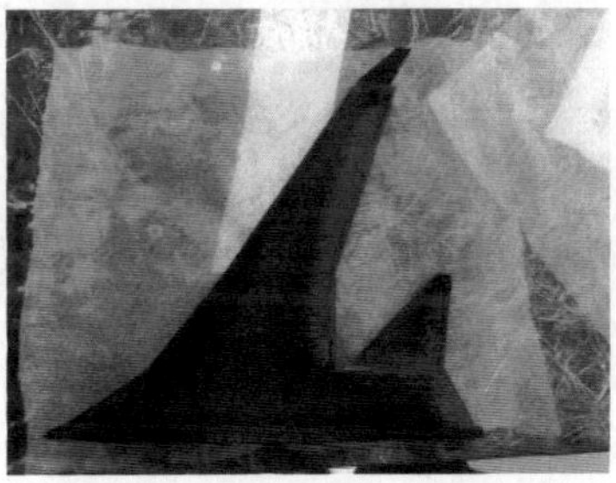

Figure 2: 3-D printed model

Figure 3: Layered with thin fibreglass

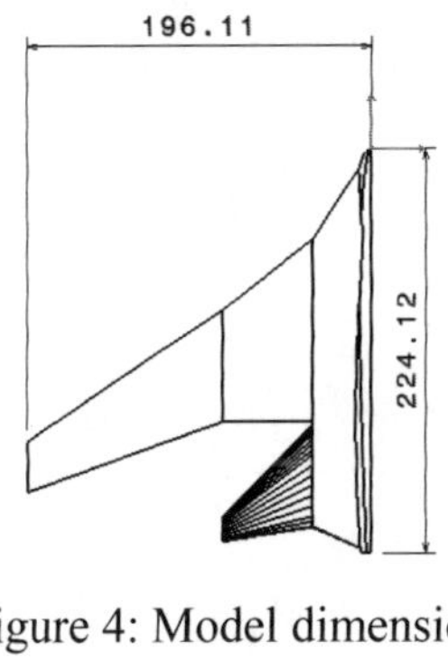

Figure 4: Model dimension

Figure 5: UPNM high speed wind tunnel

Figure 6: Wind tunnel model mounted at the energy balance

3. Results and Discussions

The data acquired are plotted to form the lift coefficient against angle of attack curve, drag coefficient against angle of attack curve, and L/D against angle of attack curve. The blockage correction is also has been made to obtained the true performance of the aircraft. As the model undergoes a wind tunnel experiment, the experimental data must be corrected using the following correction formulas [12]:

For solid blockage,

$$\Delta V = \varepsilon_{sb} V_U \tag{1}$$

$$\varepsilon_{sb} = \frac{K_1 V_b}{S^{3/2}} \tag{2}$$

where V_U is the uncorrected airspeed, K_1 for vertical model is 0.52 and S is the working section area.

For wake blockage,

$$\Delta V = \varepsilon_{wb} V_U \tag{3}$$

$$\varepsilon_{sb} = \frac{c}{2h} c_{d_u} \tag{4}$$

where c_{d_u} is uncorrected coefficient of drag, c is the model's length and h is the height of the working section.

For streamline curvature correction,

$$\alpha = \alpha_u + \frac{57.3\sigma}{2\pi}(C_{lu} + 4C_{m\,1/2u}) \tag{5}$$

$$C_l = C_{lu}(1 - \sigma - 2\varepsilon) \tag{6}$$

$$C_{m\,1/2} = C_{m\frac{1}{2u}}(1 - 2\varepsilon) + \frac{\sigma C_l}{4} \tag{7}$$

$$\sigma = \frac{\pi^2}{48}\left(\frac{c}{h}\right)^2 \tag{8}$$

Total corrected airspeed, $V = V_u(1 + \varepsilon)$ (9)

Total corrected drag, $C_{d0} = C_{d0u}(1 - 3\varepsilon_{sb} - 2\varepsilon_{wb})$ (10)

Figure 7 shows the plot of coefficient of lift, C_L against angle of attack. From the curve, it can be observed that as the angle of attack increases, the coefficient of lift is also increased. The maximum C_L obtained from the experiment is 1.114 at angle of attack 29.5°. On the other hand, the lowest C_L is at angle of attack -9.5°, which is only -0.472. From the blockage correction curve, the coefficient of lift reaches its maximum at around +28° angle of attack, which is 0.974. The minimum corrected C_L is higher compared to the uncorrected C_L, which is -0.457. The C_L values for uncorrected and corrected almost same until it reaches +13° angle of attack where the corrected C_L starts to decrease for the next angle of attack.

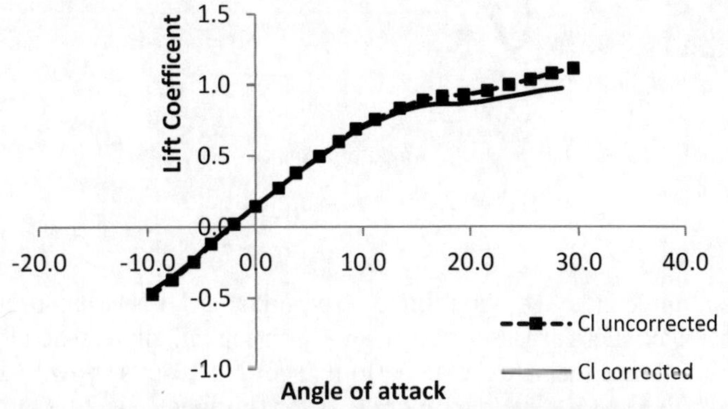

Figure 7: Lift coefficient versus angle of attack

Meanwhile, plot of coefficient of drag against angle of attack for both uncorrected and corrected for blockage data can be seen in Figure 8. The corrected coefficient of drag, C_D shows lower values compared to the uncorrected C_D data at angle around +22°. For the corrected C_D, the minimum value is 0.019 and for uncorrected data, the minimum C_D obtained is also 0.019. The maximum C_D after been corrected is 0.449 while for uncorrected data is 0.51. At angles of attack -10° to +21°, C_D is almost the same.

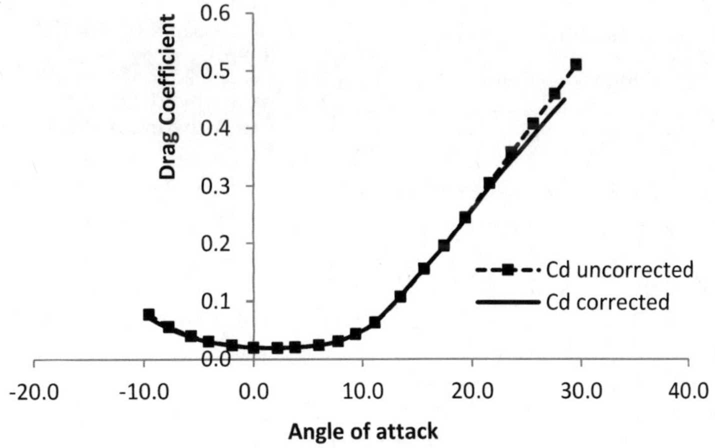

Figure 8: Lift coefficient versus angle of attack

The curve of L/D against angle of attack is shown in Figure 9. The corrected L/D shows almost the same values to the uncorrected L/D. For the corrected data, the maximum L/D achieved is 20.7 at the angle of attack +5.9° while for the uncorrected values, the L/D obtained is 20.8 at the same angle of attack. These angles of attack indicate the optimum flight configuration for Baseline V with 45° sweep tail angle.

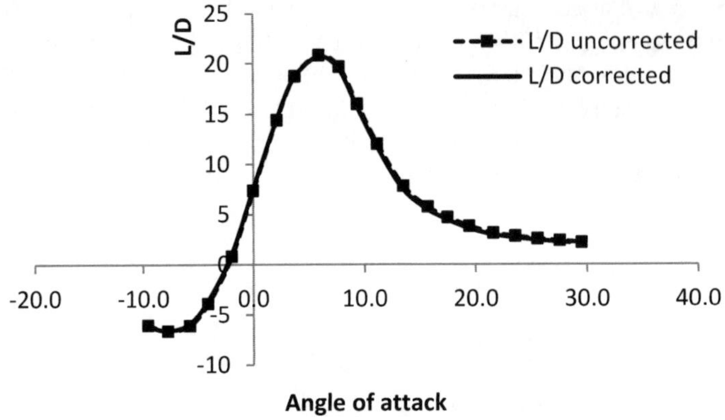

Figure 9: L/D versus angle of attack

4. Conclusion

All data obtained from the conducted wind tunnel experiment at UPNM has been studied to establish the aerodynamics characteristics for Baseline V with 45° sweep tail angle. The blockage correction has been made in order to estimate the actual data performance. The results show that Baseline V with 45° sweep tail angle will stall at +15°. The maximum L/D for Baseline V is 20.7 at +5.9° angle of attack, which is the optimum flight configuration. From the analyses that have been made, it is shown that the 45° sweep tail reduces the L/D of the Baseline V. Some improvements must be made such as moving the C.G. of the model and deflecting the tail of the model. For future study, different angles of swept tail can be tested in order to achieve the best performance of the Baseline V.

Acknowledgement

The authors would like to express their deepest gratitude to Ministry of Higher Education for granting the research fund related to this experiment (600-IRMI/MyRA 5/3/GIP (037/2017)). Our appreciation also goes to the Research Management Institute and the Faculty of Mechanical Engineering, UiTM for their various support and facilities.

References

[1] Yan W F, Wu J H and Zhang Y L 2014 *Advanced Materials Research* **1016** 354-8
[2] Potsdam M A P M A and Liebeck R H 1997 *15th Applied Aerodynamics Conference*
[3] Liebeck R H 2004 *Journal of Aircraft* **41** 10-25
[4] Qin N, Vavalle A, Le Moigne A, Laban M, Hackett K and Weinerfelt P 2004 *Progress in Aerospace Sciences* **40** 321-43
[5] Nasir R E M, Kuntjoro W and Wisnoe W 2012 *Applied Mechanics and Materials* **225** 375-84
[6] Mat I S I S, Zakaria K and Khan Z A 2013 *Advanced Materials Research* **845** 971-4
[7] Tulapurkara P E 2013 *Airplane Design (Aerodynamic)* Department of Aerospace Engineering, Indian Institute of Technology, Madras
[8] Nasir R E M, Kuntjoro W and Wisnoe W 2014 *Procedia Technology* **15** 783-91
[9] Wisnoe W, Nasir R E M, Kuntjoro W and Mamat A M I 2009 *13th International Conference on Aerospace Sciences and Aviation Technology*
[10] Kuntjoro W, Nasir R E M, Wisnoe W and Ali Z 2012 *Advanced Materials Research* **433-440** 6636-40
[11] Nasir R E M, Wisnoe W, Ramly R, Kuntjoro W and Muhammad F 2015 *Jurnal Teknologi* **12** 133-9
[12] Ross I and Altman A 2011 *Journal of Wind Engineering and Industrial Aerodynamics* **99** 523-38

IOP Conf. Series: Materials Science and Engineering **270** (2017) 012002 doi:10.1088/1757-899X/270/1/012002

Modelling the effect of changing design fineness ratio of an airship on its aerodynamic lift and drag performance

J Jalasabri, F I Romli* and M Y Harmin

Department of Aerospace Engineering, Faculty of Engineering, Universiti Putra Malaysia, 43400 Serdang, Selangor, Malaysia.

*fairuz_ir@upm.edu.my

Abstract. In developing successful airship designs, it is important to fully understand the effect of the design on the performance of the airship. The aim of this research work is to establish the trend for effects of design fineness ratio of an airship towards its aerodynamic performance. An approximate computer-aided design (CAD) model of the Atlant-100 airship is constructed using CATIA software and it is applied in the computational fluid dynamics (CFD) simulation analysis using Star-CCM+ software. In total, 36 simulation runs are executed with different combinations of values for design fineness ratio, altitude and velocity. The obtained simulation results are analyzed using MINITAB to capture the effects relationship on lift and drag coefficients. Based on the results, it is concluded that the design fineness ratio does have a significant impact on the generated aerodynamic lift and drag forces on the airship.

1. Introduction

In recent years, airships have been anticipated to make its comeback into the mainstream commercial air transportation industry. Airships used to serve as the main air transportation means since the 1930s. However, due to a series of fatal incidents that hindered their design and development, particularly the infamous Hindenburg incident that marked the end of mainstream application of airships, jet aircraft effectively took over the role as primary means of passengers' air transport until today. Subsequently, the role of airships has been effectively delegated to only tourism and advertising purposes.

The current commercial transport jet aircraft has been facing few challenges lately. Among others, these include fluctuating fuel prices and increasing environmental awareness among the public that led to stricter operational rules by the aviation authority, which are causing problems for many airlines to maintain their profitable operation. Meanwhile, with the advancement in their design and technology, airships appear to have become a superior air transportation means than jet aircraft in few operational aspects. For instance, as indicated in Ref. [1], the environmental cost per seat kilometer of an airship is just about a quarter of that for jet aircraft. With the expectation that the future air transportation rules will be stricter with regards to environmental issues such as carbon emissions, this is a big advantage to use airship in commercial air transportation. Ref. [2] further compares operational characteristics of airship against other modes of transportation as tabulated in Table 1. The presented comparison result is taken to highlight that airship is very suitable to become an alternative means of public transport. An exemplary research that expresses this idea is the currently ongoing Multi-body Advanced Airship for Transportation (MAAT) project, which is supported by the European Union [3]. Another study related to having airships as public mass transport means has been reported in Ref. [4], which envisions it as

IOP Conf. Series: Materials Science and Engineering **270** (2017) 012002 doi:10.1088/1757-899X/270/1/012002

the potential solution for the increasing road traffic congestions. Overall, it can be summarized that the airships are cheaper to fly than helicopters and aircraft, able to carry high payloads and also operate in remote and hard-to-reach areas [5].

Table 1: Comparison of airships and other transportation modes [2]

Operational Characteristics	**Airship versus Maritime**	**Airship versus Highway**	**Airship versus Railway**	**Airship versus Aerial**
Speed	Much faster	Faster	Much faster	Much slower
Load Capacity	Less capacity	Much more capacity	Less capacity	Increased capacity
Load Adaptability	Much more flexible	Less flexible	Much more flexible	More flexible
Transportation Cost	Much more expensive	More expensive	Much more expensive	Much more economical

The increased suitability of airships for use as public air transport means can also be contributed to the revolutionary new designs that are much improved than conventional or traditional airship designs. Examples of the modern airship concept designs include Atlant, Aeroscraft, SkyShuttle and Airlander airships. Of particular interest in this study is Atlant-100 airship developed by Augur RosAeroSystems company, which is believed to be suitable option for mass public transport application. Computational fluid dynamics (CFD) simulation analysis on an approximate model of the Atlant-100 airship has been conducted in this study, where the model's design fineness ratio is varied to observe the impact of its generated aerodynamic lift and drag forces. Design fineness ratio is defined as the ratio of the airship's length to its width and it is an important parameter in designing the airship due to the expected impact on both its performance and transport capacity. Knowledge on its relationship with the performance of the airship helps in making better design decision in the trade-off between performance and capacity.

2. Lift and Drag Simulation Analysis

The approximate model of the Atlant-100 airship is constructed in CATIA software and it is illustrated in Figure 1. It should be noted that this model is developed based on the available design information within the public domain and may not accurately represent the actual airship design. Nevertheless, for this study, the constructed model is taken to be adequate since the primary focus of this study is on the effects of varying design fineness ratio on the airship's lift and drag forces. Total length of the model is 100 m while its width and height are 48 m and 35 m, respectively.

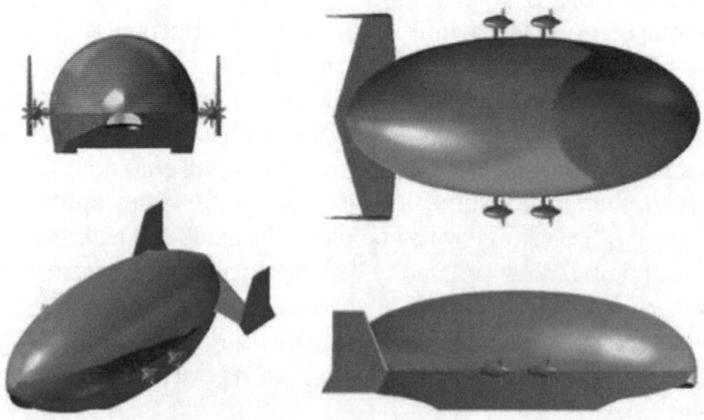

Figure 1: Constructed approximate computer-aided design (CAD) model of Atlant-100 airship

The CAD model is then used for CFD simulation analysis in Star CCM+ software. The turbulence model applied for the simulation is the Spalart-Allmaras (S-A) and polyhedral cell type is selected for the meshing. Both of these choices are made through the results of preliminary simulation studies that can be referred to in Ref. [6]. The model meshing details for the CFD simulation are tabulated in Table 2. To obtain good simulation data set for the trend analysis later, the setting for the simulation cases is made based on the full factorial design of experiment (DoE) method. In addition to the design fineness ratio, the velocity and altitude parameters in the simulation cases are also varied to represent the varying operational factors. In total, there are 36 different simulation runs executed for this study. Of important note, the variation of the design fineness ratio of the CAD model in this study is achieved by changing the airship's width while maintaining its length as constant. The CAD models corresponding to the three different design fineness ratios considered in this study are as illustrated in Figure 2, where the original design has the design fineness ratio of 2.08.

Figure 2: Different design fineness ratio models of the airship design

Examples of CFD simulation results from Star-CCM+ software for different design fineness ratios are depicted in Figure 3 to Figure 5. The velocity and pressure profile plots shown are for simulated flight at 1.5 km altitude with velocity of 100 km/h. The full simulation results are presented in Table 3.

Velocity plot (side view) Velocity plot (top view) Pressure plot

Figure 3: Results for design fineness ratio = 2.08, altitude = 1.5 km and velocity = 100 km/h

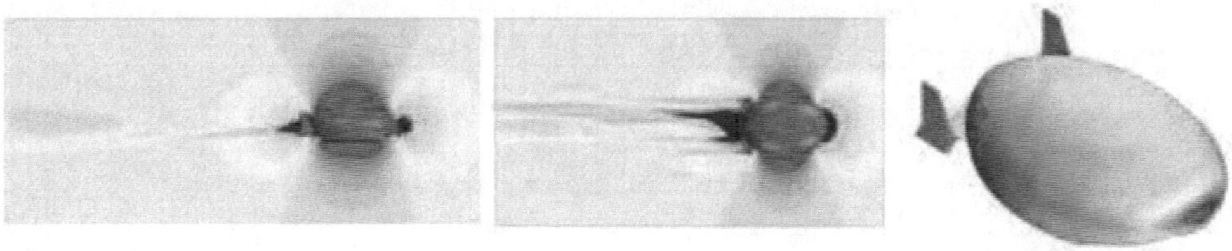

Velocity plot (side view) Velocity plot (top view) Pressure plot

Figure 4: Results for design fineness ratio = 1.39, altitude = 1.5 km and velocity = 100 km/h

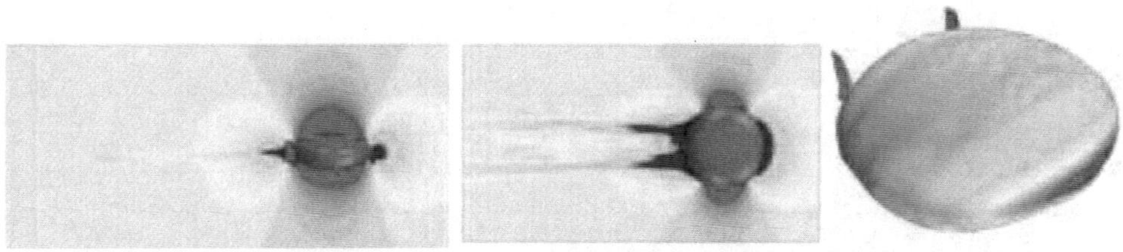

Velocity plot (side view) Velocity plot (top view) Pressure plot

Figure 5: Results for design fineness ratio = 0.93, altitude = 1.5 km and velocity = 100 km/h

IOP Conf. Series: Materials Science and Engineering **270** (2017) 012002 doi:10.1088/1757-899X/270/1/012002

Table 2: Mesh details

Mesh Model	Polyhedral Mesher
Base size	3.2m
No. Prism Layer	12
Prism layer thickness	33.33% (Default)
Growth rate	1.3 (Default)
Y-values	All + y wall treatments (Default)
Surface Size (Target)	1.6m
Tunnel Surface Size	204.8m
No. of Cells	~ 6 millions

Table 3: Simulation analysis results from Star-CCM+

Run	**Fineness Ratio**	**Altitude (m)**	**Velocity (km/h)**	C_L	C_D
1	2.08	1500	100	0.026	0.024
2	2.08	1500	140	0.039	0.026
3	2.08	1500	190	0.039	0.027
4	2.08	1500	250	0.032	0.025
5	2.08	2000	100	0.029	0.026
6	2.08	2000	140	0.025	0.023
7	2.08	2000	190	0.031	0.026
8	2.08	2000	250	0.040	0.026
9	2.08	2500	100	0.040	0.027
10	2.08	2500	140	0.022	0.025
11	2.08	2500	190	0.017	0.023
12	2.08	2500	250	0.042	0.025
13	1.39	1500	100	0.048	0.036
14	1.39	1500	140	0.044	0.032
15	1.39	1500	190	0.047	0.033
16	1.39	1500	250	0.040	0.031
17	1.39	2000	100	0.038	0.036
18	1.39	2000	140	0.047	0.034
19	1.39	2000	190	0.055	0.034
20	1.39	2000	250	0.047	0.037
21	1.39	2500	100	0.052	0.036
22	1.39	2500	140	0.055	0.036
23	1.39	2500	190	0.054	0.036
24	1.39	2500	250	0.047	0.035
25	0.93	1500	100	0.055	0.033
26	0.93	1500	140	0.089	0.037
27	0.93	1500	190	0.069	0.033
28	0.93	1500	250	0.088	0.036
29	0.93	2000	100	0.061	0.033
30	0.93	2000	140	0.070	0.035
31	0.93	2000	190	0.060	0.032
32	0.93	2000	250	0.078	0.036
33	0.93	2500	100	0.059	0.032
34	0.93	2500	140	0.058	0.032
35	0.93	2500	190	0.072	0.035
36	0.93	2500	250	0.078	0.036

IOP Conf. Series: Materials Science and Engineering **270** (2017) 012002 doi:10.1088/1757-899X/270/1/012002

3. Modelling of the Effects Trend

The data obtained from the CFD simulation runs is then processed to model the possible relationship between the design fineness ratio and the generated lift and drag forces on the airship. For this study, the interest is on the difference in the generated aerodynamic forces when the design fineness ratio has been changed. With this in mind, the data in Table 3 can be rewritten as tabulated in Table 4, with the reference benchmark case settings as the original design fineness ratio of 2.08, altitude of 1500 m and velocity of 100 km/h. The processed data in Table 4 is then used as the input into MINITAB software for the statistical analysis and modeling.

Table 4: Data input into MINITAB for statistical analysis

Run	**% Fineness Ratio Change**	**% Altitude Change**	**% Velocity Change**	**% C_L Change**	**% C_D Change**
1	0.00	0.00	0.00	0.00	0.00
2	0.00	0.00	40.00	50.00	8.33
3	0.00	0.00	90.00	50.00	12.50
4	0.00	0.00	150.00	23.08	4.17
5	0.00	33.33	0.00	11.54	8.33
6	0.00	33.33	40.00	-3.85	-4.17
7	0.00	33.33	90.00	19.23	8.33
8	0.00	33.33	150.00	53.85	8.33
9	0.00	66.67	0.00	53.85	12.50
10	0.00	66.67	40.00	-15.38	4.17
11	0.00	66.67	90.00	-34.62	-4.17
12	0.00	66.67	150.00	61.54	4.17
13	-33.17	0.00	0.00	84.62	50.00
14	-33.17	0.00	40.00	69.23	33.33
15	-33.17	0.00	90.00	80.77	37.50
16	-33.17	0.00	150.00	53.85	29.17
17	-33.17	33.33	0.00	46.15	50.00
18	-33.17	33.33	40.00	80.77	41.67
19	-33.17	33.33	90.00	111.54	41.67
20	-33.17	33.33	150.00	80.77	54.17
21	-33.17	66.67	0.00	100.00	50.00
22	-33.17	66.67	40.00	111.54	50.00
23	-33.17	66.67	90.00	107.69	50.00
24	-33.17	66.67	150.00	80.77	45.83
25	-55.29	0.00	0.00	111.54	37.50
26	-55.29	0.00	40.00	242.31	54.17
27	-55.29	0.00	90.00	165.38	37.50
28	-55.29	0.00	150.00	238.46	50.00
29	-55.29	33.33	0.00	134.62	37.50
30	-55.29	33.33	40.00	169.23	45.83
31	-55.29	33.33	90.00	130.77	33.33
32	-55.29	33.33	150.00	200.00	50.00
33	-55.29	66.67	0.00	126.92	33.33
34	-55.29	66.67	40.00	123.08	33.33
35	-55.29	66.67	90.00	176.92	45.83
36	-55.29	66.67	150.00	200.00	50.00

The effect of changes in design fineness ratio, altitude and velocity parameters towards the lift and drag coefficients of the airship model can be observed from the main effects plot. Figure 6 shows the main effects plot for the generated aerodynamic lift coefficient. It can be observed that design fineness ratio has a significant effect on the aerodynamic lift generation since its plot has a very steep slope. In contrast, altitude changes have a rather negligible impact on the lift generation while velocity is fairly impactful. Meanwhile, Figure 7 is depicting the main effects plot for the aerodynamic drag coefficient. In similar fashion, impact from changing design fineness ratio can be observed to be very significant to the generation of aerodynamic drag in comparison to almost negligible impact both from altitude and velocity.

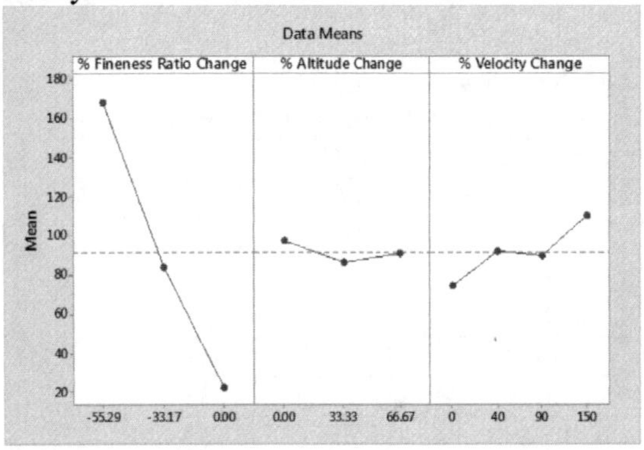

Figure 6: Main effects plot for % C_L change

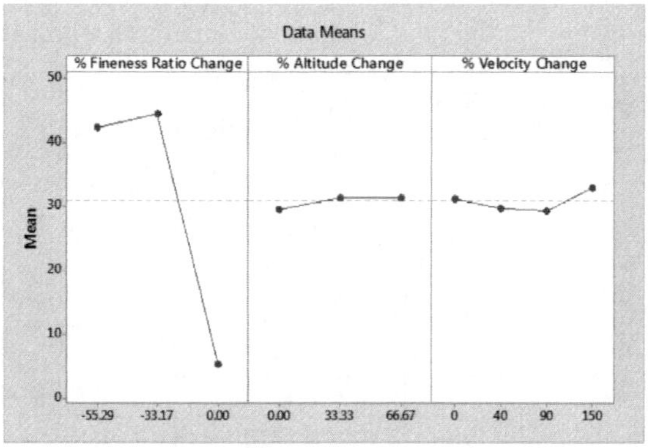

Figure 7: Main effects plot for % C_D change

In many cases, though the independent parameter itself may have a low impact, its interaction with other independent parameters can produce significant effect on the dependent parameter. Based on the interaction plots between design fineness ratio, altitude and velocity parameters in Figure 8 and Figure 9 for lift and drag, respectively, this situation is true. Looking at Figure 8, there is significant effect on aerodynamic lift from the interaction of design fineness ratio with velocity or altitude as indicated by the crossing of the plot lines. The same is observed for the generation of drag as shown in Figure 9.

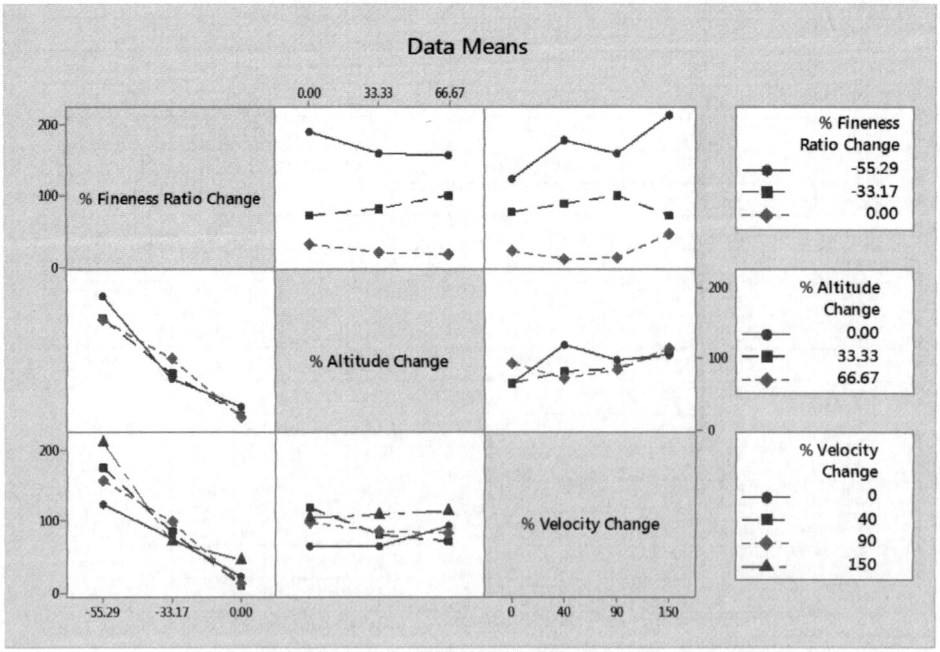

Figure 8: Interaction plot for % C_L change

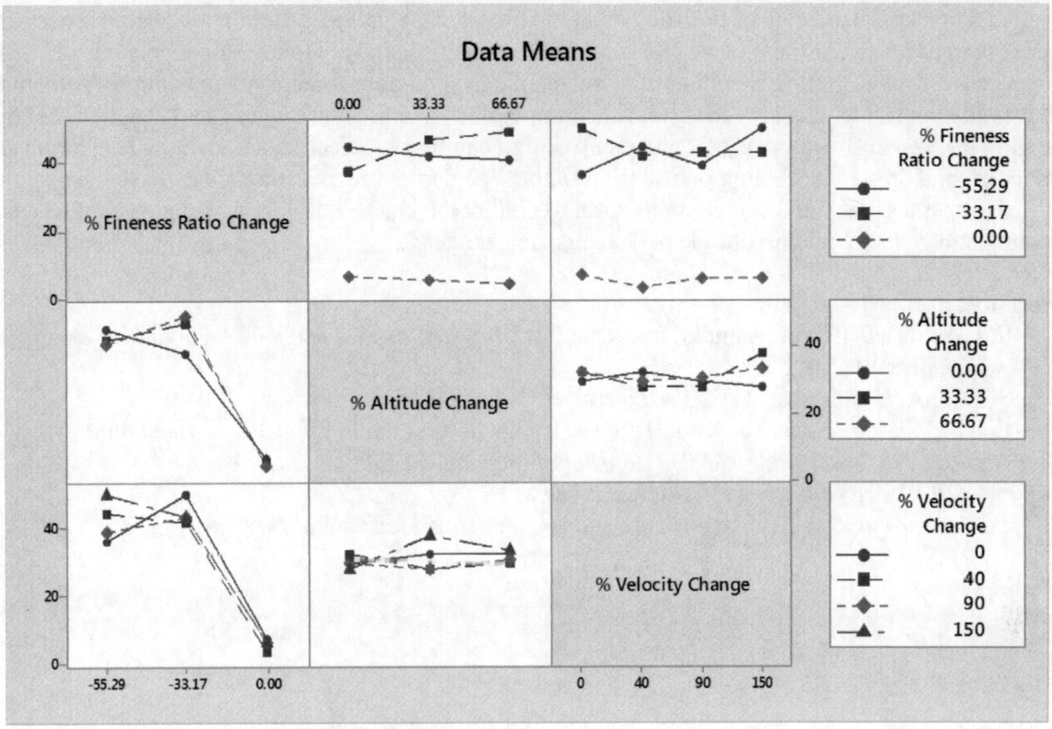

Figure 9: Interaction plot for % C_D change

Next, statistical regression model can be fitted to better capture the relationship between considered parameters and aerodynamic performance parameters of the airship design model. Using MINITAB, the fitted regression models for lift and drag forces are derived and they are indicated by Equation 1 and Equation 2, respectively, where VC is velocity change (in %), FRC is fineness ratio change (in %) and AC is altitude change (in %). Both fitted regression models have a high R-square value, which is used to measure how close the actual data are to the fitted regression line and how good the model is in capturing the variability of the response data around its mean. For Equation 1, its R-square value is 0.95, which means 95% of the effects relationship is adequately captured by the model. Meanwhile, the R-square value for Equation 2 is 0.97. Hence both of the fitted regression models can be taken to be adequately accurate in predicting the subsequent changes in the lift and drag forces on the airship design model when its design fineness ratio, altitude or velocity changes.

$$\begin{aligned}\%\ CL\ Change = {} & 0.314\ VC + 0.04454\ FRC^2 - 0.0620\ FRC{*}AC - 0.01534\ AC{*}VC \\ & - 0.001201\ FRC^2{*}AC + 0.000396\ FRC^2{*}VC + 0.000128\ FRC{*}VC^2 \\ & + 0.000101\ AC{*}VC^2 \end{aligned} \tag{1}$$

$$\begin{aligned}\%\ CD\ Change = {} & -2.036\ FRC + 0.0449\ VC - 0.02396\ FRC^2 - 0.01465\ FRC{*}AC \\ & + 0.00679\ FRC{*}VC - 0.000285\ FRC^2{*}AC + 0.000132\ FRC^2{*}VC \end{aligned} \tag{2}$$

All in all, based on the derived regression models, it can be concluded that the design fineness ratio has very significant impact on the generation of lift and drag forces for the airship design. The models in Equation 1 and Equation 2 can be used to predict the expected changes in aerodynamic lift and drag of the airship when the values of design fineness ratio is changed, along with the different operational setting of the velocity and altitude.

IOP Conf. Series: Materials Science and Engineering **270** (2017) 012002 doi:10.1088/1757-899X/270/1/012002

4. Conclusion

Aerodynamic performance characteristics of an airship are essential to determine its effectiveness and suitability in carrying out the intended mission. The aim of this study is to investigate the impact of the design fineness ratio on the generation of aerodynamic lift and drag forces on an airship. Based on the CFD simulation results and also the established regression models, it has been clearly highlighted that changing the design fineness ratio of an airship design can highly affect its aerodynamic performance, namely lift and drag forces acting on it during flight. The derived regression models have been shown to be of proper accuracy and can be easily used to predict the changes in lift and drag when the design fineness ratio, velocity and/or altitude parameters are changed.

References

[1] Stockbridge C, Cerruti A and Marzocca P 2012 *International Journal of Aeronautical and Space Sciences* **13** 170-87
[2] Schafer A W and Waltz I A 2014 *Transport Policy* **34** 1-4
[3] Ilieva G, Pascoa J, Dumas A and Trancossi M 2014 *Aerospace Science and Technology* **35** 1-14
[4] Aminian N O and Romli F I 2017 *Journal of Mechanical Engineering* **SI**
[5] Stewart W 2015 *http://www.dailymail.co.uk*
[6] Jalasabri J and Romli F I 2017 *International Review of Mechanical Engineering* **11**

IOP Conf. Series: Materials Science and Engineering **270** (2017) 012003 doi:10.1088/1757-899X/270/1/012003

Two dimensional numerical analysis of aerodynamic characteristics for rotating cylinder on concentrated air flow

M S Alias, A S Mohd Rafie*, O F Marzuki, M F Abdul Hamid and C C Chia

Department of Aerospace Engineering, Faculty of Engineering, Universiti Putra Malaysia, Serdang, Selangor, Malaysia.

*shakrine@upm.edu.my

Abstract. Over the years, many studies have demonstrated the feasibility of the Magnus effect on spinning cylinder to improve lift production, which can be much higher than the traditional airfoil shape. With this characteristic, spinning cylinder might be used as a lifting device for short take-off distance aircraft or unmanned aerial vehicle (UAV). Nonetheless, there is still a gap in research to explain the use of spinning cylinder as a good lifting device. Computational method is used for this study to analyse the Magnus effect, in which two-dimensional finite element numerical analysis method is applied using ANSYS FLUENT software to examine the coefficients of lift and drag, and to investigate the flow field around the rotating cylinder surface body. Cylinder size of 30mm is chosen and several configurations in steady and concentrated air flows have been evaluated. All in all, it can be concluded that, with the right configuration of the concentrated air flow setup, the rotating cylinder can be used as a lifting device for very short take-off since it can produce very high coefficient of lift (2.5 times higher) compared with steady air flow configuration.

1. Introduction

Magnus effect can be seen as a device that produces lifting force in perpendicular direction of the flow by providing a moving wall on its body to influence the boundary layer of fluid around it [1]. Several inventions have proven its potential applications in aerospace including the Magnus rotor wing aircraft in the early 1920s but high cost has drastically hindered its development. In the meantime, researchers continue to develop the fan-wing aircraft, which increases the lift coefficient but has some difficulties in low-speed manoeuvring and hard landing [2]. A smaller-scaled rotor wing aircraft design that uses the concept of Magnus rotor wing has been developed to prove the feasibility of flying using Magnus effect principle. The results are promising for take-off and landing in short distance, and the aircraft is able to fly at low speeds and stable at higher angles of attack [3]. Furthermore, another study involves changing the airfoil leading edge into a rotating cylinder for Moving Surface Boundary Layer Control (MSBC) and the result shows an increase in lift by 100%, stall delay from 10° to 35° and increase in lift-to-drag ratio by 167% [4]. In addition, the effect of Magnus rotor to micro air vehicle has also been successfully experimented and it has been found that the size of cylinder and payload is an important factor that contributes to the lifting of the MAV [5].

The flow dynamic of a rotating cylinder will be dramatically altered when the body is placed close to the wall. Results from a previous study show that the lift force experienced by the cylinder has been directed away from the static wall but when the gap ratio is maintained, the lift and drag increase as the

Reynolds number, *Re* increases [6, 7]. Therefore, a rotating cylinder with no wall close to the body can be a solution to create lift perpendicular to the airstream direction. A different study is carried out to investigate local flow field and rear wake of a surface-mounted finite circular cylinder in low speed wind tunnel using Particle Image Velocimetry (PIV), with $Re = 4.2 \times 10^4$ and different aspect ratios (ARs). The result has concluded that higher AR might increase turbulence and upwash that are formed with flow separation occurred at the leading edge free-end surface and a mean recirculation formation [8]. Furthermore, separation point is identified on the basis of the highest pressure coefficient on the cylinder surface [9]. Stagnation is considered as high pressure due to the existence of turbulent flow. The cylinder tends to move in y-direction due to pressure created near to the lower side of the cylinder symmetry. This is based on specific *Re*, rotational rate and angle of incidence of the air flow [10]. On the other hand, the conducted study on flow separation and effects of surface roughness at different rotating speeds has shown that the increments of lift coefficient depends on surface roughness of the cylinder [11, 12]. Moreover, the supercritical flow past a cylinder has been examined, assuming fully turbulent boundary layer [13]. The ranges of the parameters that can affect the rotating cylinder flight have been found to be $1 \geq$ Power-Law Index, $n \geq 0.2$, $0.1 \leq Re \leq 40$ and $0 \leq \alpha \leq 6$, but no lifting is produced with positive drag [14].

In summary, the research on rotating cylinder has been proven to be beneficial in many industries including aerospace. However, there is still a research gap in explaining the use of spinning cylinder as a good lift generator. The main objective of this study is to analyze the effect of steady air flow and concentrated airflow in 2D numerical simulation on rotating cylinder as a potential lifting devices for short take-off flying distance aircraft.

2. Methodology

For validation purpose of the numerical analysis, a rotating cylinder model used in the experiment in Ref. [15] is applied in this study for comparison. The ANSYS Fluent software is used to analyze the fluid flow characteristics on the rotating cylinder and the details of the meshing are tabulated in Table 1. The orthogonal quality value is close to 1, which indicates a good quality of meshes. Furthermore, as shown in Table 2, using the statistical analysis software Statistical Package for the Social Sciences (SPSS), the simulated results are indicated to be essentially 90.6% similar to the experimental results in Ref. [15]. This positive agreement between the two results is taken as a validation for the numerical setup and it can be used to simulate the fluid flow characteristics of the rotating cylinder at different α.

Table 1: Meshing statistics

Parameter	Value
Nodes	62704
Elements	62345
Min	0.590
Max	0.999
Average	0.986
Standard Deviation	2.3×10^{-3}

Table 2: Model summary for the validation study

	R	R Square	Adjusted R Square	Std. Error of the Estimate
Model	.952[a]	0.906	0.901	0.331091

[a]Predictors: Rotational speed, air velocity
Dependent variable: Rotational rate

The main interest of this study is to improve the lift coefficient of the cylinder. Therefore, several configurations to obtain high lift coefficient are considered for this study. A wall block of size 2mm x 15mm is located near the cylinder to evaluate and analyze the Magnus effect. For the concentrated air flow, the velocity inlet is placed nearby to minimize turbulent boundary layer that causes drag. For the simulations, the fixed constants are set as follows: $Re = 2.2 \times 10^4$, inlet velocity, $U_\infty = 7$m/s, $\alpha = 1.54$ and rotational speed, $U_\theta = 718.67$ rad/s (clockwise). Figure 1 illustrates the geometry model for steady and uniform air flow whereas Figure 2 is showing the geometry model on air inlet shifted near-front with different inlet opening size in front of rotated cylinder for concentrated air flow simulation.

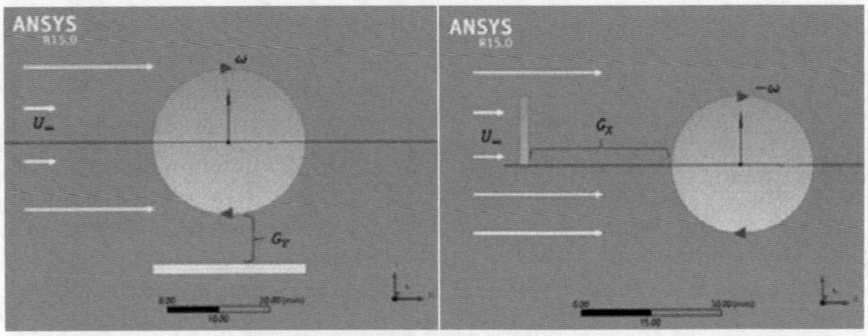

Figure 1: Geometry model for steady and uniform air flow

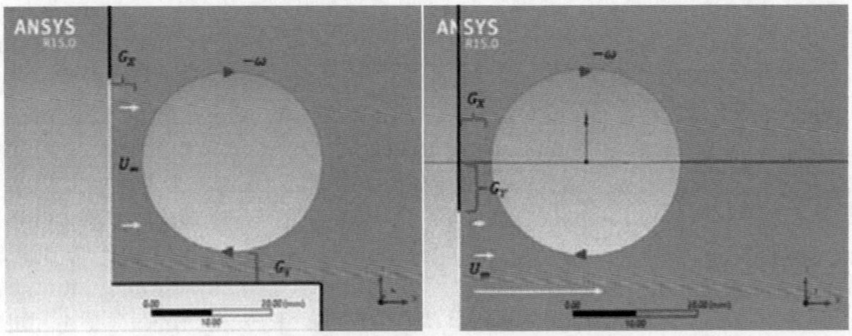

Figure 2: Geometry model for concentrated air flow

The mesh quality is important for domain discretization. The skewness is evaluated by FLUENT. Meshes have been generated for both testing where Minimum Orthogonal Quality should be close to 1 [16]. Figure 3 shows the meshing for steady and concentrated air flow approaches, where the meshes for determining optimum C_L satisfied at Minimum Orthogonal Quality of 6.148 x 10^{-1} and optimum C_D at Minimum Orthogonal Quality of 5.809 x 10^{-1}. These values are more than 0.5 and close to 1, which can be considered that a good quality of mesh has been generated.

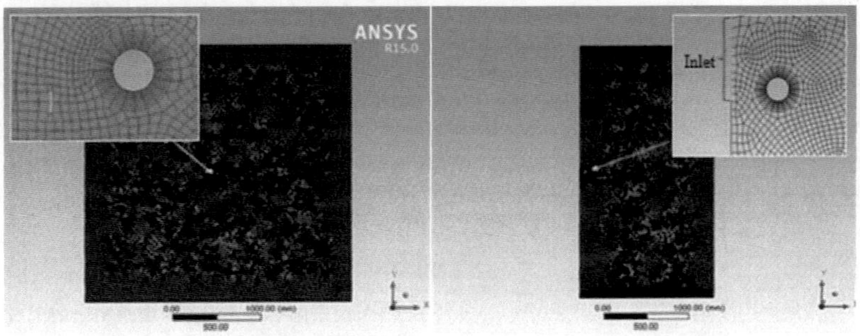

Figure 3: Domain discretization for steady and concentrated air flow

The flow behaviour near the wall is complicated and also hard to distinguish with that at different regions near the wall. Hence the concept of y^+ has been formulated, which is a dimensionless quantity that corresponds to the distance from the wall measured in the terms of viscous length. It is related to the turbulence modelling, especially mesh generation processes, and how it is going to affect the flow simulation result [16]. A non-dimensional wall distance for a wall-bounded flow can be determined by using Equation 1, where u* is the friction velocity at the nearest wall, y is the distance to the nearest wall and v is the local kinematic viscosity of the fluid. y^+ is often referred as y plus and is commonly used in boundary layer theory and in defining the law of the wall.

$$y^+ = (u^* y) / v \qquad (1)$$

For external flows, the calculation of wall height starts from the skin friction coefficient, C_f as given by Equation 2 [14].

$$C_f = 0.0508\ Re^{-0.2} \tag{2}$$

The wall shear stress, τ_w can be calculated from skin friction coefficient as in Equation 3. The value is then used to calculate the frictional velocity, U_τ as in Equation 4. Finally, the wall height is estimated by Equation 5.

$$\tau_w = 1/2 C_f\, \rho\, U^2 \tag{3}$$

$$U_\tau = \sqrt{(\tau_w/\rho)} \tag{4}$$

$$y = (y^+\, v) / U_\tau \tag{5}$$

For the simulation in this study, the fluid is assumed to be an ideal gas having constant specific heats capacity. The flow analyses are carried out using commercial CFD package, ANSYS CFX that executes the 3D RANS equations based on the finite volume numerical method. The solver type used is pressure based with absolute velocity formulation, time considered at transient on 2D space planar. The air flow is set as 'viscous' and is reflected to 'SST K-Omega, K-ω'. The values of air density, ρ of 1.225 kgm^{-3} and viscosity, μ of 1.789 x 10^{-5} are used. Moreover, for boundary conditions, the cylinder is fixed as a wall but to rotate the cylinder, momentum parameters are clarified for moving wall that is 'relative to adjacent cell zone' with 'rotational speed'. To prevent of any other disturbance, the shear condition is set to 'no slip condition' with 'wall roughness height (m)' is set to zero. Since the air flow considered is turbulent flow, the specification method is set for 'intensity' of 5% and 'viscosity ratio' of 10. The same setting is used for the outlet velocity.

The algebraic equations are solved in iterative manner, which results in velocity and pressure fields being updated together. A fully implicit coupling has been achieved by implicit discretization of the pressure gradient terms from the momentum equations, including implicit discretization of the mass flux. All procedures are linked with an algebraic multi-grid method and the set of equations is solved by a point or block using the Gauss-Seidel technique. On the other hand, to solve the equations for the unsteady form, second-order discretization scheme is used to integrate the equations in time domain. This scheme is implicit and the solving process is done with numerical stability. The time step is chosen based on the dimensionless step [16].

3. Results and Discussion

The optimization of the coefficient of aerodynamics forces, either C_L or C_D, has been carried out with a wall block is located near to the rotated cylinder body to restrict incoming air flow. It is important to see the effect in order to define the optimum gap at Y-axis, G_Y for steady source of air to increase the speed of air at the upper side of the cylinder body. Five configurations are considered in this study for high coefficient of lift while three configurations are considered for low coefficient of drag. Table 3 and Table 4 tabulate the results in a steady air flow for optimum C_L and C_D, respectively.

From Table 3, Configuration 3 has shown the highest resultant C_L=1.725 with G_Y = 10 mm, which is actually lower than to the on-design operating point C_L=2.170 (no wall block). The simulation result for this Configuration 3 is illustrated in Figure 4. The velocity distribution contour is shown in Figure 4(a), where 10% to 20% of the cylinder perimeter operating points located at the top are considered to be at high velocity while the low velocity occurred at the lower separation region. The vortex wake is identified at half of the cylinder diameter size. Furthermore, Figure 4(b) describes the upper boundary layer streamlines that is thinner than the lower boundary layer. The velocity streamlines path indicates a turbulent state at the back of the cylinder. Meanwhile, Figure 4(c) implies that 10% - 20% of the cylinder perimeter operating points located at the lower separation region are considered high pressure and the low pressure region occurred at the top of the operating cylinder.

In the meantime, from Table 4, Configuration 2 has shown the lowest result for C_D with 0.027 at G_X = 30 mm. This result is an improvement in comparison to the on-design operating point C_D=0.298 (no

IOP Conf. Series: Materials Science and Engineering **270** (2017) 012003 doi:10.1088/1757-899X/270/1/012003

wall block). The velocity distribution contour in this case is displayed in Figure 5(a), which shows at least 25% of the cylinder perimeter operating points at the upper separation and secondary boundary layer region are considered high velocity. The vortex wake is greater behind the cylinder but it rapidly decreases. Figure 5(b), on the other hand, highlights that the thinning of the upper separation region streamline has started at the boundary layer origin (stagnation point) where the air velocity is higher than the lower separation region streamlines. Moreover, Figure 5(c) shows that at least 10% - 20% of the cylinder perimeter operating points are considered high pressure at the stagnation point where the boundary layer originated. Low pressure is identified at the upper separation region.

Table 3: C_L optimization in steady air flow

Configuration	G_Y (mm)	C_L	C_D
1	5.0	1.048	0.615
2	7.5	1.296	0.607
3	10.0	1.725	0.488
4	12.5	1.118	0.485
5	15.0	0.953	0.481

Table 4: C_D optimization in steady air flow

Configuration	G_X (mm)	C_L	C_D
1	45	0.507	0.055
2	30	0.501	0.027
3	15	0.523	0.039

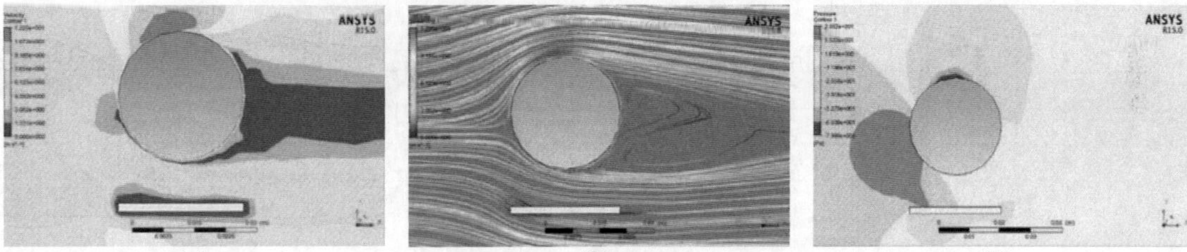

(a) (b) (c)

Figure 4: Simulation result of Configuration 3 in steady air flow for C_L optimization: (a) velocity contour, (b) streamlines, (c) pressure contour

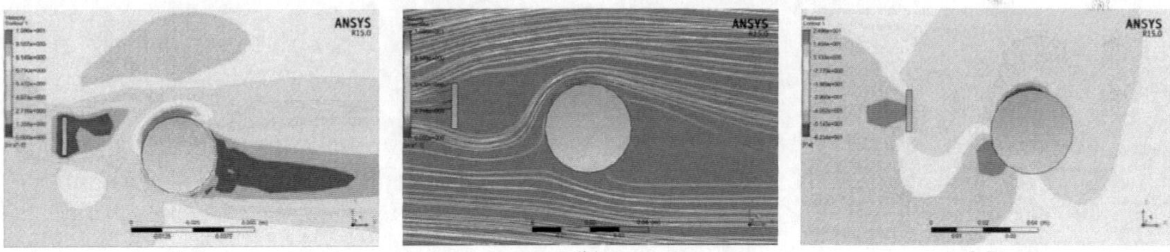

(a) (b) (c)

Figure 5: Simulation result of Configuration 2 in steady air flow for C_D optimization: (a) velocity contour, (b) streamlines, (c) pressure contour

Meanwhile, for the concentrated air flow cases, there are four considered configurations in the C_L optimization and five configurations for C_D optimization. Table 5 shows the detailed simulation results for C_L optimization study while Table 6 tabulates the obtained results in C_D optimization study. From Table 5, the highest C_L of 5.316 is obtained by Configuration 2 with G_Y=35mm and G_X=5mm. This is higher in comparison to the on-design operating point C_L of 2.170 (normal air inlet distance). Figure 6(a) shows the simulated velocity distribution contour for Configuration 2, where at least 40% - 50% of the upper separation and 10% - 15% of the lower separation cylinder perimeter operating points are considered of high velocity. In addition, Figure 6(b) shows the thinning of the streamlines at the upper separation point. The stagnation or boundary layer origin can be clearly seen to be shifted upward and deflect the air from the upper to lower regions. Figure 6(c) depicts that at least 10% - 20% of cylinders perimeter operating points with the highest pressure are located at lower separation point whereas the upper separation region is considered lower pressure. As for the optimum configuration for C_D, based on the results in Table 6, Configuration 5 appears to have the lowest C_D with 0.122 at G_Y=1492.5mm and G_X=10mm. This is an improvement compared to the on-design operating point C_D=0.298 (normal air inlet distance). The velocity distribution contour for Configuration 5 is displayed in Figure 7(a), which shows that 70% - 80% of cylinder perimeter operating points at the upper separation (which in the primary boundary layer region) is considered have the highest air velocity. Figure 7(b) shows that the upper separation region has gained the highest air velocity and the stagnation point has shifted a little below than the normal condition of Magnus effect. Last but not least, in Figure 7(c) displays that at least 10% – 15% of cylinders perimeter operating points that are considered as having the highest pressure are located at the stagnation point whereas the lowest ones are located at the upper separation region close to the wall domain.

Table 5: C_L optimization in concentrated air flow

Conf. No.	Inlet size (mm) = 1500 mm - 15 mm + G_Y	G_X (mm)	C_L	C_D
1	40.0	10	3.788	3.331
2	35.0	5	5.316	4.325
3	45.0	15	2.464	2.473
4	22.5	15	0.742	0.189

Table 6: C_D optimization in concentrated air flow

Conf. No.	Inlet size (mm) = 1500 mm - 15 mm + G_Y	G_X (mm)	C_L	C_D
1	1515.0	5	2.818	2.522
2	1507.5	5	2.245	1.765
3	1500.0	5	1.570	1.074
4	1500.0	10	1.869	0.754
5	1492.5	10	0.757	0.122

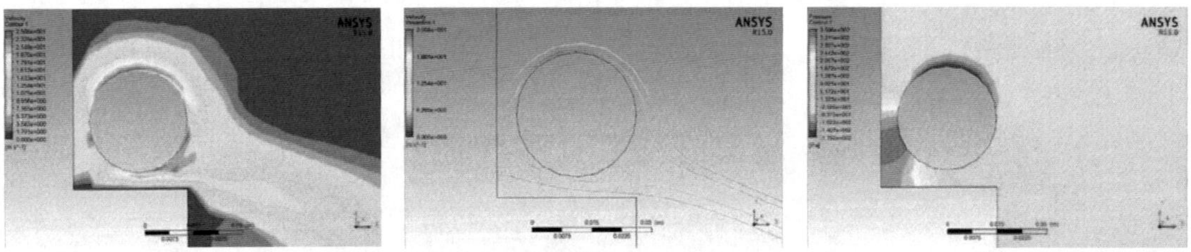

(a) (b) (c)

Figure 6: Simulation result of Configuration 2 in concentrated air flow for C_L optimization: (a) velocity contour, (b) streamlines, (c) pressure contour

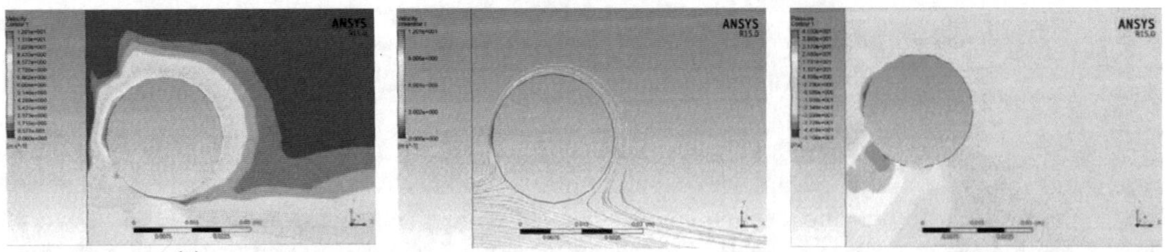

(a) (b) (c)

Figure 7: Simulation result of Configuration 5 in concentrated air flow for C_D optimization: (a) velocity contour, (b) streamlines, (c) pressure contour

4. Conclusion

The main objective of this study is to investigate the effects of steady and concentrated air flows on the rotating cylinder. Based on the simulation results for the steady air flow setup, it has been found that optimum configuration for the highest C_L produces a lower value, C_L=1.725 in comparison to that generated without the wall block configuration, C_L=2.170. In contrast, the simulated C_D result for the optimum configuration shows big improvement with C_D=0.027, which is lower than C_D=0.298 for without the block configuration. Meanwhile, for concentrated air flow setup, it is shown that the optimum C_L value of 5.316 is a great improvement compared to the on-design value of C_L = 2.170. As for the optimum C_D=0.122, the result is better than C_D = 0.298 for the on-design. All in all, it can be concluded that, with the right configuration of the concentrated air flow setup, the rotating cylinder can be used as a lifting device for very short take-off since it can produce very high coefficient of lift (2.5 times higher) compared with steady air flow configuration.

Acknowledgement
The authors like to thank Assoc. Prof. Dr.-Ing. Surjatin Wiriadidjaja for his support with the fundamental aerodynamics knowledge and application in this study. The authors also acknowledge financial support from Universiti Putra Malaysia through research grant GPB 9531600 and GP-IPB9415402.

References

[1] Seifert J 2012 *Prog. Aerosp. Sci.* **55** 17–45
[2] Collis S S, Joslin R D, Seifert A and Theofilis V 2004 *Prog. Aerosp. Sci.* **40** 237–89
[3] Carstensen S, Mandviwalla X, Vita L and Paulsen U S 2014 *Journal of Ocean and Wind Energy* **1** 41–9
[4] Gowree E R and Prince S A 2012 *28th Int. Congr. Aeronaut. Sci.*
[5] Seifert J 2012 *50th AIAA Aerospace Sciences Meeting*
[6] Rao A, Stewart B E, Thompson M, Leweke T and Hourigan K 2011 *J. Fluids Struct.* **27** 668–79
[7] Lam K M, Liu P and Hu J C 2010 *J. Fluids Struct.* **26** 703–21
[8] Rostamy N, Sumner D, Bergstrom D J and Bugg J D 2012 *J. Fluids Struct.* **34** 105–22
[9] Harichandan A B and Roy A 2012 *J. Fluids Struct.* **33** 19–43
[10] Boudaoud W, Yahiaoui T, Imine B and Imine O 2012 *EPJ Web Conf.* **25** 1104
[11] Marzuki O F, Mohd Rafie A S, Romli F I and Ahmad K A 2017 *Acta Mechanica*
[12] Marzuki O F, Mohd Rafie A S, Romli F I and Ahmad K A 2015 *ARPN Journal of Engineering and Applied Sciences* **10** 9725-9
[13] Karabelas S J, Koumroglou B C, Argyropoulos C D and Markatos N C 2012 *Appl. Math. Model.* **36** 379–98
[14] Panda S K and Chhabra R P 2010 *J. Nonnewton. Fluid Mech.* **165** 1442–61
[15] Reid E G 1924 *Tests of Rotating Cylinders* National Advisory Committee for Aeronautics
[16] ANSYS Inc. 2004 *ANSYS Modeling and Meshing Guide*

IOP Conf. Series: Materials Science and Engineering **270** (2017) 012004 doi:10.1088/1757-899X/270/1/012004

Coandă configured aircraft: A preliminary analytical assessment

M F Abdul Hamid*, E Gires, A S M Harithuddin, A R Abu Talib, A S M Rafie, F I Romli and M Y Harmin

Department of Aerospace Engineering, Faculty of Engineering, Universiti Putra Malaysia, 43400 Serdang, Selangor, Malaysia

*mohd_faisal@upm.edu.my

Abstract. The interest in the use of flow control for enhanced aerodynamic performance has grown, particularly in the use of jets (continuous, synthetic, pulsed, etc.), compliant surface, vortex-cell, and others. It has been widely documented that these active control concepts can dramatically alter the behaviour of aerodynamic components like airfoils, wings and bodies. In this conjunction, with the present demands of low-cost and efficient flights, the use of Coandă effect as a lift enhancer has attracted a lot of interest. Tangential jets that take advantage of the Coandă effect to closely follow the contours of the body have been considered to be simple and particularly effective. For this case, a large mass of surrounding air can be entrained, hence amplifying the circulation. In an effort to optimize the aerodynamic performance of an aircraft, such effect will be critically reviewed by taking advantage of recent progress. For this purpose, in this study, the design of a Coandă-configured aircraft wing will be mathematically idealized and modelled as a two-dimensional flow problem.

1. Introduction

In the era of globalization and high global concern on energy efficient flights, aircraft manufacturers have to revolutionize beyond conventional designs. Significant efforts have been devoted to the use of advanced technologies for greater energy extraction and conversion, as well as optimized propulsion systems. To fully utilize the technological elements that can contribute to innovations of future design concepts, one should explore the fundamental principles that play key roles in these technologies. One of these technologies that is of interest in this study is the circulation control wing (CCW), which is known to be a means of enhancing the bound circulation [1]. This technology is known to be superior for aerospace applications and has been extensively investigated over many years, both experimentally and numerically [2-12]. In brief, the CCW technique is executed by tangentially blowing high-velocity jet airflow over a highly curved surface. This creates a balance of the pressure and centrifugal forces, causing the boundary layer and the jet sheet to remain attached to the surface without any separation (i.e. phenomenon referred to as Coandă effect). Consequently, the rear stagnation point is repositioned toward the lower airfoil surface, further enhancing the airflow circulation around the entire airfoil. The outer irrotational flow is also turned substantially, leading to a higher value of the lift coefficient that is comparable to that achievable from the use of conventional high lift systems [2]. Figure 1 shows the introduction of a thin jet that is associated with tangential blowing over trailing edge surface, which represents one technique for lift enhancement.

IOP Conf. Series: Materials Science and Engineering **270** (2017) 012004 doi:10.1088/1757-899X/270/1/012004

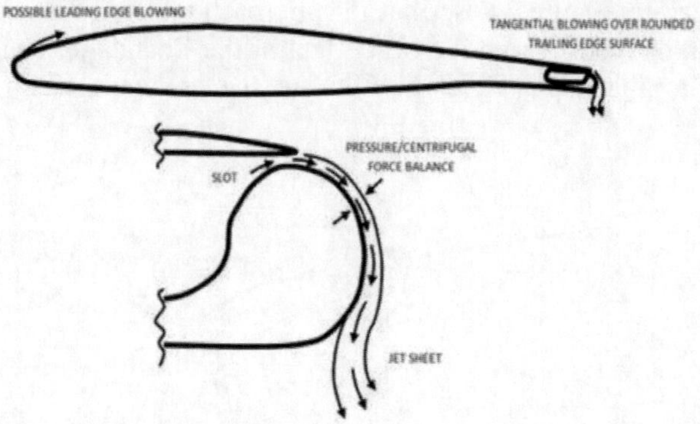

Figure 1: Coandă-jet over trailing edge surface for lift enhancement, adapted from [3]

Early CCW designs typically had a large-radius rounded trailing edge to maximize the lift benefit. These designs, however, also result in high drag penalty when the jet is turned off. One way to tackle this problem is by redesigning the shape of the trailing edge. A numerical simulation has shown that the best lift enhancing effect could be facilitated by designing the lower part of the trailing edge to be a flat surface and preserving the upper surface with a highly curved surface [4]. This curvature on the upper surface produces a large jet turning angle, leading to high generated lift. The progress of high-speed computers has made it possible to use the first-principles-based computational approaches for aerodynamic modelling of propeller blades, for example. Since these approaches are based on the laws of conservation of mass, momentum and energy, they can capture much of the physics in great details [5-7]. However, such approach also requires a considerable computational cost. Therefore, prior to the execution of a much more involved and complex computational model (e.g. via computational fluid dynamics (CFD)), a simple two-dimensional mathematical representation of the model can be drawn to pave the way. Similar approaches on modelling such designs have shown remarkable results [8-10].

In this particular work, the focus is placed on the basic airflow idealization over a two-dimensional (2D) curved rounding-off surface in subsonic flow (airfoil trailing-edge modification). A generalized mathematical model representing the momentum-jet-flow and the changes of pressure over the curved surfaces are drawn and presented for clarity. This simplified model is then analytically evaluated to determine aerodynamic performance of the Coandă-configured aircraft design. It is a kind of blended or hybrid method, combining the results from CFD and analytical models, offering a simple solution to approach such complex design problem. It should be noted that this method is not developed to replace CFD but more as a cheaper alternative to eliminate the repetitive computational cost and the reiteration of modelling procedures required.

2. Methodology

The study presented here is aimed to assist in the preliminary analysis of a Coandă configured aircraft, focusing on the aerodynamic aspect of the design. The starting point is using the 2D representations of a wing (i.e. airfoil). The method used is based on the fundamentals of physics: momentum theory [13] and Bernoulli's principle [14]. The total lift (L_T) is the sum of forces from the free-stream (L_∞) flow over an airfoil (CFD results from Ref. [8]; jet is turned off) plus the lift force generated by the Coandă effect (L_C) at the TE. This is given by Equation 1.

$$L_T = L_\infty + L_c \tag{1}$$

Moreover, the lift force from the Coandă effect consists of two parts: lift due to the jet-momentum (L_j) and lift due to the pressure differential (L_p). This turns Equation 1 into Equation 2.

$$L_T = L_\infty + \left(L_j + L_p\right) = L_\infty + \left(\int \dot{m} v_j \, ds + \int p\, r' \, ds\right) \tag{2}$$

The lift generated by Coandă effect is determined using the surface integral from pressure distribution acting on the airfoil surface. Figure 2 describes the approach taken in this work to arrive at a rounded trailing edge configuration, considering the best effect of the Coandă-jet configuration by redesigning the lower part near the trailing edge (TE) to be a flat surface, as suggested by Ref. [4].

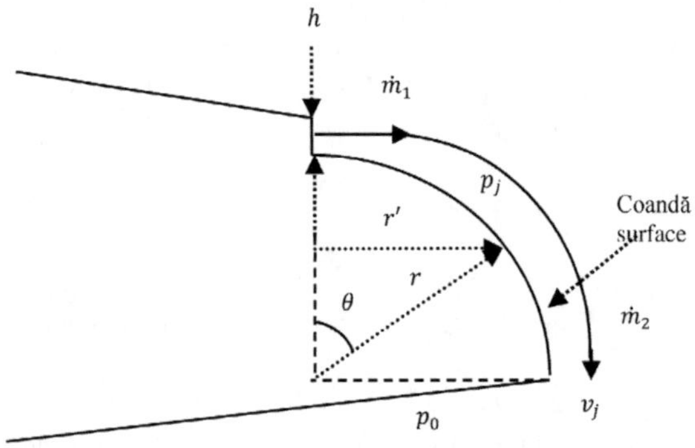

Figure 2: Construction of a rounded curved surface at the vicinity of the airfoil trailing-edge

For the first part, the lift from the momentum jet (bends of the jet flow towards the ground), L_j can be written as Equation 3. Then, based on Ref. [8], the height of jet at TE is taken to be $h = \frac{1}{50} r$. With this information, Equation 3 now becomes Equation 4.

$$L_j = \int \dot{m} v_j \, ds = \int (\rho A v_j) v_j \, ds = \rho {v_j}^2 \int_0^h dh \tag{3}$$

$$L_j = \frac{1}{50} \rho r {v_j}^2 \tag{4}$$

For the second part, the lift from pressure differential (between the lower and upper surfaces), L_p is derived from Bernoulli's equation as in Equation 5. The lift force due to pressure differential over the curved surface is then given by Equation 6.

$$p_j + \frac{1}{2} \rho {v_j}^2 = p_0 \quad \rightarrow \quad p = p_0 - p_j \tag{5}$$

$$L_p = \int p \, r' \, ds = p \int_0^{\frac{\pi}{2}} r' \, d\theta \tag{6}$$

From Figure 2, it can be observed that $r' = r \sin\theta$. Substituting this information, the lift generated by the pressure differential on the curved surface can be written as Equation 7.

$$L_p = pr \int_0^{\frac{\pi}{2}} \sin\theta \, d\theta \tag{7}$$

From CFD result of Ref. [8], $L_\infty = 17.41$ N/m. This is the lift force generated at $Re = 1 \times 10^6$ from a Coandă-modified S809 airfoil configuration when the jet flow, v_j is turned off. Inserting this value of L_∞ and substituting Equation 4 and Equation 7 into Equation 2, the component of lift forces from the jet-momentum and the pressure differential is given by Equation 8.

$$L_T = (17.41) + \left(\frac{1}{50} \rho r {v_j}^2 + \rho r {v_j}^2\right) = 17.41 + \left(\frac{51}{50} \rho r {v_j}^2\right) \tag{8}$$

The calculation of the difference or error between the CFD result (COMSOL) and the result obtained using Equation 8 can be done using Equation 9.

$$\% \text{ Error} = \frac{|\text{Calculated} - \text{Simulation}|}{\text{Calculated}} \times 100\% \tag{9}$$

IOP Conf. Series: Materials Science and Engineering **270** (2017) 012004 doi:10.1088/1757-899X/270/1/012004

3. Analysis and Discussion

It should be noted that, as to simplify the whole problem and to retain all physical aspects of the flow, few assumptions have been made in this study: the analytical formulation is modelled at the vicinity of the TE (as shown in previous Figure 2), a uniform jet velocity profile has been adopted, the jet size is constant, the jet speed or velocity is taken to be the same or larger than the local velocity at the outer edge of the boundary layer and no flow separation occurs over the curved surfaces. The tabulated results are presented in Table 1, which show the generated lift forces from two respective sources; each with specific design conditions (as indicated in Figure 3). The data in the second column of Table 1 is obtained from previous parametric study on S809 airfoil with the jet turned on, which is available in Ref. [8]. Next to it is the obtained results by combining the CFD result when jet is off [8], plus the analytical result that is obtained by Equation 8. All in all, the tabulated data in Table 1 shows that, as the Coandă-jet speed increases, the lift forces are also increased. The data in the last column shows the respective error between the two approaches at each Coandă-jet (or momentum-jet) design condition.

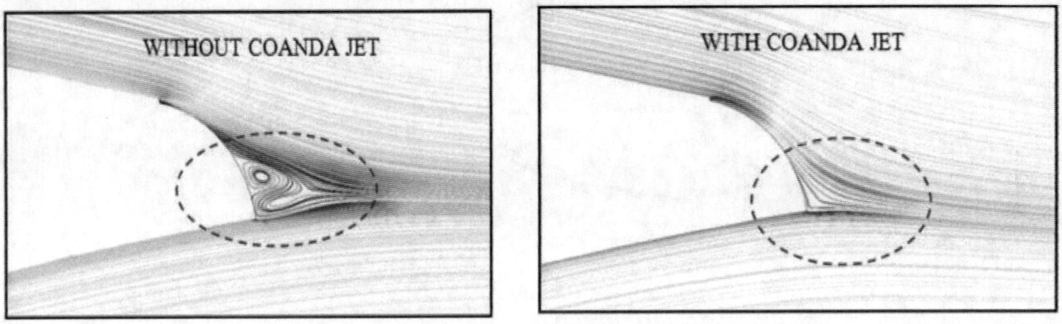

(a) (b)

Figure 3: Flow fields of S809 airfoil: (a) with, (b) without - Coandă-jet [8]

Table 1: Lift forces from the analytical and numerical results

Coandă-jet (m/s)	$\text{Lift}_{\text{COMSOL}}$ (N/m)	$\text{Lift}_{\text{Calculated}}$ (N/m)	Error (%)
15	25.47	28.66	11.11
20	41.89	37.41	11.99
25	58.50	48.65	20.24
30	75.37	62.40	20.80
35	91.71	78.64	16.62
40	106.95	97.38	9.82

Figure 4 shows that both lines of the lift forces are increasing with the Coandă-jet. The trends for both plotted lines are relatively consistent. At lower Coandă-jet values (i.e. <20 m/s), it shows that the analytical model can give a good prediction. Nonetheless, as the Coandă-jet gets higher (i.e. >20 m/s), the analytical model is shown to give lower values than the numerical model. A plausible explanation for this situation can be the fact that, as the momentum from the Coandă-jet gets higher (i.e. numerical model), the outer irrotational flow will tend to turn. This effectively promotes a substantial increase in the circulation, which eventually leads to the higher generated lift. At present, this analytical model is unable to capture the flow interaction effects from the outer irrotational flow. However, the analytical model of a Coandă-configured aircraft-wing design has presented a simple mathematical correlation to simplify the problem.

4. Conclusion

Implementation of the analytical method offers a simple modelling technique to outfit complex fluid flow problems. It can provide a valuable propulsive (i.e. lift enhancement) modelling technique when the problem is too costly to be computed. The result obtained thus far indicates that the analytical model of the Coandă configured aircraft-wing design is applicable and can be used as an alternative way to estimate the Coandă performance. This approach combines the results from CFD and analytical models, offering cheaper alternative to avoid repetitive computations. In particular, for problems that shares almost similar physical modelling design layout and setup. It is very useful, for instance, to focus on the impact of a single design parameter while retaining the rest at constant; as commonly practiced in a parametric study. A final note is that this study is only valid upon certain designs and assumptions with a limited range of Coandă-jet flow. Thus, more detailed analysis is required since the present analytical work is only valid as to assist in assessing similar models.

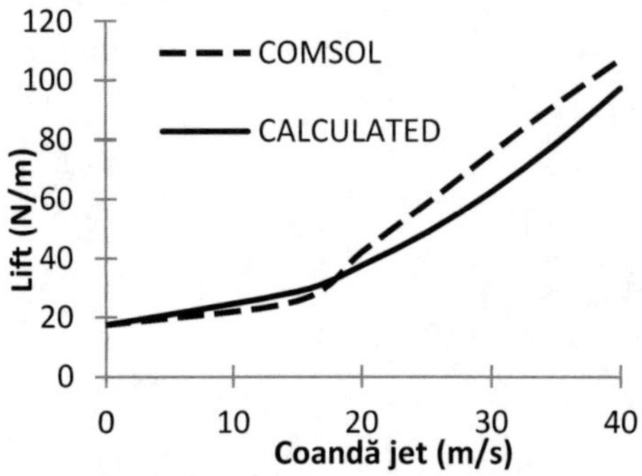

Figure 4: Comparison of the lift generated from the analytical and numerical results

Acknowledgement

The authors wish to acknowledge the contribution of Prof. Harijono Djojodihardjo in the early stages of this study. This work has been funded by Universiti Putra Malaysia through Geran Putra IPM (GP-IPM/2017/9537900).

References

[1] Gad-el-Hak M 1996 *Applied Mechanics Reviews* **49** 365-79
[2] Djojodihardjo H and Abdul Hamid M F 2013 *ASEAN Eng. J.* **3** 39-57
[3] Englar R J, Smith M J, Kelley S M and Rover R C 1994 *Journal of Aircraft* **31** 1160-8
[4] Tongchitpakdee C, Benjanirat S and Sankar L N 2006 *Journal of Solar Energy Engineering* **128** 432-44
[5] Wu J, Sankar L and Kondor S 2004 *42nd AIAA Aerospace Science Meeting and Exhibit*
[6] Guanpeng X and Sankar L N 2000 *Journal of Solar Energy Engineering* **122** 35-9
[7] Liu Y, Sankar L N, Englar R and Ahuja K 2001 *39th AIAA Aerospace Sciences Meeting*
[8] Djojodihardjo H, Abdul Hamid M F, Basri S, Romli F I and Abdul Majid D L A 2011 *IIUM Engineering Journal* **12**
[9] Schroijen M and Van Tooren M 2009 *AIAA/ASAME/SAE/ASEE Joint Propulsion Conference and Exhibit*
[10] Ahmed R I, Abu Talib A R, Mohd Rafie A S and Djojodihardjo H 2017 *Aerospace Science and Technology* **62** 136-47
[11] Englar R J, Smith M J, Kelley S M and Rover R C 1994 *Journal of Aircraft* **31** 1169-77
[12] Shojaefard M H, Noorpoor A R, Avanesians A and Ghaffarpour M 2005 *American Journal of Applied Sciences* **2** 1474–80
[13] Froude R E 1889 *Trans. Inst. Naval Architects* **30** 390
[14] Batchelor G K 1967 *An Introduction to Fluid Dynamics* Cambridge University Press

IOP Conf. Series: Materials Science and Engineering **270** (2017) 012005 doi:10.1088/1757-899X/270/1/012005

Experimental result analysis for scaled model of UiTM tailless blended wing-body (BWB) Baseline 7 unmanned aerial vehicle (UAV)

R E M Nasir, A M Ahmad*, Z A A Latif, R M Saad and W Kuntjoro

Aviation Technology Research Group, Flight Technology and Test Centre (FTTC), Faculty of Mechanical Engineering, Universiti Teknologi MARA, 40000 Shah Alam, Selangor, Malaysia

*namiahmad11@gmail.com

Abstract. Blended wing-body (BWB) aircraft having planform configuration similar to those previously researched and published by other researchers does not guarantee that an efficient aerodynamics in term of lift-to-drag ratio can be achieved. In this wind tunnel experimental study, BWB half model is used. The model is also being scaled down to 71.5% from the actual size. Based on the results, the maximum lift coefficient is found to be 0.763 when the angle is at 27.5° after which the model starts to stall. The minimum drag coefficient is 0.014, measured at zero angle of attack. The corrected lift-to-drag ratio (L/D) is 15.9 at angle 7.8°. The scaled model has a big flat surface that surely gives an inaccurate data but the data obtained shall give some insights for future perspective towards the BWB model being tested.

1. Introduction

Blended wing-body (BWB) aircraft, through its unique configuration and potential benefits, is well-suited for the role of environmental-friendly, long range, high capacity airliner [1]. However, there are issues of flight stability and control that need to be addressed and solved. The BWB tends to have poor departure characteristics due to its lower maximum lift coefficient, which results from the absence or limited number of high lift devices and also tails with long moment arm [2]. The tailless nature of the BWB aircraft with multiple elevons as control surfaces requires understanding of the impact of these elevons have towards stability. Such BWB aircraft often requires an active flight control system [3].

The study on the BWB unmanned aerial vehicle (UAV) has begun in Universiti Teknologi MARA (UiTM) since 2005, which is focused on the design, fundamental aerodynamics and flight dynamics of small BWB [4]. It is found from previous studies that a BWB aircraft having planform configuration just like the one previously being researched and published by other researchers somehow does not guarantee it to have efficient aerodynamics in terms of lift-to-drag ratio (L/D) [5]. Studies conducted by UiTM have found that each variation of BWB aircraft design has its own advantages and disadvantages [6]. The Baseline 7 BWB design, as shown in Figure 1, has been introduced to study the behavior of the control surfaces, given four elevons and without vertical stabilizer and wingtip. The challenge is that this aircraft design will have to fly with improved stability solely depending on airfoil selection and four control surfaces to stabilize the aircraft [6-7]. The tailless design shall simplify the mechanism while improving the design characteristics of Baseline 7. However, this paper only focuses on the aerodynamic behavior and characteristics of the Baseline 7 scaled model analysis inside the wind tunnel with a closed system configuration.

2. Model and Wind Tunnel Setup

The experimental test is performed in Universiti Pertahanan Nasional Malaysia (UPNM) Wind Tunnel as shown in Figure 2. The wind tunnel is a suction-type tunnel with a test section area of 0.2m x 0.2m and it is equipped with three-component force balance equipment that able to measure lift and drag. For this wind tunnel experimental setup, half model of the BWB is used. The model is scaled down to 1:7 from the actual size. The dimensions and concept of the scaled half model are shown at Figure 3 and Figure 4.

The experiment is conducted at a speed of 49.58m/s, which represents 15 m/s for full-scaled model wind tunnel test. The wing span of the scaled model had been minimised and that is 0.021m^2. The pitching angles being tested are between -10° to +30° with fixed elevator. The experiment is done from angles of attack of -10° up to +30°. The increment of 2° interval is chosen to identify the best configuration of each aerodynamic coefficient for the model. The test location has an ambient temperature of 300K with the density of air around 1.155kg/m^3.

Since the experiment is done in the wind tunnel with a closed-loop system, some corrections have to be made to achieve true values of aerodynamic coefficient [8]. The values obtained in the closed-system wind tunnel must be corrected to eliminate several effects of interference [9]. Influence of the tunnel walls and interaction between tunnel wall and model will affect the aerodynamic characteristics results [9].

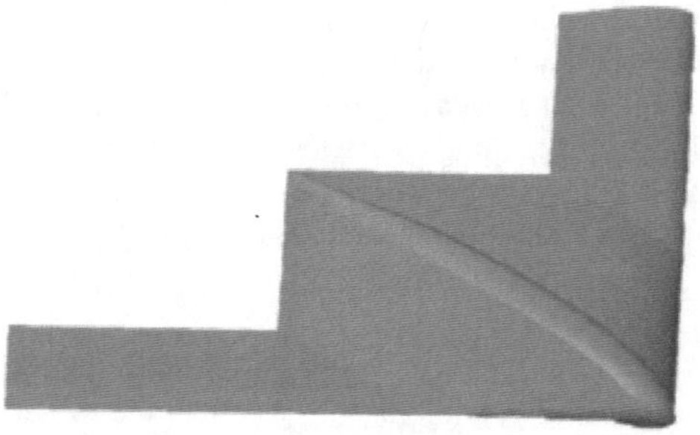

Figure 1: Baseline 7 BWB design

Figure 2: UPNM low speed wind tunnel

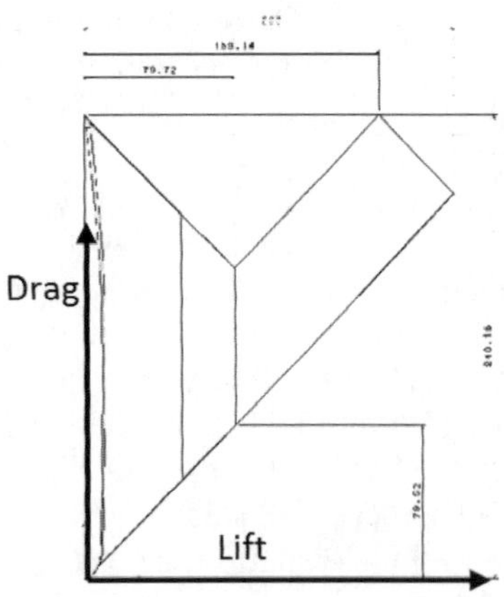

Figure 3: Configuration of wind tunnel model used from plan view

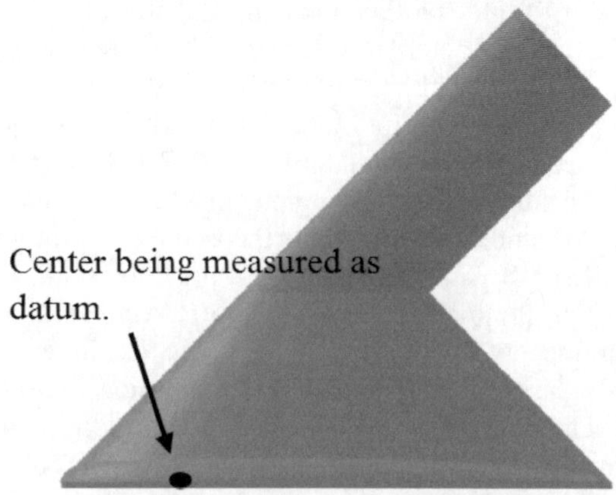

Figure 4: Half scaled wind tunnel model from the front view

The first correction is for the blockage correction. Solid blockage correction is a simple velocity correction in which the flow of air inside the wind tunnel will be in the freestream condition [10]. The solid blockage sometimes can be neglected because of its minimal correction to the velocity of the air flow affected during solid blockage correction. This is done using Equation 1 and Equation 2 where ΔV is the corrected velocity of the airspeed, ε_{sb} is the solid blockage factor, V_u is uncorrected velocity of the airflow at the leading edge, V_b is the Volume of the model, K_1 for vertical model is 0.52 and S is the working section area.

$$\Delta V = \varepsilon_{sb} V_u \tag{1}$$

$$\varepsilon_{sb} = \frac{K_1 V_b}{S^{3/2}} \tag{2}$$

The wake correction, on the other hand, carries out a handful of corrected parameters such as wake correction factor and velocity of airflow at trailing edge of the model. This is done using the following Equation 3 and Equation 4 where c_{d_u} is the uncorrected drag coefficient, c is the model's length and h is the height of working section.

$$\Delta V = \varepsilon_{wb} V_U \tag{3}$$

$$\varepsilon_{sb} = \frac{c}{2h} c_{d_u} \tag{4}$$

Furthermore, streamline curvature correction will correct the drag coefficient, lift coefficient, angle of attack and also moment coefficient. The wind tunnel wall is artificially curved to form a curvature streamline inside the system and because of this condition, the correction is required to achieve better results from wind tunnel closed system experimental procedure [11]. Equation 5 up to Equation 10 are involved in this correction procedure. In these equations, $\propto$ is the corrected angle of attack, $\propto_u$ is the uncorrected angle of attack, C_l is corrected lift coefficient, C_{lu} is uncorrected lift coefficient, $C_{m\,1/2u}$ is the uncorrected moment coefficient halved, ε is the summation of solid blockage factor and wake blockage factor, C_{dc} is corrected drag coefficient and C_{du} is uncorrected drag coefficient.

$$\sigma = \frac{\pi^2}{48} \left(\frac{c}{h}\right)^2 \tag{5}$$

$$\propto = \propto_u + \frac{57.3\sigma}{2\pi} (C_{lu} + 4C_{m\,1/2u}) \tag{6}$$

$$C_l = C_{lu}(1 - \sigma - 2\varepsilon) \tag{7}$$

$$C_{m\,1/2} = C_{m\frac{1}{2}u}(1 - 2\varepsilon) + \frac{\sigma C_l}{4} \tag{8}$$

$$V = V_u(1 + \varepsilon) \tag{9}$$

$$C_{dc} = C_{du}(1 - 3\varepsilon_{sb} - 2\varepsilon_{wb}) \tag{10}$$

3. Results and Discussion

Figure 5 shows the plot of lift coefficient versus angle of attack (i.e. from -10° to +30°). The lift coefficient is at maximum with 0.822 when the angle is at 27.5°. The model then starts to stall after the angle of maximum lift coefficient based on the plot. Nonetheless, after the wind tunnel data has been corrected, the maximum lift coefficient value is changed to 0.763 though it occurs at the same angle of attack as before. A small change in lift coefficient is observed throughout the data correction analysis. When the angle of attack is at 0°, the lift coefficient is 0.036 as shown in Figure 6.

In the meantime, Figure 7 shows the drag coefficient versus angle of attack plot. Drag coefficient is at its maximum at about 27.5° angle of attack whereas its minimum value of around 0.014 is observed at the angle of attack of 2.2°. The corrected drag coefficient data shows a slight decrease in value as it approaches the higher angle of attack. A clearer graph is observed from Figure 8 to identify the linear trendline obtained for the drag coefficient versus angle of attack plot.

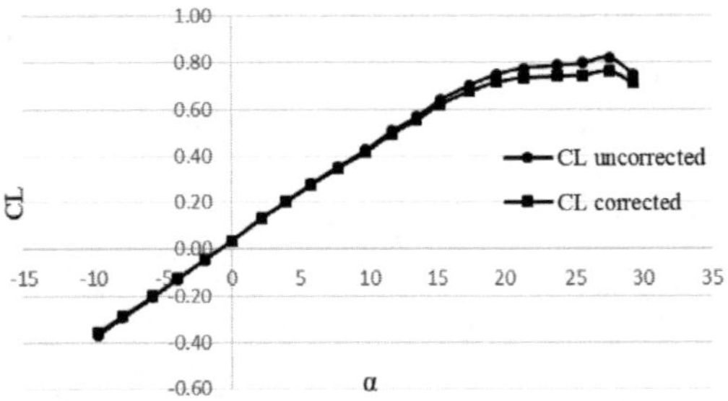

Figure 5: Lift coefficient versus angle of attack

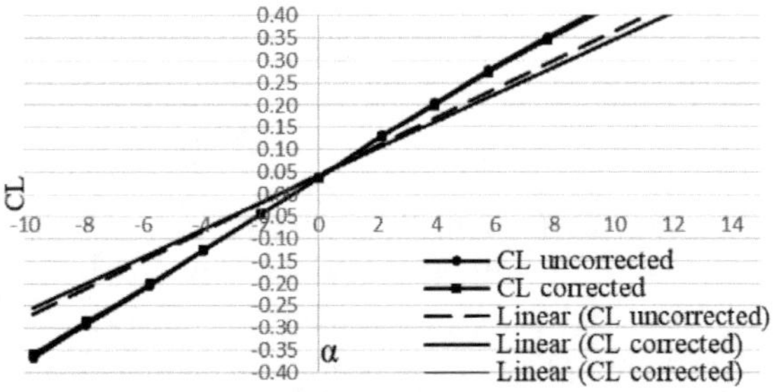

Figure 6: Lift coefficient versus angle of attack with trendline

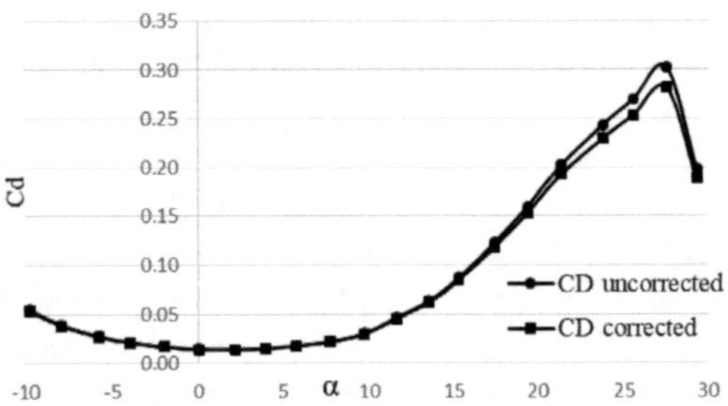

Figure 7: Drag coefficient versus angle of attack

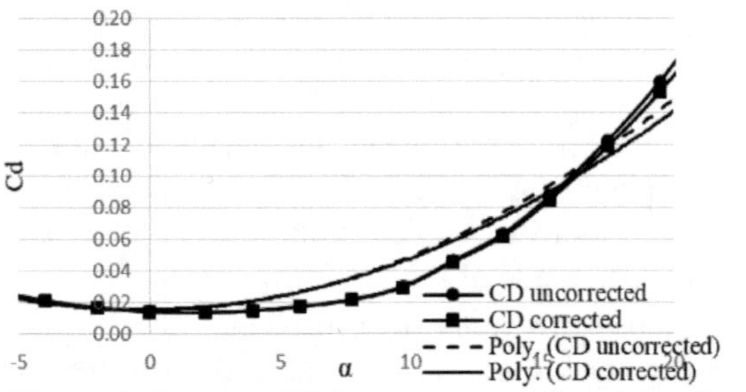

Figure 8: Drag coefficient versus angle of attack with trendline

The L/D versus angle of attack graph is presented in Figure 9. The maximum value of L/D from this experiment is achieved at angle 7.74° with a value of 16. However, the corrected L/D results in a just slightly lower value of 15.9 at angle 7.8°. At zero angle of attack, the L/D obtained is 2.6 with the corrected value. These values indicate that L/D value for Baseline 7 BWB is not that as high as L/D = 21 recorded for previous Baseline II design [5]. The higher the value of L/D of an aircraft, the better is the aerodynamic efficiency of the aircraft. In this experiment, the value of L/D for this design airplane is somewhat average compared to other previous baseline models. From the graph, the change of L/D value between uncorrected and corrected is negligible.

4. Conclusion

Data obtained from the wind tunnel experiments in this study are analyzed to obtain the aerodynamic performance characteristics for the BWB Baseline 7. The maximum lift coefficient is 0.822 at angle of attack 27.5° for uncorrected value and 0.763 at angle of attack 27.5° for corrected value. Value of drag coefficient gives a promising value of 0.014 at angle 0° for both corrected and uncorrected values. L/D value obtained from this experiment indicates a promising and good corrected value of 15.9, which is among the top of UiTM's BWB Baseline series. Table 1 shows the BWB Baseline UiTM models dated from 2005 to 2014 [4-9].

Several recommendations can be made to improve the evaluation of data in this experiment. The wind tunnel model should be fabricated with a smoother finishing such that the flow does not experience any major wake exerted at the skin of the model. In addition, the wind tunnel should be equipped with more data evaluator such as speed sensor, density sensor and more accurate force balance. The scaled model has a big flat surface that certainly gives an inaccurate data but the data obtained in this study can be used to shape the future perspective towards the model being tested.

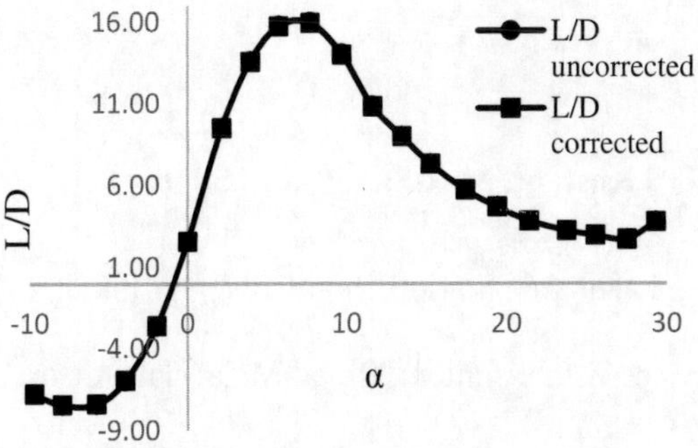

Figure 9: L/D versus angle of attack

Table 1: L/D of BWB baseline model being designed in UiTM

Year	Model	L/D Max
2005	Baseline-I	14.0
2009	Baseline-II	24.0
2012	Baseline-III	13.0
2013	Baseline-IV	15.0
2014	Baseline-V	32.0
2017	Baseline 7	15.9

Acknowledgement
The authors would like to express their deepest gratitude to Ministry of Higher Education for granting the research fund related to this experiment (FRGS/1/2016/TK09/UITM/02/1). Our appreciation also goes to the Research Management Institute and the Faculty of Mechanical Engineering, UiTM for the various support and facilities provided in this study.

References

[1] Ikeda T 2006 *Aerodynamic Analysis of A Blended-wing-body Aircraft Configuration* Bachelor Thesis, RMIT University
[2] Okonkwo P and Howard S 2016 *Progress in Aerospace Sciences* **82** 1-23
[3] Roman D, Allen J B and Liebeck R H 2000 *18th Applied Aerodynamics Conference*
[4] Kuntjoro W, Mohd Nasir R E, Wisnoe W, Mamat A M I and Abdulah M R 2008 *Aircraft Structural Design Conference*
[5] Wisnoe W, Mamat A M I, Mohd Nasir R E, Kuntjoro W and Ramly R 2008 *International Conference on Mechanical and Manufacturing Engineering*
[6] Nasir R E M, Mazlan N S C, Ali Z M, Wisnoe W and Kuntjoro W 2016 *IOP Conference Series: Materials Science and Engineering* **152** 012021
[7] Nasir R E, Kuntjoro W, Wisnoe W and Mamat A M 2009 *Journal of Mechanical Engineering* **6** 73-96
[8] Hensel R W 1951 *NACA-TN-2433*
[9] Allen H J and Vincenti W G 1944 *NACA Technical Report 782*
[10] Ross I and Altman A 2011 *Journal of Wind Engineering and Industrial Aerodynamics* **99** 523-38
[11] Peckham D H and Atkinson S A 1957 *Aeronautical Research Council Technical Report (C. P. No. 508)*

IOP Conf. Series: Materials Science and Engineering **270** (2017) 012006 doi:10.1088/1757-899X/270/1/012006

Size-scale effects of inboard Store-Y on X-plane model: A parametric study using CFD

M A Rahman[1,2], M Y Harmin[1,*], M F S Koslan[2], M R Saad[3], F I Romli[1] and M T Ahmad[3]

[1]Department of Aerospace Engineering, Faculty of Engineering, Universiti Putra Malaysia, Serdang, Selangor, Malaysia
[2]Center of Aerospace Engineering Services Establishment, Royal Malaysian Airforce, Subang Air Base, Malaysia
[3]Department of Mechanical Engineering, Faculty of Engineering, National Defence University of Malaysia, Sungai Besi, Malaysia

*myazdi@upm.edu.my

Abstract. This paper presents the computational fluid dynamics (CFD) study on the effects of the Store-Y integration to the inboard pylon of X-plane model to its aerodynamic performance. Parametric study in terms of the size-scale of the Store-Y model is simulated at various angles of attack and air velocities. The resultant lift, drag and moment force parameters are recorded in each simulation and then analysed. The findings show that the size-scale effect of Store-Y model is insignificant to the aerodynamic performance, hence offers a possibility of other types of store to be integrated at this station without affecting the overall aircraft performance.

1. Introduction

Computational fluid dynamics (CFD) has been widely used today for analyzing the characteristics of the flow that passes by an object. The development in CFD simulation is always closely tied with the enhancement in computer technology, which enables the governing complex fluid dynamic equation of Navier-Stoke to be solved in a much larger solution domain [1]. Because of this capability, CFD is becoming one of the most practical tools in aerospace engineering for the computation of the aircraft's aerodynamic performance. For instance, the CFD simulation of CREATE-AV/Kestrel Solver has been used for both F-16 and F-22, and the results are found to be in very good correlation with the original equipment manufacturer (OEM) data in terms of aerodynamic performance characteristics [2]. On the other hand, the 3D wing of NACA 2412 has also been simulated using a well-known CFD software of ANSYS FLUENT and the results are comparable with the theory of lift generation [3].

There are many other studies that have been performed to verify the CFD simulation results with the experimental data. These include, among others, the use of 3D RANS CFD software to simulate the canard for SAAB passenger aircraft [4], ANSYS CFX CFD software for different wing design of remote control aircraft [5], General Zonal Euler Solver (ZEUS) to simulate the supersonic missile of SM-2 Block IVA [6] and the use of CFD++ software to study the drag force on a different geometry of nacelle [7]. All these results have been verified with the wind tunnel testing and they are found to be in good agreement. Therefore, the CFD tool has been proven as reliable in computing the aerodynamic performance of the considered model.

One of the main challenges in CFD simulation is to acquire an effective meshing while assuring the accuracy of the results [8]. As the complexity of the model increases, the overall meshing procedure is also becoming more complicated, leading in increment of number of elements as well as resulting to a much higher computational time. In addition, the CFD solution may become stiffer to be solved and, in the worst case scenario, the numerical integrator might even fail to converge. Having this in mind, an effective strategy for the setup of CFD environment is required to achieve computational efficiency of the CFD simulation. In this work, the CFD simulation of ANSYS FLUENT is used to compute the aerodynamic performance of the X-plane wind tunnel model. The CFD environment is set up based on the wind tunnel operational envelope and conditions, along with the use of standard k-ε model as the turbulence model.

2. Simulation Modelling

A scaled down model of 1:50 ratio of the X-plane aircraft is considered in this study. The dimension of the model is 0.3m in length, 0.11m in span and 0.45m in height. This studied model represents the X-plane wind tunnel model in a wind tunnel test section of 1m × 1m × 2m in width, height and length, respectively. Initially, the physical scaled down model of the X-plane is scanned using the 3D scanner in order to produce a Computer Aided Design (CAD) model that is compatible with CAD environment of SOLIDWORK. Following this modelling process, the assembly procedure is done to integrate the Store-Y to the inboard pylon of the X-plane. They are then merged together, forming a single solid body of the model. Figure 1 shows CAD models of the clean X-plane and the X-plane with the stores. Note that, only the stores that are located at the inboard pylon of the wing are considered in this study.

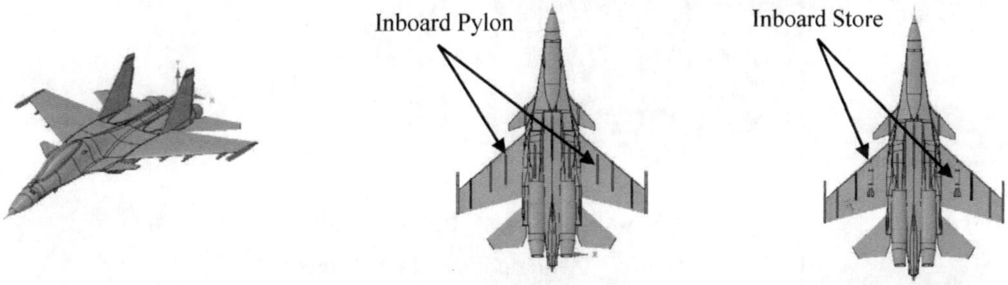

(a) Isometric view (b) Clean X-plane (c) X-plane with Store-Y

Figure 1: X-plane CAD model

The CAD model is then imported to the CFD environment of ANSYS FLUENT, where meshing is done and boundary conditions are defined. Note that, the size of the computational domain (or known as an enclosure) is equivalent to the size of the test section and the boundary conditions have been set based upon the wind tunnel operational envelope and conditions. During the transition from the CAD to the CFD environment, compatibility issues might arise that can cause the model to loss some of its features. Under those circumstances, the repair procedure needs to be carried out to identify any faulty occurrence in terms of edges, faces and gap in the model. Once all necessary repairs have been made, the Boolean operation is then employed for subtracting the solid body from the enclosure, leaving only the respective fluid domain. This operation enables simplification of the meshing process, especially involving the complex region.

The meshing procedure begins by considering a coarse mesh density setting with the employment of the tetrahedron mesh. To ensure a good approximation of the solution, a mesh convergence analysis is carried out by increasing the mesh density of the fluid domain until the solutions have satisfactorily reached its convergence. If the solution is unable to acquire its convergence, the minimum element size, curvature size and skewness parameters have to be redefined accordingly. Figure 2 provides a sample of the mesh convergence analysis for the X-plane with the Store-Y, where the solutions are normalized with the corresponding solutions of mesh density of 1×10^6 elements. It can be seen that the solutions begin to reach its convergence at mesh density of 6×10^5 elements and it is noted that beyond this point, the solutions only differ by no more than 0.02%. Furthermore, Figure 3(a) and 3(b) provide the CFD representation of the surrounding wall and the surrounding model meshing of the considered study, respectively.

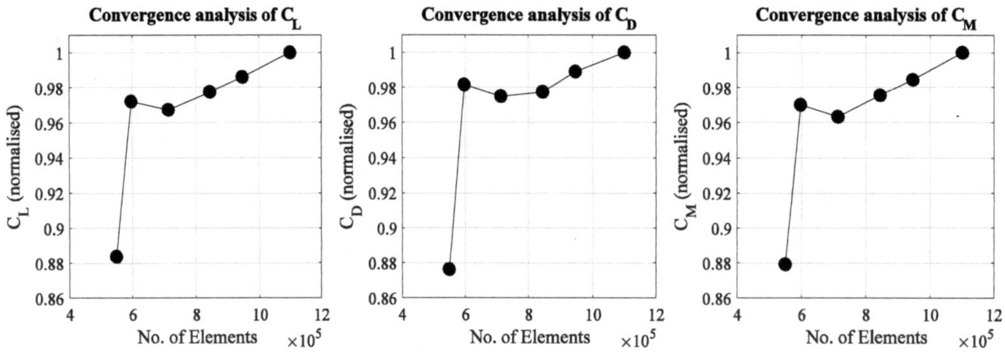

Figure 2: Mesh convergence analysis

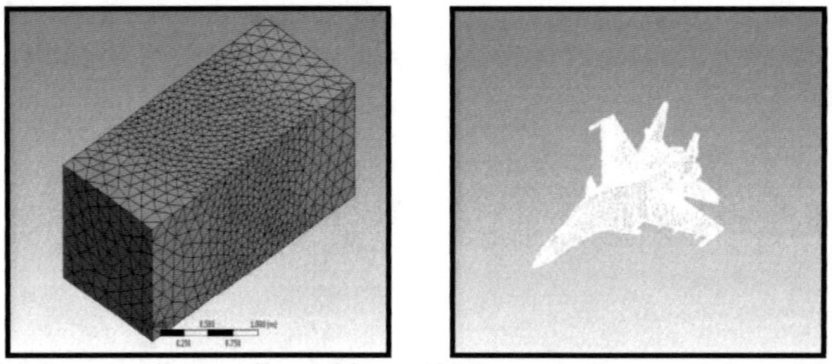

(a) Surrounding wall meshing (b) Surrounding model meshing

Figure 3: X-plane in CFD environment

Once all CFD environment parameters have been satisfactorily defined, the X-plane model is then simulated. A parametric study in terms of store dimensions is carried out under the selected range of velocities and angles of attack, α. A turbulence model of k-ε is employed as it can offer the simplest level of closure, resulting into relatively low computational time [9]. The fluid solver of sweeping and semi-direct methods are utilized to iterate the governing fluid equations to the specified convergence criterion [10].

3. Results and Discussion

Due to the confidentiality of publishing the technical data to the general public, only the plot trends are provided in this paper. The results are presented in terms of lift, drag and moment against independent variable of angle of attack, α with corresponding velocities of namely V1, V2 and V3. The study is intended to provide initial assessment on the effect of the store size to the aerodynamic performance.

Figure 4 shows a comparison in aerodynamic performance between the clean X-plane and X-plane with Store-Y. In terms of the lift trend, the increment in lift magnitude with increasing α is due to the simulation conditions that are below the stall region where the lift curve slope, $\delta L/\delta\alpha$ is positive value. Note that, at a much higher α, the lift begins to lose its performance with of a soft stall characteristic. Meanwhile, in terms of the drag trend, an increase in drag magnitude is mainly due to the increment of the frontal area and boundary layer thickness. Lastly, the negative slope of moment trend indicates the pitching magnitude of nose up condition. Overall, the aerodynamic performance of the X-plane with Store-Y is slightly underperformed as compared to the clean X-plane. This is mainly due to the flow

IOP Conf. Series: Materials Science and Engineering **270** (2017) 012006 doi:10.1088/1757-899X/270/1/012006

over the wing being disturbed by the stores, resulting in increase of boundary layer separation. Since the store is located near to the fuselage, very minimal degradation in aerodynamic performance should be expected.

On the other hand, Figure 5 provides a comparison of the aerodynamic performance of the X-plane for three different size-scales of the Store-Y with scaling ratio of 0.75:1, 1:1 and 1.25:1. Note that the aerodynamic impact due to the store size is still very minimal and hence, the previous discussion is also applicable here as well. Table 1 presents the standard deviation of the simulation results, which describes the level of dispersion among the parametric results of the store models for each considered α and velocity. It can be observed that the standard deviations of the aerodynamic performance are very small, with a deviation of at most 1.01 in magnitude. This shows an insignificant impact of store sizes for the store located at the inboard pylon to the aerodynamic performance.

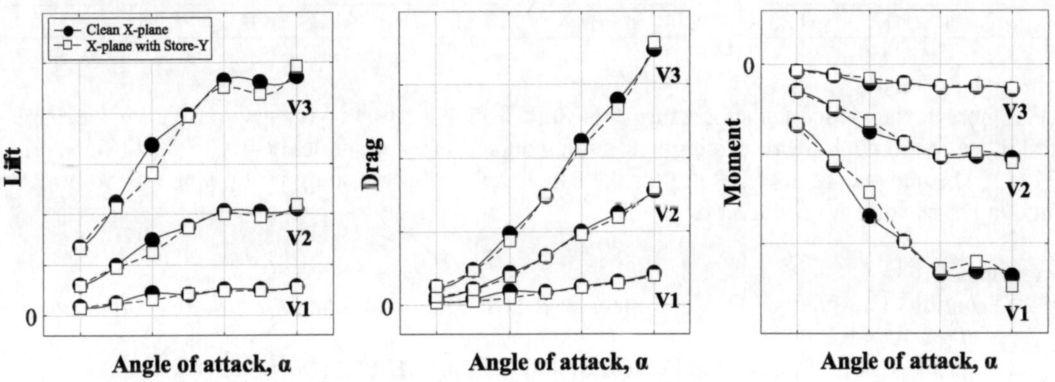

Figure 4: Aerodynamic performance of clean X-plane and X-plane with Store-Y

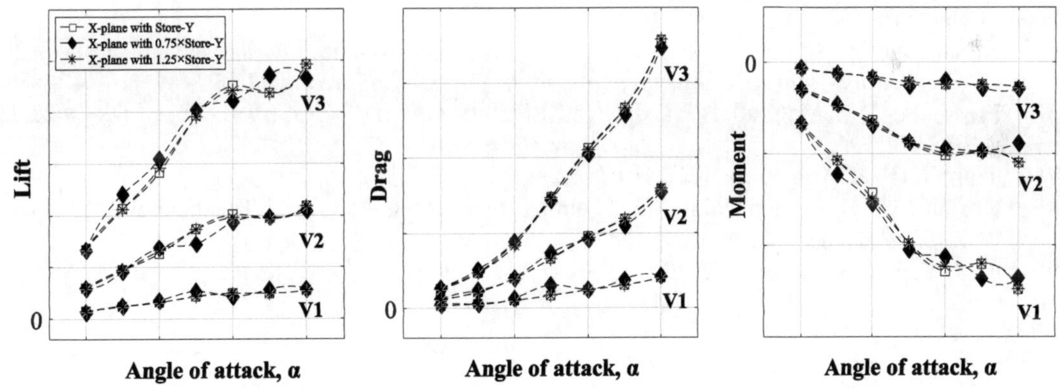

Figure 5: Aerodynamic performance of X-plane with various size scale of Store-Y

4. Conclusion

The X-plane wind tunnel model with different size-scale of Store-Y is modeled and simulated on the CFD simulator. The finding is encouraging, where the size-scale effects of the inboard Store-Y on the X-plane model is found to be insignificant. This provides a possibility for another type of stores to be installed at this particular station without affecting the overall aerodynamic performance of the aircraft. Further investigation should be concentrated in validating the findings of this study through wind tunnel testing.

Table 1: Standard deviation of the simulation results on the size-scale effects of Store-Y

α	Standard Deviation								
	Lift			Drag			Moment		
	V1	V2	V3	V1	V2	V3	V1	V2	V3
α-1	0.06	0.01	0.02	0.04	0.02	0.04	0.05	0.02	0.04
α-2	0.03	0.09	0.84	0.00	0.07	0.12	0.02	0.08	0.36
α-3	0.10	0.29	0.68	0.08	0.01	0.14	0.13	0.25	0.57
α-4	0.29	0.83	0.36	0.40	0.23	0.10	0.11	0.02	0.18
α-5	0.17	0.36	0.73	0.02	0.10	0.19	0.13	0.21	0.45
α-6	0.21	0.07	1.01	0.22	0.27	0.26	0.09	0.13	0.40
α-7	0.07	0.25	0.74	0.08	0.11	0.29	0.04	0.51	0.35

Acknowledgement
The authors acknowledge the support for this work from Malaysian Ministry of Science, Technology and Innovation through the research grant funding under ScienceFund RMK-10: Project No. 04-01-04-SF1243. The authors are also grateful for the support given in this study by Lt. Col. (R) Mohd Ramly Ajir and Dr. Md Amzari Md Zhahir.

References

[1] Tannhill J C, Anderson D A and Pletcher R H 1997 *Computational Fluid Mechanics and Heat Transfer* CRC Press
[2] Dean J P, Clifton J D, Bodkin D J and Ratclif C J 2011 *AIAA Aerospace Sciences Meeting*
[3] Triet N M, Viet N N and Thang P M 2015 *VNU Journal of Science: Mathematics-Physics* **31** 68-75
[4] Ghoreyshi M, Darragh R, Harrison S, Lofthouse A J and Hamilton P E 2016 *AIAA Applied Aerodynamics Conference*
[5] Krishnamurthy S, Jayashankar S, Rao S V, Rochen Krishna T S and Nyamanavar S 2014 *International Journal of Mechanical and Production Engineering* **2** 63-8
[6] Frostbutter D A, McGrath B E and Roger R P 2001 *Johns Hopkins Apl Technical Digest* **22** 289-301
[7] Trapp L G and Argentieri H G 2010 *J. Aerosp. Technol. Manag.* **2** 145-54
[8] Koslan M F S, Ahmad Zaidi A M, Othman M Z, Abdullah S and Tanakodi S 2013 *Modern Applied Science* **7** 23-8
[9] Tu J, Yeoh G H and Liu C 2012 *Computational Fluid Dynamics* Elsevier
[10] Kohnke P 1994 *ANSYS Theory Reference Release 5.6*

IOP Conf. Series: Materials Science and Engineering **270** (2017) 012007 doi:10.1088/1757-899X/270/1/012007

Dynamics and control of quadcopter using linear model predictive control approach

M Islam, M Okasha* and M M Idres

Department of Mechanical Engineering, International Islamic University Malaysia, Kuala Lumpur, Malaysia

*mokasha@iium.edu.my

Abstract. This paper investigates the dynamics and control of a quadcopter using the Model Predictive Control (MPC) approach. The dynamic model is of high fidelity and nonlinear, with six degrees of freedom that include disturbances and model uncertainties. The control approach is developed based on MPC to track different reference trajectories ranging from simple ones such as circular to complex helical trajectories. In this control technique, a linearized model is derived and the receding horizon method is applied to generate the optimal control sequence. Although MPC is computer expensive, it is highly effective to deal with the different types of nonlinearities and constraints such as actuators' saturation and model uncertainties. The MPC parameters (control and prediction horizons) are selected by trial-and-error approach. Several simulation scenarios are performed to examine and evaluate the performance of the proposed control approach using MATLAB and Simulink environment. Simulation results show that this control approach is highly effective to track a given reference trajectory.

1. Introduction

The interest in quadcopter has gradually increased among researchers due to its structural simplicity and flexibility in flight, in addition to its versatile applications in both civil and military areas [1]. In conjunction to this, it is vital to operate the quadcopter with proper control to gain better performance from it. An effective controller ensures its smooth and collision-free flight in the complex environment considering aerodynamic drag and moments [2]. In literatures, the quadcopter's control problem has been widely investigated using several control approaches including proportional-integral-derivative (PID), linear quadratic regulator (LQR) and H-infinity for linear control system. In the meantime, for nonlinear control system, approaches like backstepping, feedback linearization and model predictive control are applied. The comparison between PID and LQR control techniques on micro quadcopter has been demonstrated in a previous study and the system is shown to be stabilized around the hover position but PID shows poor performance at different operating points [3, 4]. Moreover, another study has applied the backstepping control approach based on Lyapunov theory to stabilize the quadcopter to track a given desired position and attitude. In that work, an under-actuated subsystem is introduced to control the horizontal position through roll and pitch angles while a fully-actuated subsystem is used to control the vertical position through yaw and a propeller subsystem to control propeller forces [5]. Additionally, another study used feedback linearization for trajectory tracking to control rotational and translational dynamics [6].

Model Predictive Control (MPC) becomes one of the widespread controllers nowadays because of its capability in working with constraints and disturbances, predictive behaviour, simplicity in tuning and advanced performance with multi-variable at the same time. It is considered as a nonlinear control system that works on predicting future states and errors [7]. MPC has been used to track the reference trajectory considering disturbances and nonlinear H-infinity to obtain the robustness of the system in quadcopter [8]. In a previous study, MPC is applied to gain robust performance from the system under wind-gust disturbance condition for attitude reference tracking in quadcopter [9]. In that work, MPC has successfully tracked the reference point using a single MPC technique on the quadcopter platform that considers external disturbances in the system and constraints for the actuators saturation at control inputs. The study presented in this paper investigates the quadcopter's dynamics and control using the MPC approach.

2. System Design

Quadcopter basically holds a rigid cross-linked structure that has four independent rotors with fixed pitched propellers. Among the four propellers, two are rotating in clockwise direction while the other two rotate in anti-clockwise direction, as shown in Figure 1. The control of quadcopter is obtained by changing the angular speed of the propellers Ω_i (i = 1, 2, 3, 4). The rotational movement of quadcopter along X, Y and Z axes can be described by roll (ϕ), pitch (θ) and yaw angle (ψ).

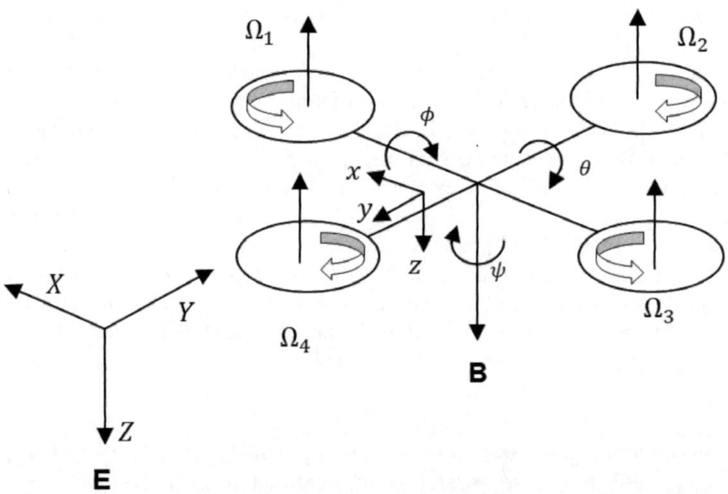

Figure 1: Configuration of quadcopter, where **B** and **E** denote the body fixed frame and the Earth fixed frame, respectively

Every controller input has an effect on a certain movement such as u_2 affects on roll movement, u_3 affects on pitch movement, u_4 affects on yaw movement and u_1 has an effect on upward movement along Z axis. Here, as '+' (plus) configuration is chosen for this quadcopter, the control inputs produce the effects on the system as described in Equation 1 to Equation 4. The initial conditions and nominal parameters for simulation of the quadcopter are shown in Table 1 [10].

$$u_1 = k_f(\Omega_1^2 + \Omega_2^2 + \Omega_3^2 + \Omega_4^2) \tag{1}$$

$$u_2 = k_f(\Omega_4^2 - \Omega_2^2) \tag{2}$$

$$u_3 = k_f(-\Omega_3^2 + \Omega_1^2) \tag{3}$$

$$u_4 = k_M(\Omega_1^2 - \Omega_2^2 + \Omega_3^2 - \Omega_1^2) \tag{4}$$

Table 1: Parameters and initial conditions for simulation

Symbol	Description	Value	Unit
I	Moment of inertia	$\begin{pmatrix} 7.5e-3 & 0 & 0 \\ 0 & 7.5e-3 & 0 \\ 0 & 0 & 1.3e-2 \end{pmatrix}$	kg.m^2
l	Arm length	0.23	m
I_r	Inertia of motor	6e-5	kg.m^2
k_f	Thrust coefficient	3.13e-5	Ns^2
k_M	Moment coefficient	7.5e-7	Nms^2
m	Mass of quadcopter	0.65	kg
g	Gravity	9.81	ms^2
k_t	Aerodynamic thrust drag coefficient	$\begin{pmatrix} 0.1 & 0 & 0 \\ 0 & 0.1 & 0 \\ 0 & 0 & 0.1 \end{pmatrix}$	Ns/m
k_r	Aerodynamic moment drag coefficient	$\begin{pmatrix} 0.1 & 0 & 0 \\ 0 & 0.1 & 0 \\ 0 & 0 & 0.1 \end{pmatrix}$	Nm.s

The mathematical model of the quadcopter is given by Equation 5 to Equation 10 [11-13].

$$\ddot{x} = \frac{-1}{m}[k_{t_x}\dot{x} + u_1(sin\phi sin\psi + cos\phi cos\psi sin\theta)] \tag{5}$$

$$\ddot{y} = \frac{-1}{m}[k_{t_y}\dot{y} + u_1(sin\phi cos\psi - cos\phi sin\psi sin\theta)] \tag{6}$$

$$\ddot{z} = \frac{-1}{m}[k_{t_z}\dot{z} - mg + u_1 cos\phi cos\theta] \tag{7}$$

$$\dot{p} = \frac{-1}{I_x}[k_{r_x}p - lu_2 - I_y qr + I_z qr + I_r q\omega_r] \tag{8}$$

$$\dot{q} = \frac{-1}{I_y}[-k_{r_y}q + lu_3 - I_x pr + I_z pr + I_r p\omega_r] \tag{9}$$

$$\dot{r} = \frac{-1}{I_z}[u_4 - k_{r_z}r + I_x pq - I_y pq] \tag{10}$$

Another kinematic relationship is required between Euler rates, $[\dot{\phi}, \dot{\theta}, \dot{\psi}]^{\mathrm{T}}$ on the earth fixed frame and angular velocity, $[p,\ q, r]^{\mathrm{T}}$ of the quadcopter to describe the whole complete system. This is given by Equation 11 to Equation 13 [14].

$$\phi = p + rcos\phi tan\theta + qsin\phi tan\theta \tag{11}$$

$$\dot{\theta} = qcos\phi - rsin\phi \tag{12}$$

$$\dot{\psi} = r\frac{cos\phi}{tan\theta} + q\frac{sin\phi}{cos\theta} \tag{13}$$

Therefore, the complete dynamic model of the quadcopter can be described by four control inputs, $u = [u_1 u_2 u_3 u_4]^T$ and 12 state vectors, $x_s = [x\ y\ z\ \dot{x}\ \dot{y}\ \dot{z}\ \phi\ \theta\ \psi\ p\ q\ r]^T$.

3. Controller Design

MPC, or known as receding horizon control (RHC), is a control approach that comprises a systematic algorithm where the dynamic model of the system is solved under a finite, moving horizon and closed control problem. It has the ability to use constraints in both control inputs and outputs on the system during design process. It basically predicts a number of outputs of the system such that it can generate an optimized control effort for the system to reach the reference trajectory. The optimization problem is solved for a predefined time interval that is also known as prediction horizon at each sampling time interval. The immediate optimized control signal is applied in the system until the next sampling time interval and this process is repeated for each sampling time interval [15].

3.1. Plant Model and Prediction Horizon

A nonlinear system can be written as in Equation 14, where $x(t)\epsilon R^n$ denotes the states of the system and $u(t)\epsilon R^m$ denotes system inputs.

$$\dot{x} = f((x), u(t)) \tag{14}$$

The quadcopter's dynamic model is linearized at hover condition as in Equation 15 to Equation 18 [15, 17], where the nominal states and control inputs are x_T and u_T, respectively, k is the sample time, $A \epsilon R^{n\times n}$ is the state matrix, $B \epsilon R^{n\times m}$ is input matrix, $y \epsilon R^p$ is system outputs, $C \epsilon R^{p\times n}$ is output matrix and $D \epsilon R^{p\times m}$ is feedforward matrix. For this system, $n = 12$ and $m = 4$ are considered.

$$\Delta x_{k+1} = A\Delta x_k + B\Delta u_k \tag{15}$$

$$\Delta y_k = C\Delta x_k + D\Delta u_k \tag{16}$$

$$\Delta x_k = x_k - x_T \tag{17}$$

$$\Delta u_k = u_k - u_T \tag{18}$$

A prediction horizon has to be determined such that the controller can predict a number of future states to reach the desired states. A state observer with an estimator is also required to be implemented in the controller to predict the future states while the estimator predicts future behaviour of the system. A Linear Quadratic Estimator is applied for the algorithm. By expanding Equation 15 and Equation 16 up to $k + N$, the future states and outputs can be achieved depending on initial states and future inputs as given by Equation 19 and Equation 20, respectively.

$$\Delta x_{k+N} = A^N\Delta x_k + A^{N-1}B\Delta u_k + A^{N-2}B\Delta u_{k+1} + \cdots + AB\Delta u_{k+N-2} + B\Delta u_{k+N-1} \tag{19}$$

$$\Delta y_{k+N} = CA^N\Delta x_k + C(A^{N-1}B\Delta u_k + A^{N-2}B\Delta u_{k+1} + \cdots + AB\Delta u_{k+N-2} + B\Delta u_{k+N-1}) \tag{20}$$

The equations can also be written in matrix form as shown by Equation 21 and Equation 22.

$$\begin{pmatrix} \Delta x_k \\ \Delta x_{k+1} \\ \Delta x_{k+2} \\ \vdots \\ \Delta x_{k+N-1} \end{pmatrix} = \begin{pmatrix} I \\ A \\ A^2 \\ \vdots \\ A^{N-1} \end{pmatrix} \Delta x_k + \begin{pmatrix} 0 & 0 & \cdots & 0 & 0 \\ B & 0 & \cdots & 0 & 0 \\ AB & B & \cdots & 0 & 0 \\ \vdots & \vdots & \ddots & \vdots & \vdots \\ A^{N-2}B & A^{N-3}B & \cdots & B & 0 \end{pmatrix} \begin{pmatrix} \Delta u_k \\ \Delta u_{k+1} \\ \Delta u_{k+2} \\ \vdots \\ \Delta u_{k+N-1} \end{pmatrix} \tag{21}$$

$$\begin{pmatrix} \Delta y_k \\ \Delta y_{k+1} \\ \Delta y_{k+2} \\ \vdots \\ \Delta y_{k+N-1} \end{pmatrix} = \begin{pmatrix} C & 0 & 0 & \cdots & 0 \\ 0 & C & 0 & \cdots & 0 \\ 0 & 0 & C & \cdots & 0 \\ \vdots & \vdots & \vdots & \ddots & \vdots \\ 0 & 0 & 0 & \cdots & C \end{pmatrix} \begin{pmatrix} \Delta x_k \\ \Delta x_{k+1} \\ \Delta x_{k+2} \\ \vdots \\ \Delta x_{k+N-1} \end{pmatrix} + \begin{pmatrix} D & 0 & 0 & \cdots & 0 \\ 0 & D & 0 & \cdots & 0 \\ 0 & 0 & D & \cdots & 0 \\ \vdots & \vdots & \vdots & \ddots & \vdots \\ 0 & 0 & 0 & \cdots & D \end{pmatrix} \begin{pmatrix} \Delta u_k \\ \Delta u_{k+1} \\ \Delta u_{k+2} \\ \vdots \\ \Delta u_{k+N-1} \end{pmatrix} \tag{22}$$

As D = 0 in most of the cases, the two equations can be rewritten as in Equation 23 and Equation 24 in a shorter form.

$$\Delta X_k = A_m \Delta x_k + B_m \Delta u_k \tag{23}$$

$$\Delta Y_k = C_m \Delta x_k \tag{24}$$

3.2. Control Design

The MPC technique is helped by cost function in its control algorithm to calculate the optimal solution at every sampling time interval. The cost function is generally designed in a way that the predicted outputs are directed towards the desired states as described in Equation 20 while the control efforts are minimized as well. In this study, the cost function is minimized by the norm of the difference between the current outputs and desired trajectory and the norms of motor inputs as in Equation 25, where $\widehat{W}_u$ and $\widehat{W}_y$ are given by Equation 26 and Equation 27, respectively [16].

$$J(\Delta x, \Delta u) = (\Delta u_k)^T \widehat{W}_u^{\,2} (\Delta u_k) + (\Delta Y_k - \Delta Y_k^r)^T \widehat{W}_y^{\,2} (\Delta Y_k - \Delta Y_k^r) \tag{25}$$

$$\widehat{W}_u = \begin{bmatrix} W_{u|0,1} & 0 & \cdots & 0 & \cdots & 0 & 0 & \cdots & 0 \\ 0 & W_{u|0,2} & \cdots & 0 & \ddots & 0 & 0 & \ddots & 0 \\ \vdots & \vdots & \ddots & \vdots & \vdots & \vdots & \vdots & \cdots & \vdots \\ 0 & 0 & \cdots & W_{u|0,m} & \cdots & 0 & 0 & 0 & 0 \\ \vdots & \vdots & \vdots & \vdots & \ddots & \vdots & \vdots & \vdots & \vdots \\ 0 & 0 & \cdots & 0 & \cdots & W_{u|N-1,1} & 0 & \cdots & 0 \\ 0 & 0 & \cdots & 0 & \cdots & 0 & W_{u|N-1,2} & \cdots & 0 \\ \vdots & \vdots & \ddots & \vdots & \ddots & \vdots & \vdots & \ddots & \vdots \\ 0 & 0 & \cdots & 0 & \cdots & 0 & 0 & \cdots & W_{u|N-1,m} \end{bmatrix} \tag{26}$$

$$\widehat{W}_y = \begin{bmatrix} W_{y|0,1} & 0 & \cdots & 0 & \cdots & 0 & 0 & \cdots & 0 \\ 0 & W_{y|0,2} & \cdots & 0 & \ddots & 0 & 0 & \ddots & 0 \\ \vdots & \vdots & \ddots & \vdots & \vdots & \vdots & \vdots & \cdots & \vdots \\ 0 & 0 & \cdots & W_{y|0,m} & \cdots & 0 & 0 & 0 & 0 \\ \vdots & \vdots & \vdots & \vdots & \ddots & \vdots & \vdots & \vdots & \vdots \\ 0 & 0 & \cdots & 0 & \cdots & W_{y|N-1,1} & 0 & \cdots & 0 \\ 0 & 0 & \cdots & 0 & \cdots & 0 & W_{y|N-1,2} & \cdots & 0 \\ \vdots & \vdots & \ddots & \vdots & \ddots & \vdots & \vdots & \ddots & \vdots \\ 0 & 0 & \cdots & 0 & \cdots & 0 & 0 & \cdots & W_{y|N-1,m} \end{bmatrix} \tag{27}$$

3.3. Quadratic Programming

As the cost function is in quadratic form, a quadratic programming is chosen to solve the optimization problem. The main purpose of the quadratic programming is to reduce the cost function $J(\Delta x, \Delta u)$ by finding out a feasible search direction, Δu.

3.4. Input and Constraint Handling

During the design of the quadcopter, it is important to apply constraint at the force of each motor. This is to enable the motors to be operated between maximum and minimum rotations per minute (rpm). There is an upper bound, u_{ub} and a lower bound, u_{lb} at the control inputs where $u_{lb} \leq u_{k+i} \leq u_{ub}$ for $i = 0, 1, 2, \ldots, N-1$. As the dynamic model is linearized around a certain operating point, the MPC approach solves the perturbed control inputs for the linearized model. The constraints can be described in matrix form as in Equation 28, where $I_{m \times m}$ is an identity matrix.

$$\begin{bmatrix} \mathrm{I}_{\mathrm{m\times m}} \\ -\mathrm{I}_{\mathrm{m\times m}} \end{bmatrix} \Delta \mathrm{u}_{\mathrm{k+i}} \le \begin{bmatrix} \mathrm{u}_{\mathrm{ub}} - \Delta \mathrm{u}_{\mathrm{T}} \\ -(\mathrm{u}_{\mathrm{lb}} - \Delta \mathrm{u}_{\mathrm{T}}) \end{bmatrix} \tag{28}$$

After rearranging, Equation 28 can be rewritten as Equation 29, where $\boldsymbol{I_u}$ is given by Equation 30.

$$I_u \Delta u_k \le \Delta u_b \tag{29}$$

$$I_u = \begin{bmatrix} \begin{bmatrix} I_{m\times m} \\ -I_{m\times m} \end{bmatrix} & 0 & \cdots & 0 \\ 0 & \begin{bmatrix} I_{m\times m} \\ -I_{m\times m} \end{bmatrix} & \cdots & 0 \\ \vdots & \vdots & \ddots & \vdots \\ 0 & 0 & \cdots & \begin{bmatrix} I_{m\times m} \\ -I_{m\times m} \end{bmatrix} \end{bmatrix}, \Delta u_b = \begin{bmatrix} \begin{bmatrix} u_{ub} - \Delta u_T \\ -(u_{lb} - \Delta u_T) \end{bmatrix} \\ \begin{bmatrix} u_{ub} - \Delta u_T \\ -(u_{lb} - \Delta u_T) \end{bmatrix} \\ \vdots \\ \begin{bmatrix} u_{ub} - \Delta u_T \\ -(u_{lb} - \Delta u_T) \end{bmatrix} \end{bmatrix} \tag{30}$$

It is also necessary to consider a limit on the angles to avoid kinematic singularities because of the limitations of the model. The angles are limited within the bounds and the bounds are given as follows for roll, pitch and yaw, respectively: $-\pi \le \phi \le \pi$, $-\frac{\pi}{2} \le \theta \le \frac{\pi}{2}$ and $-\pi \le \psi \le \pi$.

If a specific output is to be constrained, it can be described as Equation 31 and the constraints can be represented by Equation 32, where z_{lb} and z_{ub} are denoted as lower bound and upper bound for the outputs.

$$\Delta z_{lb} = C_z \Delta x_k \tag{31}$$

$$z_{lb} \le C_z x_{k+i} \le z_{ub};\ i = 0, 1, 2, 3, \ldots., N-1 \tag{32}$$

Similarly, it also can be shown in matrix form as in Equation 33.

$$\begin{bmatrix} C_z \\ -C_z \end{bmatrix} \le \begin{bmatrix} z_{ub} - C_z x_T \\ -(z_{lb} - C_z x_T) \end{bmatrix} \tag{33}$$

From Equation 23 and Equation 24, the constraints can be described as in Equation 34 where Δx_k is substituted and Γ_z is given in Equation 35.

$$\Gamma_z (A_m \Delta x_k + B_m \Delta u_k) \le \Delta z_b \tag{34}$$

$$\Gamma_z = \begin{bmatrix} \begin{bmatrix} C_z \\ -C_z \end{bmatrix} & 0 & \cdots & 0 \\ 0 & \begin{bmatrix} C_z \\ -C_z \end{bmatrix} & \cdots & 0 \\ \vdots & \vdots & \ddots & \vdots \\ 0 & 0 & \cdots & \begin{bmatrix} C_z \\ -C_z \end{bmatrix} \end{bmatrix}, \ \Delta z_b = \begin{bmatrix} \begin{bmatrix} z_{ub} - C_z x_T \\ -(z_{lb} - C_z x_T) \end{bmatrix} \\ \begin{bmatrix} z_{ub} - C_z x_T \\ -(z_{lb} - C_z x_T) \end{bmatrix} \\ \vdots \\ \begin{bmatrix} z_{ub} - C_z x_T \\ -(z_{lb} - C_z x_T) \end{bmatrix} \end{bmatrix} \tag{35}$$

The control input and output constraints can be described by one single Equation 36, where Π is given in Equation 37 [17].

$$\Pi \Delta u_k \le \Upsilon \tag{36}$$

$$\Pi = \begin{bmatrix} M_u \\ \Gamma_z B_m \end{bmatrix}, \Upsilon = \begin{bmatrix} \Delta u_b \\ \Delta z_b - \Gamma_z A_m \Delta x_k \end{bmatrix} \tag{37}$$

4. Simulated Results

As the dynamics of the quadcopter is nonlinear, it has been linearized at a certain point and considered as $[0, 1m, 0, 0, 0, 0, 0, 0, \frac{5\pi}{180}, 0, 0, 0]^T$. For this study, prediction horizon, N = 30 and control horizon, M = 2 are selected after several initial simulations where these two values provided the most efficient performance under the consideration of settling time and overshoot. The effects of different N along x, y and z axes on settling time and overshoot are shown in Figure 2. It can be observed from the figure that the settling time is increasing when N increases while the overshoot is decreasing with increasing N along the x, y and z axes.

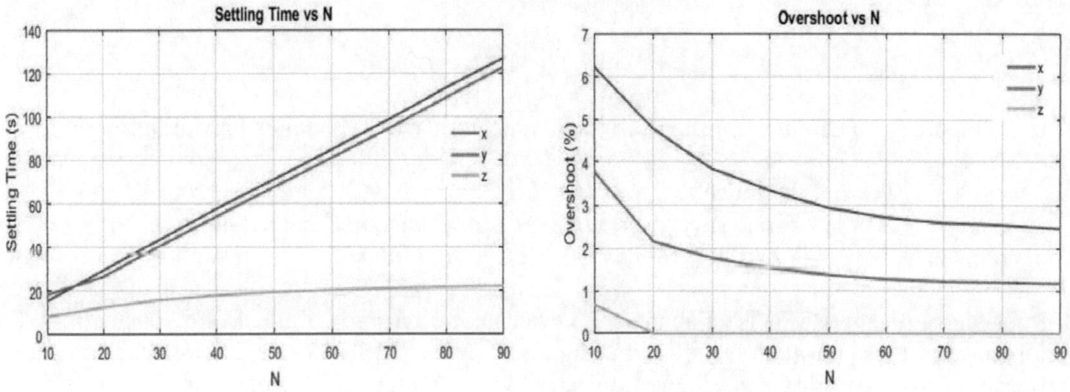

Figure 2: Effects of N on settling time and overshoot

Sample time for the model is chosen as 0.25s after the confirmation of system stability where the pole of the system has been achieved: $[0, 0, 0, -0.1538, -0.1538, -0.1538, 0, 0, 0, 0, 0, 0]^T$. Figure 3 illustrates the efficiency of MPC controller that can reject disturbances without affecting the control outputs. In this scenario, x = 3m, y = 2m and z = -5m is selected as the reference point to show the comparison among the disturbances that range from 0.1 to 1. The RMSE is found to remain exactly the same along x, y and z axes with 5.13%, 5.11% and 3.09%, respectively. This means that the external disturbance has no effect on the system. Moreover, Figure 4 shows the control effort against time for better understanding.

Three different trajectories are chosen for tracking under certain constraints at control inputs and addition of disturbances in the system where the angular velocity of each motor is taken as 848 rad/s. This helps to find the constraints as: $0 < u_1 < 90$, $-22.52 < u_2 < 22.52$, $-22.52 < u_3 < 22.52$ and $-1.08 < u_4 < 1.08$, and the disturbances considered for the four control inputs are [0.1, 0.1, 0.1, 0.1] as in Figure 5.

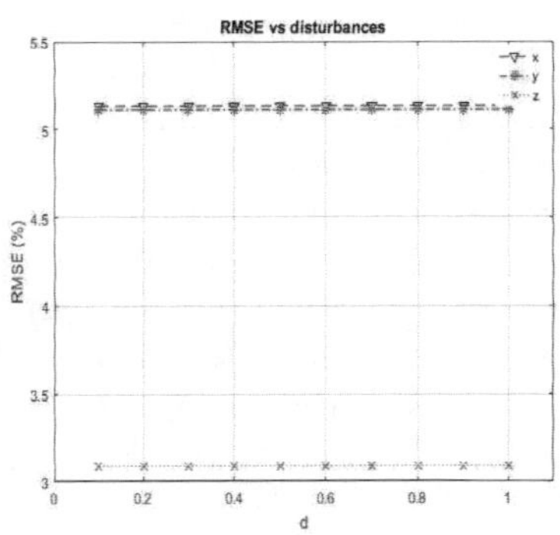

Figure 3: Effects of different disturbances on RMSE

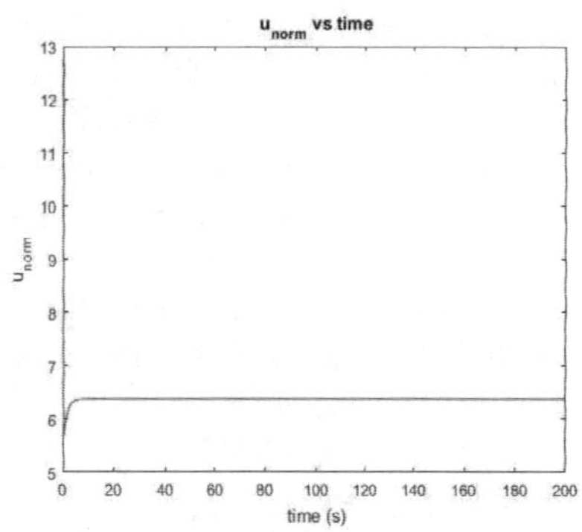

Figure 4: Control efforts (u_{norm}) against time

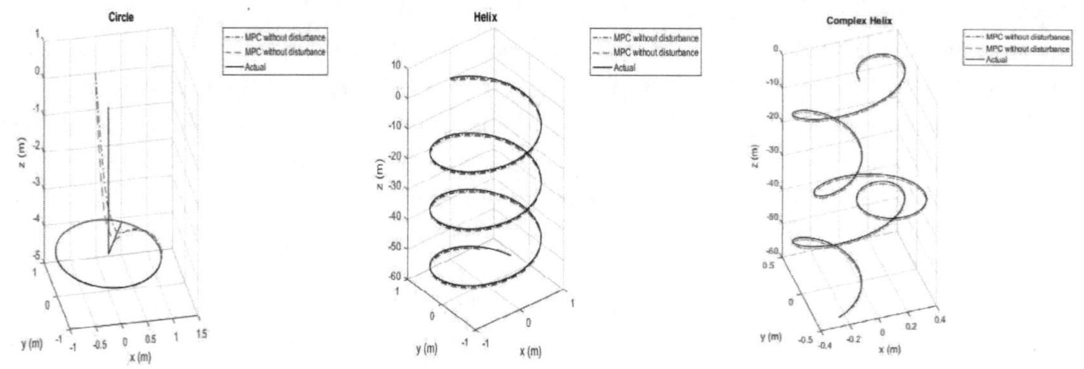

(a) (b) (c)

Figure 5: (a) Circular trajectory, (b) helix trajectory and (c) complex helix trajectory

For circular trajectory, $x = sin(0.1t), y = cos(0.1t), z = -4m$

For helix trajectory, $x = sin(0.1t), y = cos(0.1t), z = -0.3t$

For complex helix, $x = cos(0.05t) - cos^3(0.05t),\ y = sin(0.05t) - sin^3(0.05t), z = -0.3t$

Root-Mean-Square (RMS) is an approach to evaluate the accuracy of the data by comparison. The performance of the controller is evaluated using RMS error (RMSE). Table 2 shows the comparison between reference trajectory and achieved trajectory values with respect to time for circular, helix and complex helix trajectory under disturbance and without disturbance. From Table 2, it is found that the RMSE is less than 5% for the three trajectories, which is considered tolerable.

Table 2: RMSE for circle, helix and complex helix trajectories with and without disturbances

	RMSE for with disturbance			**RMSE for without disturbance**		
	x (%)	**y (%)**	**z (%)**	**x (%)**	**y (%)**	**z (%)**
Circle	2.3320	2.8844	0.8866	2.3321	2.8878	1.2029
Helix	2.2526	1.2278	2.1067	2.2524	1.2281	2.9558
Complex Helix	1.0628	2.5308	1.3508	1.0628	2.5308	1.3508

5. Conclusion

The presented work shows the use of Linear Model Predictive Control (LMPC) approach for different trajectories (i.e. circle, helix and complex helix trajectory) under disturbances. It is designed with the help of MPC toolbox in Simulink. The main advantage of MPC controller that make it different from other controllers such as PID, LQR, H-infinity or feedback linearization is the optimization of control inputs and outputs under the consideration of disturbances, noise and constraints. This has been shown to help achieve proper inputs and outputs under certain requirements of the system although the noise factor is not considered in this work. The most crucial issues to design a MPC model include choosing proper prediction horizon, control horizon and sample time because they all affect the system stability. After the confirmation of system stability, the system tracking can be improved by tuning to the proper gains. This study has successfully demonstrated a proper tracking with minimal RMSE under different disturbances that have been shown to be of negligible effect to the system outputs. In future, nonlinear Model Predictive Control (NMPC) approach will be designed that is expected to be more suitable for nonlinear quadcopter model.

IOP Conf. Series: Materials Science and Engineering **270** (2017) 012007 doi:10.1088/1757-899X/270/1/012007

References

[1] Zhao W and Go T H 2014 *Journal of the Franklin Institute* **351** 1335-55
[2] Tanveer M H, Hazry D, Ahmed S F, Joyo M K, Warsi F A, Razlan Z M, Wan K and Hussain A T 2014 *Australian Journal of Basic and Applied Sciences* **8** 428-31
[3] Bouabdallah S, Noth A and Siegwart R 2004 *IEEE/RSJ International Conference on Intelligent Robots and Systems*
[4] Pounds P E I 2007 *Design, Construction and Control of A Large Quadrotor Micro Air Vehicle* Australian National University
[5] Madani T and Benallegue A 2006 *IEEE/RSJ International Conference on Intelligent Robots and Systems*
[6] Bonna R and Camino J 2015 *International Symposium on Dynamic Problems of Mechanics*
[7] Kendoul F 2012 *Journal of Field Robotics* **29** 315-78
[8] Raffo G V, Ortega M G and Rubio F R 2008 *International Federation of Automatic Control World Congress*
[9] Alexis K, Nikolakopoulos G and Tzes A 2010 *18th Mediterranean Conference on Control and Automation*
[10] Bouabdallah S 2007 *Design and Control of Quadrotors with Application to Autonomous Flying* Thesis, Ecole Polytechnique Federale de Lausanne
[11] Efe M Ö 2011 *IEEE Transactions on Industrial Informatics* **7** 354-61
[12] Khan H S and Kadri M B 2015 *International Conference on Information and Communication Technologies*
[13] Elkholy H M 2014 *Dynamic Modeling and Control of a Quadrotor using Linear and Nonlinear Approaches* Thesis, American University in Cairo
[14] Bolandi H, Rezaei M, Mohsenipour R, Nemati H and Smailzadeh S M 2013 *Intelligent Control and Automation* **4** 335
[15] Tule C 2014 *Trajectory Generation and Constrained Control of Quadrotors* Thesis, University of Texas Arlington
[16] Mathworks *https://www.mathworks.com/help/mpc/ug/optimization-problem.html*
[17] Ru P and Subbarao K 2017 *Aerospace* **4** 31

IOP Conf. Series: Materials Science and Engineering **270** (2017) 012008 doi:10.1088/1757-899X/270/1/012008

Enhancing the stabilization of aircraft pitch motion control via intelligent and classical method

H Lukman[1], S Munawwarah[1], A Azizan[2], F Yakub[1,*], S A Zaki[1] and Z A Rasid[1]

[1]Malaysia-Japan International Institute of Technology, Universiti Teknologi Malaysia, 54100 Kuala Lumpur, Malaysia
[2]Advanced Informatic Systems, Universiti Teknologi Malaysia, 54100 Kuala Lumpur, Malaysia

*mfitri.kl@utm.my

Abstract. The pitching movement of an aircraft is very important to ensure passengers are intrinsically safe and the aircraft achieve its maximum stability. The equations governing the motion of an aircraft are a complex set of six nonlinear coupled differential equations. Under certain assumptions, it can be decoupled and linearized into longitudinal and lateral equations. Pitch control is a longitudinal problem and thus, only the longitudinal dynamics equations are involved in this system. It is a third order nonlinear system, which is linearized about the operating point. The system is also inherently unstable due to the presence of a free integrator. Because of this, a feedback controller is added in order to solve this problem and enhance the system performance. This study uses two approaches in designing controller: a conventional controller and an intelligent controller. The pitch control scheme consists of proportional, integral and derivatives (PID) for conventional controller and fuzzy logic control (FLC) for intelligent controller. Throughout the paper, the performance of the presented controllers are investigated and compared based on the common criteria of step response. Simulation results have been obtained and analysed by using Matlab and Simulink software. The study shows that FLC controller has higher ability to control and stabilize the aircraft's pitch angle as compared to PID controller.

1. Introduction

Aircraft technology has transformed human life by enabling regional and international transportation of people and goods. Aviation has essentially reduced the travelling time and cost, and revolutionized travelling experience with introduction of in-flight luxuries that are not commonly available in other modes of transportation like ground vehicles, trains or boat. However, an aircraft has to be controlled with a different control system to guarantee flight passengers are safe and the aircraft is in fly-worthy conditions. There are many critical parts in an aircraft that need to be perfectly controlled and one of them is the elevator. The elevator influences aircraft motion by affecting the pitch movement [1, 2]. There are many intelligence methods that are widely used for aviation control such as proportional, integral and derivatives (PID), linear quadratic control and fuzzy logic control (FLC) [3-5]. The FLC system has become one of today's successful technology to control sophisticated systems. It addresses system complexity by resembling the human decision making and is able to generate precise solutions given certain approximate input information. It closes the gaps in engineering control design that has been limited by rigid mathematical techniques such as linear manipulate design.

The development of automated control system has played a crucial role in the growth of civil and military aviation. Aircraft and missile structures are being assembled with control structures that offer stability, disturbance attenuation and reference signal tracking, even as their aerodynamic coefficients vary over a huge dynamic range due to large Mach-altitude fluctuations and uncertainties resulting from inaccurate wind tunnel measurements [6, 7]. The aggregate of nonlinear dynamics, modelling uncertainties and parameter variant in characterizing an aircraft and its operating environment are the major problems in flight control system [8]. To reduce the complexity of analysis, the aircraft is often assumed as an inflexible body and the aircraft's motion includes a small deviation from equilibrium flight condition [7, 9]. One of the objectives of a pitch control machine is to manipulate or assist the pilot to steer the plane by guiding the pitch steadily. With a pitch controller, the aircraft will go back to the desired attitude within a reasonable duration of time after a pitch disturbance. Moreover, the pitch controller can quickly make a pitch change after a given command. There are many problems faced in developing a pitch controller for an aircraft longitudinal motion. Researchers have varying opinion on which controller to be applied to ensure adequate suitability and minimum error when simulations are implemented [10]. By using a simulation tool, it can be observed whether the output is following the desired input or not. In order to get the best results, the best controller needs to be selected to minimize the errors in the simulation.

A simulation on the behaviours of the pitch motion of an aircraft is implemented in this study. In addition, a control system is developed to enhance the performance of previous control system, using intelligence method for the pitch angle control. Throughout this paper, the aircraft dynamic system is modelled with mathematical equations. Mathematical modelling involves the process of describing the dynamics of the system in a set of differential equations. By applying the physical law governing a particular system, which is from Newton's Laws, the system dynamics for many mechanical systems can be obtained. The motions of an airplane consist of longitudinal and lateral motions. However, to narrow down the scope of study, only equation of longitudinal motion is considered. The modelling derivation states that the summation of all external forces acting on the aircraft's body is identical to the time fee of exchange of the momentum of aircraft's frame. Furthermore, the summation of the outside moments acting on the aircraft's frame is identical to the time fee of exchange of the moment of angular flight momentum. In this paper, two proposed controllers are designed, namely PID and FLC. The proposed controller is then enhanced by combining the controllers which are proportional (P), proportional and derivatives (PD), PID and FLC to improve the aircraft response. Analyses are implemented on each type of controllers, whether it has fulfilled the criteria needed for the system. Several mathematical models are used to perform the simulations.

2. Pitch Motion of an Aircraft System

The pitch of aircraft is controlled through an elevator that is commonly positioned at the rear of the plane, parallel to the wing that houses the ailerons [11]. Pitch rate manipulation typically responds very fast to the adjustments in control input compared to pitch angle. Pitch control is a longitudinal challenge and the design of the autopilot system that controls the pitch of an aircraft is complex. By moving the elevator backwards, the pilot controls the elevator up (a function of bad camber) resulting in the downwards pressure on the horizontal tail to expand. Angle of attack on the wings increases, so the aircraft nose is pitched up and lift is usually produced. In micro-lighting fixtures and grasp gliders, the pitch movement is reversed and the pitch control device is much less complicated. Hence, whilst the pilot manipulate backwards the elevator, it produces a nostril-down pitch and the angle of attack at the wing is reduced. The pitch angle of an aircraft is manipulated via adjusting the angle and therefore boosting the force of the rear elevator. Figure 1 shows the pitch angle and the direction of an aircraft.

A few assumptions need to be made before the modelling process is done. First, the aircraft is at a steady state cruising at constant altitude and velocity, hence the thrust and drag cancel out each other while the lift and weight balance out each other. Second, the change in pitch angle does not change the speed of an aircraft under any circumstance. Moreover, the atmosphere in which the plane is flying is assumed to be undisturbed, thus forces and moment due to atmospheric disturbance are taken as zero.

Besides that, the pitch angle is assumed not to be more than 90 degree or 1.6 radians. If the pitch angle goes beyond that limit, it will damage the whole operation of the aircraft. The angle of deflection also has its limitation that need to be considered. The input for the system is from zero to the limited value in the system whereas the output for the system is also limited at a certain range of angles.

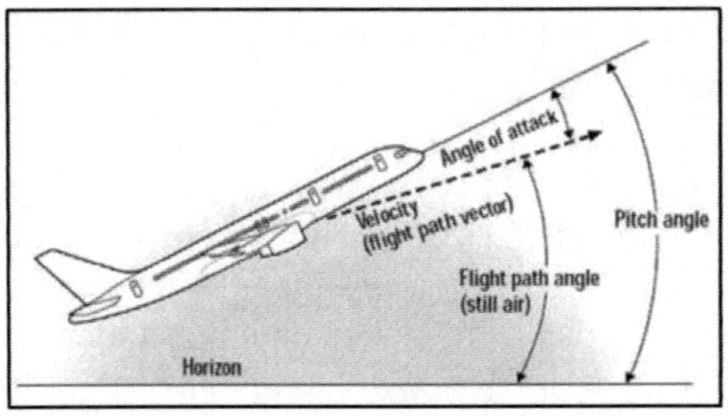

Figure 1: The pitch angle with the direction of an aircraft

The operational framework of this study includes the equation from free body of an aircraft taking the lift at certain pitch angle, which are derived to form three equations [12, 13]. Those equations are in first order system before being converted into a transfer function. The following dynamic equations in Equation 1 to Equation 3 describe the longitudinal dynamics involving the pitch motion of a typical aircraft, where *u*, *v* and *w* represents forward, side and lateral velocities, respectively, while *p*, *q* and *r* are roll, pitch and yaw rates, respectively.

$$F_x = m(\dot{u} + qw - rv) \tag{1}$$

$$F_z = m(\dot{w} + pv - qu) \tag{2}$$

$$M_y = I_y\dot{q} + rp(I_x - I_z) + I_{xz}(p^2 - r^2) \tag{3}$$

It is necessary to solve completely the aircraft problem with the following assumptions for the aircraft motion, as in Equation 4 to Equation 9.

$$\text{Rolling rate: } p = \dot{\phi} - \psi \sin\theta \tag{4}$$

$$\text{Yawing rate: } q = \dot{\theta} \cos\phi + \dot{\psi}\cos\theta \sin\phi \tag{5}$$

$$\text{Pitching rate: } r = \psi \cos\theta \cos\phi - \dot{\theta} \sin\phi \tag{6}$$

$$\text{Pitch angle: } \dot{\theta} = q \cos\phi - r \sin\phi \tag{7}$$

$$\text{Roll angle: } \dot{\phi} = p + q\sin\phi\tan\theta + r\cos\phi\tan\theta \tag{8}$$

$$\text{Yaw angle: } \dot{\Psi} = (q\sin\phi + r\cos\phi)\sec\theta \tag{9}$$

The equations are linearized by adding perturbation equation and small disturbance theory to replace the nonlinearities in the aircraft general dynamic equations [14]. The gravitational force components along the x, y and z –axis can be easily shown to be as in Equation 10 to Equation 12, respectively.

$$F_{xg} = -mg \sin\theta \tag{10}$$

$$F_{yg} = mg \cos\theta \sin\phi \tag{11}$$

$$F_{zg} = mg \cos\theta \cos\phi \tag{12}$$

Then, Equation 1 to Equation 3 are deduced to Equation 13 to Equation 15, respectively.

$$F_x = m(\dot{u} + qv - rv) + mg \sin\theta \tag{13}$$

$$F_z = m(\dot{w} + pv - qu) - mg\cos\theta\cos\phi \tag{14}$$

$$M_y = I_y\dot{q} + rp(I_x - I_z) + I_{xz}(p^2 - r^2) \tag{15}$$

The equations above can be transformed into the state-space form as stated by Equation 16 to Equation 18. The model in state-space form is used for controller design. Moreover, the parameters used for the domestic aircraft are tabulated in Table 1.

$$\begin{bmatrix} \dot{\Delta\alpha} \\ \dot{\Delta q} \\ \dot{\Delta\theta} \end{bmatrix} = \begin{bmatrix} \frac{z_\alpha}{u_0} & 1 & 0 \\ M_\alpha + \left[M_{\alpha}.\frac{z_\alpha}{u_0}\right] & M_q + M_{\dot{\alpha}} & 0 \\ 0 & 1 & 0 \end{bmatrix} \begin{bmatrix} \Delta\alpha \\ \Delta q \\ \Delta\theta \end{bmatrix} + \begin{bmatrix} -\left[\frac{z_\alpha}{u_0}\right] \\ -\left[M_\alpha + \left[M_{\dot{\alpha}}.\frac{z_\delta}{u_0}\right]\right] \\ 0 \end{bmatrix} \Delta\delta \tag{16}$$

$$\begin{bmatrix} \dot{\Delta\alpha} \\ \dot{\Delta q} \\ \dot{\Delta\theta} \end{bmatrix} = \begin{bmatrix} -2.02 & 1 & 0 \\ -6.9868 & -2.9476 & 0 \\ 0 & 1 & 0 \end{bmatrix} \begin{bmatrix} \Delta\alpha \\ \Delta q \\ \Delta\theta \end{bmatrix} + \begin{bmatrix} 0.232 \\ 0.0203 \\ 0 \end{bmatrix} \Delta\delta e \tag{17}$$

$$Y = \begin{bmatrix} 0 & 0 & 1 \end{bmatrix} \begin{bmatrix} \Delta\alpha \\ \Delta q \\ \Delta\theta \end{bmatrix} + [0] \tag{18}$$

Table 1: Parameters of domestic aircraft [2]

Longitudinal Derivatives	**Components**		
	Dynamics Pressure and Dimensional Derivative $Q = 36.8\ lb/ft^2$, $QS = 6771lb$, $QS\overline{C} = 38596ft.lb$, $(\overline{c}/2u_O) = 0.016s$		
	X-Force (S^{-1})	**Z-Force (Z^{-1})**	**Pitching Moment (FT^{-1})**
Rolling velocities	$X_u = -0.045$	$Z_u = -0.369$	$M_u = 0$
Yawing velocities	$X_w = 0.036$, $X_{\dot{w}} = 0$	$Z_w = -2.02$, $Z_{\dot{w}} = 0$	$M_w = -0.05$, $M_{\dot{w}} = -0.051$
Angle of attack	$X_\alpha = 0$, $X_{\dot{\alpha}} = 0$	$Z_\alpha = -355.42$, $X_{\dot{\alpha}} = 0$	$M_\alpha = -8.8$, $X_{\dot{\alpha}} = -0.8976$
Pitching rate	$X_q = 0$	$Z_q = 0$	$M_q = -2.05$
Elevator deflection	$X_{\delta e} = 0$	$Z_{\delta e} = 0$	$M_{\delta e} = 0$

3. Controller Design

Two types of controller are implemented in this study: PID and FLC controllers. The PID acts as the benchmark while the FLC controller is to enhance the performance of PID controller. Software used for the simulation are Matlab and Simulink version 2016b.

The most common industrial controller is based totally on the PID set of regulations. PID is known as control loop feedback mechanism that is applied on top of structured things. PID controller attempts to compensate the error between a measured variable and its desired reference value by mathematical calculations. The corrective values from PID will be the input into the controller, producing stabilized controlled movements. The PID controller calculation involves three separate parameters that consist of proportional, integral and derivatives values, as indicated by Equation 19 and Figure 2. Proportional value decides the reaction to the current error whereas the integral value is based on the sum of recent errors. On the other hand, derivatives component determines the reaction to the cost at which the error has since changed.

$$u(t) = K_p\, e(t) + K_i \int_0^t e(\tau)d\tau + K_d\, {de(t)}/{dt} \tag{19}$$

The gain is known as the critical advantage whereas the frequency of oscillation is known as crucial frequency. There are other conventional tuning methods such as the Cohen-Coon method that is used

IOP Conf. Series: Materials Science and Engineering **270** (2017) 012008 doi:10.1088/1757-899X/270/1/012008

when the main design criterion is related to the disturbance rejection but it may only be used for first order models consisting of massive manner delays [15]. Such relay auto-tuning technique eliminates the possibility of flying the aircraft near the steadiness limit.

In the meantime, FLC is a superior approach for sorting and when dealing with statistics. It has also been proven to be the step forward for most control system applications since it mimics the human manipulative common sense. It can be embedded into anything from small hand-held products to the massive computerized manipulative systems. It makes use of imprecise but very descriptive language to address input statistics similar to a human operator. It is very robust and forgiving of operator and facts input, and usually works when first applied with little or no tuning. A novel technique for FLC controller is employed for stabilization of the pitch control in this study [7, 11]. Figure 3 shows the overall closed-loop system for FLC control systems for the pitch control of an aircraft. The input to the fuzzy controller is the error, which measures the system performance, and the rate at which the error changes. The output is the change to be applied to the control signal. From Figure 3, the error is computed by comparing the desired point with the aircraft output. The change of error is generated by deriving the error. The error and change of error are fed to the fuzzy controller through a multiplexer.

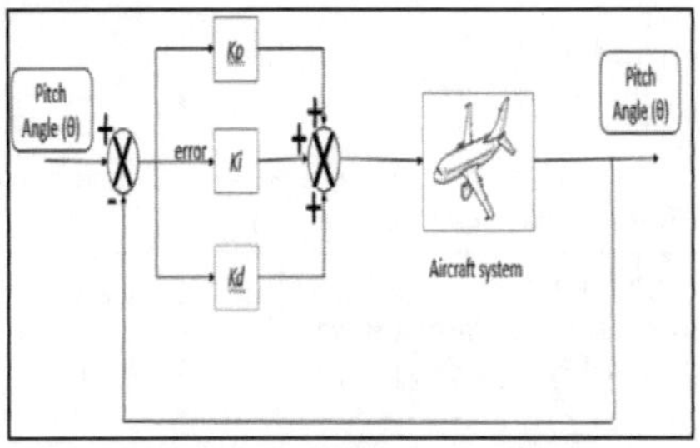

Figure 2: PID controller system

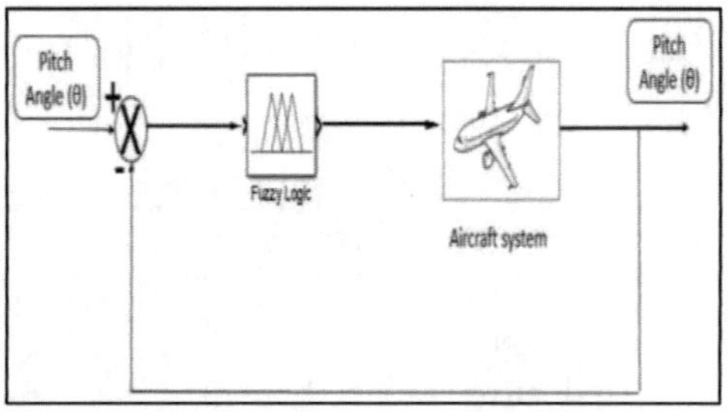

Figure 3: FLC controller system

FLC has two inputs (i.e. error change and error) and one output (i.e. control signal) as shown in Figure 4, Figure 5 and Figure 6, respectively. The variables that implies inputs and output have been classified as N(Negative), Z(Zero), P(Positive). Since there are three fuzzy variables and each fuzzy variable has three membership functions, the fuzzy controller for pitch control of an aircraft has a total on nine membership functions. Each membership function is constrained to be triangular such that each has three parameters. Table 2 shows the fuzzy rule employed in this study that is based on 3 x 3.

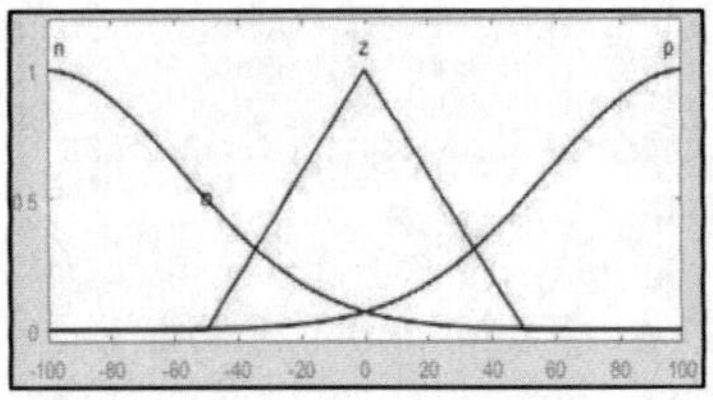
Figure 4: Fuzzy set for the input "error"

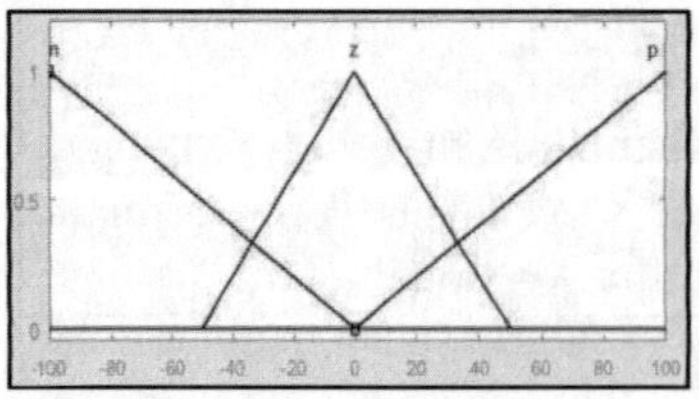
Figure 5: Fuzzy set for the input "Δ_{error}"

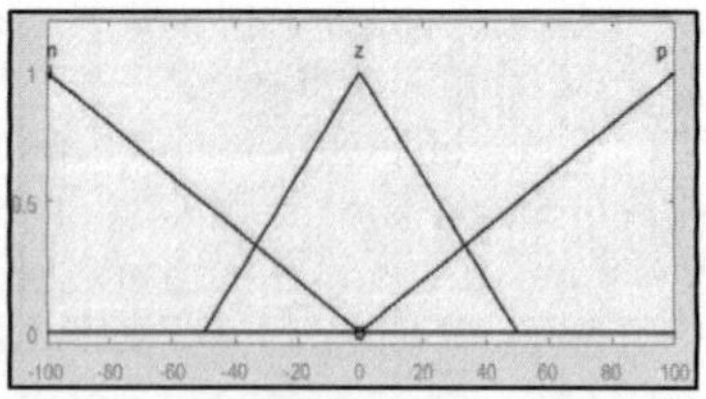
Figure 6: Fuzzy set for the output "output"

Table 2: The fuzzy rule base

Δ_{error} (de)	**error (e)**		
	N	**Z**	**P**
Negative (N)	N	N	P
Zero (Z)	N	Z	P
Positive (P)	N	P	P

4. Result and Discussion

The simulations are analysed based on the mathematical model for the aircraft pitch motion system. Firstly, the performance of the controllers are observed on the pitch angle of the aircraft without any feedback. Then, the results obtained are analysed using the classical approach and intelligent method, which is PID and FLC, respectively. From Figure 7, it can be seen that the pitch control system is not stable without a controller. The curve of the pitch angle is approaching infinity as the time increases. Therefore, a feedback controller needs to be designed in order to stabilize the system. In this study, a unit step command is required in order for the pitch angle to follow a reference value of 0.2 radian. The controller output is made proportional to the error where the proportionality constant called the proportional gain, K_P is introduced. The K_P will effectively reduce the rise time but does not eliminate the steady-state error. This controller is capable of maintaining the output steady-state value at desired value as shown in Figure 8. From Figure 8, K_P is set to be 1.5. Increasing the value K_P will decrease the rise time but at the same time increase the overshoot of the system. When the K_P is increased, it will produce large oscillations before the system reaches the steady-state value. The output response gives fast response with a rise time of 0.388 seconds, settling time of 1.871 seconds, percentage of overshoot of 4.8% and percentage of steady-state error of 0.05%.

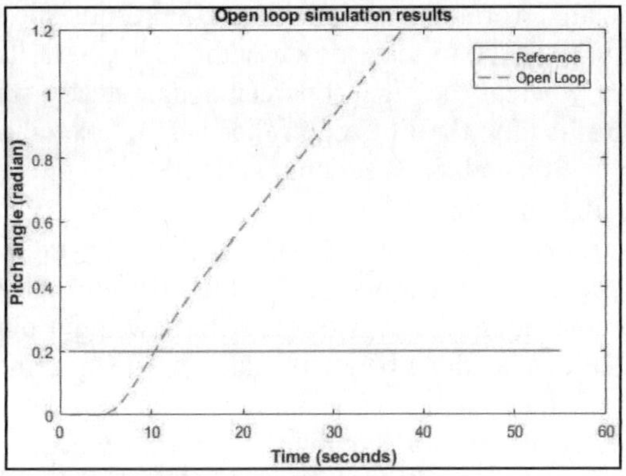

Figure 7: Without controller

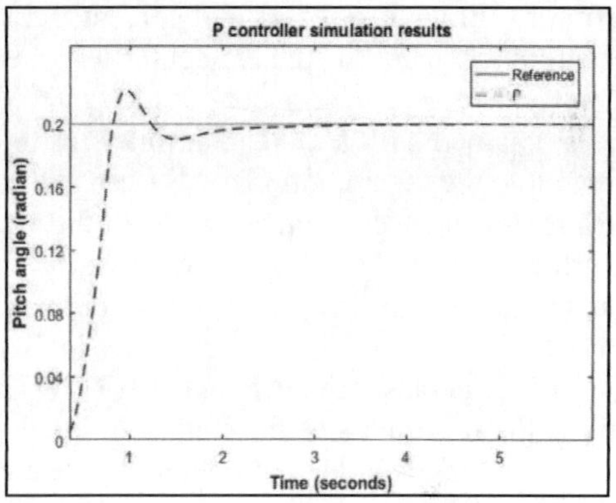

Figure 8: With P controller

Figure 9 illustrate the response of PD controller for pitch angle to the unit step reference input. The derivatives component is used to create damping in dynamic system and thus stabilizing its behaviour. The derivatives action is proportional to the rate of change of error measurement, which compensates the errors by changing the measurement and improving the system transient response. This increases the stability of the system. The PD controller provides a very fast response with settling time of 1.098 seconds and rise time of 0.23 seconds, but the response has a sharp peak. The transient response has an overshoot around 2.3%. In addition, it also gives a percentage of steady-state about 0.05%. Increasing K_D will make the response become a little bit slower, hence a decrease of the overshoot.

The value of K_P, K_I and K_D are dependent on each other. Changing one of them can affect the other two. Effects of each controller gain K_P, K_I and K_D on a closed-loop system will make a system more stable. K_P will have the effect of reducing rise time but it will never eliminate the steady-state error. K_I will have the effect of eliminating the steady-state error but it will make the transient response worse. K_D will have the effect in increasing the stability of the system by reducing overshoot and improving the transient response. Using Zieglar-Nicols method and after several trials and errors, the parameters value for the PID controller are determined as K_P= 12.45, K_I = 1.75 and K_D= 0.12. The simulation results for PID controller are shown in Figure 10. It can be observed that the response of PD controller is a little bit slower than PID controller in term of rise time and settling time, which both values for the latter controller are 0.16 seconds and 1.89 seconds, respectively.

Meanwhile, Figure 11 shows the output response of pitch control by using the FLC approach where the two inputs and one output have been applied to the system. The FLC provides good performance in term of settling time, percentage of overshoot, percentage of steady-state error and rise time. From the results obtained, it shows that the pitch angle follows the reference value. This controller is able to provide a better response than the PID controller and does not produce any overshoot. The response is comparatively faster where the settling time is about 0.96 seconds and the rise time is 0.16 seconds.

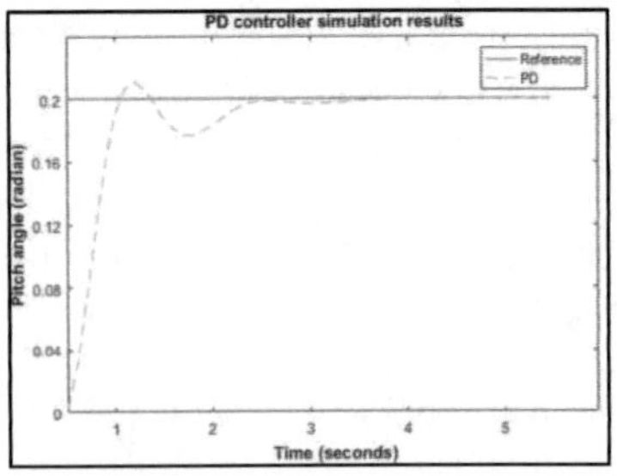

Figure 9: PD controller

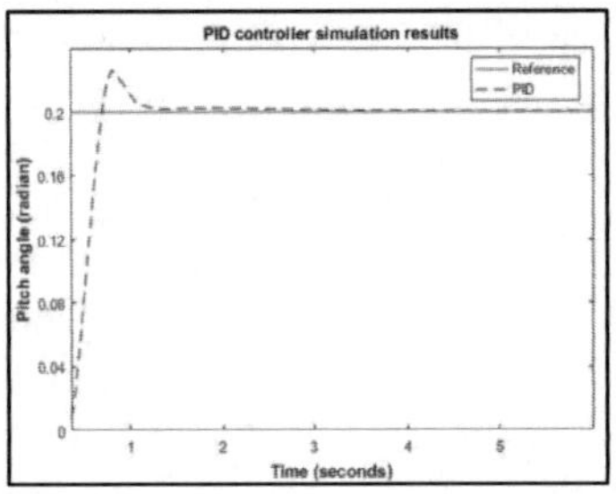

Figure 10: PID controller

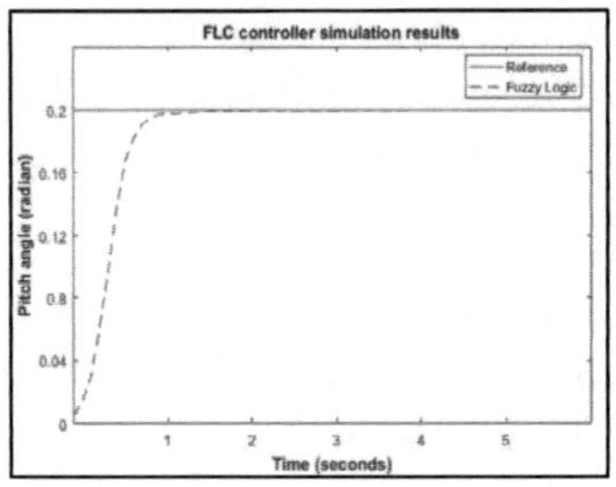

Figure 11: FLC controller

Moreover, it can be observed that the intelligent based approach gives the best output performance in term of settling time, overshoot percentage, steady state error and rise time. From the results shown in Figure 8 to Figure 11, the PD controller takes the longest time to settle down with 3.5 seconds. The highest percentage of overshoot goes to the P controller whereas the highest percentage of steady-state error is obtained by P and PD controller for 0.05. Significantly, the FLC controller has 0% overshoot compared to the other controllers. Based on these results, it can thus be concluded that FLC is the best controller to enhance the pitch motion of the aircraft compared to the PID.

Furthermore, in order to know whether the PID controller is robust or not, a new input pitch angle is set and a disturbance is applied. The disturbance is reflected on the cross wind effect that will hit the plane while on air in pitch motion scenario. The PID controller system is tested with new input pitch angle of 0.1 radians. Meanwhile, the 0.3 step input that acts as disturbance is added to the system to examine the robustness of the controller. The results obtained are shown in Figure 12 and Figure 13, which show that the PID controller still give a good performance by achieving the criteria needed. The PID controller settles down and follows the desired input of 0.1 radians. It is concluded that the PID controller is robust due to its performance from different input response and with applied disturbance.

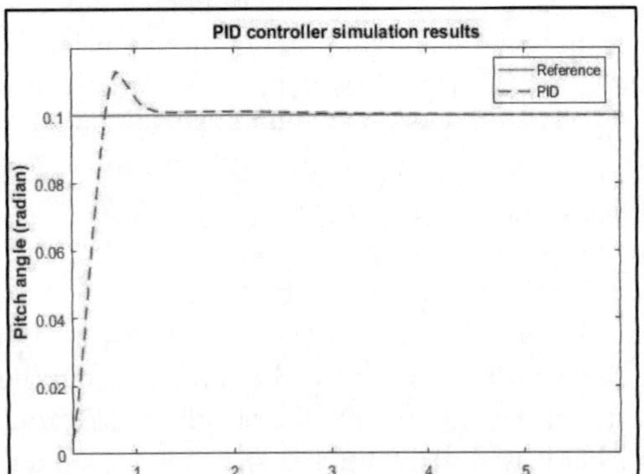

Figure 12: PID controller for input 0.1 radians and 0.3 step input disturbance effect

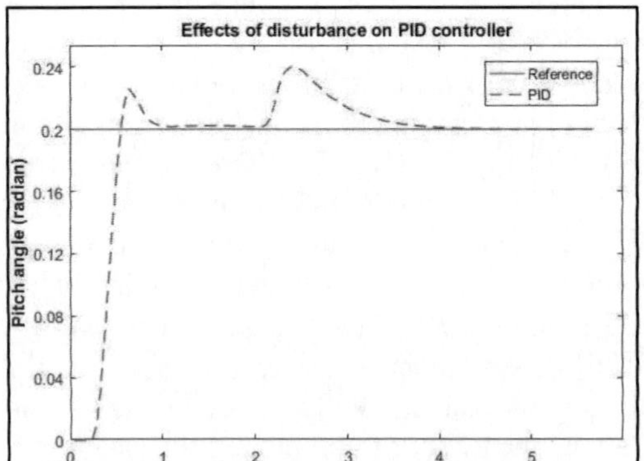

Figure 13: PID controller for input 0.2 radians and 0.3 step input disturbance effect

The FLC controller is also tested with similar disturbance criteria as the PID to see its robustness. Similar to PID controller, the chosen value for new input pitch angle is 0.1 radians and 0.3 step input is applied as disturbance on both chosen values of pitch angle. From Figure 14, it shows that the FLC controller still give a good performance by achieving the criteria required. The FLC controller settles down and follows the desired input 0.1 radians. When there is a disturbance, the steady-state error is

IOP Conf. Series: Materials Science and Engineering **270** (2017) 012008 doi:10.1088/1757-899X/270/1/012008

observed because it does not reach the settling time for 0.2 radians and 0.1 radians. Some tunings need to be done to improve the performance of the FLC when applied with a 0.3 step input disturbance for both input.

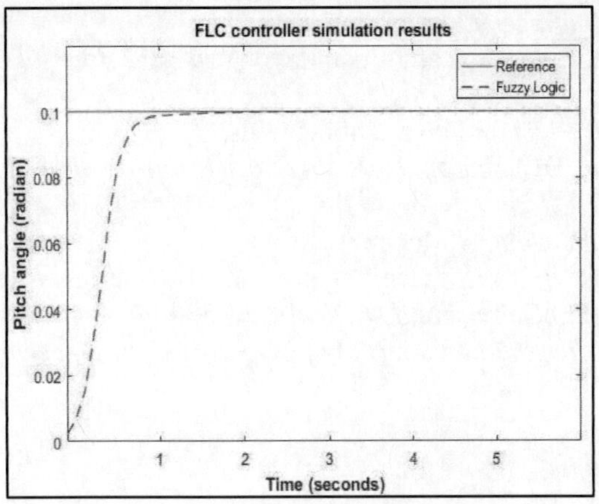

Figure 14: FLC controller for input 0.1 radians and 0.3 step input disturbance effect

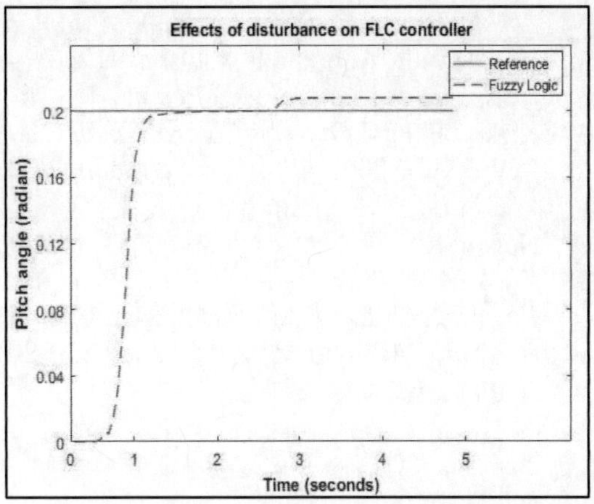

Figure 15: FLC controller for input 0.2 radians and 0.3 step input disturbance effect

5. Conclusion

The validated model of pitch control of an aircraft is very helpful in developing the control strategy for an actual system. Pitch control of an aircraft requires a pitch controller to maintain the pitch angle at the desired value. This can be achieved by reducing the error signal, which is the difference between the output angle and the desired angle. In this study, two methods in designing a controller have been explored: PID and FLC controllers. It is difficult to design a third order system, thus for PID controller method, the rate feedback is added to improve the system performance. From the results, it can be concluded that the controllers designed are all capable to control the pitch angle of an aircraft using a linearized system. Simulation and analysis of the results have shown that FLC controller gives the best performance in relative to the PID controller. However, further improvement needs to be done on the FLC controller in order to improve its performance such that it becomes more robust and much better response can be achieved. The limitation in the FLC controller design should be eliminated by adding more membership functions to the controller.

Acknowledgement

The authors like to thank Dr. Mohamed Sukri Mat Ali and Dr. Shamsul Sarip for their comments and advices for this study. This work has been financially supported by the research grant of PAS UTM under Grant Number PY/2016/07732.

References

[1] Osheku C A, Adetoro M A L, Agboola F A, Kisabo A B and Funmilayo A A 2012 *J. Sci. Res. Reports* **1** 1–16

[2] Hess R 1981 *AIAA Guidance and Control Conference*

[3] Singh A R and Giri V K 2012 *International Journal of Engineering Research & Technology* **1** 1–6

[4] Volosencu C 2012 *Tuning Fuzzy PID Controllers* in *Introduction to PID Controllers - Theory, Tuning and Application to Frontier Areas* InTech

[5] Wahid N and Hassan N 2012 *Third International Conference on Intelligent Systems, Modelling and Simulation*

[6] Levant A, Pridor A, Gitizadeh R, Yaesh I and Ben-Asher J Z 2000 *AIAA J. Guid. Control Dyn.* **23** 586–94
[7] Torabi A, Ahari A A, Karsaz A and Kazemi S H 2014 *J. Math. Comput. Sci.* **8** 113–27
[8] Aguilar-Ibanez C, Sossa-Azuela H and Suarez-Castanon M S 2015 *International Conference on Mechatronics, Electronics and Automotive Engineering*
[9] Kisabo A B, Agboola F A, Osheku1 C A, Adetorol M A L and Funmilayo A A 2012 *Journal of Scientific Research and Reports* **1** 1-16
[10] Kirk D E 2004 *Optimal Control Theory: An Introduction* Dover Publications
[11] Lavresky E 2005 *http://www.cds.caltech.edu/archive/help/uploads/wiki/files/136/Aircraft_Pitch _Roll _Dynamics.pdf*
[12] Nelson R C 1998 *Aircraft Stability and Automatic Control* McGraw-Hill
[13] Cook M V 2007 *Flight Dynamics Principles* Elsevier Aerospace Engineering Book Series
[14] Levant A, Pridor A, Gitizadeh R and Yaesh I 2000 *Control and Dynamics* **23** 586-94
[15] Salem M, Ashtiani M A and Sadati S H 2013 *International Journal of Scientific & Engineering Research* **4** 34–8

IOP Conf. Series: Materials Science and Engineering **270** (2017) 012009 doi:10.1088/1757-899X/270/1/012009

Gain scheduled linear quadratic control for quadcopter

M Okasha*, J Shah, W Fauzi and Z Hanouf

Department of Mechanical Engineering, International Islamic University Malaysia, Kuala Lumpur, Malaysia.

*mokasha@iium.edu.my

Abstract. This study exploits the dynamics and control of quadcopters using Linear Quadratic Regulator (LQR) control approach. The quadcopter's mathematical model is derived using the Newton-Euler method. It is a highly manoeuvrable, nonlinear, coupled with six degrees of freedom (DOF) model, which includes aerodynamics and detailed gyroscopic moments that are often ignored in many literatures. The linearized model is obtained and characterized by the heading angle (i.e. yaw angle) of the quadcopter. The adopted control approach utilizes LQR method to track several reference trajectories including circle and helix curves with significant variation in the yaw angle. The controller is modified to overcome difficulties related to the continuous changes in the operating points and eliminate chattering and discontinuity that is observed in the control input signal. Numerical non-linear simulations are performed using MATLAB and Simulink to illustrate to accuracy and effectiveness of the proposed controller.

1. Introduction

Quadcopters have significantly impacted the commercial and industrial fields with their multifaceted applications. They are ideal platforms for aerial photography, inspection, surveillance, and search and rescue missions in complex and dangerous environments due to their low cost, small size, superior mobility and hover capability. Many researchers are currently working on enhancing the quadcopter's capability and performance, especially when it comes to stability and control issues [1].

In literatures, researchers have designed various linear, nonlinear or even hybrid control techniques such as Proportional-Integral-Derivative (PID), Linear Quadratic Regulator (LQR), Sliding Mode, Backstepping, Model Predictive Control (MPC) and Robust Control techniques [2-8]. Most of these techniques have been successfully applied only for near hovering conditions. However, under the wide range of operating conditions or aggressive manoeuvres, it is difficult to control the quadcopter since it is a highly manoeuvrable, nonlinear, coupled and under-actuated system (i.e. it has four input forces with six output coordinates, so having an under-actuated degree of two).

This paper is focused on improving the difficulties associated with LQR controller by considering actuator saturation, continuous changes of operating points with high manoeuvrability and chattering of the control input signal. Moreover, it utilizes an accurate mathematical model of quadcopter based on the Newton-Euler nonlinear detailed formulation. This mathematical model includes the gyroscopic rotor dynamics and aerodynamics drag and moments that affect the quadcopter.

2. Dynamic Model

Quadcopter has four rotors arranged in a planar fashion. Two of these rotors are opposite to each other rotate in clockwise direction and the other to rotate in counter-clockwise direction. Changing the rotor speeds results in changing the overall lift an moments produced by the quadcopter. The quadcopter X configuration used in this study is shown in Figure 1, with body frame denoted by (x_b, y_b, z_b) and the world frame denoted by (x, y, z) [7-8].

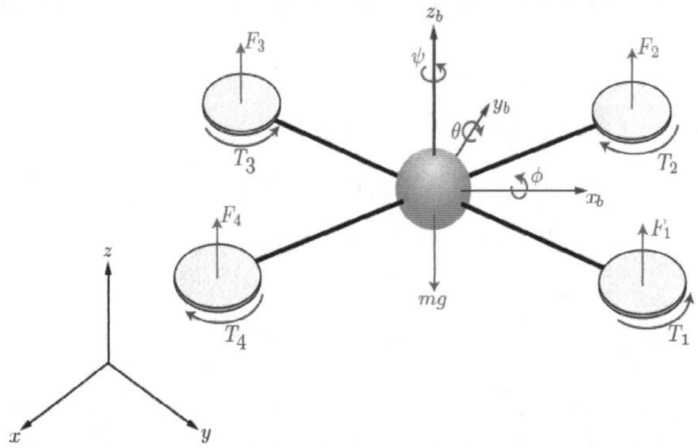

Figure 1: Quadcopter X configuration with coordinate frames (adopted from reference [8])

Using the Newton-Euler method, the equations of motion (translational and rotational) are derived as Equation 1 to Equation 9, where (x, y, z), $(\dot{x}, \dot{y}, \dot{z})$ and $(\ddot{x}, \ddot{y}, \ddot{z})$ are components of position, velocity and acceleration vectors, respectively, in world coordinate frame, (s, c, t) denote *sin, cosine* and *tangent* trigonometric function, respectively, g is gravitational constant, m is mass of the quadcopter, (J_x, J_y, J_z) are diagonal components of inertia matrix, J_r is the rotors' inertia, ω_r is the rotors' relative speed, (p, q, r) are the body's angular velocity components, (φ, θ, ψ) are roll, pitch and yaw Euler angles, respectively, $(K_{t_x}, K_{t_y}, K_{t_z})$ are diagonal components of constant aerodynamic translational drag matrix, $(K_{r_x}, K_{r_y}, K_{r_z})$ are diagonal components of constant aerodynamic rotational drag matrix and (u_1, u_2, u_3, u_4) are the components of the control input vector defined by Equation 10 to Equation 13, respectively.

$$\ddot{x} = -(K_{t_x}\dot{x} - mu_1(s\varphi s\psi + c\varphi s\theta c\psi))/m \tag{1}$$

$$\ddot{y} = -(K_{t_y}\dot{y} + mu_1(s\varphi c\psi - c\varphi s\theta s\psi))/m \tag{2}$$

$$\ddot{z} = -(gm + K_{t_z}\dot{z} - mu_1 c\theta c\varphi)/m \tag{3}$$

$$\dot{\varphi} = p + rc\varphi t\theta + qs\varphi t\theta \tag{4}$$

$$\dot{\theta} = qc\varphi - rs\varphi \tag{5}$$

$$\dot{\psi} = rc\varphi/c\theta + qs\varphi/c\theta \tag{6}$$

$$\dot{p} = -(K_{r_x}p - J_x u_2 - J_y qr + J_z qr + J_r q\omega_r)/J_x \tag{7}$$

$$\dot{q} = (-K_{r_y}q + J_y u_3 - J_x pr + J_z pr + J_r p\omega_r)/J_y \tag{8}$$

$$\dot{r} = (-K_{r_z}r + J_z u_4 + J_x pq - J_y pq)/J_z \tag{9}$$

$$u_1 = (F_1 + F_2 + F_3 + F_4)/m \tag{10}$$

$$u_2 = (-F_1 + F_2 + F_3 - F_4)l/J_x \tag{11}$$

$$u_3 = (-F_1 - F_2 + F_3 + F_4)l/J_y \tag{12}$$

$$u_4 = (T_1 - T_2 + T_3 - T_4)/J_z \tag{13}$$

For Equation 10 to Equation 13, l is the force moment arm while F_i and T_i are the aerodynamic force and moment produced by i^{th} rotor, respectively, that can be simplified into Equation 14 and Equation 15. Here, K_f and K_m are the aerodynamic force and moment constants, respectively, whereas ω_i is the angular velocity of rotor i.

$$F_i = K_f \omega_i^2 \tag{14}$$

$$T_i = K_m \omega_i^2 \tag{15}$$

To have an accurate and realistic dynamics model, detailed gyroscopic and aerodynamic effects, (drag forces and moments acting on the quadcopter) are considered in the derivation. Form Equation 1 to Equation 9, it is evident that this system is under actuated since it has four inputs (u_1, u_2, u_3, u_4) with six coordinate outputs $(x, y, z, \varphi, \theta, \psi)$. In next section, the linearized model is obtained and LQR controller is employed to track several reference trajectories with a constant or varying heading angle (x, y, z, ψ).

3. LQR Controller Design

The nonlinear dynamics can be compactly written as in Equation 16, where **x** and **u** are the quadcopter state and control input vectors, respectively. These vectors are defined as in Equation 17 and Equation 18.

$$\dot{\mathbf{x}} = \mathbf{f}(\mathbf{x}, \mathbf{u}) \tag{16}$$

$$\mathbf{x} = (x, y, z, \dot{x}, \dot{y}, \dot{z}, \varphi, \theta, \psi, p, q, r) \tag{17}$$

$$\mathbf{u} = (u_1, u_2, u_3, u_4) \tag{18}$$

The steady state equilibrium points of the quadcopter satisfy Equation 19. By solving Equation 19, the results are given by Equation 20 and Equation 21.

$$\mathbf{0} = \mathbf{f}(\mathbf{x}_{ss}, \mathbf{u}_{ss}) \tag{19}$$

$$\mathbf{x}_{ss} = (x_{ss}, y_{ss}, z_{ss}, 0,0,0,0,0, \psi_{ss}, 0,0,0) \tag{20}$$

$$\mathbf{u}_{ss} = (g, 0,0,0) \tag{21}$$

In order to design LQR controller for the nonlinear dynamics, the linearized model is obtained about the steady state values as indicated by Equation 22, where $\delta\mathbf{x} = \mathbf{x} - \mathbf{x}_{ss}$ and $\delta\mathbf{u} = \mathbf{u} - \mathbf{u}_{ss}$ while the Jacobian matrix **B** and **A** are found as in Equation 23 and Equation 24, respectively.

$$\delta\dot{\mathbf{x}} = \mathbf{A}\delta\mathbf{x} + \mathbf{B}\delta\mathbf{u} \tag{22}$$

$$\mathbf{B} = \frac{\partial \mathbf{f}}{\partial \mathbf{u}}_{(\mathbf{x}_{ss}, \mathbf{u}_{ss})} = \begin{bmatrix} 0 & 0 & 0 & 0 \\ 0 & 0 & 0 & 0 \\ 0 & 0 & 0 & 0 \\ 0 & 0 & 0 & 0 \\ 0 & 0 & 0 & 0 \\ 1 & 0 & 0 & 0 \\ 0 & 0 & 0 & 0 \\ 0 & 0 & 0 & 0 \\ 0 & 0 & 0 & 0 \\ 0 & 1 & 0 & 0 \\ 0 & 0 & 1 & 0 \\ 0 & 0 & 0 & 1 \end{bmatrix} \tag{23}$$

$$\mathbf{A} = \frac{\partial \mathbf{f}}{\partial \mathbf{x}}_{(\mathbf{x}_{ss},\mathbf{u}_{ss})} = \begin{bmatrix} 0 & 0 & 0 & 1 & 0 & 0 & 0 & 0 & 0 & 0 & 0 & 0 \\ 0 & 0 & 0 & 0 & 1 & 0 & 0 & 0 & 0 & 0 & 0 & 0 \\ 0 & 0 & 0 & 0 & 0 & 1 & 0 & 0 & 0 & 0 & 0 & 0 \\ 0 & 0 & 0 & -\frac{K_{t_x}}{m} & 0 & 0 & gs\psi & gc\psi & 0 & 0 & 0 & 0 \\ 0 & 0 & 0 & 0 & -\frac{K_{t_y}}{m} & 0 & -gc\psi & gs\psi & 0 & 0 & 0 & 0 \\ 0 & 0 & 0 & 0 & 0 & -\frac{K_{t_z}}{m} & 0 & 0 & 0 & 0 & 0 & 0 \\ 0 & 0 & 0 & 0 & 0 & 0 & 0 & 0 & 0 & 1 & 0 & 0 \\ 0 & 0 & 0 & 0 & 0 & 0 & 0 & 0 & 0 & 0 & 1 & 0 \\ 0 & 0 & 0 & 0 & 0 & 0 & 0 & 0 & 0 & 0 & 0 & 1 \\ 0 & 0 & 0 & 0 & 0 & 0 & 0 & 0 & 0 & -\frac{K_{r_x}}{J_x} & 0 & 0 \\ 0 & 0 & 0 & 0 & 0 & 0 & 0 & 0 & 0 & 0 & -\frac{K_{r_y}}{J_y} & 0 \\ 0 & 0 & 0 & 0 & 0 & 0 & 0 & 0 & 0 & 0 & 0 & -\frac{K_{r_z}}{J_z} \end{bmatrix} \quad (24)$$

This linear parameter varying system is observable and controllable of rank 12 as found in Ref. [8] with output feedback, $\mathbf{y} = (x, y, z, \psi)$. The control input vector can be computed using LQR controller as in Equation 25 where $\mathbf{u} = \delta\mathbf{u} + \mathbf{u}_{ss}$ and $\mathbf{K}$ is the LQR gain matrix used to minimize cost function in Equation 26. In Equation 26, $\mathbf{Q}$ and $\mathbf{R}$ are symmetric positive definite weighting matrices for state convergence and control action respectively. Increasing the elements of $\mathbf{Q}$ results in a quicker state convergence while increasing elements of $\mathbf{R}$ limits the control input action.

$$\delta\mathbf{u} = -\mathbf{K}\delta\mathbf{x} \quad (25)$$

$$\mathrm{J} = \int_0^{\infty} (\delta\mathbf{x}^T \mathbf{Q} \delta\mathbf{x} + \delta\mathbf{u}^T \mathbf{R} \delta\mathbf{u}) \mathrm{dt} \quad (26)$$

The gain matrix $\mathbf{K}$ can be computed using the following Matlab command, $\mathbf{K} = \mathbf{lqr}(\mathbf{A}, \mathbf{B}, \mathbf{Q}, \mathbf{R})$. As observed in Equation 24, matrix $\mathbf{A}$ is a parameter linear varying matrix in the heading angle, ψ. To maintain stability, three LQR controllers are implemented and tested to track a time varying reference trajectories (i.e. circular and helical) with varying heading angle for the following cases: (1) fixed-gain control, where the controller gain is constant and the linearized model is evaluated at constant nominal heading angle, (2) gain-scheduled control, which the controller gains are switching based on heading angle reference value described by Equation 27 as shown in Figure 2, and (3) continuous gain control, which the controller gains are evaluated continuously as function of the reference heading angle.

$$\mathbf{u}(\mathbf{x}) = \begin{cases} \mathbf{u}_1\left(\mathbf{x}; \frac{\pi}{12}\right), & 0 \le \psi(t) \le \frac{\pi}{6} \\ \mathbf{u}_2\left(\mathbf{x}; \frac{\pi}{4}\right), & \frac{\pi}{12} \le \psi(t) \le \frac{\pi}{3} \\ & \vdots \\ & \vdots \\ \mathbf{u}_{12}\left(\mathbf{x}; \frac{23\pi}{4}\right), & \frac{11\pi}{6} \le \psi(t) \le 2\pi \end{cases} \quad (27)$$

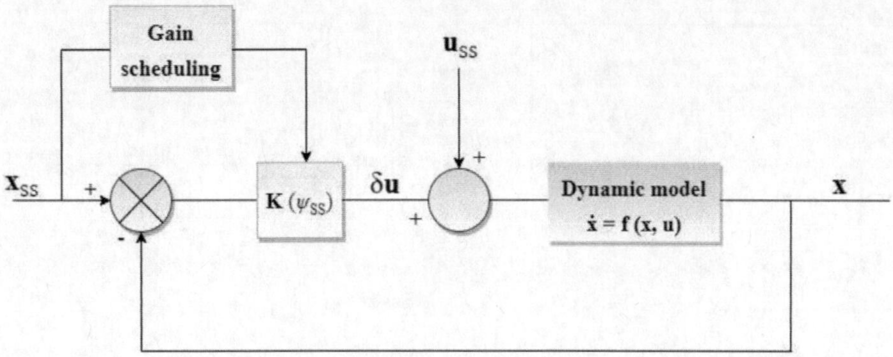

Figure 2: Block diagram of the gain-scheduled controller

4. Results and Discussion

To assess the performance of the developed controller, Matlab and Simulink environments are applied in the numerical simulation. The quadrotor's parameters are taken from Ref. [2] and Ref. [7] based on the OS4 hardware. For physical implementation purposes, actuators' saturation limits are considered in the simulation. The numerical values of **Q** and **R** matrices are tuned by trial and error method. These matrices are diagonal and have the following values: **Q** = diag(10,10,10,1,1,1,1,1,10,1,1,1) and **R** = diag(0.001,.001,0.001,.001).

Simulation results are presented for the three cases mentioned in the previous section: (1) tracking circular trajectory with variable heading angle using LQR fixed-gain control, (2) tracking circular and helical trajectories with variable heading angle using LQR gain-scheduled control and (3) tracking circular and helical trajectories with variable heading angle using LQR continuous gain control. Figure 3 shows the 3D and 2D time history of the states (blue lines) and control input to track a circular reference trajectory (red dotted lines) with variable heading angle. In this figure, the LQR fixed-gain control is computed based on the linearized model at constant heading angle ($\psi = 0$). It is observed that this controller is no longer guaranteed stability as the heading angle deviates away from the value about the linearization is taken. To correct this issue, the LQR gain-scheduled control is employed and tested to track circular and helical trajectories with varying heading angles as shown in Figure 4 and Figure 5. In these figures, the controller is stable and able to track the reference signals.

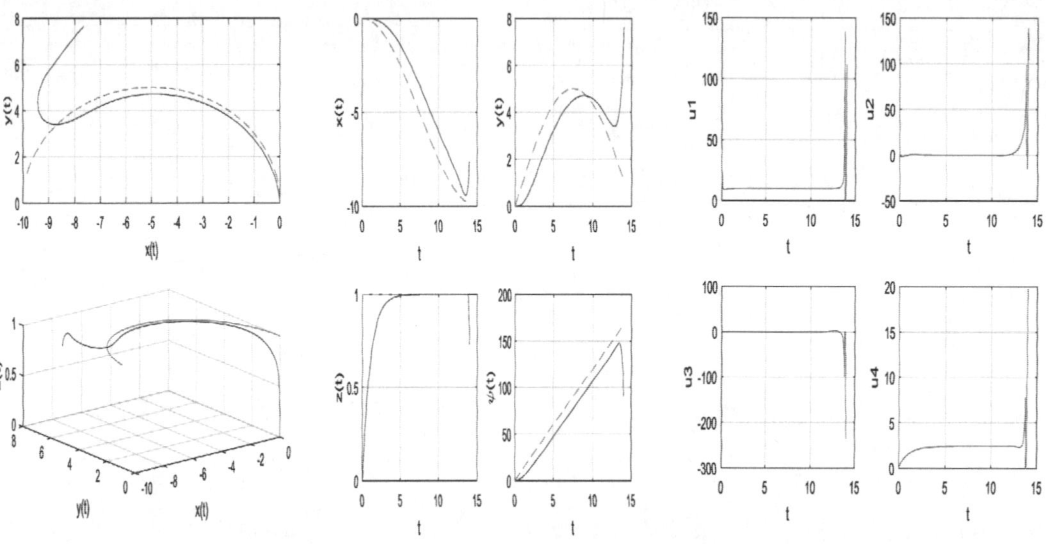

Figure 3: 3D and 2D trajectory, states and control inputs for the fixed gain circular trajectory

IOP Conf. Series: Materials Science and Engineering **270** (2017) 012009 doi:10.1088/1757-899X/270/1/012009

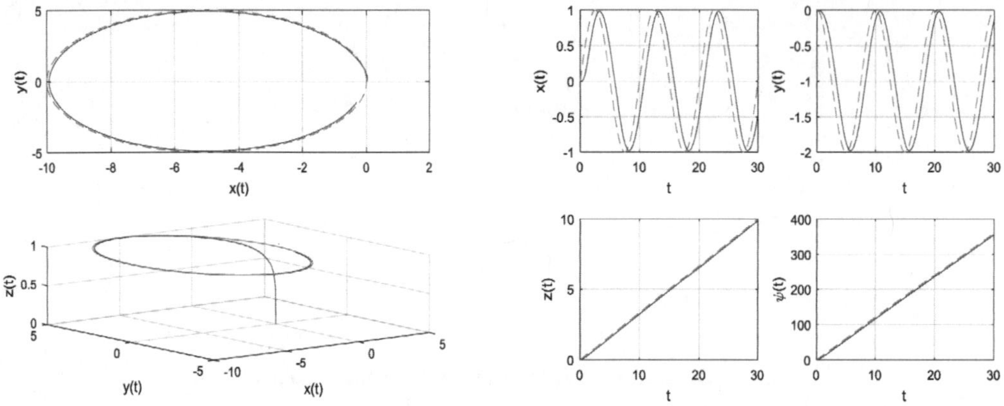

Figure 4: 3D and 2D trajectory and states for the gain-scheduled circular trajectory

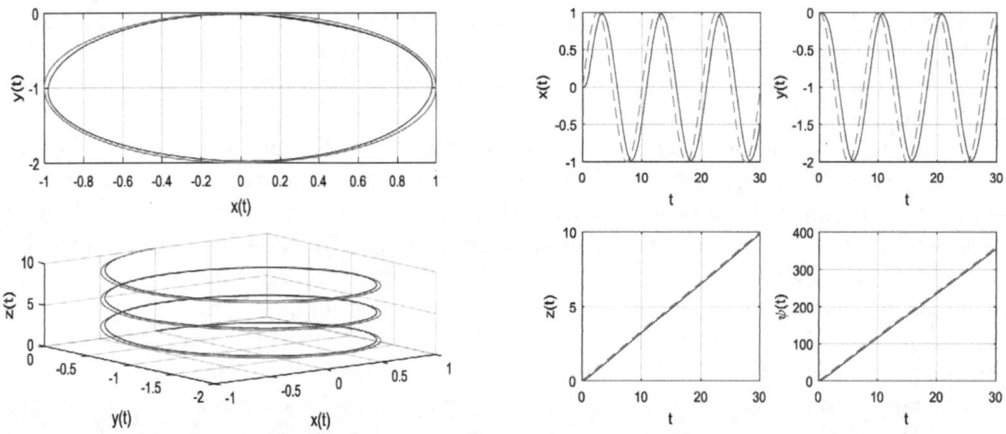

Figure 5: 3D and 2D trajectory and states for the gain-scheduled helical trajectory

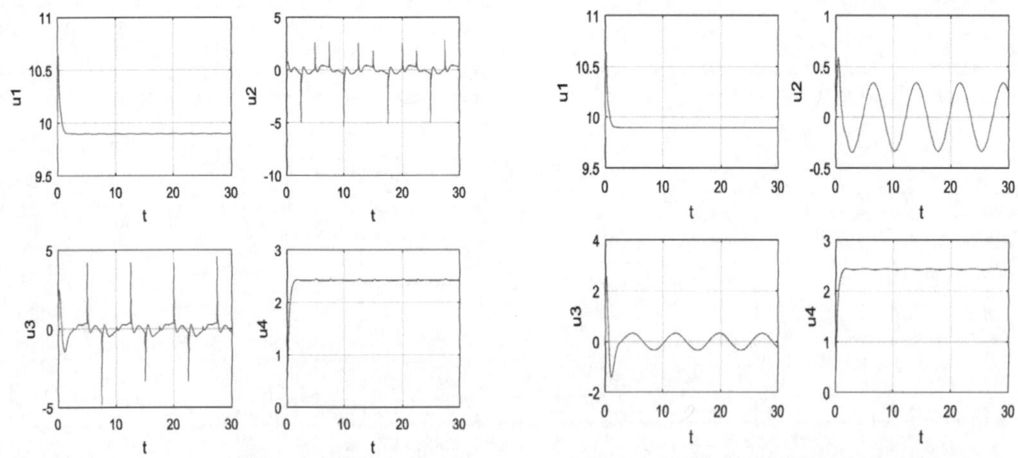

(a) (b)

Figure 6: Control input for the helical trajectory: (a) gain-scheduled and (b) continuous gain

IOP Conf. Series: Materials Science and Engineering **270** (2017) 012009 doi:10.1088/1757-899X/270/1/012009

However, while the gain-scheduled control is successful in tracking various reference trajectories, the physical limitations of the control efforts need to be considered during implementations. Figure 6(a) shows the control efforts to track the helical trajectory for gain-scheduled case. As a result of switching discretely between operating points of gain-scheduled control, it is observed that there are discontinuity and chattering in the control input signals, which affect physical implementation of this control law. It is found that these limitations can be eliminated by employing continuous gain control techniques as shown in Figure 6(b). Furthermore, it is noted on all simulated cases that there are steady state errors in tracking reference signals. These errors can be reduced through proper selection of **Q** and **R** weighting matrices. To have zero state error, the LQR method should be modified to include and integral action in the control law.

5. Conclusion

In this paper, a detailed mathematical model of quadcopter is derived using Newton-Euler method, including aerodynamic moments and gyroscopic nonlinear terms. Then, a parameter varying linearized model is obtained by linearizing the nonlinear model around the steady state values. The LQR control method is employed with fixed and variable gain control law to track various reference signals (circular and helical curves). It is observed, for conventional fixed gain control law, the controller is inadequate for tracking reference signal with significant variation in the heading angle. The gain-scheduled control law has been designed to successfully resolve this issue. Continuous gain control law is able to overcome scattering and discontinuity problems of control input signals. In future, an integral action term will be introduced in the LQR control design to eliminate the steady state errors that is observed in the tracking of reference signals. Furthermore, special attention will be given to the **Q** and **R** weighting matrices to determine how they should be varied to best achieve various design objectives and shape the dynamic response. Finally, these laws will be tested and implemented on a quadcopter hardware in the loop real time simulation.

References

[1] Kendoul F 2012 *J. Field Robotics* **29** 315-78
[2] Bouabdallah S, Noth A and Siegwart R 2004 *IEEE/RSJ International Conference on Intelligent Robots and Systems*
[3] Li J and Li Y 2011 *International Conference in Mechatronics and Automation*
[4] Bouabdallah S and Siegwart R 2005 *IEEE International Conference in Robotics and Automation*
[5] Nicol C, Macnab C J B and Ramirez-Serrano A 2011 *Mechatronics* **21** 927-38
[6] Mueller W M and D'Andrea R 2013 *European Control Conference*
[7] Elkholy H 2014 *Dynamic Modeling and Control of A Quadrotor using Linear and Nonlinear Approaches* Thesis, American University in Cairo
[8] Sawyer S 2015 *Gain-Scheduled Control of a Quadcopter UAV* Thesis, University of Waterloo

IOP Conf. Series: Materials Science and Engineering **270** (2017) 012010 doi:10.1088/1757-899X/270/1/012010

Quadrotor trajectory tracking using PID cascade control

M Idres*, O Mustapha and M Okasha

Department of Mechanical Engineering, International Islamic University Malaysia, Jalan Gombak, 53100 Kuala Lumpur, Malaysia.

*midres@iium.edu.my

Abstract. Quadrotors have been applied to collect information for traffic, weather monitoring, surveillance and aerial photography. In order to accomplish their mission, quadrotors have to follow specific trajectories. This paper presents proportional–integral–derivative (PID) cascade control of a quadrotor for path tracking problem when velocity and acceleration are small. It is based on near hover controller for small attitude angles. The integral of time-weighted absolute error (ITAE) criterion is used to determine the PID gains as a function of quadrotor modeling parameters. The controller is evaluated in three-dimensional environment in Simulink. Overall, the tracking performance is found to be excellent for small velocity condition.

1. Introduction

Recently, a growing interest in unmanned aerial vehicles (UAVs) has been shown among the research community. Quadcopters are flying vehicles that can be equipped with cameras and sensors to perform many complex tasks. Due to their high maneuverability and small size, quadrotors have been widely used. Their potential applications include search-and-rescue, homeland security, military surveillance, and earth sciences [1-3]. In general, quadrotors are naturally unstable, under-actuated, under damped, coupled and nonlinear systems that need to be controlled. Several control techniques can be utilized to control a quadrotor including proportional–integral–derivative (PID) [4, 5], linear quadratic regulator (LQR) [6], and backstepping and sliding mode control [7, 8]. Because of its simple structure and good stability, PID plays an important role in quadrotor control.

In this study, a PID controller is used for attitude and position control. Assuming small acceleration and small attitude angles, the corresponding control laws are derived. A Simulink model is developed where the performance of the controller is tested for different path-tracking cases.

2. Methodology

Figure 1 illustrates the quadrotor dynamic model with world (earth fixed) frame (x_W, y_W, z_W), body-fixed frame (x_B, y_B, z_B), rotor forces, F_i and rotor moments, M_i. The corresponding dynamic model is represented by Equation 1 to Equation 5 [4, 9], where r is the position vector of the center of mass of the quadrotor, F_i is the thrust force from rotor 'i', R is the transformation matrix from body frame to earth frame, p, q and r are the three components of angular velocity corresponding to roll, pitch and yaw motions, L is the distance from the center of mass to the axis of rotation of rotors, M_i is moment of rotor 'i' about center of mass, I is the inertia matrix, ω_i is rotor angular speed, ω_{des} is the desired rotor angular speed and k_F, k_M and k_m are constants.

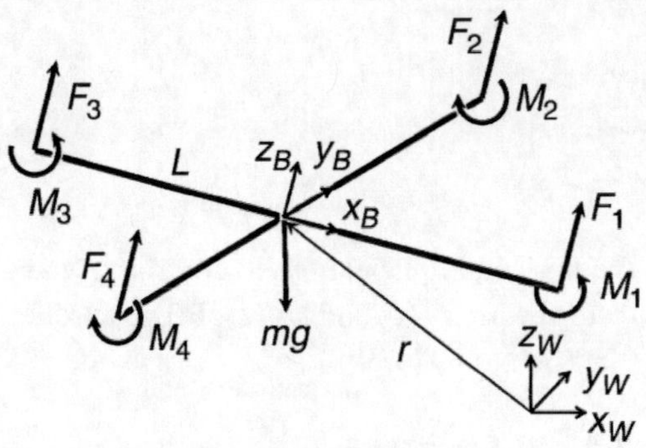

Figure 1: Quadrotor dynamic model representation

$$m\ddot{\boldsymbol{r}} = \begin{bmatrix} 0 \\ 0 \\ -mg \end{bmatrix} + R \begin{bmatrix} 0 \\ 0 \\ \sum F_i \end{bmatrix} \tag{1}$$

$$I \begin{bmatrix} \dot{p} \\ \dot{q} \\ \dot{r} \end{bmatrix} = \begin{bmatrix} L(F_2 - F_4) \\ L(F_3 - F_1) \\ M_1 - M_2 + M_3 - M_4 \end{bmatrix} - \begin{bmatrix} p \\ q \\ r \end{bmatrix} \times I \begin{bmatrix} p \\ q \\ r \end{bmatrix} \tag{2}$$

$$I = \begin{bmatrix} I_{xx} & 0 & 0 \\ 0 & I_{yy} & 0 \\ 0 & 0 & I_{zz} \end{bmatrix} \tag{3}$$

$$F_i = k_F\,\omega_i^2; \quad M_i = k_M\omega_i^2 \tag{4}$$

$$\dot{\omega}_i = k_m(\omega_i^{des} - \omega_i) \tag{5}$$

Figure 2 shows the cascade control loops [4, 9]. Position controller determines the required change in motor speed, $\Delta\omega_F$ to follow the hovering height, $r_{T,3}(t)$and the required attitude angles: ϕ^{des}, θ^{des} and ψ^{des} based on position error, $r_T(t) - r(t)$. The attitude controller then determines the required change in the motor speeds: $\Delta\omega_\phi$, $\Delta\omega_\theta$ and $\Delta\omega_\psi$ in order to achieve the desired attitude angles. Thus the quadrotor will follow the specified trajectory, $r_T(t)$.

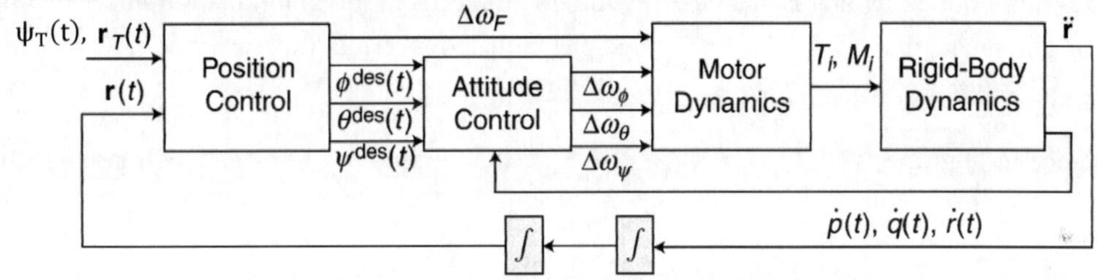

Figure 2: Nested control loops for position and attitude control

The vector of desired rotor speeds can be written as a linear combination of four terms as in Equation 6 [4, 9], where ω_h is the nominal rotor speed required to hover in steady state: $\omega_h = \sqrt{m\,g/(4\,k_F)}$ and the deviations from this nominal value are $\Delta\omega_F$, $\Delta\omega_\phi$, $\Delta\omega_\theta$ and $\Delta\omega_\psi$. $\Delta\omega_F$ results in a net force along the z-axis direction in the earth frame while $\Delta\omega_\phi$, $\Delta\omega_\theta$ and $\Delta\omega_\psi$ produce moments causing roll, pitch and yaw, respectively.

$$\begin{bmatrix} \omega_1^{des} \\ \omega_2^{des} \\ \omega_3^{des} \\ \omega_4^{des} \end{bmatrix} = \begin{bmatrix} 1 & 0 & -1 & 1 \\ 1 & 1 & 0 & -1 \\ 1 & 0 & 1 & 1 \\ 1 & -1 & 0 & -1 \end{bmatrix} \begin{bmatrix} \omega_h + \Delta\omega_F \\ \Delta\omega_\phi \\ \Delta\omega_\theta \\ \Delta\omega_\psi \end{bmatrix} \tag{6}$$

Equation 7 is obtained by linearizing Equation 2 about the hovering operating point (assuming small attitude angles) [4, 9].

$$\dot{p}^{des} = \frac{4\,k_F\,L\,\omega_h}{I_{xx}}\,\Delta\omega_\phi\,, \quad \dot{q}^{des} = \frac{4\,k_F\,L\,\omega_h}{I_{yy}}\,\Delta\omega_\theta\,, \quad \dot{r}^{des} = \frac{8\,k_M\omega_h}{I_{zz}}\,\Delta\omega_\psi \tag{7}$$

Near the hovering operating point, it is assumed that $\dot{\phi} \approx p, \dot{\theta} \approx q$ and $\dot{\psi} \approx r$. Subsequently, the deviations from hovering required for attitude PD control are given by Equation 8. Since the position PID controller will generate angle references for the attitude controller, integral action is not necessary

IOP Conf. Series: Materials Science and Engineering **270** (2017) 012010 doi:10.1088/1757-899X/270/1/012010

for the latter. The integral term is needed to ensure steady state error, which is not required for the internal attitude loop.

$$\begin{aligned}\Delta\omega_\phi &= k_{p,\phi}(\phi^{des}-\phi)+k_{d,\phi}(p^{des}-p)\\ \Delta\omega_\theta &= k_{p,\theta}(\theta^{des}-\theta)+k_{d,\theta}(q^{des}-q)\\ \Delta\omega_\psi &= k_{p,\psi}(\psi^{des}-\psi)+k_{d,\psi}(r^{des}-r)\end{aligned} \tag{8}$$

Pitch and roll angles are used to control position in the horizontal plane, $\Delta\omega_\psi$ to control yaw angle and $\Delta\omega_F$ to control hover height. Assume a tracking trajectory with $r_T(t)$ and $\psi_T(t)$. PID controller is used to calculate the commanded acceleration ($\ddot{r}_i^{des}$) as in Equation 9 [4, 9].

$$(\ddot{r}_{i,T}-\ddot{r}_i^{des})+k_{p,i}(r_{i,T}-r_i)+k_{i,i}\int(r_{i,T}-r_i)dt+k_{d,i}(\dot{r}_{i,T}-\dot{r}_i)=0 \tag{9}$$

For near hover condition, $\ddot{r}_{i,T}\approx\dot{r}_{i,T}\approx 0$. Therefore, this algorithm is only successful in following a track with small velocity and acceleration. Linearizing Equation 1 about the hovering operating point leads to Equation 10 [4, 9].

$$\begin{aligned}\phi^{des} &= (\ddot{r}_1^{des}\sin\psi_T-\ddot{r}_2^{des}\cos\psi_T)/g\\ \theta^{des} &= (\ddot{r}_1^{des}\cos\psi_T+\ddot{r}_2^{des}\sin\psi_T)/g\\ \Delta\omega_F &= m\,\ddot{r}_3^{des}/(8\,k_F\omega_h)\end{aligned} \tag{10}$$

By linearizing Equation 1 and Equation 2 around hovering operating point, plant transfer functions for position, roll, pitch, and yaw by $P(s)=\frac{K}{s^2}$ are approximated [4, 9]. In this case study, $K=1/m$ for position, $K=\frac{4\,k_F\,L\,\omega_h}{I_{xx}}$ for roll and pitch, and $K=\frac{8\,k_M\omega_h}{I_{zz}}$ for yaw.

In reference to Figure 3, with $C(s)=K_p+\frac{K_i}{s}+K_d s$, the closed loop transfer function is given by Equation 11.

$$\frac{y(s)}{r(s)}=\frac{K(k_d s^2+k_p s+k_i)}{s^3+K(k_d s^2+k_p s+k_i)} \tag{11}$$

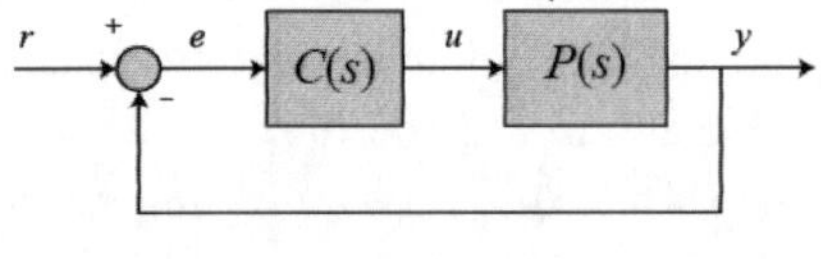

Figure 3: Transfer function for PID tuning

PID controller is designed using minimization of integral of time-weighted absolute error (ITAE) criterion [10]. The ω_n of the closed loop system is selected based on the desired settling time ($T_s=\frac{4}{\xi\omega_n}$ and $\xi\approx 0.8$). PID parameters are then determined by comparing the characteristic equation of the closed loop system with the optimum characteristic equation for ITAE as shown in Equation 12 [10]. The resulting PID controller parameters serve as the starting values for later tuning by trial and error.

$$s^3+1.75\,\omega_n s^2+2.15\,\omega_n^2 s+\omega_n^3 \tag{12}$$

3. Results and Discussions

The quadrotor parameters are tabulated in Table 1 and the controller gains are listed in Table 2, with ω_h = 109.68 rev/s. The attitude controller (roll and pitch) is made ten times faster than the position controller by appropriate choice of settling time.

Table 1: Quadrotor parameters [4]

Parameter	Value	Units
m	1.08	Kg
L	0.22	m
I_{xx}	0.1045	Kg.m^2
I_{yy}	0.1045	Kg.m^2
I_{zz}	0.2090	Kg.m^2
k_F	2.2×10^{-4}	N/(rev/s)^2
k_M	5.4×10^{-6}	N m / (rev/s)^2
k_m	20	s^{-1}

Table 2: Controller parameters

Controller	Gain K	Settling time T_s	Frequency ω_n	k_p $2.15\,\omega_n^2/K$	k_i ω_n^3/K	k_d $1.75\,\omega_n/K$
Position (PID)	0.952	6 s	0.83	1.567	0.607	1.53
Roll (PD)	0.203	0.6 s	8.3	734.9	0	71.81
Pitch (PD)	0.203	0.6 s	8.3	734.9	0	71.81
Yaw (PD)	0.005	6 s	0.83	291.55	0	298.37

Figure 4 shows the step response for step input of 2m in *x*, *y* or *z*. Desired track and obtained track are plotted, where an overshoot of 15% (*x* and *y*) and 25% (*z*). The rise time is 2s and the settling time is 15 s. A compromise is made for fast response (small rise time) at the expense of high overshoot. The tracking of the ramp trajectory in *x*, *y*, or *z* with a slope (velocity) equals to 1 is demonstrated in Figure 5. The Root Mean Square Error (RMSE) is about 0.0001.

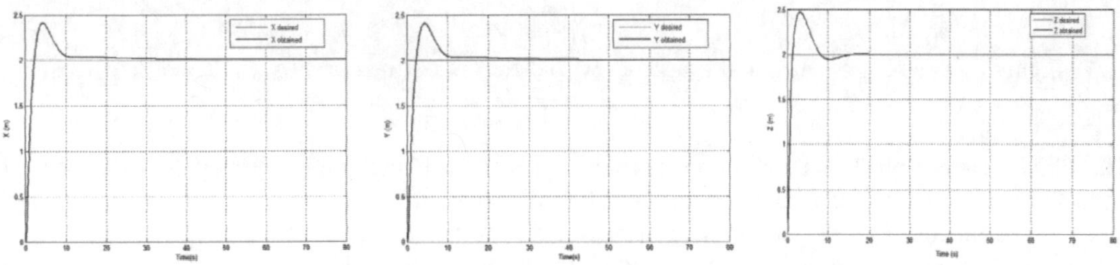

(a) (b) (c)

Figure 4: Step input response in (a) *x*, (b) *y* and (c) *z*

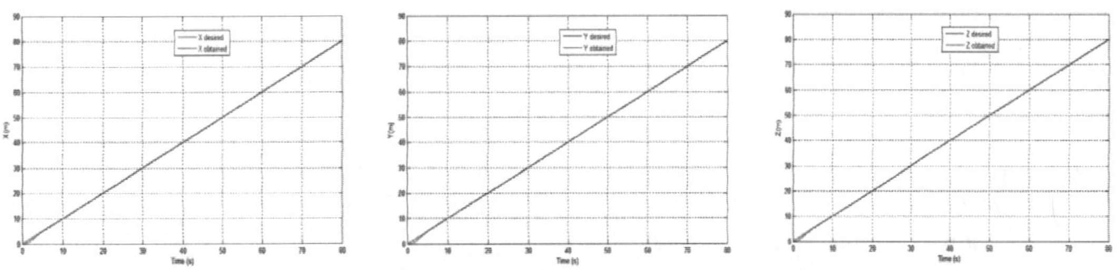

(a) (b) (c)

Figure 5: Tracking ramp trajectory in (a) *x*, (b) *y* and (c) *z*

The quadrotor is also commanded to move in a circle (radius = 6m and velocity = 2m/s) in *x-y* plane, *x-z* plane or *y-z* plane as shown in Figure 6. It has been shown that the tracking performance is excellent with RMSE of 0.0002.

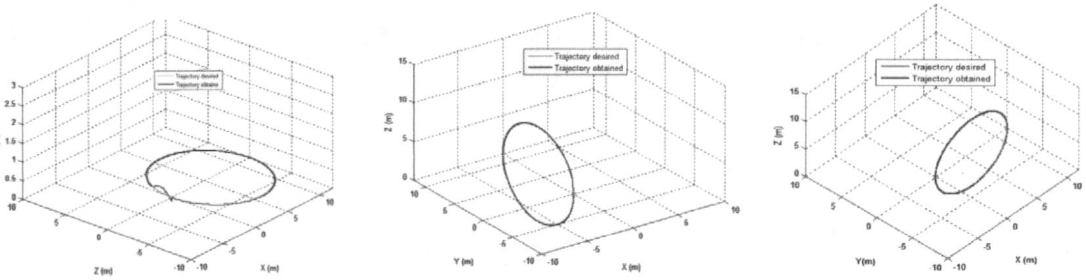

(a) (b) (c)

Figure 6: Tracking of a circular trajectory in (a) *x-y* plane, (b) *x-z* plane and (c) *y-z* plane

The quadrotor has been commanded to move in a sinusoidal path in one direction (*x*, *y* or *z*) with 1m amplitude and different time periods (80s, 20s and 10s) as shown in Figure 7. The RMSE increases as the period decreases (velocity increases) from 0.0002 for period = 80s to 0.045 for period = 10s. This demonstrates the limitation of the current algorithm to the small velocity trajectories as it is an assumption in the development of this control method.

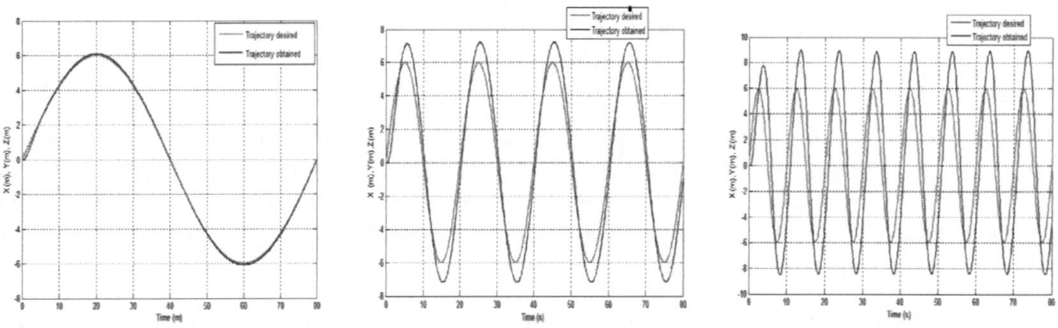

(a) (b) (c)

Figure 7: Tracking a sinusoidal trajectory with different periods (a) 80s, (b) 20s and (c) 10s

4. Conclusions

The PID cascade control method is implemented for quadrotor tracking of a specified trajectory. The controllers are based on near hover condition. The PID gains are obtained as the function of quadrotor modeling parameters using ITAE criterion. Tracking performance has been tested for ramp, circular and sinusoidal trajectories. The performance is found to be acceptable as long as velocity is small. For the future work, tracking of large angle and large velocity trajectories will be considered, which entail the modification of control algorithm.

Acknowledgement

The authors acknowledge the financial support for this study from the International Islamic University Malaysia (IIUM).

References

[1] Ghazbi S N, Aghli Y, Alimohammadi M and Akbari A A 2016 *International Journal of Smart Sensing and Intelligent Systems* **9** 309-33
[2] Gupte S, Mohandas P and Conrad J 2012 *IEEE Southeastcon*
[3] Sarris Z 2001 *9th IEEE Mediterranean Conference on Control and Automation*

IOP Conf. Series: Materials Science and Engineering **270** (2017) 012010 doi:10.1088/1757-899X/270/1/012010

[4] Mellinger D 2012 *Trajectory Generation and Control for Quadrotors* PhD Thesis, University of Pennsylvania
[5] Kugelberg I 2016 *Black-Box Modeling and Attitude Control of a Quadcopter* Master Thesis, Linköping University
[6] Hoffmann G, Rajnarayan D G, Waslander S L, Dostal D, Jang J and Tomlin C J 2004 *23rd Digital Avionics Systems Conference*
[7] Bouabdallah S and Siegwart R 2005 *IEEE International Conference of Robotic and Automatic*
[8] Hao N D, Mohamed B and Rafaralah H 2014 *European Control Conference*
[9] Mustapha O 2017 *Modeling, Control and Navigation of Quadrotor* Master Thesis, International Islamic University Malaysia
[10] Dorf R C and Bishop R H 2011 *Modern Control Systems* Prentice Hall

IOP Conf. Series: Materials Science and Engineering **270** (2017) 012011 doi:10.1088/1757-899X/270/1/012011

Electromagnetic calibration system for sub-micronewton torsional thrust stand

J K Lam[1], S C Koay[1] and K H Cheah[2,*]

[1]School of Engineering, Taylor's University, Malaysia
[2]School of Engineering and Physical Science, Heriot-Watt University, Malaysia

*k.cheah@hw.ac.uk

Abstract. It is critical for a micropropulsion system to be evaluated. Thrust stands are widely recognised as the instrument to complete such tasks. This paper presents the development of an alternative electromagnetic calibration technique for thrust stands. Utilising the commercially made voice coils and permanent magnets, the proposed system is able to generate repeatable and also consistent steady-state calibration forces at over four orders of magnitude (30 – 23000 μN). The system is then used to calibrate a custom-designed torsional thrust stand, where its inherent ability in ease of setup is well demonstrated.

1. Introduction

Small satellites have found increasing applications in communications and space weather study [1]. This is because they are simpler, can be developed in shorter time frame and inherently cost effective [2]. In particular, nanosatellites (typically <10 kg) have gained more interests. They can be developed and launched into space within only two years, which is highly advantageous for demonstrating new and innovative ideas in outer space exploration. One of the major sub-systems in a nanosatellite is the micropropulsion system. It makes use of microthrusters to perform tasks like attitude control, station keeping, drag compensation and orbital transfer [3]. Throughout the years, various microthrusters have been developed with the help of microelectromechanical systems (MEMS) that have been proven as a promising approach [4].

In light of this, the accurate and precise performance characterisation of a microthruster becomes critical [5]. As the forces produced by microthrusters are extremely low (in the order of micronewton), resolving them requires highly sensitive measuring instruments. The pendulum thrust stand is widely regarded as the more efficient and suitable method to measure constant and impulsive thrust forces produced by the microthrusters [6]. Essentially, it is a spring-mass-damper system in which a structure holding the microthruster is supported by a torsional spring. The structure will oscillate as the forces are applied. By analysing the oscillation, the forces produced by the microthrusters can be determined. There are three main configurations for pendulum thrust stand, i.e. hanging [7, 8], inverted [9, 10] and torsional [11–13], with each of them having their own advantages and limitations.

As a measuring instrument, it is necessary for the pendulum thrust stand to be calibrated. Basically, the calibration establishes the relationship between the thrust stand response in terms of displacement (measured by the sensor) and the forces applied. There are two categories of calibration techniques: contact and non-contact. The contact calibration techniques include string-pulley-weight system [7], impact hammer [14] and impact pendulum [15]. These are older methods that have become obsolete

[16] as their performance pales in comparison to that of the modern non-contact methods. Non-contact calibration systems include gas dynamic [16, 17], electrostatic (ES) [14, 18–20] and electromagnetic (EM) [12, 21–22]. The gas dynamic calibrators are reliable in producing calibration forces between nanonewton and sub-micronewton [17]. In contrast, ES calibrators are able to provide a wider range of calibration forces, typically between hundreds of nanonewton and thousands of micronewton [20]. As for EM calibrators, though the reported calibration forces are sub-micronewton and above, they also exhibit good consistency and repeatability. Besides, EM calibrators are much easier to be implemented compared to gas dynamic and ES. They mostly consist of an electromagnet (solenoid) coupled with permanent magnet, current-carrying copper wire, or metal conductor.

This study explores the feasibility of using the commercially available voice coil to develop EM calibration system for a sub-micronewton torsional thrust stand. The developed calibration system is very compact as the voice coil is rather small in size. This is highly beneficial for integration with small testing facilities, in particular the vacuum chamber, in which the setup cost is proportional with the overall size. The selected voice coil is first tested with different permanent magnets to investigate the characteristics of the EM calibration force generated. The system is then implemented onto the thrust stand to demonstrate its performance as a calibrator.

2. Torsional Thrust Stand Setup

A thrust stand is designed and constructed based on the working principle of a torsional pendulum [6]. Generally, the torsional thrust stand has a good balance in measuring high and low vibrational noise sensitivities. The existing thrust stand consists of a 60 cm torsional arm made of U-shaped aluminium beam. It is lightweight (210 g only) yet sufficiently stiff to support external loadings mounted onto it. The torsional arm is supported by a single-ended flexural pivot (F-20, C-Flex) that acts as the torsional spring. The pivot is clamped and connected to a heavy rectangular aluminium base (3 kg). Four anti-vibration mounts (126-3904, RS Pro) were installed underneath the corners of the base to enhance its stability against external vibration. A strong permanent magnet is placed in close proximity under the torsional arm to induce an eddy current braking force in order to dampen the oscillation of the arm. A high resolution (0.5 μm) laser displacement sensor (HL-G103-S-J, Panasonic) is positioned at one end of the arm to measure the deflection of the torsional arm. The EM calibrator is installed at the other end of the arm. A transparent acrylic casing is used to enclose the entire assembly in order to minimise the effect of ambient disturbance on the thrust stand response. This setup is shown in Figure 1.

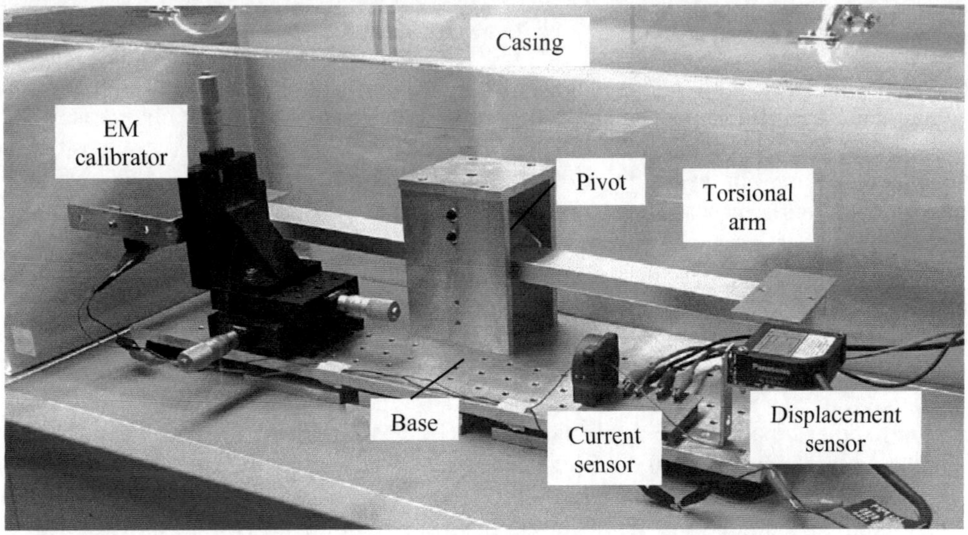

Figure 1: Setup of the torsional thrust stand for existing study

3. Electromagnetic Calibration System

Unlike ES calibration system, which is well established and proven, EM calibration for torsional thrust stand is a relatively new idea. It can be further improved in terms of performance as well as simplified for its implementation. While the fundamental working principles of EM calibrators remain the same, effective yet commercially available components can be utilised in order to further innovate EM as an alternative technique for thrust stand calibration.

3.1. Voice Coil and Permanent Magnet as Electromagnetic Calibrator

The EM calibration system used in this study consists of a voice coil and a permanent magnet. Voice coil, as shown in Figure 2, is readily available in different diameters and is often used in loudspeaker. It is essentially a solenoid, whereby an electromagnetic field is generated when the electrical current passes through the coil. Operation of voice coil is governed by Ampere's Law as stated in Equation 1, where B is the electromagnetic field strength, μ is the permeability of vacuum, N is the number of turns of wire and I is the amount of current flow.

$$B = \mu NI \tag{1}$$

In this study, electrical current of various levels are supplied to the voice coil in order to generate electromagnetic field of different strengths. The voice coil is then engaged to a permanent magnet to induce interactions between their magnetic fields. They are arranged in a way such that they repel each other. As a result, it produces the electromagnetic force needed for the calibration of the thrust stand.

Figure 2: Voice coils of various diameter and two different permanent magnets

3.2. Electromagnetic Force Measurement using Weighing Balance

The amount of electromagnetic force generated by the EM calibration system is first measured through an experiment. After a few trials, a 25.5 mm diameter voice coil is selected as it provides the range of calibration force required for this study. A voice coil of this size is compact, thus it is easy to be set up. Two different types of permanent magnet, i.e. ferrite disk magnet (weaker magnetic field strength) and neodymium disk magnet (stronger magnetic field strength) are utilized. By using magnets of different levels of magnetic field strength, the range of forces generated is wider. As such, the torsional thrust stand can be calibrated to suit different types of micropropulsion systems.

The setup for electromagnetic force measurement is shown in Figure 3 and Figure 4. A weighing balance (HR250AZ, A&D Weighing) with a resolution of 0.1 mg (0.981 μN) is applied to measure the electromagnetic force generated. The permanent magnet is fixed onto the weighing balance while the voice coil is fixed externally to the mechanical stage (PT3/M, Thorlabs) for position and engagement adjustment. A power supply (MP303-3, Meguro) is used to supply electrical current to the voice coil. A current sensor (CTSR1-P, LEM) with a resolution of 10 mA and a digital oscilloscope (DS1102E, Rigol) are used to measure the amount of current that flows through the coil. Electrical current ranges from 0.01 A to 0.4 A are supplied and the corresponding readings from the weighing balance are then recorded.

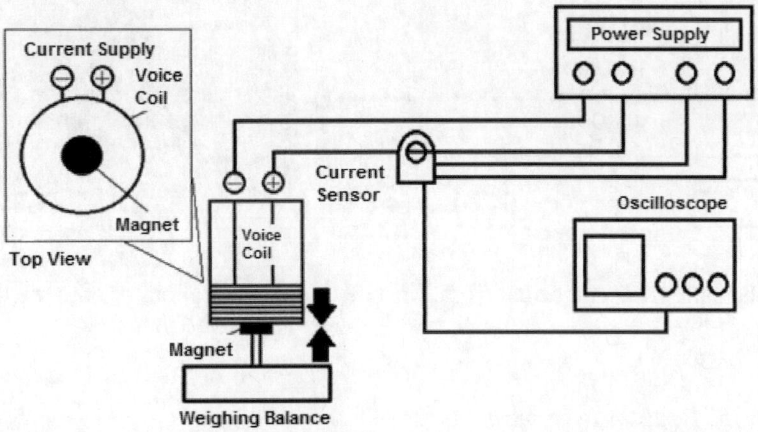

Figure 3: Schematics of EM force measurement setup and arrangement of voice coil and magnet

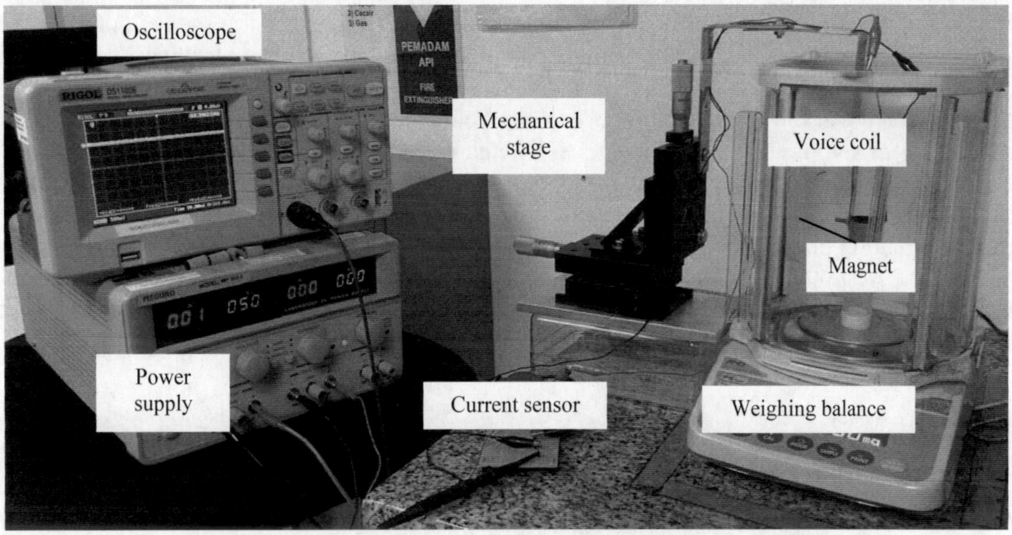

Figure 4: Actual experimental setup for EM force measurement

The effect of the engagement distance between the voice coil and the magnet on the stability of the electromagnetic force is studied. This is to identify the acceptable range of engagement distance for generation of consistent and repeatable electromagnetic force. For this purpose, the current is fixed at three distinct levels (0.1 A, 0.25 A and 0.4 A), which represents low, mid and high range of forces, respectively. The engagement distance is varied from -5 mm to 5 mm using the mechanical stage. Similarly, the corresponding readings from the weighing balance are recorded.

3.3. Implementation onto Thrust Stand

After the force measurement, the EM calibrator is installed onto the thrust stand as shown in Figure 5. The permanent magnet is fixed to the other end of the torsional arm as opposed to the end with the laser displacement sensor. The voice coil is placed externally and its positioning and engagement to the magnet are adjusted using the mechanical stage. The measured electromagnetic forces are applied to the torsional arm and the deflections of the arm are recorded using the linear displacement sensor.

IOP Conf. Series: Materials Science and Engineering **270** (2017) 012011 doi:10.1088/1757-899X/270/1/012011

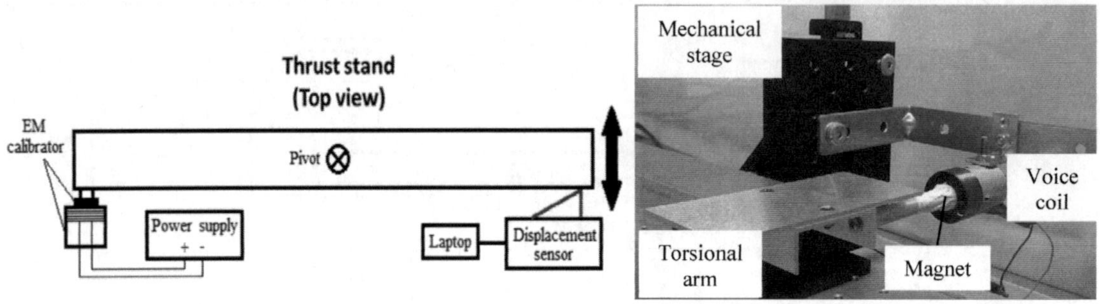

Figure 5: Schematic (left) and actual (right) installation of EM calibrator onto the thrust stand

4. Results and Discussion

4.1. Electromagnetic Force Measurement

Figure 6 shows the electromagnetic forces generated by the EM calibration system at varying levels of the electrical current. Overall, the neodymium magnet based system generates much higher force than that with ferrite magnet. This is predominantly due to the difference in their magnetic field strengths. Neodymium magnet is well known for its strong magnetism. Thus, given the same amount of current through the voice coil, the magnet with stronger field strength repels harder, leading to a higher force generated, and vice versa.

It can be noted that the relationship between electromagnetic force generated and electrical current supplied is linear and directly proportional, as dictated in Equation 1. Such relationship indicates that the electromagnetic force generated using the proposed EM calibrator is predictable and reliable. This is crucial for its applications onto the torsional thrust stand later on.

The combination of these two magnets has an extended range of electromagnetic forces that cover four orders of magnitude, i.e.30 – 2200 μN and 920 – 23000 μN, for ferrite magnet based and neodymium magnet based calibrator, respectively. This is advantageous as the thrust stand can be calibrated to suit a wider range of applications.

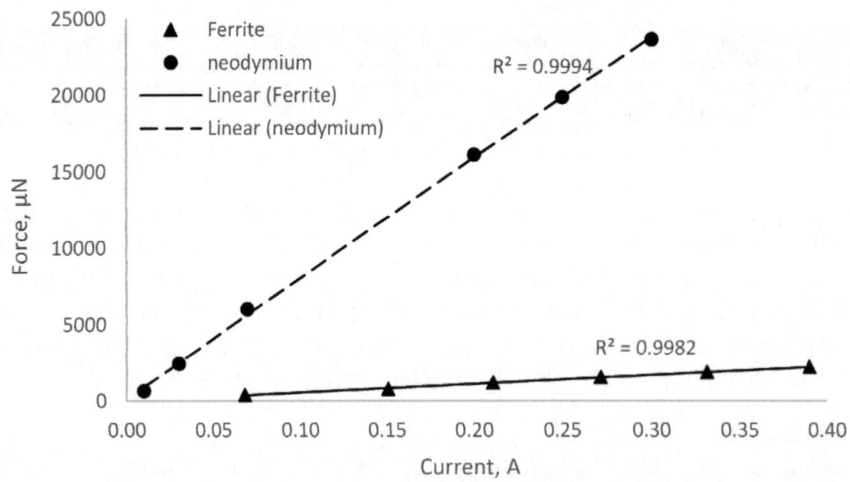

Figure 6: Electromagnetic force generated by ferrite and neodymium magnets at varying currents

Meanwhile, Figure 7 shows the change in the electromagnetic force with the engagement distance for both ferrite magnet based (top) and neodymium magnet based (bottom) calibrators. It should be noted that 0 mm engagement distance represents that the magnet is placed at the edge of the voice coil; whereas positive engagement distance means that the magnet is moved further away from the voice coil, and vice versa for negative engagement distance. Regardless of the polarity of the engagement distance, the electromagnetic force generated decreases as the magnet moves away or into the voice coil. The decreasing trends are due to the reduced magnetic field interaction between the voice coil and the magnet as the engagement distance changes. However, the order of deviation is different. The experimental measurement shows that the electromagnetic force decreases in a less drastically manner as the magnet is moved further away (positive engagement distance) from the coil.

Existing study reveals that beyond the range of ±2 mm of engagement distance, the electromagnetic force generated has decreased too much (~14 %) which has a negative impact on its repeatability and consistency. The effect is more apparent when the electrical current is at the higher level. The obtained results serve as an important guideline for installing the EM calibrator onto the thrust stand. The engagement distance should be kept within ±2 mm for reproducible and consistent calibration force generation.

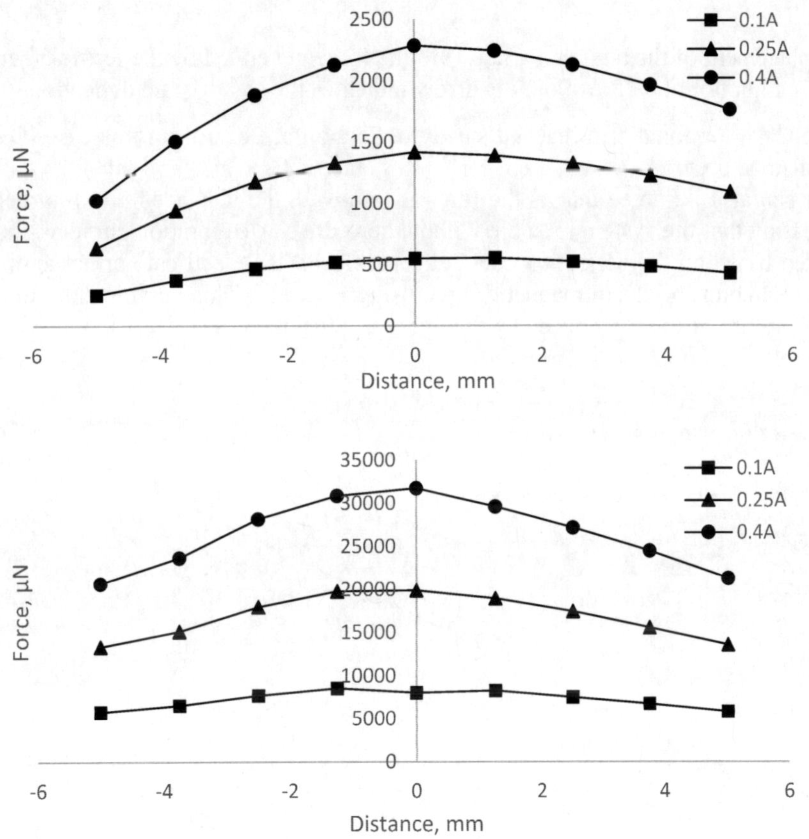

Figure 7: Electromagnetic force generated by ferrite (top) and neodymium (bottom) magnets corresponding to varying engagement distances at three current levels

4.2. Calibration for Steady-state Force

After the EM calibrator is installed, steady-state (constant) electromagnetic forces are generated and applied to the torsional arm. A similar range of electrical current (0.01 A to 0.4 A) is supplied and the corresponding deflection of the arm is recorded. Figure 8 depicts the sample response of the torsional arm due to the force applied.

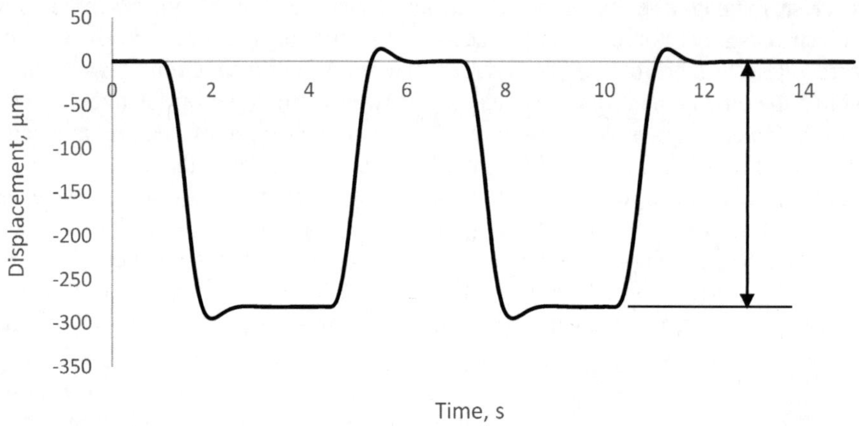

Figure 8: Displacement of the torsional arm due to the force generated by the ferrite magnet based EM calibrator at 0.25 A. Pointing arrow indicates the steady state deflection.

As observed, the torsional arm reached steady deflection at 282 µm in this case. Damping of the response can also be observed i.e. at ~2 s, ~5.5 s, ~8 s, and ~11 s. This shows that the damper installed is effective. It can also be noted that as the force is removed, the displaced arm returned to its initial position, indicating that the system is stable without zero drift. A calibration curve as shown in Figure 9 is then plotted based on the deflection data collected. Here, the electrical current applied is mapped to its equivalent amount of electromagnetic force using Figure 6. This curve can be directly used for measuring the amount of force produced by the micropropulsion system.

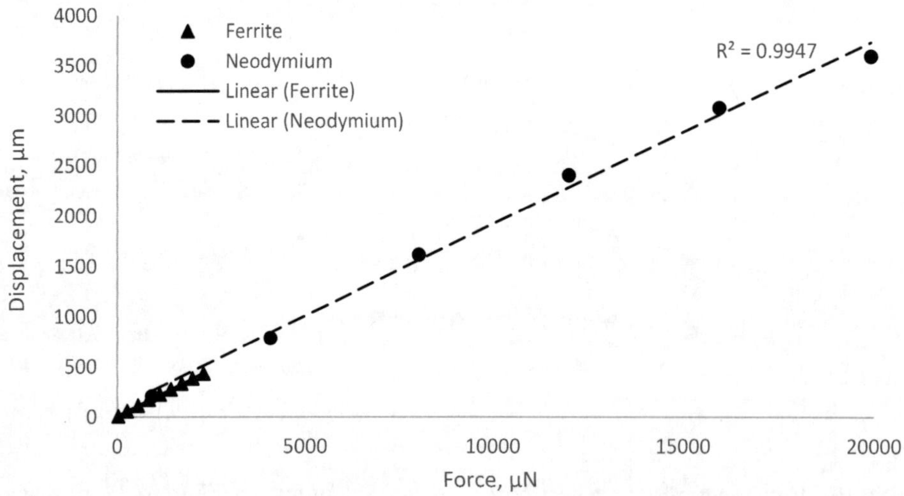

Figure 9: Calibration curve for steady-state force measurement

By using the combined ferrite and neodymium magnet based EM calibrator, the thrust stand has been calibrated to be able to resolve steady-state forces at four order of magnitude i.e. 30 – 23000 µN. From Figure 9, it is also observed that the two calibrators overlap at ~920 – 2200 µN, in which more in-depth study is needed for this overlapping region to obtain accurate calibration results.

5. Conclusion

The proposed EM calibrator is considered feasible as an alternative thrust stand calibration technique. With the combination of commercially available voice coils and various magnets, calibration force over four orders of magnitude ranging at 30 – 23000 µN is produced. Furthermore, the existing EM calibration system also exhibits good repeatability and linearity in generating the forces. The system is also inherently compact, thus easy for set up. Nevertheless, more detailed analysis of EM calibrators are in the pipeline and more extensions of its utility such as producing impulsive calibration forces are to be carried out.

Acknowledgement
The authors acknowledge the funding for this project from Taylor's Research Grant Scheme (Project Code: TRGS/MFS/1/2016/SOE/001).

References
[1] Shiroma W, Martin L, Akagi J, Akagi J, Wolfe B, Fewell B and Ohta A 2011 *Cent. Eur. J. Eng.* **1** 9-15
[2] Cheah K H and Low K S 2016 *IOP Conf. Ser. Mater. Sci. Eng.* **152** 1-8
[3] Cheah K H and Low K-S 2015 *J. Micromechanics Microengineering* **25** 1-10
[4] Mueller J, Marrese C, Polk J, Yang E, Green A, White V, Bame D, Chadraborty I and Vargo S 2003 *Acta Astronaut.* **52** 881–95
[5] Gamero-Castaño M, Hruby V and Martínez-Sánchez M 2001 *27th International Electric Propulsion Conference*
[6] Polk J E, Pancotti A, Haag T, King S, Walker M, Blakely J and Ziemer J 2017 *J. Propuls. Power* **33** 539–55
[7] Polzin K A, Markusic T E, Stanojev B J, Dehoyos A and Spaun B 2006 *Rev. Sci. Instrum.* **77** 105108
[8] Moeller T and Polzin K A 2010 *Rev. Sci. Instrum.* **81** 115108
[9] Xu K G and Walker M L R 2009 *Rev. Sci. Instrum.* **80** 55103
[10] Jacobson D T and Manzella D H 2003 *39th AIAA/ASME/SAE/ASEE Joint Propulsion Conference and Exhibit*
[11] White H, March P, Lawrence J, Vera J, Sylvester A, Brady D and Bailey P 2016 *J. Propuls. Power* **33** 830-41
[12] Hathaway G 2015 *Rev. Sci. Instrum.* **86** 105116
[13] Lun J and Law C 2014 *Meas. Sci. Technol.* **25** 1-8
[14] Pancotti A P, Gilpin M and Hilario M S 2012 *Rev. Sci. Instrum.* **83** 35109
[15] Koizumi H, Komurasaki K and Arakawa Y 2004 *Rev. Sci. Instrum.* **75** 3185
[16] Selden N P and Ketsdever A D 2003 *Rev. Sci. Instrum.* **74** 5249
[17] Jamison A J, Ketsdever A D and Muntz E P 2002 *Rev. Sci. Instrum.* **73** 3629–37
[18] Cheah K H, Low K S, Tran Q V and Lau Z 2015 *IEEE Trans. Instrum. Meas.* **64** 3467–75
[19] Yan A H, Appel B C and Gedrimas J G 2009 *47th AIAA Aerospace Sciences Meeting*
[20] Chen S and Pan S 2011 *Meas. Sci. Technol.* **22** 45104
[21] Tang H, Shi C, Zhang X, Zhang Z and Cheng J 2011 *Rev. Sci. Instrum.* **82** 35118
[22] He Z, Wu J, Zhang D, Lu G, Liu Z and Zhang R 2013 *Rev. Sci. Instrum.* **84** 55107

IOP Conf. Series: Materials Science and Engineering **270** (2017) 012012 doi:10.1088/1757-899X/270/1/012012

Sustainability assessment of turbofan engine with mixed exhaust through exergetic approach

S Saadon* and M S Mohd Redzuan

Department of Aerospace Engineering, Faculty of Engineering, Universiti Putra Malaysia, 43400 Serdang, Selangor, Malaysia

*mimisaadon@upm.edu.my

Abstract. In this study, the theory, methods and example application are described for a CF6 high-bypass turbofan engine with mixed exhaust flow based on exergo-sustainable point of view. To determine exergetic sustainability index, the turbofan engine has to undergo detailed exergy analysis. The sustainability indicators reviewed here are the overall exergy efficiency of the system, waste exergy ratio, exergy destruction factor, environmental effect factor and the exergetic sustainability index. The results obtained for these parameters are 26.9%, 73.1%, 38.6%, 2.72 and 0.37, respectively, for the maximum take-off condition of the engine. These results would be useful to better understand the connection between the propulsion system parameters and their impact to the environment in order to make it more sustainable for future development.

1. Introduction

The civil aviation industry plays a major role in the current world economy with more than 2 billion people travel around the world annually and 23,000 aircraft in commercial industry [1]. Such a rapid growth of this industry also causes an unavoidably increase in the aviation greenhouse gas emissions. One of the proposed solutions to encounter this is by using energy and exergy with better efficiency, in hopes for better reduction of the pollutant emissions. Exergy analysis is a useful performance analysis method and the determination of the irreversibility rates is based on the calculation of isentropy. By increasing the efficiency, it is hoped that there will be less demand of fuel and consequently less waste to be released. Both engine fuel consumption and environment impact due to aircraft play an important role to attain sustainable development with environmental protection. Hence, extra efforts with many approaches have been applied in aviation industry over the years to better understand and improve the performance characteristics of aircraft.

An aircraft is a complicated system that consists of several parts that are exposed to environmental conditions during operation, which are often varying and unstable. The aviation industry has started to construct remarkably complex aircraft designs with many mixed systems that create demands for more data acquisition to estimate the overall system performance. Many methods have been proposed since the very beginning of this industry to study the flight performance but it is agreed that a common basis must be applied when assessing different approaches in aircraft design, particularly when constructing new systems. The application of exergy analysis method will be very helpful to do comparative study between the different system designs and this is shown by the increasing number of papers concerning aircraft engines in the last four years [2].

The studies concerning energy and exergy analysis of turbofan engine have already began since a decade ago. About 47% of exergy-based analyses have been linked to aircraft engines, mostly deal on turbofan engines and around 22% of them discussed on turbojet engines [2]. The first few conducted studies have used hydrogen as fuel replacing kerosene to run the engine [3,4], which is shown to have a great impact in reducing Specific Fuel Consumption (SFC) rate by as much as 39.2% [3]. Moreover, the exergy efficiency is inversely proportional with altitude due to decreasing value of temperature [4]. In the meantime, several studies have also been conducted using either Jet-A1 or JP8 as the fuel [1, 5-9]. Most of these studies cover not only the thermodynamic analysis part, but also the economic [5, 7] and sustainability factors [1, 9]. However, all of them take a turbofan engine with separate exhaust as the case study. According to one of the studies, turbofan engine with mixed exhaust and no afterburner would give the optimized value of SFC [10]. In another study, fan and core engine exhausts are shown to have the highest potential improvement rate based on exergy analysis performed, which means that the focus in future development should be placed on these two components since they still have higher chances to be improved for better performance [6].

Sustainability has been linked to issues of environmental, economic and social dimensions toward responsible management of resources. A higher sustainability index indicates a low waste exergy ratio and low environmental effect factor. In this study, a turbofan engine with mixed exhaust is introduced and, its performance and sustainability index based on exergy is analyzed. The results are compared to the exergy analysis of turbofan with separate exhaust engine done by previous researchers.

2. System Description

The schematic of a turbofan engine that consists of a fan, a low-pressure compressor (LPC), a high-pressure compressor (HPC), an annular combustion chamber (CC), a low-pressure turbine (LPT) and a high-pressure turbine (HPT) is illustrated in Figure 1. The fan, compressors and turbines are taken to be adiabatic in this study. The first drive shaft is the high pressure (HP) system that links the HPT and HPC. Meanwhile, the second shaft is the low pressure (LP) system that links the LPT to the LPC and the fan. Because of these two systems, this engine is described as a two-spool turbofan engine. In the turbofan with mixed exhaust engine, after initially compressed in the fan, the "cold air" is compressed again with the "hot air" coming out of the turbine. Eventually both flows of air are mixed in a mixer to attain the same static pressure before entering the nozzle for expansion.

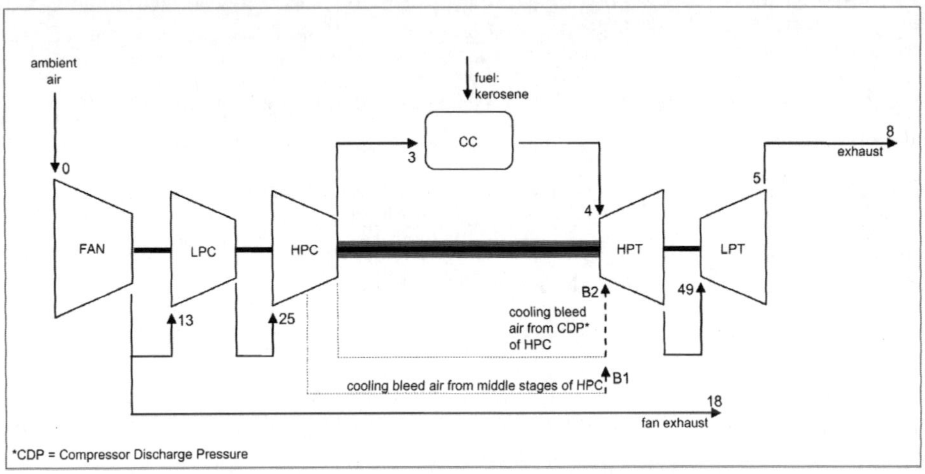

Figure 1: Schematic of CF6 turbofan engine [6]

The intake air mass flow rate of 679.2kg/s (or 16.6%) is used in the combustion process along with 2.2kg/s of fuel. The "cold air", which flows through the fan and exhaust, constitutes around 82.4% of the intake air flow. Meanwhile, the bleed air flowing out from HPC to HPT quantifies for 1% of this air flow, which is mainly used for cooling of the first stage of HPT. The gas flows in the engine are assumed to behave ideally. The maximum engine take-off thrust is 185.05kN with an inlet air speed of 147m/s. For this study, kerosene is used and it has a specific chemical exergy of 45.8MJ/kg [11]. The air composition at the engine inlet is taken from Ref. [12]. The combustion equation involved can be expressed in terms of mole fractions of each gas as in Equation 1.

$$0.013174C_{12}H_{23} + 3.963210(0.7748N_2 + 0.2059O_2 + 0.0003CO_2 + 0.019H_2O) \rightarrow 0.159277CO_2 + 0.226802H_2O + 0.582186O_2 + 3.070695N_2 \quad (1)$$

3. Exergy Analysis of Turbofan Engine

Only physical and chemical exergies are considered inside each component, excluding the exhaust part due to the high velocity of discharging flows at the exit. Moreover, potential exergy is also neglected since there is no significant height difference between intake and exit of the engine. Specific physical exergy is defined by Equation 2 [13], where $\bar{h}$ and $\bar{s}$ are the specific molar enthalpy and specific molar entropy, respectively, with the reference temperature is set as the ambient.

$$\overline{ex_{ph,k}} = \overline{h_k} - \overline{h_0} - T_0(\overline{s_k} - \overline{s_0}) \quad (2)$$

Furthermore, the chemical exergy is defined as in Equation 3 [13]:

$$\overline{ex_{ch,k}} = \sum_k x_k \overline{ex_k^{ch}} + \bar{R}T_0 \sum_k x_k ln x_k \quad (3)$$

Finally, the kinetic exergy is normally defined by Equation 4:

$$\overline{ex_{ke,k}} = \dot{m}_a \frac{V_{aircraft}^2}{2} \quad (4)$$

Some other exergy-based measures are also useful for the system evaluation, which include the exergy efficiency, exergy destruction rate, relative exergy destruction, fuel depletion ratio productivity lack, fuel and product exergy factors, and the rate of exergetic improvement potential. They are described as follow:

- Exergy efficiency of each component is the ratio of the product exergy to the fuel exergy, as given by Equation 5:

$$\varepsilon_k = \frac{\dot{P}_k}{\dot{F}_k} \quad (5)$$

- Exergy destruction rate is the rate of exergy destruction due to losses within the components and it is calculated by Equation 6:

$$\dot{Ex}_{des,k} = \dot{F}_k - \dot{P}_k \quad (6)$$

- Relative exergy destruction gives the exergy destruction of each component as a ratio of the total exergy destruction. This value is normalised and hence permits comparison of performance among different components of the system. It is given in Equation 7:

$$X_k = \frac{\dot{Ex}_{des,k}}{\dot{Ex}_{des,tot}} \quad (7)$$

- Fuel depletion ratio and productivity lack describe the exergy destruction for each component as a percentage of the total fuel and product exergy, respectively, given by Equation 8(a) and Equation 8(b).

$$\delta_k = \frac{\dot{Ex}_{des,k}}{\dot{F}_{tot}} \quad (8a)$$

$$\xi_k = \frac{\dot{Ex}_{des,k}}{\dot{P}_{tot}} \quad (8b)$$

- Fuel and product exergy factors estimate the parts of the fuel and product exergy for a system unit as fraction of the total fuel and product exergy values, respectively. They can be calculated using Equation 9(a) and Equation 9(b).

$$f_k = \frac{\dot{F}_k}{\dot{F}_{tot}} \tag{9a}$$

$$p_k = \frac{\dot{P}_k}{\dot{P}_{tot}} \tag{9b}$$

- Rate of exergetic improvement potential defines the percentage of potential improvement for each component. It helps to define the component that has to be improved. This can be calculated using Equation 10.

$$IP_k = (\dot{1} - \varepsilon_k)(\dot{F}_k - P_k) \tag{10}$$

The main component of the turbofan engine that will be reviewed in this study is the exhaust, which is different from other previous studies. The exergy inlet, outlet and efficiency equations for the exhaust are given by Equation 11(a), Equation 11(b) and Equation 11(c), respectively.

$$\dot{E}x_{in,exh} = \dot{E}x_{ph,exh} + \dot{E}x_{ch,exh} \tag{11a}$$

$$\dot{E}x_{out,exh} = \dot{E}x_{ke,exh} \tag{11b}$$

$$\varepsilon_{exh} = \frac{\dot{E}x_{out,exh}}{\dot{E}x_{in,exh}} \tag{11c}$$

4. Exergetic Sustainability Assessment for Turbofan Engine

The computed exergetic parameter is then used to deduce the exergy-based sustainability parameters. This study considers take-off condition at sea level. To calculate the exergy efficiency of the overall turbofan engine, the useful product of the engine (i.e. thrust power) and the resources needed (i.e fuel and inlet air chemical exergy) have to be defined. Thrust for a mixed flow and un-chocked turbofan engine is calculated using Equation 12, where $m_{\dot{e}xh}$ and V_{exh} are the mass flow rate and aircraft velocity at the exhaust, respectively, f is the ratio of mass flow rate of the fuel to the total mass flow of rate of air, β is the bypass ratio and V_0 is the aircraft velocity at intake fan.

$$T = m_{\dot{e}xh}\{(1 + f + \beta)V_{exh} - (1 + \beta)V_0\} \tag{12}$$

Thrust power and exergy efficiency are derived using Equation 13 and Equation 14, respectively.

$$\dot{T} = TxV_0 \tag{13}$$

$$\varepsilon = \frac{\dot{T}}{F_{tot}} \tag{14}$$

Several exergy-based sustainability parameters are evaluated for a turbofan engine based on Ref. [14]. First, the waste exergy ratio is the ratio of total waste exergy output to total exergy input, as given in Equation 15. In a turbofan engine studied, all waste exergy is unused as the exhaust emissions cannot be recovered.

$$Waste\ exergy\ ratio = \frac{F_{t\dot{o}t} - \dot{T}}{F_{t\dot{o}t}} \tag{15}$$

Next, recoverable exergy amount given by Equation 16 is null in this case study due to unrecoverable exhaust emissions to the atmosphere.

$$r_{re} = \frac{Recoverable\ exergy}{F_{t\dot{o}t}} \tag{16}$$

Furthermore, the exergy destruction factor indicates the reduction in the positive effect of the engine based on exergy sustainability. It is defined as the ratio of total exergy destruction to total fuel exergy, which is given by Equation 17.

$$f_{exd} = \frac{\dot{Ex}_{des,tot}}{\dot{F}_{tot}} \tag{17}$$

Meanwhile, the environmental effect factor is defined as in Equation 18.

$$r_{eef} = \frac{Waste\ exergy\ ratio}{\varepsilon} \tag{18}$$

Finally, the sustainability assessment of a turbofan engine through exergy method can be addressed by applying the exergetic sustainability index. It is the inverse of the environmental effect factor, as given by Equation 19, and varies from 0 to 1.

$$\Theta_{esi} = \frac{1}{r_{eef}} \tag{19}$$

5. Results and Discussion

In this study, the exergy efficiency, exergy destruction, fuel depletion ratio, productivity lack, fuel and product exergy factors, and the improvement potential rate of each component of CF6 turbofan engine are defined and evaluated using sea level data. This type of engine has been used in multiple aircraft including Airbus A300/310/330 and Boeing B767/747. It has a relatively high by-pass ratio at around 4.5. The data used in this analysis such as mass flow rate, molar specific enthalpy and entropy are all tabulated in Table 1 [15], along with the results obtained for the exergy rates of different points in the turbofan engine. The obtained results are compared to the previous study in Ref. [6], which concerns a turbofan engine with separated exhaust.

Table 1: Thermodynamics parameters for the turbofan engine

Component	**Mass flow rate (kg/s)**	**Molar specific enthalpy (kJ/kmol)**	**Molar specific entropy (kJ/kmol.K)**	**Physical exergy rate (MW)**	**Chemical exergy rate (MW)**	**Kinetic exergy rate (MW)**	**Total exergy rate (MW)**
Fan inlet	679.18	8939.63	200.04	0.00	-1.28	-	-1.28
Fan outlet	679.18	10508.52	200.34	34.60	-1.28	-	33.32
LPC inlet	119.54	10508.52	200.34	6.09	-0.23	-	5.86
LPC outlet	119.54	11557.58	200.91	9.70	-0.23	-	9.47
HPC inlet	119.54	11557.58	200.91	9.70	-0.23	-	9.47
HPC outlet	116.14	25805.49	203.13	63.78	-0.22	-	63.56
CC inlet	112.74	25805.49	203.13	61.92	-0.21	-	61.71
CC outlet	114.94	49321.70	225.56	129.12	0.28	-	129.40
HPT inlet	114.94	49321.70	225.56	129.12	0.28	-	129.40
HPT outlet	121.74	33864.62	226.02	71.24	0.30	-	71.54
LPT inlet	121.74	33864.62	226.02	71.24	0.30	-	71.54
LPT outlet	121.74	24013.09	227.05	28.54	0.30	-	28.84
Mixed exhaust inlet	681.38	-	-	85.92	1.67	-	87.59
Mixed exhaust outlet	681.38	-	-	-	-	26.83	26.83
Bleed air 1	3.40	21036.40	203.70	1.29	-0.01	-	1.28
Bleed air 2	3.40	25805.49	203.15	1.86	-0.01	-	1.86
Fuel	2.20	7253.00	-	-	100.97	-	100.97

IOP Conf. Series: Materials Science and Engineering **270** (2017) 012012 doi:10.1088/1757-899X/270/1/012012

Figure 2 and Figure 3 show that the turbine has the greatest exergy efficiency while the exhaust is found to have the lowest one. The high efficiency in the fan, compressors and turbines are due to their high isentropic efficiencies, which is around 80-95%. This implies that the highest exergy loss for the system would be from the exhaust, followed by the combustion chamber. The loss of exergy from the exhaust is believed to be caused by high-quality energy that is discharged from the component. In the meantime, the loss in combustion chamber can be contributed to the internal irreversibility within the component. From both figures, it is also observed that all turbomachinery components exhibit a high value of exergy efficiency except for LPC. This is understandable since LPC is used to avoid irregular flow, stall and blocking, and to achieve high-pressure ratio that subsequently leads to irreversibilities. Because of this, some parts of the compressor are required to operate at rather low rotation per minute (RPM). Comparing these results to the those with separated exhaust in Ref. [6], exergy efficiencies for fan and core engine exhaust are found to be around 12%, which are much lower than the those for a mixed exhaust engine studied in this paper. Consequently, the total exergy loss in fan and core engine exhaust is higher than the that in the mixed exhaust alone.

In Figure 4, exhaust has the largest fuel depletion ratio and productivity lack with values of 9.97% and 12.39%, respectively. It is followed by the combustion chamber at 7.29% and 9.06%, respectively, where chemical energy is converted to heat. Comparing with the exergy analysis for turbofan engine with separate exhaust in Ref. [6], the fuel depletion ratio and productivity lack of the mixed exhaust engine are higher. This might be due to additional heating of the secondary air exiting the fan by the primary air from core engine when they are mixed to attain the same temperature before exiting the exhaust. Furthermore, fuel and product exergy factors in Figure 5 indicate that combustion chamber is the primary consummator of the total fuel and product exergy with 35.65% and 36.77%, respectively. It is followed by HPT with 29.04% and 37.05%, respectively. These results are anticipated since both the conversion of chemical energy to heat and the conversion of mechanical to electrical power occur within these two components. Meanwhile, it is good to acknowledge that the sum of fuel and product exergy factors of both fan and core engine exhaust when they are in separated exhaust as reported in Ref. [6] is lower than those in the mixed exhaust.

Figure 6 highlights that more than 50% of the total exergy destruction rate occurs in the exhaust. Nevertheless, this value is found to be lower than the total relative exergy destruction in fan and core engine exhaust separately found in the previous study, which is 66% [6]. This is totally expected since the exergy loss rate of both components separately is higher than the loss in mixed exhaust alone. The results of potential improvement rate in Figure 7 implies that the mixed exhaust and the combustion chamber still deserve more focus and research than the other components. Although recent advances in combustion have been initiated, it is still restricted to the laws of thermodynamics.

Lastly, the most significant contribution of this study is the exergetic sustainability performance for the engine, which is shown in Figure 8. This part of the study has never been done for a CF6 engine.

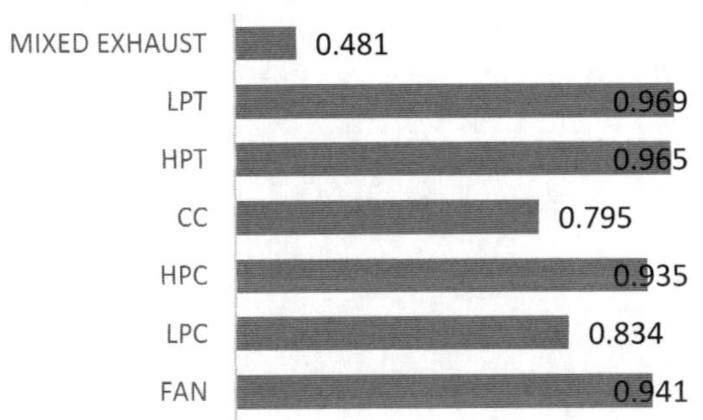

Figure 2: Exergy efficiency of each component of the turbofan engine

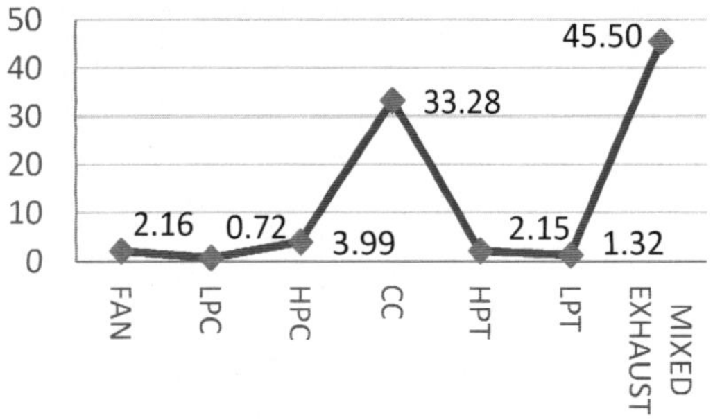

Figure 3: Exergy destruction of each component of the turbofan engine

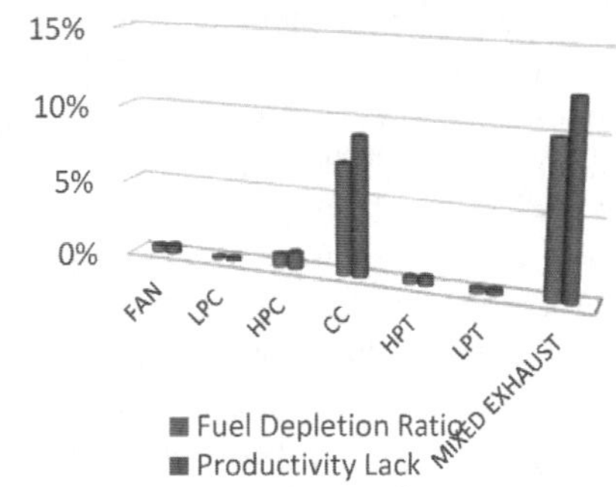

Figure 4: Productivity lack and fuel depletion ratio of each component of the turbofan engine

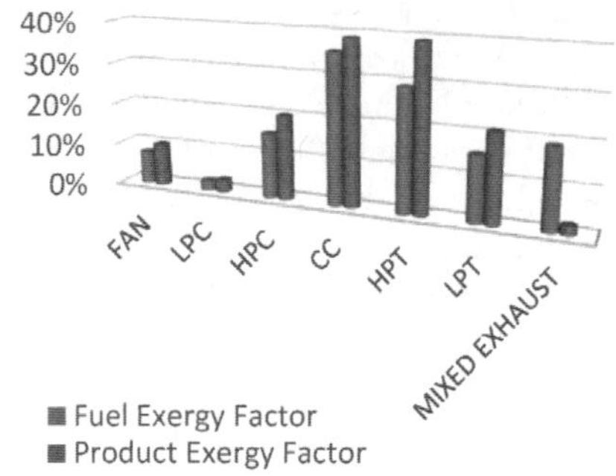

Figure 5: Product and fuel exergy factors of each component of the turbofan engine

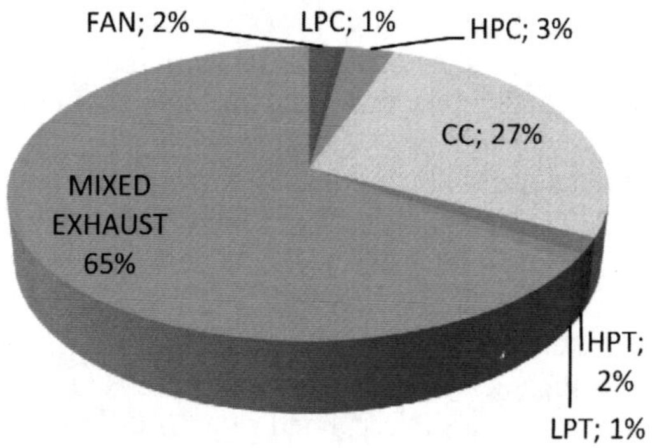

Figure 6: Relative exergy destructions of each component of the turbofan engine

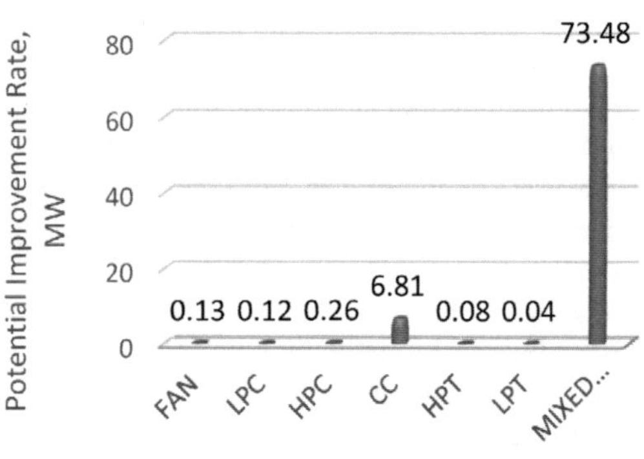

Figure 7: Potential improvement rate of each component of the turbofan engine

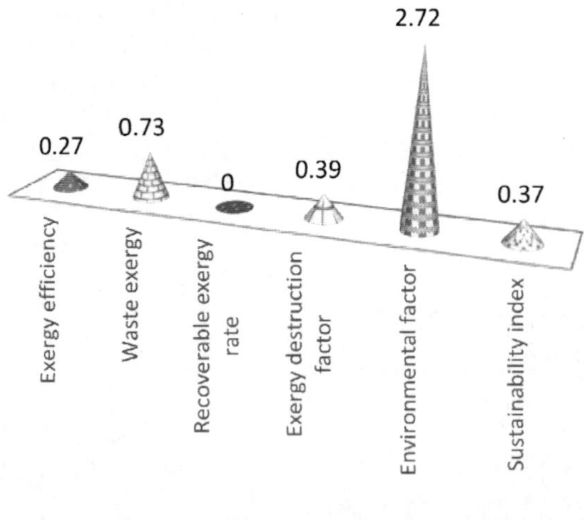

Figure 8: Exergetic sustainability performance of the turbofan engine

In Figure 8, the obtained result for waste exergy ratio is slightly higher than theoretically suggested (approximately 70%) and this implies that significant amount of provided exergy rate to the system is wasted. Subsequently, the exergetic sustainability indicator for the mixed exhaust engine is rather low, which is 0.37. Meanwhile, the higher the value of the sustainability index, the better it is for practical

IOP Conf. Series: Materials Science and Engineering **270** (2017) 012012 doi:10.1088/1757-899X/270/1/012012

applications. The other notable factor is the exergy destruction factor, which is calculated to be 0.39. The amount of recoverable exergy is zero because the emissions released cannot be recovered in the engine. To improve the environmental sustainability, it is necessary not only to utilize sustainable or renewable energy sources, but also to use the natural sources more efficiently while minimizing the environmental damage.

6. Conclusion

In this paper, exergy analysis and sustainability assessment based on exergy for a turbofan engine with mixed exhaust during take-off condition at sea level have been performed. This study aims to present the possibilities and advantages that the turbofan engine with mixed exhaust and sustainability exergy-based study might offer to the aircraft propulsion system. The main conclusions that can be outlined from this study are as follows:

- The exergy efficiency for the mixed exhaust component is calculated to be higher (48.1%) than the one for fan and core engine separately (12.94% and 12.71%, respectively).
- The overall exergy efficiency for turbofan engine with mixed exhaust is 26.9%.
- The other important parameter that shows a drop in positive effect is the exergy destruction factor, which is found to be 0.39.

Sustainability index based on exergy analysis has been shown to be an important method to assess the sustainability of the turbofan engine. Exergy aids in the determination of improvement potential rates and reduction in the irreversibility of the process. In the future, it is advisable to perform an exergo-environmental analysis and life-cycle assessment to this turbofan engine to assess the environmental impact of the engine.

References

[1] Aydin H, Turan O, Hikmet Karakoc T and Midilli A 2014 *Int. J. Exergy* **14** 388-412

[2] Sohret Y, Ekici S, Altuntas O, Hepbasli A and Hikmet Karakoc T 2016 *Progress in Aerospace Sciences* **83** 57-69

[3] Karakoc T H, Turgut E T and Hepbasli A 2007 *3rd International Green Energy Conference*

[4] Turgut E T, Karakoc T H and Hepbasli A 2007 *Int. J. Energy Res.* **31** 1383-97

[5] Turgut E T, Karakoc T H and Hepbasli A 2009 *Int. J. Exergy* **6** 277-94

[6] Turgut E T, Karakoc T H, Hepbasli A and Rosen M A 2009 *Int. J. Exergy* **6** 181-99

[7] Tona C, Raviolo P A Pellegrini L F and de Oliveira Jr S 2010 *Energy* **35** 952-9

[8] Turan O, Aydin H, Karakoc T H and Midilli A 2014 *J. Autom. Control Eng.* **2** 110-4

[9] Turan O 2015 *Energy* **86** 722-36

[10] Yadav R, Jugseniya Chandrakant B, Pashilkar Abhay A 2005 *ASME Turbo Expo 2005: Power for Land, Sea and Air*

[11] Dagaut P and Cathonnet M 2006 *Progress in Energy and Combustion Science* **32** 48-92

[12] Bejan A ,Tsatsaronis G and Moran M 1996 *Thermal design and optimization* Wiley, USA

[13] Szargut J M, Morris D R and Steward F R 1988 *Exergy analysis of thermal, chemical and mettalurgical processes* Hemisphere, New York

[14] Midilli A, Kucuk H and Dincer I 2012 *Environmental Progress & Sustainable Energy* **31** 604-11

[15] General Electric 1983 *General Electric CF6-80 Engine Student Book*

IOP Conf. Series: Materials Science and Engineering **270** (2017) 012013 doi:10.1088/1757-899X/270/1/012013

Critical needs for piston engine overhaul centre in Malaysia

M H Khairuddin*, M Y Yahya and M K Johari

Universiti Kuala Lumpur – Malaysian Institute of Aviation Technology, Lot 2891, Jalan Jenderam Hulu, 43900 Dengkil, Selangor, Malaysia.

*hafis@unikl.edu.my

Abstract. Piston engine overhaul centre (PEOC) is the branch of aviation Maintenance, Repair and Overhaul (MRO) providers, which plays a pivotal role in maintaining the fleet of business and commercial aircraft in the world. The centre typically offers three main MRO capabilities: airframe, engine and component services. Companies holding a PEOC(s) are all subjected to stringent procedures and conditions regulated and audited by the International Civil Aviation Organization. Currently, piston engine maintenance and repair for Asian countries is conducted only in Singapore. The focus of this study is to establish the needs for a PEOC in Malaysia, which will cater almost all small local aircraft companies such as transport and flying school companies.

1. Introduction

Today, there are still many aviation companies that fly piston engine aircraft for their core business. Depending on their nature of business, some aircraft companies do have a significant preference in operating piston-powered or turboprop aircraft in their fleet. As a further indication on the continued relevancy of the piston engine use, there are also extensive studies on piston engine that are still being conducted, such as the study on reducing CO_2 gas emission from the aircraft exhaust [1]. The ongoing use of piston engine creates a need for its maintenance and repair services, which are provided by the piston engine overhaul centre (PEOC). This is a branch of aviation Maintenance, Repair and Overhaul (MRO) providers that typically offers three main MRO capabilities: airframe, engine and component services.

The Malaysian aviation industry has progressed immensely over the last few decades. Originally with only a single airline monopolizing the local industry, Malaysia now has at least six commercial airlines excluding charter flight companies and airfreight carriers [2]. These main local international airlines include AirAsia, AirAsiaX, Malaysia Airlines and Malindo Air while the other two, FireFly and MASWings focus on domestic destinations only. Aside from the commercial airline companies, the Malaysia's Royal Military Air Force fleet is also consisted of turbine-powered jets. Whilst these major airlines utilize turbine engines, many smaller private aviation companies and clubs actually have piston engine aircraft that are often being used for recreational purposes and activities. Table 1 shows the number of piston engine aircraft owned by several local companies and also their corresponding maintenance costs (by PEOC in Singapore). This data is obtained from interview sessions with the management team of the companies. Note that a pseudonym is used for each company in Table 1 to hide their actual names.

Table 1: Piston engine distribution in Malaysia and their maintenance cost

Name of Company	**No. of Piston Engine**	**Repair Cost from Service at PEOC in Singapore per Engine**			**Transportation Cost per Engine**
		4-cylinder	**6-cylinder**	**8-cylinder**	
Company M	32	RM 35000	RM 45000	RM 65000	RM 4000
Company H	25				RM 7000
Company A	41				RM 7000
Company S	9				RM 5000
Company A2	12				RM 5000
Company F	6				RM 2000
Company I	13				RM 14,000
Company C	6				RM 4000
Company A3	3				RM 5000
Company P	4				RM 4000
Others	57				RM 5500

It can be observed from Table 1 that the total number of piston engine in Malaysia is 207. Based on this significant number of engines and their relatively high maintenance costs, it is believed that there exists a crucial need for the establishment of a PEOC in Malaysia. Currently, there is no such centre locally or in other Asian countries except Singapore. For small aviation companies, having their own manpower for piston engine MRO is not practical due to the high cost. However, prior to getting the approval from the Malaysia's Department of Civil Aviation (DCA), there is a need for a study to be conducted to examine the existing needs of PEOC among local aviation companies that operate piston engine aircraft, which will support its establishment in Malaysia

2. Maintenance of Piston Engine

It has been stated that the MRO of complex capital goods is afflicted with challenges with regard to several processes [3]. Complex capital goods include diesel locomotives, aircraft engines, industrial gas turbines and wind turbines. The employees working in this field are highly qualified and specially accredited. Therefore, the supply of new employees is limited. At the same time, the wages for such labour are high. Consequently, the capacities have to be planned as accurately as possible.

This is one key challenge for MRO companies because the planning of regeneration processes is much more complex than that for customary production processes due to high diversity of parts and damage. The process cannot be standardized as easily because every new regeneration job is unique in its damage set. Thus, the workload of upcoming jobs is unknown beforehand, which makes them even harder to plan. The capacity planning in the field of MRO has hardly been studied scientifically and currently, no practical relevant, rule-based and thus transparent method or software application that is suitable for planning or controlling the regeneration of complex capital goods exists. The regeneration process for an aircraft engine begins after its removal from the wing. The engine is then transported to the designated facility, where an incoming inspection is carried out to determine the damage to the engine and the necessary work to be carried out. After disassembling the engine into its parts, the parts are inspected again to define the exact damage and the processing of the parts.

A decision has to be made as to whether the part is serviceable, repairable or has to be purchased anew. The reassembly of the stored, repaired and purchased parts takes place as soon as the necessary parts are available. The regeneration ends when a test run of the reassembled engine is successful. The overall summary of the regeneration process is shown in Figure 1.

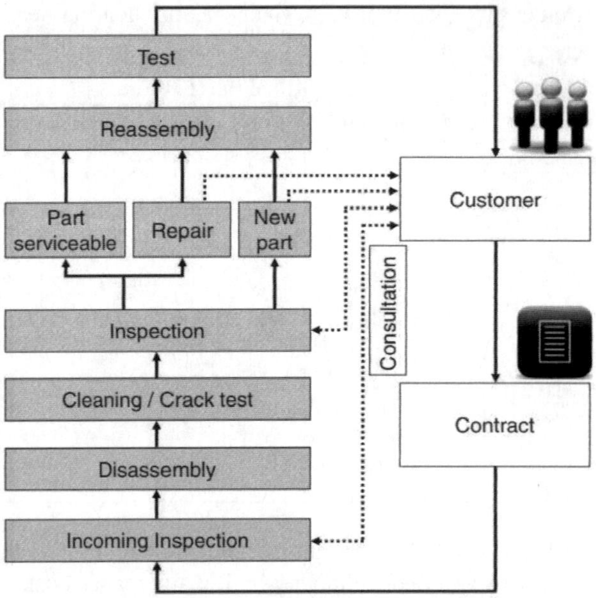

Figure 1: Regeneration process flow [3]

A significant challenge in aircraft engine is quantifying its residual life to overhaul. The algorithm described by Piancastelli et al. (2012) calculates the residual life of a petrol piston engine with a good level of reliability [4]. This method has been tested on few small, latest-generation, naturally-aspirated aircraft and racing piston engines, and has been demonstrated to be effective in several experiments. This method is implemented directly on the electronic control system of the engine with very few lines of C-code. The method can also be used in many industrial engines. This innovative method assumes that only two main factors (power level and wear) affect engine durability or time between overhaul. These two factors are considered separately and combined with the worst case criteria. The wear is assumed to follow the logarithmic law and a formula similar to the Miner's law for material fatigue is used, making it possible to calculate the power-level curve with the knowledge of only two points. The wear curve is also related to elapsed engine cycles. All in all, the algorithm is very simple and can be implemented with just a few lines of software code accessing data collected from existing sensors. The system is currently used to evaluate actual residual life of racing engines.

3. Comparison Analysis

Based on the data in Table 1, several comparison analyses can be made to justify the proposal for the establishment of PEOC in Malaysia. Like any other businesses, Malaysian aviation companies have always strived to offer the best services for their clients while acquiring high profits at the same time. However, the operational costs have been challenging their business model for years now, primarily because of several factors such as manpower, currency difference and transportation costs that include movement of their piston engines into another country.

The overhaul and repair services are important segments of the remanufacturing industry. Such services are traditionally characterized by complicated process plans, stochastic operations and the usage of routable inventory [5]. This includes, for examples:

- highly variable operations including disassembly that disassembles an asset into parts, part repair operations and assembly that assembles the repaired parts back into an asset. The disassembly operation is often time consuming and performed in conjunction with a "discovering" inspection process. The uncertain service scope is often a combined result of customer requirements and the inspection process
- highly variable processing time for various operations

IOP Conf. Series: Materials Science and Engineering **270** (2017) 012013 doi:10.1088/1757-899X/270/1/012013

- existence of serial-number-specific-parts (SNS) that are required to be assembled into the original assets they belonged to, rotable parts that are refurbished parts satisfying certain qualifications for general use, and the concomitant rotable inventory

With this in mind, PEOC has just the right touch as business opportunity in Malaysia: when there are customer needs, there are business needs [6]. This situation eventually reflects the Target Platform Needs and Certification Requirements at applicable standard for delivering aircraft with piston engine. This is called System Level Requirements (SLR) as shown in Figure 2.

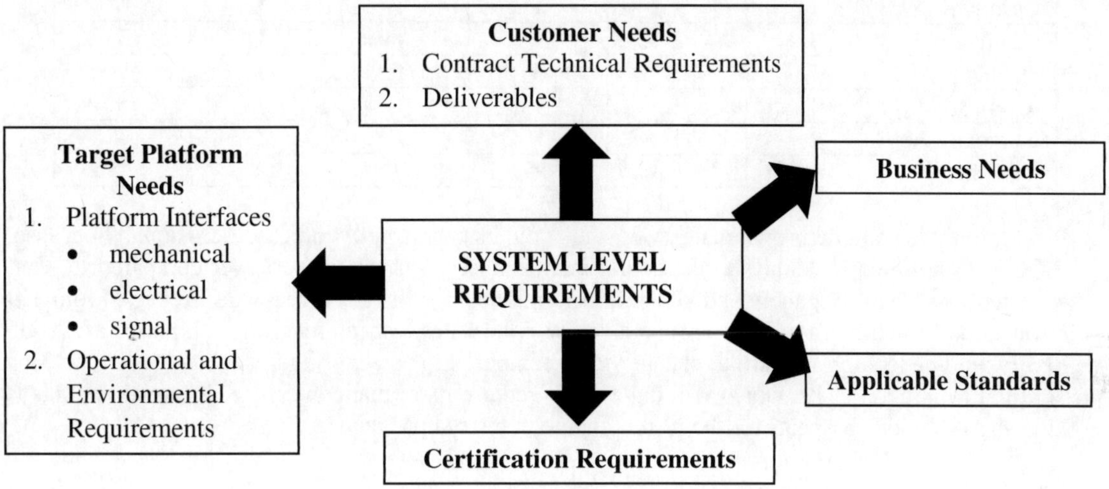

Figure 2: System Level Requirements (SLR) [6]

For SLR, the main contributors are target platforms' need (the interfaces as well as operational and environmental conditions), customers' need, business needs, applicable standards and the certification basis, if applicable. Table 2 tabulates the overview of elements forming the engine system requirement set. The initial phase of the project agreement among the stakeholders is the most affordable and a significant assignment to be accomplished. Project's incorporator makes an effort to reach agreement among the stakeholders on what to be delivered, how the product should comply with the customer requirement and when it should be accomplished considering the acceptance criteria of the product. Project stakeholders perform brainstorming, interface analysis, interviews, observation, requirements workshop, surveys and questionnaire. The main purpose of these activities is to obtain and understand project business needs. In this way, technical requirements of contract are established. Also, customer defines all relevant deliverables to the contractor. At this stage, agreement is accomplished, technical requirements of contract are set and deliverables are declared between the engine developer and the customer. This model can be applied when PEOC needs to be established by adjusting the model in the form requirements in Malaysia. Hence, the relevant companies will no longer have to comply with additional Singapore's aviation law requirements as they can simply proceed with their current local Department of Civil Aviation (DCA) regulations.

With the piston engine maintenance, the main focus of Target Platform Needs is definitely from the mechanical perspectives. Besides having sufficient manpower for the tasks, it is crucial to note that doing the maintenance work locally in Malaysia is also considerably easier especially in assembling and re-assembling engine parts, not just because of the geographical advantages but also due to similar operational and environment requirements by the local DCA. For instance, in Malaysian Institute of Aviation Technology, piston engines experts are not only experienced trainers but they are also fully-certified by DCA for maintenance work. Apart from that, the aviation companies previously listed in Table 1 already have their own in-house manpower for maintenance as tabulated in Table 2. Despite having these numbers, the companies still have to outsource the maintenance to Singapore because of the absence of a PEOC here in Malaysia.

Table 2: Available manpower for piston engine maintenance by company

Name of Company	Engineer	Technician	Mechanics	Manager	Finance Manager	Total
Company M	2	15	6	1	1	35
Company H	2	12	5	1	1	21
Company A	3	20	6	1	1	31
Company S	1	3	3	1	1	9
Company A2	1	2	1	1	1	6
Company F	1	1	1	1	1	5
Company I	1	3	3	1	1	9
Company C	1	1	1	1	1	5
Company A3	1	1	-	1	1	4
Others	6	6	15	9	9	35

In addition to having their own manpower, the numbers can actually increase occasionally because, if PEOC is established in Malaysia, the actual maintenance costs would be lower compared to what other countries such as Singapore and USA are offering because of the currency differences. From that point onwards, it will be more competitive as more companies, especially from Southern East Asian regions, would be looking forward to cheaper rates compared to current Singapore's prices. Hence, the possibility of more job tasks, which will undeniably require more manpower, is substantial and this is nothing short of good prospects for the piston engine industry in Malaysia.

4. Conclusion

Currently, piston engine maintenance and repair for Asian countries is conducted only in Singapore. It has been highlighted in this paper that the demand for maintenance and repair of piston engine aircraft, either from local or international companies, is significant especially from the perspectives of cost and logistics. In addition, the readiness of local manpower for the work tasks seems to be adequate. Hence, it can be taken that the needs for a PEOC in Malaysia has been established.

References

[1] Weclas M 2010 *Journal of Thermodynamics* **2010** 1-39
[2] Malaysia Airports Holdings Berhad 2017 *http://www.klia.com.my*
[3] Eickemeyer S, Steinkamp S, Schuster B, Bodenhage F, Nyhuis P 2014 *Procedia CIRP* **23** 252-7
[4] Piancastelli L, Frizziero L, Morganti E and Pezzuti E 2012 *Walailak Journal of Science and Technology* **9** 425-31
[5] Luh P B, Yu D, Soorapanth S, Khibnik A I and Rajamani R 2005 *IEEE Transactions on Automation Science and Engineering* **2** 145-57
[6] Kivilcim E and Banu B 2014 *Applied Mechanics and Materials* **598** 215-8

IOP Conf. Series: Materials Science and Engineering **270** (2017) 012014 doi:10.1088/1757-899X/270/1/012014

Preliminary studies on readiness of biojet fuel for commercial aviation: The feasibility and potential in Malaysia

H Mohd Noh[1,*], M N Mahammad Taher[1], G A Rodrigo[2], N A Abdul Rahman[1], J Othman[3] and N H R Yahaya[1]

[1]Universiti Kuala Lumpur – Malaysian Institute Aviation Technology, Dengkil, Selangor, Malaysia
[2]Air Transport Department, ETSIAE, Universidad Politecnica De Madrid, Pz Cardenal Cisneros, Madrid, Spain
[3]Universiti Kuala Lumpur – Malaysian France Institute, Bandar Baru Bangi, Selangor, Malaysia

*hazariah@unikl.edu.my

Abstract. This paper demonstrates the need for a new alternative energy using biojet fuel in commercial aviation. The demand of air travels leads the authority, airlines and government in seeking for new renewable and sustainable energy for aircraft operation in the future. This study looks into the level of readiness in using biofuel. 40 personnel who are working in the aviation industries have participated and completed the survey questionnaires. The preliminary findings suggest that the impact towards this new fuel will lead to a better environment, less cost, better maintenance and energy sustainability. The usage of biojet fuel seems possible to be pursued in Malaysia.

1. Introduction

Many new engines today have much better fuel efficiencies than previous ones, which helps to reduce the fuel consumption. While this facilitates the airlines in dealing with the instability of fuel prices, the problem with carbon dioxide (CO_2) emissions remains. Many research have been focused on finding alternative energy sources to replace the fossil fuel, which will also aid in reducing the maintenance and cost for airlines [1-3]. The current growth in aviation sector has been highlighted by the statistics in the annual report of International Air Transport Association (IATA) [4-9]. It is estimated that about 2.3 billion passengers use air transportation to travel around the world since it is the shortest and also the fastest way of traveling. Based on the future projection, air traffic is expected to increase at 4.8% per year until 2036, with the Asia/Pacific region as the primary contributor to this increment [10]. The rise in air traffic also increases the environmental concerns for the region.

Researchers in the Sustainable Way for Alternative Fuel and Energy in Aviation (SWAFEA) have previously claimed that the use of alternative fuels such as 50% SPK can provide significant reduction in engine soot and sulphur oxides (SOx) emissions due to reduced content of aromatics and sulphur [11]. In line with a similar notion, more research have been carried out to promote the usage of biojet fuel to reduce emissions [12]. The purpose of this investigation is to carry out a preliminary study in identifying the basic knowledge of biojet fuels among aviation personnel, feasibility of biojet fuel used in aviation and the impact it has on the aviation industry.

IOP Conf. Series: Materials Science and Engineering **270** (2017) 012014 doi:10.1088/1757-899X/270/1/012014

2. Research Method

In the big picture, this research will cover several phases as indicated in Figure 1. The four phases are essentially adapted from Ref. [13]. First phase of this research is already completed and the findings reveal the awareness, feasibility and impact of biojet fuel on the aviation industry. Phase 2, which is the focus of this study, the survey questionnaire [14] is applied to understand the readiness of aviation personnel for adaptation of biojet fuel in the aviation industry and aircraft maintenance in particular. A brief guide has been introduced at the beginning of the questionnaire.

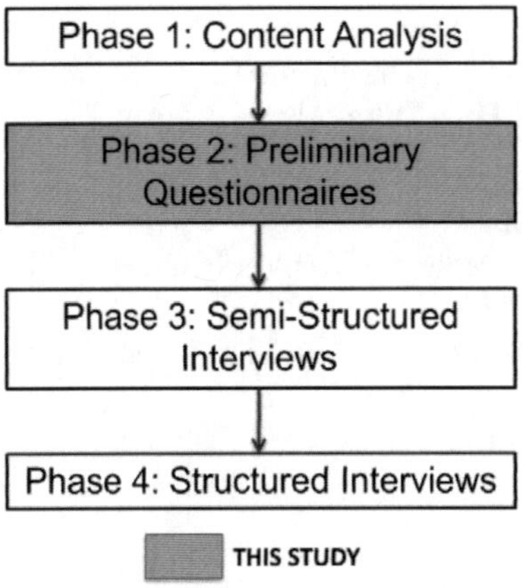

Figure 1: Research phase

According to Ref. [15], "the higher the need for control, the more critical it be that an interviewer is present and possibly that the questionnaire is done on paper". The collected data in this Phase 2 is the content analysis data, which has been gathered in previous Phase 1 [16]. The respondents are chosen based on convenience sampling. The questionnaire has been divided into four main sections: personal (biographical) data, general knowledge on biojet fuels, feasibility of biojet fuels in aviation and impact of the usage of biojet fuels. All respondents' response to the questionnaire is obtained through face-to-face manner.

3. Results and Discussion

3.1. Section A: Personal (biographical) Data

The participants are asked to indicate their gender. As tabulated in Table 1, 28 respondents (70%) are female while the remaining 12 participants (30%) are male. Historically, engineering has been a male-dominated profession although more females are joining this profession in current trends. On the other hand, Table 2 shows the frequency for different groups of age for the respondents. Both age groups of 35-40 and 45-50 have the highest frequency with 18% while the lowest is 5% for the age groups of 20-25, 55-60 and 60-65. From the distribution of the age groups, it seems that majority of the respondents have an experience and knowledge about aviation sector with more than five years in this industry. It should be noted that the targeted group in this study is the aviation personnel who have more than four years of experience in handling aircraft, particularly on engine or aircraft systems.

Table 1: Gender of respondents

Gender	Frequency
Female	28
Male	12

Table 2: Age of respondents

Age Group	20-25	26-30	30-35	36-40	40-45	46-50	50-55	56-60	60-65	Total
Frequency	2	4	5	7	5	7	6	2	2	40

Participants are also asked to indicate their position in the industry by writing the relevant answers. As illustrated in Figure 2, 17 (43%) respondents used to be License Aircraft Engineer from various airlines. The other answered positions include apprentice engineer (7%), aircraft technician (12%) and aviation lecturer (30%) and students (8%). Aviation students chosen were those in their final year of study and were currently on-the-job training (OJT). In the meantime, the aviation lecturers are among those closer to the industries and current new technology. Furthermore, the respondents are also asked to put the relevant numbers of experiences. As shown in Figure 3, the majority of the respondents (i.e. 17 respondents or 42%) reported having between 16 to 20 years of experience in the aviation industry and other three respondents (8%) had between 26 to 30 years of industry experience. The experienced participants have tremendous impact on this study for the practicality and accuracy of its contribution.

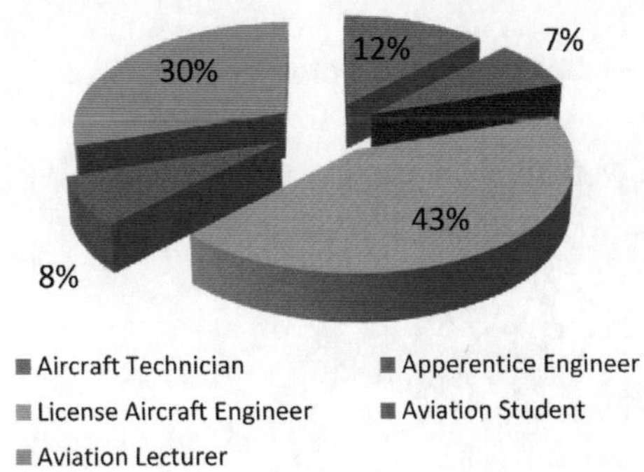

Figure 2: Respondents' position in the industries

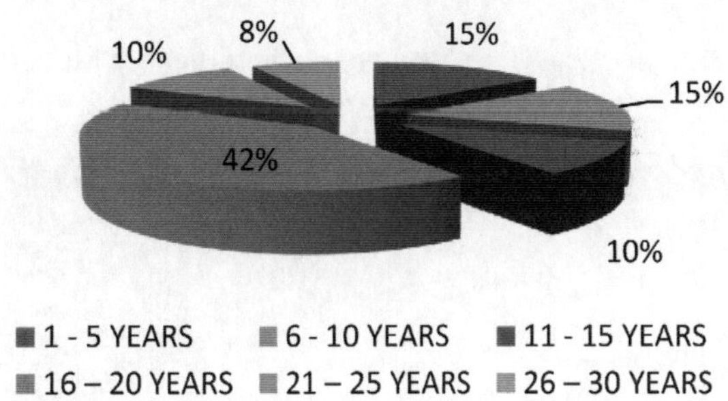

Figure 3: Respondents' length of experience in the industries

3.2. Section B: Personal Knowledge about Biojet Fuels in Aviation

The participants' responses on questions B1 to B5 reflect their personal knowledge about biojet fuels in aviation, and its usage and impact to aviation industries. The responses are tabulated in Table 3. The answers for question B1 shows that majority of the respondents are moderately knowledgable about biojet fuels usage in aviation. The knowledge about biojet fuels to the environment, to aircraft engines, and reducing fuel cost by using biojet fuels needs to be educated to all personals in aviation industries through training. In addition, based on average response to question B2, the participants' knowledge on the environment is very high. It shows that the respondents highly agree that biojet fuels are good for the environment. Furthermore, Question B3 shows that the respondents are only moderately agreed that biojet fuel is good for aircraft engine. This indicates that the knowledge on what biojet can do to the aircraft engine has not been properly relayed. Moreover, question B4 shows that the respondents are also moderately agreed that biojet fuel is good for reducing the fuel cost. The financial aspect is not being shared to the respondents but they understood that the cost of biojet fuels as alternative fuels is much cheaper than currently used fuels once it is adequately established. Last but not least, answers for question B5 show that the respondents are moderately knowledgeable on potential profits in using biojet fuels in aviation industries as alternative fuels. This due to the knowledge of cheaper alternative fuels that can be applied on most engines, especially those in the automotive vehicles.

All in all, the questions in this Section B are focused on how good biojet fuels to the environment, to aircraft engines, and to reduce fuel cost. Figure 4 highlights the summary of responses for Section B. Questions B1, B3 and B5 have the highest number of respondents (i.e. 50%) answering "moderate" knowledge. This shows that the average respondents know about biojet fuel usage in aviation industry. However, it seems that the respondents have a poor knowledge in biojet fuel.

Table 3: Responses for Section B of the questionnaire

Question	Responses					
	Very High	**High**	**Moderate**	**Low**	**Very Low**	**Unsure**
B1: Do you know about biojet fuels usage in aviation?	1	5	20	8	5	1
B2: Do you think biojet fuel is good to the environment?	30	0	10	0	0	0
B3: Do you think biojet fuel is good to aircraft engine?	2	8	20	2	3	5
B4: Do you think biojet fuel is good for reduced fuel cost?	5	15	10	1	0	9
B5: Rate your thought on how profitable biojet fuel to airlines/ aviation company?	4	6	20	0	0	10

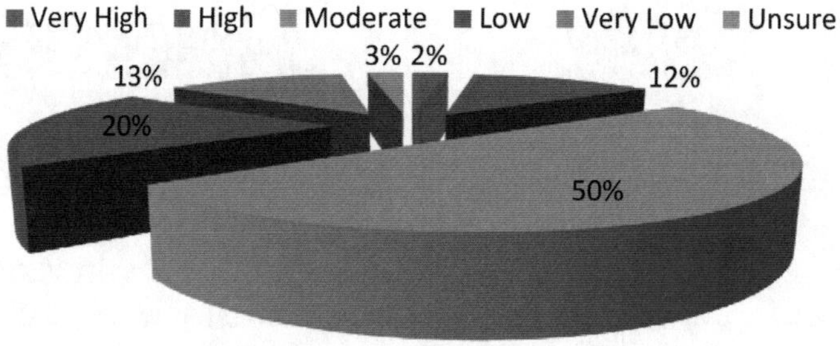

Figure 4: Respondents' knowledge

3.3. Section C: The Feasibility of Biojet Fuels in Aviation

Question C1 is designed to probe whether the biojet fuel is considered safe for aviation industry. Most of the respondents (77%) have agreed with that statement while only nine respondents (23%) are not sure on that matter. For the second until fourth questions (i.e. C2, C3 and C4), the respondents have averagely answered "moderate". These questions are more on the technical side when the biojet fuels are used in aircraft engines. The questions are asking whether any modification will be required to the engines if biojet fuel is used, whether the adaptation level to biojet fuel will be easier when that takes place, and whether the required changes from normal fuel (Jet A/Jet A-1) to biojet fuel will be of low technical feasibility. In the last question C5 however, nine respondents' (23%) agree that biojet fuels will be widely used in future. Figure 5 summarizes the responses for Section C of the questionnaire.

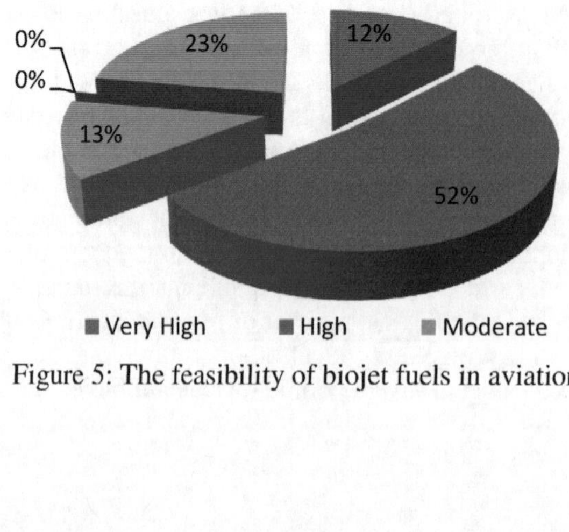

Figure 5: The feasibility of biojet fuels in aviation

3.4. Section D: Impact from the Usage of Biojet Fuels

Figure 6 highlights the responses obtained for Section D of the questionnaire. The first question D1 is asking the respondents whether they think that biojet fuels lead to a better environment, with 13 of the respondents has rated it as very high (32%). The second question D2, on the other hand, is asking if the biojet fuels provide better fuel cost and the highest answer is "moderate" by 13 respondents (32%). For questions D3 and D4, the respondents averagely answered as "moderate", with 55% and 37%, respectively. The questions are about the increase of engine performance and the saving in airlines fuel cost if biojet fuel is used. Finally, the last question D5 is asking whether sustainable biojet fuels can be achieved. Most of the respondents have rated this question as moderate (42%).

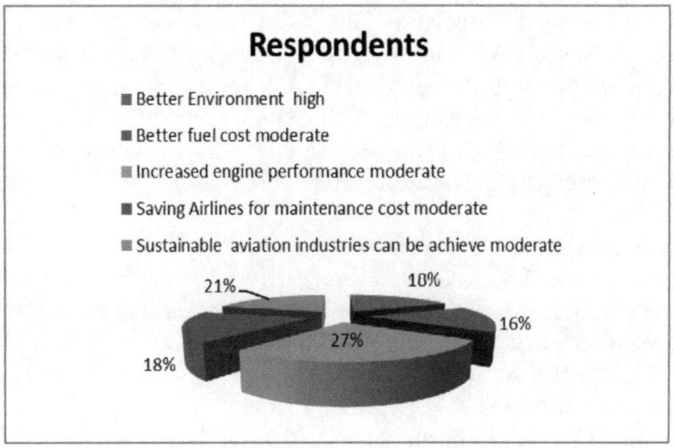

Figure 6: The impact of using biojet fuel based on 40 respondents

Overall, in measuring the readiness level, the respondents are aware that biojet fuels are safe to be used in aviation. This is supported by the results of more than 25 airlines already tested or conducted biojet fuel flights [17, 18]. However, the level of knowledge on the new green technology, especially biojet fuels, is still at moderate level as shown in Figure 4 where 36% of the respondents are unsure of biojet fuel. Overall, majority of the respondents have acknowledged the good impact of using biojet fuels to the environment, aircraft safety and fuel. The minimal harmful emissions of gas release from biofuels as compared to the existing fossil fuels will help in producing a greener environment. On top of that, the safety and high specification of fuel quality are important in the aviation sector and biojet fuels are offering them. The technical requirements for aviation biojet fuels are high-performance fuel (that can withstand a range of operational conditions), does not compromise safety, directly substitute traditional jet fuel aviation, and meets the required performance target [19]. The feasibility of biojet fuel adaptation in aviation industries is also given by the responses of 52% at the high rate as shown in Figure 5.

4. Conclusion

Challenges remains in adapting the biojet fuel in aviation industry as it requires high-quality standard requirements in measuring the impact on the environment as well as the fuel cost. The cost has a high impact for new alternative fuel as increase in cost will create a huge hurdle to the aviation industries. Aviation industries will be reluctant to adopt this new energy if there is an increase in fuel operation. The investigation done in this study has shown that the basic knowledge of biojet fuels among aviation personnel is still at a moderate level. The potential feasibility of biojet fuel used in aviation industries in Malaysia can be further explored and its impact towards environment, better fuel cost, increase in engine performance and maintenance can be better assessed for more sustainable aviation industries. The readiness situation in Malaysia for this purpose can be projected through the research's feasibility studies, which is expected to create more local opportunities in the perspective of social, economic and operation.

Acknowledgement
The researchers would like to thank Ministry of Higher Education, Malaysia for the research funding under the Fundamental Research Grant Scheme: FRGS/1/2015/TK10/UNIKL/03/1.

References

[1] Noh H, Alonso G, Nair S and Dahdi Y 2015 *WIT Transactions on Ecology and the Environment* **206** 103–11
[2] Arushi and Drews S 2011 *Aviation and Environment* Centre for Science and Environment, India
[3] International Civil Aviation Organization 2010 *ICAO Environment Report*
[4] International Air Transport Association 2009 *IATA Environment Report*
[5] International Air Transport Association 2007 *IATA Report on Alternative Fuels*
[6] International Air Transport Association 2010 *IATA Report on Alternative Fuels*
[7] International Air Transport Association 2011 *IATA Report on Alternative Fuels*
[8] International Air Transport Association 2012 *IATA Report on Alternative Fuels*
[9] International Air Transport Association 2012 *IATA Report on Alternative Fuels*
[10] International Air Transport Association 2010 *International Air Transport Association Annual Report*
[11] Air Transport Action Group 2012 *Powering the Future of Flight*
[12] Noh H M, Benito A and Alonso G 2016 *Transp. Res. Part D Transp. Environ.* **46** 298–316
[13] Henk H, Kramer R, Sesink L and Joris Z V 2006 *Project Management Handbook* DANS
[14] Bradburn N, Sudman S and Wansink B 1990 *Comput. Environ. Urban Syst.* **14** 72
[15] Harrell M C and Bradley M A 2009 *Data Collection Methods Semi-Structured Interviews and Focus Groups* National Defense Research Institute
[16] Noh H M, Rodrigo G A and Rahman N A A 2016 *IOP Conf. Ser. Mater. Sci. Eng.* **152** 12043
[17] International Air Transport Association 2015 *Fact Sheet: Alternative Fuels*
[18] Mohd Noh H, Mohd Tahir M N, Nur Halimatun Radhiah Y, Abdul Rahman N A, Rodrigo G A and Othman J 2017 *Sci. Int.* **29** 1–15
[19] European Biofuels Technology Platform 2014 *http://www.biofuelstp.eu/aviation-biofuels.html*

IOP Conf. Series: Materials Science and Engineering **270** (2017) 012015 doi:10.1088/1757-899X/270/1/012015

An algorithm on simultaneous optimization of performance and mass parameters of open-cycle liquid-propellant engine of launch vehicles

M A Eskandari[1,*], H K Mazraeshahi[1], D Ramesh[1], E Montazer[2], E Salami[2] and F I Romli[3]

[1]Faculty of Aerospace Engineering, K. N. Toosi University of Technology, Iran
[2]Department of Mechanical Engineering, Faculty of Engineering, University of Malaya, Malaysia
[3]Department of Aerospace Engineering, Faculty of Engineering, Universiti Putra Malaysia, Malaysia

[*]mae.eskandari@yahoo.com

Abstract. In this paper, a new method for the determination of optimum parameters of open-cycle liquid-propellant engine of launch vehicles is introduced. The parameters affecting the objective function, which is the ratio of specific impulse to gross mass of the launch vehicle, are chosen to achieve maximum specific impulse as well as minimum mass for the structure of engine, tanks, etc. The proposed algorithm uses constant integration of thrust with respect to time for launch vehicle with specific diameter and length to calculate the optimum working condition. The results by this novel algorithm are compared to those obtained from using Genetic Algorithm method and they are also validated against the results of existing launch vehicle.

1. Introduction

The need to increase the specific impulse of rocket launcher engines has led design engineers to start the development of more complex engine cycles starting from simple gas pressure fed engines to the very complex staged combustion cycle engines. Currently, a wide range of different cycles exist and they are not only differ in terms of performance, but also in terms of mass, cost, reliability, etc., which in general makes it difficult to quickly determine the one that is best suited for certain mission or task. Ernst (2014) has applied the modular approach where engine components are sized using performance, dimension and mass model, making use of corrected ideal rocket theory and empirical relations [1]. His work focused mainly on the methodology and the construction of the models, and further includes the optimization of an upper stage and verification, validation and uncertainty and sensitivity analysis of the tool and the optimization to assess its accuracy, precision and applicability. Consequently, the research develops analysis modules for classifying components of liquid rocket engine and computing performance and weight of each component. Each module analyzes with combinations of input and output during system analysis. In order to increase its fidelity, the modules are based on the empirical formulas and practical data [2].

Furthermore, Lee and Roh (2013) presented system analysis program that can compute the weight and specific impulse used as performance indicator of the propulsion system, and the performance of each component for given values of propellant, required thrust and combustion chamber pressure [2]. Analysis of today's liquid-propellant engine of launch vehicles is a challenging and complicated task, which requires advanced tools and algorithms. In general, an optimum design of engine itself will not result in the optimum operation of launch vehicle since the interaction between the rocket's subsystems must also be taken into consideration. There are several software packages that have been designed to determine the optimum performance parameters of the engine for specific inputs such as REDTOP2 [3]. Furthermore, there are also software packages like Fordu that are able to determine the optimum parameters of the propulsion subsystem including engine mass rate, pumps' input pressure and engine mass [4, 5]. However, it has been noted that the mass parameters are not typically considered together in the optimization process with the performance parameters.

It is believed in this study that simultaneous optimization of both performance and mass parameters of the rocket subsystems can lead to better results. Hence, a new algorithm is introduced in this study, which uses semi-empirical relations to calculate the mass of the propulsion subsystems such as engine, tanks and pressurizing subsystem, and this describes the propulsion subsystem with a better accuracy. The main advantage and novelty of this proposed algorithm is the coupling between the performance and mass parameters of rocket subsystems, which are optimized simultaneously instead of separately like in most existing algorithms.

2. Performance Characteristics Algorithm for Propulsion Subsystem

For the optimization of performance parameters, the following algorithm is proposed. In the first step, some constraints have to be defined as inputs. In this algorithm, the combustion chamber's pressure, burning time, thrust force, ratio of mass flow of oxidizer to the mass flow of fuel (O/F) and type of propellant are taken as inputs. Propulsion subsystem is considered open-cycled for this paper and its schematic is shown in Figure 1.

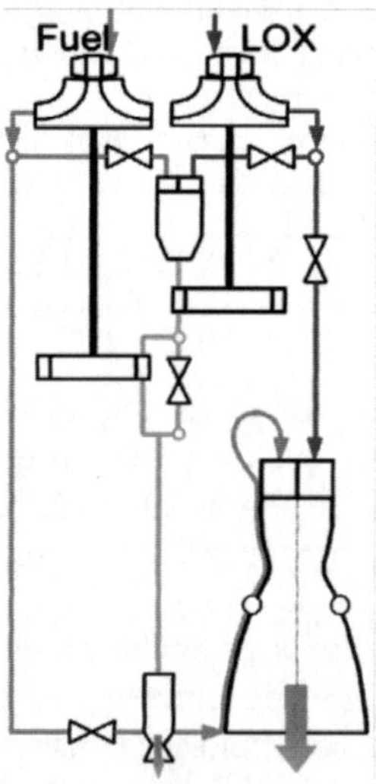

Figure 1: Simplified schematics of open-cycle liquid-fueled rocket [6]

According to the preliminary thrust force and specific impulse, the mass flow rate of the propulsion can be calculated from Equation 1.

$$\dot{m}_{eng} = \frac{T_{eng}}{I_{SP_{eng}} g_0} \tag{1}$$

Then, each component mass flow rate than can be driven from the following Equations 2 and 3:

$$\dot{m}_{ox_{eng}} = \dot{m}_{eng} \frac{(O/F)_{eng}}{1+(O/F)_{eng}} \tag{2}$$

$$\dot{m}_{fe_{eng}} = \dot{m}_{eng} \frac{1}{1+(O/F)_{eng}} \tag{3}$$

Next, the calculation of pressure drop in the whole propulsion subsystem is done. To simplify these calculations, pressure drop including cooling channel, injectors, pipes and orifices can be considered as a factor of combustion chamber's pressure. In the designing process of engine, the effort is focused on minimizing these waste as much as possible. Usually, pressure drop in the pipes and control valves is between 35 to 50 percent of the combustion chamber's pressure. For regenerative cooling system, this drop is normally about 20 to 30 percent of the combustion chamber's pressure. The pressure drop in injectors depends on their shape and type. However, in this study, this drop percentage is taken to be about 30 percent of the combustion chamber's pressure. Also, pressure drop is usually about 6 to 10 percent in chamber's pressure for orifices [7]. Therefore, the discharge pressure of oxidizer and fuel pumps can be calculated using Equation 4 and Equation 5, respectively.

$$P_{P_{ox}} = 1.5939 P_{cc} \tag{4}$$

$$P_{P_{fu}} = 2.2248 P_{cc} \tag{5}$$

In order to decrease the pressure drop error, geometric factors have been introduced to optimize the discharge pressure of the pumps. This enables the geometry of the pipes to be taken as constant in the process of optimization. These geometric factors can be found from Equation 6 and Equation 7.

$$K_{P_{ox}} = \frac{\rho_{ox}\left(P_{P_{ox}} - P_{cc}\right)}{\dot{m}_{ox_{eng}}} \tag{6}$$

$$K_{P_{fu}} = \frac{\rho_{fu}\left(P_{P_{fu}} - P_{cc}\right)}{\dot{m}_{fu_{eng}}} \tag{7}$$

To obtain mass flow rate of the turbine, power of the pumps is calculated according to Equation 8 and Equation 9. In this study, the pumps' input pressure has been chosen as an algorithm input, where ∇P_{pox} and ∇P_{pfu} defined by Equation 10 and Equation 11, respectively, are the pressure difference of the pumps. In the meantime, *i* and *e* indicate the input and exit section of the pumps, respectively.

$$P_{req pox} = \frac{g_o \dot{m}_{oxeng} \nabla P_{pox}}{\eta_{pox}} \tag{8}$$

$$P_{req pfu} = \frac{g_o \dot{m}_{fueng} \nabla P_{pfu}}{\eta_{pfu}} \tag{9}$$

$$\nabla P_{pox} = P_{poxe} - P_{poxi} \tag{10}$$

$$\nabla P_{pfu} = P_{pfue} - P_{pfui} \tag{11}$$

Moreover, the efficiency of the pumps has a drastic effect in producing more power and consuming less mass flow rate. This heavily depends on their manufacturing technology and these parameters are also considered as constant inputs. For this case, η_P has been chosen equal to 0.65.

The adiabatic work of the turbine depends on the expansion ratio and temperature of gases leaving the gas generator. The expansion ratio for open-cycle engines typically ranges from 20 to 50 and the temperature is based on the tolerable temperature of the blades. The efficiency of turbine also depends on its manufacturing and in this work, it is taken as 0.4.

$$L_{ad} = RT_{gg}\left[1-\left(\frac{1}{P_{tr}}\right)^{\frac{(\gamma_{gg}-1)}{\gamma_{gg}}}\right] \quad (12)$$

$$\dot{m}_g = \frac{P_{req_{ox}} - P_{req_{fu}}}{L_{ad}\eta_{turbine}} \quad (13)$$

After the mass flow rate and O/F ratio of the gas generator have been determined, the mass flow rate of each propellant component is calculated using Equation 14 to Equation 18. The O/F ratio of the gas generator is determined with regards to the thermal resistant limitation of its material. In this case study, the ratio is equal to 0.16 that shows a rich-fueled gas generator.

$$\dot{m}_{fu_{gg}} = \frac{\dot{m}_{gg}}{1+(O/F)_{gg}} \quad (14)$$

$$\dot{m}_{fu_{gg}} = \dot{m}_{gg}\frac{(O/F)_{gg}}{1+(O/F)_{gg}} \quad (15)$$

$$\dot{m}_{ox_{cc}} = \dot{m}_{ox_{eng}} - \dot{m}_{ox_{gg}} \quad (16)$$

$$\dot{m}_{fu_{cc}} = \dot{m}_{fu_{eng}} - \dot{m}_{fu_{gg}} \quad (17)$$

$$\dot{m}_{cc} = \dot{m}_{fu_{cc}} + \dot{m}_{ox_{cc}} \quad (18)$$

Therefore, the ratio of oxidizer to fuel for combustion chamber can be calculated using Equation 19.

$$(O/F)_{cc} = \frac{\dot{m}_{ox_{cc}}}{\dot{m}_{fu_{cc}}} \quad (19)$$

At this point, the necessary means to calculate specific heat ratio and produce the gas characteristic speed within thermal equilibrium has been made available. The Astra software for the computation of the produced gas is used. As a result, theoretical diameter of throat can be obtained as in Equation 20.

$$D_t = \sqrt{\frac{4c^*\dot{m}_{cc}}{\pi P_{cc}}} \quad (20)$$

This is the first step for the determination of the throat diameter. After this, the throat diameter of the combustion chamber is kept constant. This leads to more accurate answers in the optimization space. In the next following steps, for each chamber's pressure, for a constant O/F ratio and throat diameter, the performance and mass parameters are calculated. In the end, the one corresponding to the highest specific impulse to gross mass will be selected as the optimum point of design.

To find the mass of tanks, the amount of propellant in the launch vehicle is required. This amount depends on the burning duration of the propellant. To further simplify the calculations, the integral of thrust with respect to time is kept constant in all of the iterations. As a result, the duration of burning will be known after determination of thrust. For specific amount of mass flow rate and chamber's pressure, the specific heat ratio and velocity of produced gas will be calculated. Therefore, the total mass flow rate of chamber, specific impulse and the produced thrust of the chamber can be derived from Equation 21 and Equation 22 [8, 9], where λ is the efficiency of nozzle that is usually taken to be around 0.9 to 0.98.

$$\dot{m}_{cc} = \frac{P_{cc} A_t}{c^*_{cc}} \tag{21}$$

$$I_{sp_{cc}} = \frac{c^*_{cc}\lambda}{g_0}\left[\gamma_{cc}\sqrt{\frac{2}{\gamma_{cc}-1}\left(\frac{2}{\gamma_{cc}+1}\right)^{\frac{\gamma_{cc}+1}{\gamma_{cc}-1}}\left(1-\frac{P_e}{P_{cc}}\right)^{\frac{\gamma_{cc}-1}{\gamma_{cc}}}}\right] \tag{22}$$

$$T_{cc} = I_{sp_{cc}}\dot{m}_{cc} g_0 \tag{23}$$

The gas generator mass flow rate is to be defined next. Accordingly, specific amount for mass flow rate is initially chosen – usually about 4 percent of chamber's mass flow rate – and check whether the produced power is equal to the consumed power by pumps. The process is then iterated by changing the amount of the mass flow rate until both the produced and consumed power are equal to each other. Using the same procedure, the mass flow of gas generator can be found [10].

$$\nabla \mathrm{P}_{\mathrm{pox}} = \frac{\mathrm{k}_{\mathrm{poxe}}\dot{m}^2_{oxeng}}{\rho_{\mathrm{ox}}} + \mathrm{P}_{\mathrm{cc}} - \mathrm{P}_{\mathrm{poxi}} \tag{24}$$

$$\nabla P_{p_{fu}} = \frac{k_{p_{fu_e}}\dot{m}^2_{fu_{eng}}}{\rho_{fu}} + P_{cc} - P_{p_{fu_i}} \tag{25}$$

$$P_{reqox} + P_{reqfu} = \frac{g_0\dot{m}\nabla \mathrm{P}_{\mathrm{p}}}{\eta_p} = \eta_T \dot{m}_{gg} c_p T_{emi}\left[1-\left(\frac{1}{P_{t-ratio}}\right)^{\frac{\gamma_{cc}-1}{\gamma_{cc}}}\right] \tag{26}$$

Using both previous Equations 14 and 15, the mass flow rate of oxidizer and fuel in gas generator can now be determined. Subsequently, the total mass flow rate of fuel and oxidizer can be calculated as in Equation 27 to Equation 30.

$$\dot{m}_{fu_{eng}} = \dot{m}_{fu_{cc}} + \dot{m}_{fu_{gg}} \tag{27}$$

$$\dot{m}_{ox_{eng}} = \dot{m}_{ox_{cc}} + \dot{m}_{ox_{gg}} \tag{28}$$

$$(O/F)_{eng} = \frac{\dot{m}_{ox_{eng}}}{\dot{m}_{fu_{eng}}} \tag{29}$$

$$\dot{m}_{eng} = \dot{m}_{ox_{eng}} + \dot{m}_{fu_{eng}} \tag{30}$$

Finally, the total thrust force and specific impulse are calculated using Equation 31 to Equation 33. It should be noted that the specific impulse of gas generator depends on the type of the propellant.

$$T_{gg} = I_{sp_{gg}}\dot{m}_{gg} g_0 \tag{31}$$

$$T_{eng} = T_{cc} + T_{gg} \tag{32}$$

$$I_{sp_{eng}} = \frac{T_{eng}}{\dot{m}_{eng} g_0} \tag{33}$$

3. Mass Characteristics Algorithm of Propulsion Subsystem

The mass characteristics of propulsion system mainly depend on the regime parameters. Hence mass of the different parts of the propulsion system can be described as a function of the regime parameters. The propulsion subsystem's mass characteristics primarily consist of the mass of the chamber, turbo-pump, engine and tank. For each of these characteristics, there are already some empirical functions that depend on the performance parameters of subsystems. They can be used to compute the total mass of propulsion subsystem. The mass of the turbo-pump can be obtained from the following Equation 34 and Equation 35.

$$D = \sum_{i=ox,fu} \frac{\dot{m}_i}{\omega} \left(\frac{P_{p_e,i} - P_{p_i,i}}{\rho_i} \right)^{1.5} \tag{34}$$

$$m_{T,p} = \begin{cases} 6.29 + 0.981 \times 10^{-3} D & 1170 < D < 3.22 \times 10^4 \\ 21 + 0.54 \times 10^{-3} D & 3.22 \times 10^4 < D < 7.52 \times 10^5 \end{cases} \tag{35}$$

The number of pump's revolution, ω can be calculated from Equation 36 and Equation 37, where $C_{cav.\max}$ is the cavitation maximum characteristic velocity that ranges from 3000 to 5000m/s while $\Delta h_{cav.res.}$ is the cavitation reserve factor of pumps that ranges between 10 to 30 J/kg.

$$\Delta h_{cav.} = \frac{P_{pi} - P_s - \rho \Delta h_{cav.res.}}{\rho} \tag{36}$$

$$\omega = C_{cav.\max} \frac{\Delta h_{cav.}^{0.75}}{248.Q^{0.5}} \tag{37}$$

Furthermore, the mass of the chamber can be found using Equation 38 to Equation 47, in which β_a is the nozzle wall's imperfection at the exit and it is normally taken to be between 2 to 6 degrees.

$$\gamma_{T_{cc}} = 3.03 \frac{P_{cc} \times 10^{-6}}{\sqrt{d_t}} \qquad 1.58 \times 10^6 \le \frac{P_{cc}}{\sqrt{d_t}} \le 5.86 \times 10^6 \tag{38}$$

$$\gamma_n = -23.58 + 5.894 \times 10^{-2} \left(\frac{P_{cc}}{\sqrt{\varepsilon D_t}} \right)^{0.475} \qquad 0.0632 \times 10^6 \le \frac{P_{cc}}{\sqrt{\varepsilon D_t}} \le 0.316 \times 10^6 \tag{39}$$

$$\bar{s}_c = \frac{s_c}{A_t} = 3.54 L^* \sqrt{\frac{P_{cc} \bar{q}_k}{\dot{m}_{cc}}} - 2 \frac{1}{\sqrt{\bar{q}_k c^*}} - \sqrt{\bar{q}_k c^*} + 1 \tag{40}$$

$$\bar{s}_{sub} = \frac{s_{sub}}{A_t} = \frac{2}{\bar{q}_k c^*} + \frac{0.818}{\sqrt{\bar{q}_k c^*}} - 0.974 \tag{41}$$

$$\bar{q} = \frac{\dot{m}_{cc}}{P_{cc} A_{cc}} \tag{42}$$

$$z = 1 - \left[\frac{\sin(\beta_a)}{0.6 - (0.018\gamma - 0.0175)\left(\sqrt{\bar{A}} + 24\right)} \right]^{\frac{4}{3}} \tag{43}$$

$$\bar{s}_{\text{sup}} = \frac{s_{\text{sup}}}{A_t} = \bar{s}_0 \left[1 - \left(1.415 - \frac{0.27}{\sqrt[4]{\bar{A}}} \right) \right] f(z) \tag{44}$$

$$f(z) = 1 - e^{-(1-z)^{\frac{1}{3}}} \tag{45}$$

$$\bar{s}_0 = (32 - 10\gamma)\left(\sqrt{\bar{A}} - 1\right) + \left(2.1 + 1.6\gamma^4\right)\left(\sqrt{\bar{A}} - 1\right)^{2.25} \tag{46}$$

$$m_{Tc} = \frac{\dot{m}c^*}{P_{cc}} \left[\gamma_{Tcc}\left(\bar{s}_c + \bar{s}_{sub}\right) + \gamma_n \bar{s}_{\text{sup}} + \frac{163}{\left(P_{cc} c^* \dot{m}\right)^{0.25}} - \frac{8.5g}{P_{cc}} \right] \tag{47}$$

Now, by already knowing the mass of turbo-pump and combustion chamber, the total mass of the engine can be derived using Equation 48.

$$m_{eng} = \begin{cases} N \times m_{T_C} + \sum m_{Tp} + 0.217\frac{T}{g} + 57.5 & 14.7 \le \frac{T}{g} \le 981 \\ N \times m_{T_C} + \sum m_{Tp} + 0.377\frac{T}{g} - 93.1 & 982 \le \frac{T}{g} \le 1680 \end{cases} \tag{48}$$

On the other hand, the mass of the tanks can be calculated using Equation 49 and Equation 50. It can be noted that this mass depends on type, pressure and mass of propellant.

$$m_{Tank_{fu}} = \frac{m_{prop_E}}{\left[(O/F)+1\right]\rho_{fu}} \left(\frac{2.1 f_{s,fu} \rho_{Tank,fu}}{\sigma_{y,Tank,fu}} P_{Tank,fu} + \frac{4.2\delta_{iso,fu}\rho_{iso,fu}}{D_{Tank,fu}} \right) + 0.649 \frac{\rho_{Tank,fu} D^3_{Tank,fu}}{\sigma_{y,B,fu}} f_{s,fu} P_{Tank,fu} \tag{49}$$

$$m_{Tank_{fu}} = \frac{m_{prop_E}(O/F)}{\left[(O/F)+1\right]\rho_{ox}} \left(\frac{2.1 f_{s,ox} \rho_{Tank,ox}}{\sigma_{y,Tank,ox}} P_{Tank,ox} + \frac{4.2\delta_{iso,ox}\rho_{iso,ox}}{D_{Tank,ox}} \right) + 0.649 \frac{\rho_{Tank,ox} D^3_{Tank,ox}}{\sigma_{y,B,ox}} f_{s,ox} P_{Tank,ox} \tag{50}$$

Lastly, the final gross mass of the propulsion subsystem consisting of the mass of engine, tanks and propellant can be written as the following Equation 51 and Equation 52.

$$m_{inert} = m_{eng} + \sum m_{Tank,i} + m_{other} \tag{51}$$

$$m_{gross} = m_{prop_e} + m_{inert} \tag{52}$$

4. Case Study for Validation of the Proposed Algorithm

In the process of designing a liquid fuel engine, one of the most important issues is the achievement of basic knowledge regarding the mass and functional aspects of the engine and the optimal parameters of it. In this paper, various massive and functional dimensions are tried and found before designing a liquid fuel engine. This will be a great assistance to the designers of the propulsion system in order to select appropriate design parameters like thrust, specific impulse, etc. To see whether this algorithm is effective, its results for the sample case study are compared to that obtained using Genetic Algorithm method. In general, the Genetic Algorithm has been designed to maximize the objective function and search for the optimized working condition. The main inspiration behind Genetic Algorithm comes from Darwin's evolution theory [11, 12]. This optimization technique is largely based on the idea of natural selection and reproduction.

The case study that is used to validate and demonstrate the effectiveness of the proposed algorithm is the optimization of third stage of Tsyklon-3 launch vehicle shown in Figure 2. The input variables for the procedure are tabulated in Table 1.

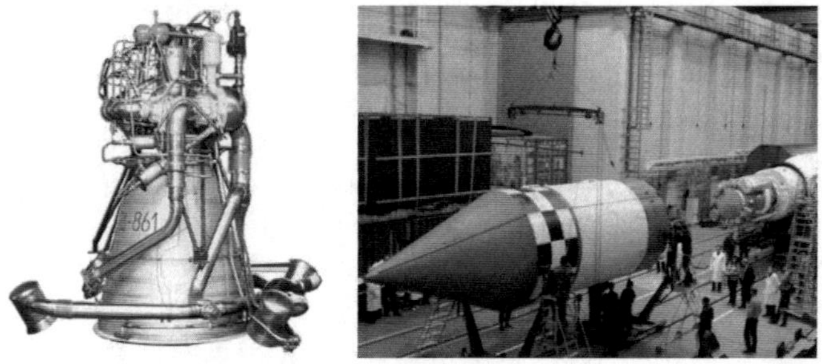

Figure 2: Third stage of Tsyklon-3 with RD-861 engine as its engine

Table 1: Input variables from Tsyklon-3 third stage engine

Variable	Value
Engine Type	Open-Cycle
Propellant Type	UDMH-N2O4
Chamber's Pressure	88.8 bar
Burning Time	125 s
O/F Ratio	2.1
Engine's Exit Pressure	0.2 bar
Specific Impulse	317 s
Thrust	78.708 kN

The process flowchart of the proposed algorithm is illustrated in Figure 3. The results obtained are presented in Table 2 and Table 3. From Table 2, it can be observed that the performance results for specific impulse and thrust force are fairly acceptable. The largest error recorded is for the engine O/F ratio, which can be contributed to the difference between the gas generators defined in the algorithm and the real model. Meanwhile, in Table 3, one of the largest errors is corresponding to empty mass of the rocket. The primary reason for this condition is the fact that some components' mass are neglected in the modelling.

Table 2: Results for the performance parameters by the proposed algorithm

Parameters	Actual Value for Tsyklon-3 Third Stage	Results from the Proposed Algorithm	Percentage Error
Specific Impulse	317 s	317.82 s	0.26%
Thrust	78.708 kN	78.74 kN	0.04%
$(O/F)_{eng}$	2.1	2.18	3.80%

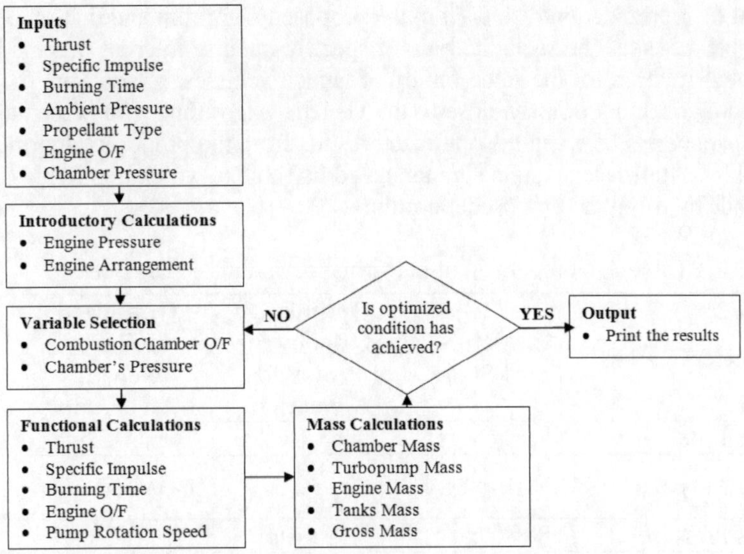

Figure 3: Simple flowchart of the proposed algorithm

Table 3: Results for the mass parameters by the proposed algorithm

Parameters	Actual Value for Tsyklon-3 Third Stage	Results from the Proposed Algorithm	Percentage Error
Propellant Mass	3139 kg	3297.3 kg	5.04%
Engine Mass	120 kg	118.8 kg	0.97%
Empty Mass	1407 kg	1344.3 kg	4.45%
Gross Mass	4666 kg	4760.0 kg	2.01%

Figure 4 depicts the relation between the ratio of specific impulse to the subsystem's mass and both the mass flow rate and the chamber's pressure. It can be observed that there exists a peak for specific impulse to mass ratio, which is the optimum operation point for the given data inputs.

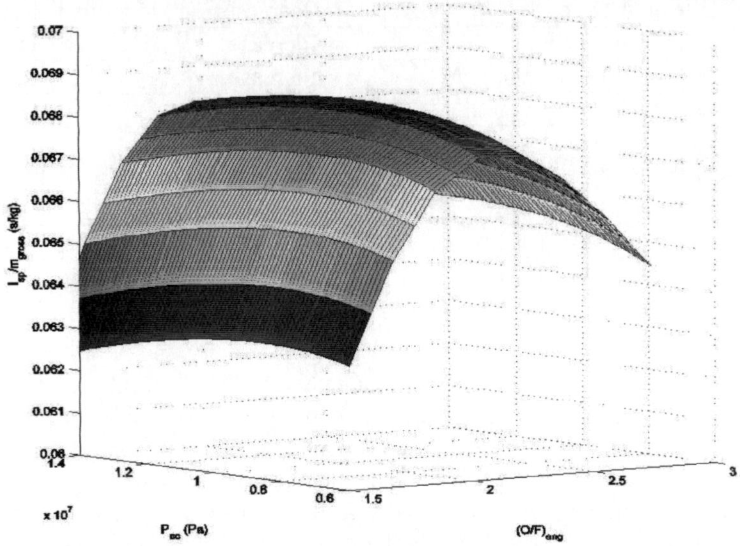

Figure 4: Ratio of specific impulse to gross mass against mass flow rate and chamber's pressure

IOP Conf. Series: Materials Science and Engineering **270** (2017) 012015 doi:10.1088/1757-899X/270/1/012015

Comparison of the optimization results from the proposed algorithm and that of Genetic Algorithm is tabulated in Table 4. As can be seen, the ratio of specific impulse to gross mass is increased, which shows an optimized situation for the same family of launch vehicles. In addition, the obtained results from the proposed algorithm are fairly close to the Genetic Algorithm. The percentage differences in the values of the parameters between the optimized results from the proposed algorithm and the actual values in the Tsyklon-3 third stage engine are tabulated in Table 5, which highlights the improvements that have been made by using the proposed algorithm.

Table 4: Comparison of performance and mass parameters

Parameters	Tsyklon-3 Third Stage	Optimized Model by the Proposed Algorithm	Optimized Model by Genetic Algorithm	Percentage Error (%)
Rocket Type	Open-Cycle	Open-Cycle	Open-Cycle	
Propellant Type	UDMH-N2O4	UDMH-N2O4	UDMH-N2O4	
Chamber's Pressure	88.80 bar	104 bar	103.93 bar	0.07
Burning Time	125 s	110.26 s	110.35 s	0.08
Specific Impulse	317 s	318.23 s	318.30 s	0.02
Thrust	78.71 kN	93.067 kN	92.995 kN	0.08
O/F Ratio	2.1000	2.0784	2.1541	3.64
Propellant Mass	3139 kg	3287 kg	3286.42 kg	0.02
Engine Mass	120 kg	129.07 kg	129.02 kg	0.04
Empty Mass	1407 kg	1354.40 kg	1353.81 kg	0.04
Gross Mass	4666 kg	4641.40 kg	4640.24 kg	0.02
Specific Impulse to Gross Mass	0.0675	0.0686	0.068594	0.01

Table 5: Improvements by the proposed algorithm from the actual Tsyklon-3 third stage engine

Parameters	Increment / Decrement	% Changes
Chamber's Pressure	Increase	11.25
Burning Time	Decrease	1.16
Specific Impulse	Increase	3.78
Thrust	Increase	1.60
O/F Ratio	Decrease	1.02
Propellant Mass	Increase	2.42
Engine Mass	Increase	14.02
Empty Mass	Decrease	19.04
Gross Mass	Decrease	3.10
Specific Impulse to Gross Mass	Increase	0.14

5. Conclusion

Finding the optimum operation point in designing launch vehicles is a real challenge. As mentioned before, the energetic optimization of the propulsion subsystem such as specific impulse and thrust is not enough. The mass of the system also has to be considered for optimum condition as well. The most prominent feature of this presented work is to include the mass of the system into consideration during the optimization process through the new proposed algorithm. The results from the proposed algorithm have been shown to be good and fairly in line with other established optimization procedures such as Genetic Algorithm. The obtained ratio of specific impulse to the subsystem's gross mass ratio with

IOP Conf. Series: Materials Science and Engineering **270** (2017) 012015 doi:10.1088/1757-899X/270/1/012015

constant integral of thrust to time has been discussed and better results have been obtained. Moreover, this viewpoint for optimization can also be extended to other liquid-propellant launch vehicles such as multiple gas generators and closed-cycle systems. This algorithm can be applied in preliminary and conceptual design of liquid-propellant launch vehicles to reduce the cost of experiment and time.

References

[1] Ernst R 2014 *Liquid Rocket Analysis (LiRA): Development of a Liquid Bi-Propellant Rocket Engine Design, Analysis and Optimization Tool* Master Thesis, TU Delft
[2] Lee S and Roh T 2013 *49TH AIAA/ASME/SAE/ASEE Joint Propulsion Conference*
[3] Bradford J, Charania A and Germain B S 2004 *40th AIAA/ASME/SAE/ASEE Joint Propulsion Conference and Exhibit*
[4] Montazer E, Mirzaei M, Salami E, Ward T A, Romli F I and Kazi SN 2016 *IOP Conference Series: Materials Science and Engineering* **152** 012023
[5] Ariff O K, Salami E, Ward T A and Romli F I 2015 *53rd AIAA Aerospace Sciences Meeting*
[6] Haidn O J 2008 *http://www.rto.nato.int*
[7] Larson W J, Henry G N and Humble R W 1995 *Space Propulsion Analysis and Design* McGraw-Hill
[8] Karimi H, Nassirharand A and Beheshti M 2003 *Journal of Propulsion and Power* **19** 938-44
[9] Karimi H and Nassirharand A 2006 *Aircraft Engineering and Aerospace Technology* **78** 132-7
[10] Ramesh D and Aminpoor M 2007 *43rd AIAA/ASME/SAE/ASEE Joint Propulsion Conference and Exhibit*
[11] Goldberg D 1989 *Genetic Algorithms in Optimization, Search and Machine Learning* Addison-Wesley
[12] Elbeltagi E, Hegazy T and Grierson D 2005 *Advanced Engineering Informatics* **19** 43-53

IOP Conf. Series: Materials Science and Engineering **270** (2017) 012016 doi:10.1088/1757-899X/270/1/012016

Design of an adaptive CubeSat transmitter for achieving optimum signal-to-noise ratio (SNR)

F D Jaswar[1,*], T A Rahman[1], M N Hindia[1] and Y A Ahmad[2]

[1]Department Wireless Communication Centre, Faculty of Electrical Engineering, Universiti Teknologi Malaysia, Johor, Malaysia
[2]Kulliyah of Engineering, International Islamic University, Kuala Lumpur, Malaysia.

*fitridewi268@gmail.com

Abstract. CubeSat technology has opened the opportunity to conduct space-related researches at a relatively low cost. Typical approach to maintain an affordable cubeSat mission is to use a simple communication system, which is based on UHF link with fixed-transmit power and data rate. However, CubeSat in the Low Earth Orbit (LEO) does not have relative motion with the earth rotation, resulting in variable propagation path length that affects the transmission signal. A transmitter with adaptive capability to select multiple sets of data rate and radio frequency (RF) transmit power is proposed to improve and optimise the link. This paper presents the adaptive UHF transmitter design as a solution to overcome the variability of the propagation path. The transmitter output power is adjustable from 0.5W to 2W according to the mode of operations and satellite power limitations. The transmitter is designed to have four selectable modes to achieve the optimum signal-to-noise ratio (SNR) and efficient power consumption based on the link budget analysis and satellite requirement. Three prototypes are developed and tested for space-environment conditions such as the radiation test. The Total Ionizing Dose measurements are conducted in the radiation test done at Malaysia Nuclear Agency Laboratory. The results from this test have proven that the adaptive transmitter can perform its operation with estimated more than seven months in orbit. This radiation test using gamma source with 1.5krad exposure is the first one conducted for a satellite program in Malaysia.

1. Introduction

The cubeSat platform has been used in space-related researches as technology demonstration. It is a relatively low cost option as compared to other space technology platforms like geostationary satellite that requires development cost up to more than USD300 million for the smallest single-satellite and several billion dollars for a multi-satellite global system [1]. The typical cubeSat employs a simple communication system using ultra high frequency (UHF) band with fixed transmit power and data rate to maintain the development cost at an affordable price [2]. However, a satellite that is orbiting the low earth orbit (LEO) does not have relative motion with the earth rotation. This LEO condition will create variability of distance for the transmission signal from the satellite to the ground station. The satellite communication field of view will be available when it is passing the ground station at above of 0° seen by the ground station's antenna elevation angle, which ranges from 5° to 90°. The satellite distance will be maximum when the antenna elevation angle is minimum. For a simple communication system, the transmission power is fixed at the minimum power that could ensure the link margin is available at the most extended satellite distance, which is at the minimum value of the ground station's antenna elevation angle [3]. A transmitter with adaptive capability is proposed to adapt to the variable receiving antenna elevation angle hence, improving and optimising the satellite link.

For a typical satellite mission, the link budget is analysed at the worst channel condition and at the minimum receiving antenna elevation angle to achieve the link margin. The satellite transmitter will have a fixed radio frequency (RF) transmit power and data transmission rate. The RF transmit power is set to fixed amount that could make a transmission for the longest satellite distance available, which means the Free Space Loss (FSL) is at maximum value. The satellite transmitter will maintain the RF transmit power even though at lower FSL. Based on conducted studies, RazakSAT satellite in LEO is consistently transmitting signal at antenna elevation angle that varies from 10° to 23° [4]. The received power signals are varied and not at the optimum signal-to-noise ratio (SNR).

CubeSat data rate is selected based on the amount of collected data in the mission and the average of satellite contact time. An effective data rate will depend on the SNR, which is composed of other factors such as the RF transmit power and the antenna elevation angle, which will also determine the satellite contact time. The satellite payload capabilities will be limited and scientific duty cycle will be reduced by using maximum downlink rate [5]. Satellite data rate is primarily limited by the available link power budget rather than the available radio electronics technology [6]. This finding shows that even though the maximum data rate would allow more data to be transmitted to the ground station, the link power budget will limit the cumulative data collected. A trade-off study of the transmission data budget with satellite contact duration using the satellite simulation toolkit for downloading needs to be performed. Using a transmitter with multi-mode operation, the SNR will be at the optimum level since the adaptive function will be determined by the ground station's antenna elevation angle.

There are several CubeSats carrying a payload that uses Software Defined Radio (SDR) solutions [7]. SDR is implemented due to its flexibility that allows it to be configured. For example, SDR allows receiver frequency in the payload to be tunable as demonstrated in GOMX1 developed by Gomspace [8]. Other than SDR, System On Chip (SOC) solution is also capable of providing adaptive function for transceivers. This solution is widely used in the mobile computing technology due to its highly integrated circuit and low power consumption compared to SDR [9]. In this project, the Automated Gain Control (AGC) amplifier is using the SOC solution that allows the RF transmit power and data rate to be reconfigured.

This paper presents the adaptive UHF transmitter as a solution to overcome the variability of path loss in the propagation path. The transmitter is designed to have four selectable modes of operation with different data rates and RF transmission power to achieve optimum SNR and efficient power consumption. The adaptive capability will influence the amount of data downlink to suit the satellite operation. The transmitter is developed with the capability to be integrated with other CubeSat bus sub-systems. The prototype is designed and tested to be compatible with the interface of the other sub-systems such as power modules, On Board Computer (OBC), antenna, satellite structure, electrical grounding and satellite thermal condition.

2. Methodology

The primary target of this research is to develop a CubeSat transmitter with adaptive capability. The link budget and data trade-off analyses are performed to set the transmitter multi-mode specification. The RF simulation using the Advance Device System (ADS) software is conducted to an optimised transmitter design. The circuitry and printed circuit board (PCB) designs are developed using Protel DXP. The PCB is then fabricated using an in-house manufacturing process.

2.1 Link Budget and Data Transmission Analysis

The transmitter is developed as one of the subsystems in the TiGA-U CubeSat mission [10]. In the satellite mission study, the satellite altitude is 700 km and the ground station antenna's elevation angle varies from 0° to 90°. The satellite is using an omni-directional antenna with a gain of 0dBi while the ground station uses directional tracking Yagi antenna with 15.9 dBi gain. The RF transmit power from 0.5W to 2W, with an increment of 0.5W for each step, is used in the link budget analysis. Additional

IOP Conf. Series: Materials Science and Engineering **270** (2017) 012016 doi:10.1088/1757-899X/270/1/012016

input that represents the adaptive RF transmit power is also included in the analysis to show that the optimum SNR is achievable. Satellite Concepts of Operation (CONOPS) and data transmission trade-off are used to generate the adaptive multi-mode specification. From the analysis, the input and output configuration of the adaptive controller is created.

2.2 RF System Design Simulation

The design process is started using the standard fixed transmitter architecture before using frequency synthesiser with the AGC. Then, the transmitter is set using a cascade gain analysis to determine the gain required at each stage of the system until it reaches the required transmitter output power. Using ADS software, system level design simulation has been performed to achieve the preliminary system specification. Figure 1 shows the Gain Budget Analysis simulation setup for the three stages of the amplifier: AGC amplifier, preamp and High Power Amplifier (HPA) at 2W of RF transmit power. The simulations are also repeated for 0.5W and 1W of RF transmit power.

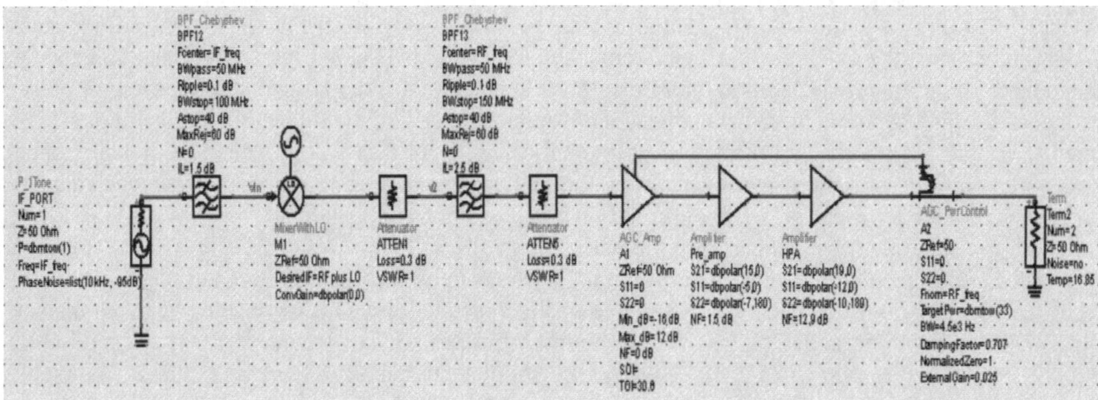

Figure 1: Transmitter system level design

From the system level analysis and circuit design simulation results, the transmitter specification is produced as tabulated in Table 1. The physical size of the transmitter is using the standard size of CubeSat module with PCB connector location that allows for stacking with other CubeSat subsystem modules. The transmitter is designed with an interface to the other subsystems such as antenna, power module and OBC as shown in Figure 2. The adaptive capability only involves the synthesiser module where the data rate and gain of the AGC amplifier can be configured. Other modules such as pre-amp and HPA are designed with a fixed configuration.

Table 1: UHF transmitter specification

Item	Specification
Frequency	430 - 440 MHz (UHF band)
Transmit Output Power	0.5 to 2 W
Data Rate	9.6kbps, 19.2kbps
Phase Noise at 10kHz Offset	<-95 dBc/Hz
Interface	UART, SMA connector
Power Supply	3.3V, 5V
Current (max)	0.6 A
Power Consumption	<5W
Dimension	91mm x 96mm x 15mm

IOP Conf. Series: Materials Science and Engineering **270** (2017) 012016 doi:10.1088/1757-899X/270/1/012016

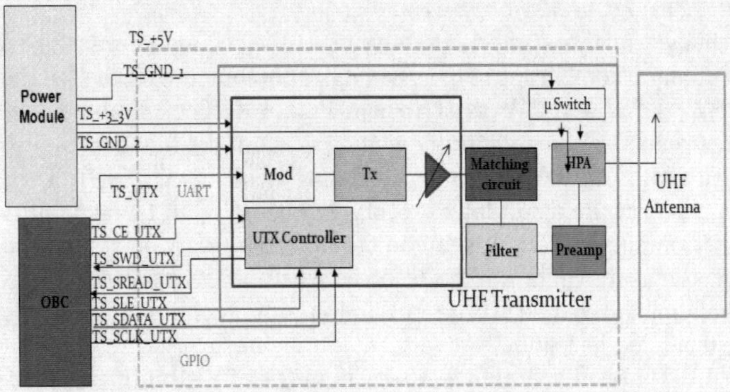

Figure 2: Block Diagram of UHF Transmitter

2.3 Prototype Development and Functional Test

From system level simulation, ideal components specification is replaced with available components in the market and the transmitter design is then optimized. The parts are then purchased and tested separately. The PCB design is fabricated using Computer Numerical Control (CNC) milling machine with two-layer copper PCB. The major sections of the upper layer are the voltage source with most of signals track. The bottom layer consists of ground signals and interconnected signals. Three prototypes are developed for the adaptive transmitter and each module is tested for functionality. The functional test measurements are performed in the lab environment to benchmark the specifications by using a spectrum analyser with counter, power measurement and phase noise measurement functions.

2.4 Space Environmental-Radiation Test

For the radiation test reference, the MicroMAS system expected mission dose is used. The CubeSat is designed for a one-year mission in LEO with a nominal orbit of 402 x 424 km and 51.6° inclination. Based on estimates from SPENVIS tool [11], a one-year mission dose is approximately 1.2 krad(Si). The Total Ionizing Dose (TID) test is conducted at the Malaysia Nuclear Agency- Gamma laboratory in Kajang, Selangor as shown in Figure 3. The high dose rate gamma source, Cobalt 60 is used in this TID test. The transmitter is exposed to the gamma source in 15-minute intervals and each exposure is at 100rad. The total exposure duration is 233 minutes and the estimated TID is 1553rad.

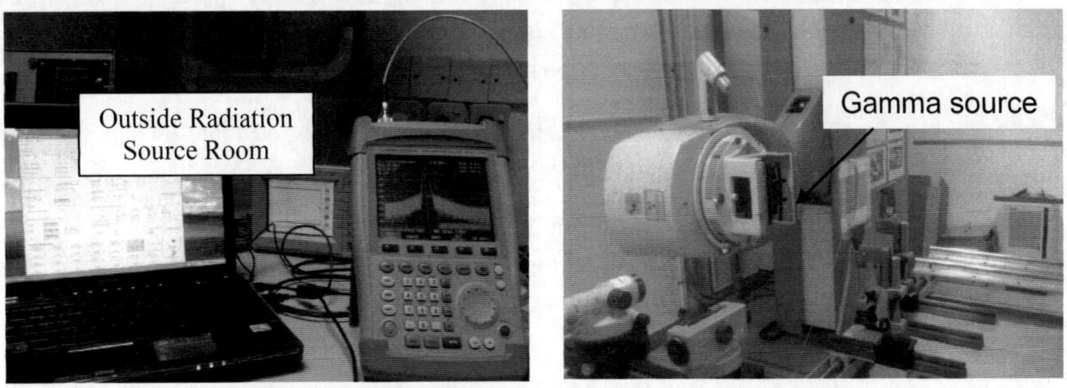

Figure 3: Radiation test setup at Malaysia Nuclear Agency

3. Results and Discussion

In this section, each of the methods implemented during development of the transmitter is discussed. Results that are collected from the link budget analysis until the environment test are described.

3.1 Link Budget and Data Transmission Analysis

Based on the link budget analysis, the SNR outputs at the receiving antenna are plotted for various elevation angles that range from 5° until 90°. The plot is combined with the RF transmission power that is fixed at 0.5W, 1W, 1.5W and 2W as shown in Figure 4. The minimum threshold to establish a transmission link between satellite and ground station is set at the required SNR based on the data rate with additional 3dB margin at the preliminary satellite design stage [12]. The SNR outputs for the adaptive RF transmit power are also plotted to highlight that, with the capability of choosing the RF transmit power, SNR outputs are optimised compared to the transmitter with fixed RF output power. From the link budget analysis, data rate trade of between 9,600bps, 19,200 bps, and 38,400 bps are recorded, and the relationship between data rate, minimum receiving antenna elevation angle and RF transmit power are tabulated in Table 2.

The satellite in LEO typically provides 10-15 minutes of visibility from the ground station. The receiving antenna elevation angle during the passing of the satellite as viewed by the ground station can be predicted and computed using a simulation program [13]. By knowing the elevation angle in advance, the transmitter can adapt to the required RF transmit power and the data rate for optimum SNR. The satellite contact duration simulation is performed using Software Toolbox Kit software. The maximum data transmission during satellite contact duration simulated by STK is calculated and listed in Table 2.

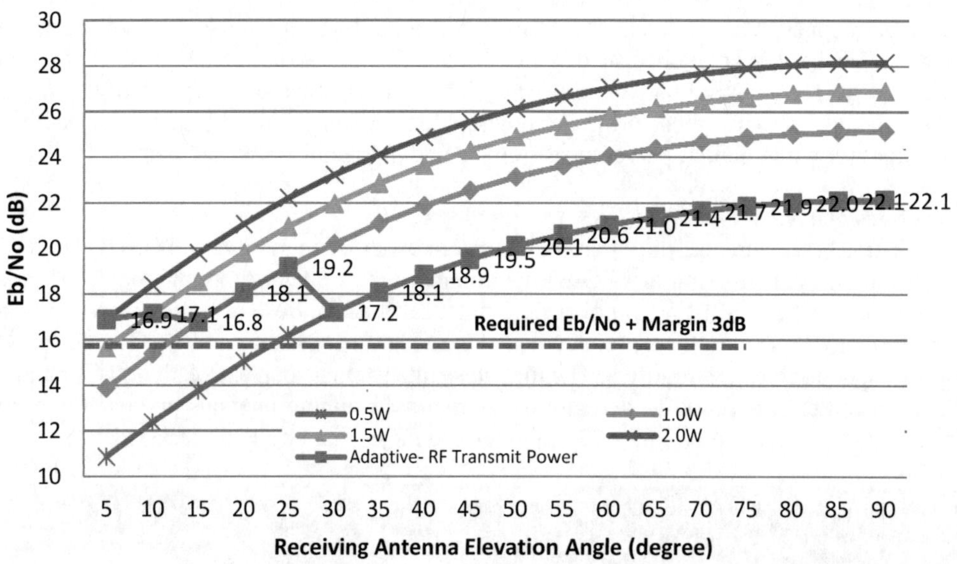

Figure 4: Link budget analysis results for 19,200bps data rate

Table 2: Relationship between data rate, minimum receiving antenna elevation angle and RF transmit power

Transmission data rate (bps)	Maximum data transmit during satellite contact duration (kilobyte)	Minimum Receiving antenna elevation angle (°)	RF Transmit power (W)
9600	469	10	0.5
9600	594	5	1
19200	933	15	1
19200	1,426	5	2
38400	3,027	25	2

From Table 2, the inputs that will determine the selection of RF transmit power and data rate are simplified as shown in Table 3. The receive antenna elevation angle predictions for each satellite path are available in OBC based on input from the Attitude Determination and Control (ADCS) subsystem. The input from power module or battery condition is also available in the OBC or direct input from the power module. The nominal power condition means that the satellite is in optimum condition and is capable of performing any satellite operation. The safe mode is determined as the satellite is in low power. Therefore, only housekeeping data can be sent to the ground station. Data transmission input is available from OBC based on the satellite operational condition (i.e. whether it is on Payload Data Downloading or only housekeeping data).

Table 3: Input and output configuration for multi-mode operation

Input Selection			**Controller Output**	
Minimum Receiving Antenna Elevation Angle (°)	**Power Module / Battery Condition**	**Data Transmission**	**RF Transmit Power (W)**	**Data Rate (bps)**
5	Nominal	Big Data	2	19,200
15	Nominal	Big Data	1	19,200
5	Nominal	Nominal	1	9,600
10	Safe Mode / Eclipse	Nominal	0.5	9,600

3.2 Simulation Results

From system level until the circuit and PCB layout design, RF parameters such as RF transmit power, phase noise, spurious and amplifier gain are monitored. These parameters are optimised before the actual circuit is designed. Figure 5 and Figure 6 show the simulation results, which are summarized in Table 4.

Meas_Name	A1	HPA	Pre_amp	A2
Cmp_S21_dB	4.419	19.000	15.000	0.000
NF_RefIn_dB	2.500	3.995	3.104	3.995
OutPGain_dB	0.256	33.927	15.589	33.927
OutPwr_dBm	-0.744	32.927	14.589	32.927
OutNPwrTotal_dBm	-171.210	-134.987	-155.162	-134.987
OutTOI_dBm	30.600	46.151	36.273	45.771

Figure 5: Cascade gain budget simulation results

freq	Mix(1)	Mix(2)	Spur_dBc	spec_dBm
0.0000 Hz	0	0	-184.981	-151.768
70.00 MHz	1	0	-152.732	-119.520
140.0 MHz	2	0	-156.093	-122.880
150.0 MHz	-3	1	-190.836	-157.624
210.0 MHz	3	0	-230.264	-197.051
220.0 MHz	-2	1	-243.071	-209.859
290.0 MHz	-1	1	-163.196	-129.984
360.0 MHz	0	1	-141.229	-108.016
430.0 MHz	1	1	0.000	33.213
500.0 MHz	2	1	-137.762	-104.549
570.0 MHz	3	1	-162.553	-129.340
580.0 MHz	-2	2	-182.797	-149.585
650.0 MHz	-1	2	-235.977	-202.764
720.0 MHz	0	2	-155.320	-122.108
790.0 MHz	1	2	-153.472	-120.259
860.0 MHz	2	2	-180.885	-147.672

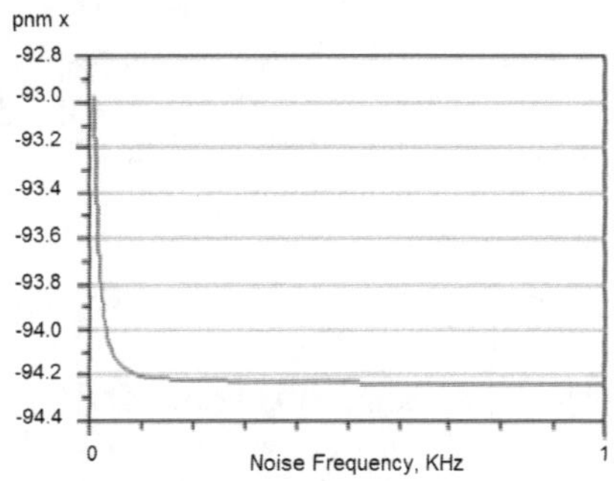

Figure 6: Spurious and phase noise simulation results

Table 4: Transmitter simulation results

Item	Specification
Frequency	430 - 440 MHz (UHF band)
Transmit Output Power	0.5 to 2 W
Data Rate	9.6kbps, 19.2kbps
Phase Noise at 10kHz Offset	-95 dBc/Hz
Spurious	-137 dBc

3.3 Prototype Development and Functional Test Results

The PCB is designed for two-layer PCB layout, which is limited by the capability of the prototyping milling machine available in-house. The PCB casing is also designed and fabricated to support thermal conductivity for HPA for long hours of transmission. Figure 7 shows the PCB layout with mechanical chasing design. After completing the PCB fabrication, electronics components are soldered and tested for their functionality. Figure 7 also depicts the Qualification Model (QM), which is produced as the third prototype. The functional test results are summarized in Table 5.

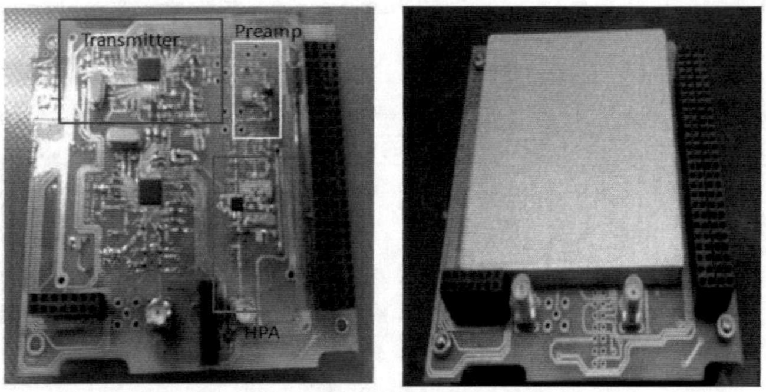

Figure 7: Qualification model (QM)

Table 5: Functional test results

Test Item	Requirement	Performance	Compliance
Frequency	437.325 MHz (UHF band)	437.325 MHz (UHF band)	Fully Comply
Transmit Output Power	0.5 to 2 W (27 to 33 dBm)	27.0 dBm 30.0 dBm 31.5 dBm 33.0 dBm	Fully Comply
Phase Noise at 10kHz offset	<-85 dBc/Hz	-95 dBc/Hz	Fully Comply

3.4 TID Test Results

The transmitter is turned ON before and after the chamber is closed to make sure that everything is in excellent condition before the gamma exposure is started. It takes about four hours to complete the 1.5krads of total gamma exposure. The transmitter is also turn ON and OFF when chamber is closed. The recorded RF power reading is +7.5dBm before deducting the losses and attenuation, with an equivalent value of 23dB. Therefore, the result is +30 dBm. This result is taken when the adaptive power control function is configured at 1W or +30dBm. The transmitter is fully functioning before and after the TID test. The ON and OFF of RF transmit power has been configured during the exposure of gamma source and does not appear to have any difficulty to transmit at the expected RF output power. The transmitter is also turned ON and reconfigured at the desired downlink frequency and manage to oscillate to the required result as shown in Table 6.

Table 6: Radiation test results

Item	Specification	Before Radiation Test (0 krad)	After Radiation Test (1.53 krad)	Verification	Compliance
Visual Inspection	-	Components and soldering pad in good condition.	Components and soldering pad in good condition.	Inspection	Fully Comply
Downlink Frequency	UHF-band, ~437.325 MHz	UHF-band, ~437.325 MHz	UHF-band, ~437.325 MHz	Test	Fully Comply
Downlink RF Transmit Power	30 dBm ± 0.5	30.1 dBm	30.1 dBm	Test	Fully Comply
Current Reading at 3V @ Transmitter SOC	26 mA	26mA	26mA	Test	Fully Comply

4. Conclusion

The UHF transmitter with adaptive multi-mode function is designed and developed in this research. This paper focuses on the design, validation methodology in the space radiation condition and the analysis of the adaptive multi-mode function. The connection between the RF transmit power, receive antenna elevation angle and data rate is investigated through the link budget analysis and CONOPS consideration. From the results, the adaptive multi-mode operation for the transmitter is designed with the capability to reconfigure the data rate and RF transmit power. The proof of concept, EM and QM models have been developed and validated. The transmitter uses industrial grade components that are fully in compliance with the RF output power and frequency stability requirement in a wide operating temperature range for the space environment. The multi-mode control function has also been tested and demonstrated to be functioning in the radiation test. The results show potential improvement in the CubeSat performance with regards to SNR and power consumption by adopting multi-mode control function.

References

[1] Evans B G 2000 *Satellite Communication Systems* The Institution of Electrical Engineers
[2] Klofas B 2013 *10th Annual CubeSat Developers' Workshop*
[3] Ekpo S and George D 2013 *IEEE System Journal* **7** 642-8
[4] Nuroddin A C M, Ismail A F, Badron K, Zulkurnain N F, Ismail M and Salim H 2013 *IEEE International Conference on Space Science and Communication (IconSpace)*
[5] Selva D and Krejci D 2012 *Acta Astronautica* **74** 50-68
[6] Bouwmeest J and Guo J 2010 *Acta Astronautica* **67** 854–62
[7] Alvarez J L, Rice M, Samson J R and Koets M A 2016 *IEEE Aerospace*
[8] Lars K A, Kaas K and Bisgard M 2014 *28th Annual AIAA/USU Conferences on Small Satellite*
[9] Vladimirova T and Bridges C 2008 *NASA/ESA Conference Adaptive Hardware System*
[10] Jaswar F D 2014 *InnoSAT2: TiGA-U Telecommunication Subsystem* ATSB
[11] Wertz J R 2006 *Space Mission Analysis and Design* Microcosm Inc.
[12] Kingsbury R, Schmidt F, Cahoy K and Sklair D 2013 *IEEE Radiation Effects Data Workshop*

IOP Conf. Series: Materials Science and Engineering **270** (2017) 012017 doi:10.1088/1757-899X/270/1/012017

Design and testing of shape memory alloy actuation mechanism for flapping wing micro unmanned aerial vehicles

N F Kamaruzaman and E J Abdullah*

Department of Aerospace Engineering, Faculty of Engineering, Universiti Putra Malaysia, 43400 Serdang, Selangor, Malaysia.

*ermira@upm.edu.my

Abstract. Shape memory alloy (SMA) actuator offers great solution for aerospace applications with low weight being its most attractive feature. A SMA actuation mechanism for the flapping micro unmanned aerial vehicle (MAV) is proposed in this study, where SMA material is the primary system that provides the flapping motion to the wings. Based on several established design criteria, a design prototype has been fabricated to validate the design. As a proof of concept, an experiment is performed using an electrical circuit to power the SMA actuator to evaluate the flapping angle. During testing, several problems have been observed and their solutions for future development are proposed. Based on the experiment, the average recorded flapping wing angle is 14.33° for upward deflection and 12.12° for downward deflection. This meets the required design criteria and objective set forth for this design. The results prove the feasibility of employing SMA actuators in flapping wing MAV.

1. Introduction

Unmanned aerial vehicles (UAVs) can be defined as a "device used or intended to be used for flight in the air that has no on-board pilot" [1]. Current generations of UAVs "can be as small as an insect or as large as a charter flight"[2]. They are typically launched from a road or a small vehicle, and are large enough to house cameras, sensors or other information-gathering equipment. The recent researches on UAVs are looking into the design and control of the tilt-rotor UAVs, which is a hybrid aircraft that can perform high-speed flights like fixed-wing aircraft and can also operate in hover conditions on top of performing vertical take-off and landing (VTOL) like helicopters and quadrotors. Most of them have been equipped with fixed or rotary wings, and switch between helicopter and fixed-wing aircraft flight modes by tilting their rotors. UAVs have the advantages over manned vehicles since the functions and operations can be implemented at a much lower cost, faster and safer.

SMA wire actuator is a stepping stone towards further development of flapping wing technology. SMA materials are metallic alloys that undergo martensitic transformation when they are subjected to applied thermo-mechanical loads. SMA can be activated by external heating (i.e. ambient temperature) or electrical heating (i.e. Joule heating with electrical current). SMAs are very attractive for actuation purposes because of their large excitation forces and displacements. With their characteristics of high power-to-weight ratio, large recovery strain and low driving voltage, SMA actuators have been widely used in a variety of applications including aircraft wing control, robotic gripper, active endoscope and automotive mirror actuator.

Micro Air Vehicle (MAV) is a miniature aircraft with maximum wing span of 15cm (i.e. about 6 inches), as defined by the Defense Advanced Research Programs Agency (DARPA) [3]. This physical size limitation puts MAVs at least an order of magnitude smaller than any operational UAVs. MAVs that are developed by the University of Florida, in particular, vary in wingspan from 4 to 6 inches and can carry a payload up to 15g [4]. There are three main categories of MAVs: fixed wing, rotary wing and flapping wing. The use of MAVs for military operations has also become more common today. A new trend in the MAV community is to take inspirations from flying insects or birds to achieve better flying capabilities [5]. For instance, a 50g mechanical insect known as Entomopter is designed using Reciprocating Chemical Muscle to actuate four wings for its flapping flight. The wings are created by using stereolithography and fused deposition modelling techniques, and produce a flapping frequency of 70Hz [6]. The study of flexible flapping-wing Micro Air Vehicles (FMAVs) has fascinated many researchers because of their aerodynamic benefits, particularly in low Reynolds number and small size [7]. Owing to their advantages of small size, low velocity, high agility and manoeuvrability, micro or small ornithopters are being potentially considered for applications in rescue missions, remote sensing, and spy and reconnaissance operations in small or closed media. The flapping flyers, inspired from the biological flappers, have a flexible wing structure. They use the root cyclic driving input (i.e. cranking moment) and a passive bending and twisting motion due to flexibility in both span-wise and chord-wise directions. The flapping of the flexible wings is characterized by the simultaneous generation of lift and thrust forces. Moreover, characteristics of loads acting on them, including aerodynamic, inertia and internal elastic loads, can vary substantially over one flapping cycle [8]. Therefore, understanding the aerodynamic, structural and flight dynamic attributes of flapping is necessary to design the high-performance and robust FMAVs [9]. There are few variables to consider when conducting researches regarding tandem wing configuration such as gap distance between the paired wings, phase angle of each wing, aerodynamic performance [10], and endurance of the wings subjected to harsh conditions. To produce high performance membrane wing structures, aerodynamic-featured wing profile needs to be inherent in the design of the membrane wing

An effective mechanical flapping system is important to produce the sufficient lift and thrust forces required for flying a flapping wing MAV. When concerning a smaller size ornithopter [11] or insects with ornithopter-like flapping wing kinematics like butterfly, a smaller and more compact mechanical flapping system configuration than a four-bar crank rocker mechanism is required. Therefore, a slider crank is introduced [12]. The extensive usage of piezoelectric materials is limitless due to their robust deformation. A flexible piezoelectric actuator is mounted underneath the wings of an ornithopter-type MAV to gain camber control during flapping flight and when an electrical field is fed, the actuator will deform and control the camber of the wings with its deformation [13].

2. Shape Memory Alloy

SMAs are a unique group of materials that have the capability to recover from large deformation well beyond the normal elastic strain limit of metals. This behaviour is associated with the thermoelastic martensitic phase transformation from the parent phase austenite (A) to the phasemartensite (M). This property renders the materials the capability to produce mechanical work output without requiring any complex mechanical design. Furthermore, the deformation recovery might occur spontaneously upon the unloading of the stress that causes the deformation (referred to as the pseudoelasticity) or delayed from the time of deformation until the material is heated at a later stage at a different location (referred to as the shape memory effect). Owing to these special features, SMAs have been used in a wide range of innovative designs and smart structures like actuator sensors, medical devices, miniature devices, damping absorbers and fibre reinforced composite structures [14]. SMAs have also been proposed in robotics as an alternative to the conventional electric, hydraulic and pneumatic motors because of the increasing need to develop smaller and lighter robots. Indeed, one of the advantages of this technology is the lighter actuator with bigger power/weight ratio, therefore making SMAs a valid alternative to conventional actuators [15].

SMA has aroused continuous attention mainly because of its two distinctive characteristics: shape memory effect (SME) and pseudoelasticity (SE). Shape memory effect refers to the behaviour of alloy that undergoes a phase transition under the influence of temperature, pressure, mechanical effect and other factors. Self-adapting wing structure of shape memory alloy refers to the base material, sensing element, driving element and electronic control system that are all integrated together. According to external changing environment and their own situation, they can adapt based on self-diagnosis of the structure adjustment. If the aircraft wings have self-adapting deformation ability, they can improve the performance of the system to a great extent, especially in improving the radius of gyration, increasing endurance, increasing the load and improving the obvious advantages of maximum speed [16]. Several research has been carried out in related fields such as structural design of stealth fighter using SMA to drive the folding wing. Recently, Chengdu aircraft group has also introduced manufacturing process of memory alloy technology [17].

It can be said that a big potential can be seen for shape memory actuator in aerospace industry but its application has yet to be widely spread presently. Boeing has also embarked in a project to employ the SMA technology for morphing aircraft with component and system level optimization at multiple flight conditions [18]. A particular attention is paid to flapping wing using this SMA since they are the least developed so far and their characteristics are not widely known. It can be seen that SMA actuator can replace the motor and significantly reduce the system's weight. Due to limited previous study on the feasibility of employing SMA on flapping wing, this work is carried out to explore its potential by designing a flapping wing MAV actuating mechanism using SMA actuator "muscles" to actuate the wing.

3. Methodology

The aim of the experiments is to understand the behaviour of the selected two different types of SMA samples. The result is then used to select and determine the suitable spring type or wire type of SMA to be applied in the proposed flapping wings, and to identify the design parameters. The two types of SMA sample (i.e. spring and wire) as shown in Table 1 are used as the guidelines for comparison with the properties from the manufacturer. Behaviour of the SMA with regards to voltage activation power supply, time-to-contract and few other parameters are recorded for further analysis. Figure 1 shows the experiment setup.

Table 1: Material properties of SMA samples

Type	Spring	Wire
Material	Nickel Titanium (NiTi)	Nickel Titanium (NiTi)
Resistance	Not stated	12.2 Ohm/meter
Length	2 cm	35 cm
Maximum Deformation	15 cm	Not stated
Activation Temperature	45-50 °C	70-90 °C
Number of Windings (Spring)	21	N/A
Outer Diameter (Wire)	0.75 mm	0.31 mm
Outer Diameter (Winding)	6.5 mm	N/A

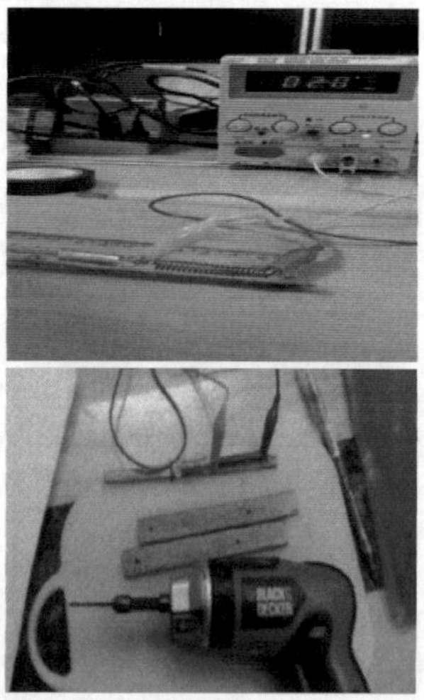

Figure 1: Experiment setup layout

The results of the experiment are tabulated in Table 2, which indicate that the SMA wire began to activate and contracted when 1.5V and 0.19A current power is supplied. The maximum recorded SMA

IOP Conf. Series: Materials Science and Engineering **270** (2017) 012017 doi:10.1088/1757-899X/270/1/012017

wire contraction recorded is 2cm at supplied power of 3.5V and average 0.51A. Further voltage and current increment is found to give no additional contraction to the SMA wire. On the other hand, the SMA spring is fully activated and contracted as much as 3.7cm at supplied power of 7V and average 2.23A. The SMA spring sample began to activate and contracted when 2V and 0.28A current power is supplied. Based on these results, wire type SMA has given the fastest contraction as compared to the spring type SMA. For the cooling process, wire type SMA has also given the fastest time to cool as compared to the spring type SMA. It is decided to convert the SMA wire into a spring and it has been performing better than the manufactured SMA spring.

The schematic diagram is shown in Figure 2 illustrates the use of four SMA wire-spring to provide the force required to flap the wing. When the power supply connected to the circuit on top of the wing is turned ON, the wire reacts and the wing flaps upwards. When the top circuit is switched OFF while the bottom circuit is switched ON, the wing flaps downwards. From the conducted experiments, it is observed that the SMA wire-spring has taken a short time to cool. This can be contributed to the faster subsequent movements.

The fabrication of the flapping wing is depicted in Figure 3. The body of the wing is made from bamboo while the wing is made from aluminium plate. The prototype is tested using DC power supply to verify the functionality of the flapping mechanism. Figure 4 shows the setup for the testing. The experiment is conducted continuously with increasing power using the same sample. The displacement of wing tip is measured manually in order to determine the angle of deflection.

Table 2: Comparison between spring and wire SMA

SMA Type	Activation Voltage (Volts)	Maximum Compression (cm)
Spring	7.0	3.7
Wire	3.5	2.0

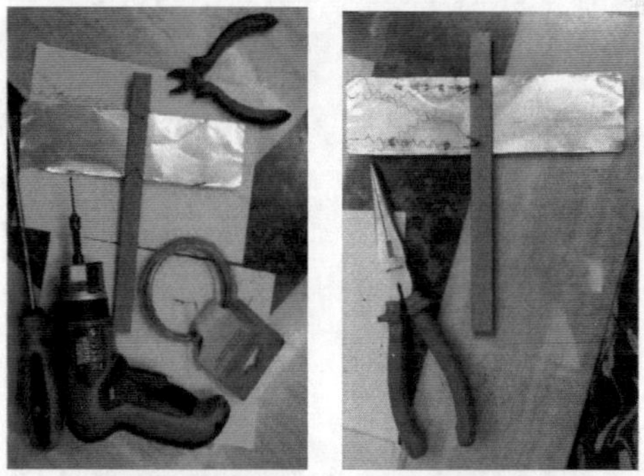

Figure 3: Fabrication of prototype

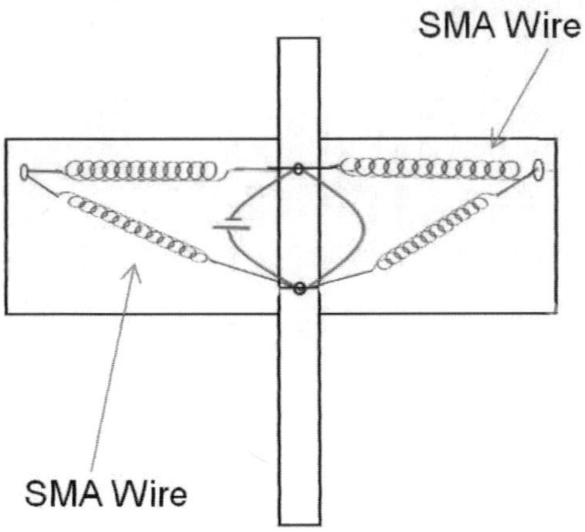

Figure 2: Schematic diagram

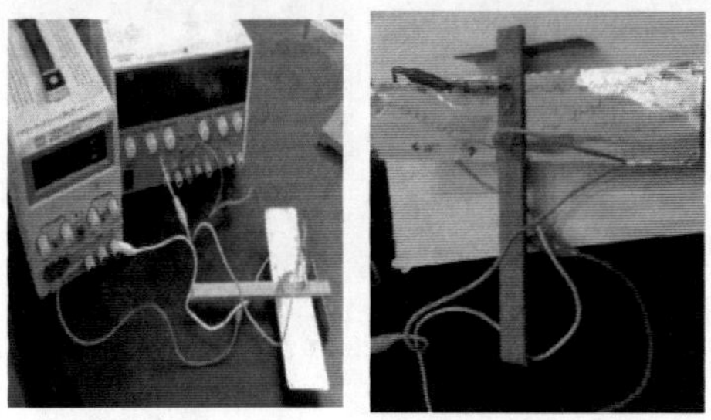

Figure 4: Lab testing for prototype

Electrical activation is calculated using Equation 1 to Equation 3 for the resistance, voltage and the power, respectively, where ρ is the SMA resistivity, L is the free length of element and A is the cross sectional area.

$$R = \frac{\rho . L}{A} \quad (1)$$

$$V = IR \quad (2)$$

$$W = I.V = I^2.R \quad (3)$$

IOP Conf. Series: Materials Science and Engineering **270** (2017) 012017 doi:10.1088/1757-899X/270/1/012017

The activation requirements for the wire and spring type SMA are used in the design selection. The design of the actuator has been determined by investigating the selected materials of both types using experimental analysis result. The angle of deflection can be calculated by Equation 4.

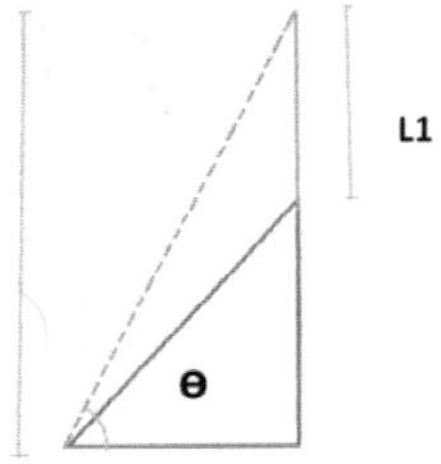

$$\tan^{-1}\theta = \frac{L2-L1}{1/2\ wing\ span} \tag{4}$$

4. Results and Discussion

The final design chosen for this study is as shown in Figure 5 and Figure 6. The engineering drawing is done using CATIA V5 software. During the assembly design, the operation of the mechanism can be tested to verify that it functions as expected. From the results of SMA behaviour experiments, the prototype testing is conducted to see the movement of the wings using wire-spring type SMA. The data obtained is tabulated in Table 3 when the wing is flapped upwards and Table 4 when it is flapped downwards. From the data, the flapping angle is observed to generally increase as the supplied power is increased.

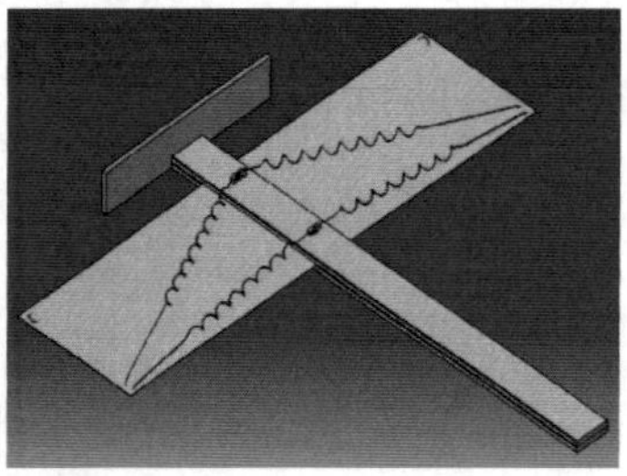

Figure 5: CATIA drawing of selected design

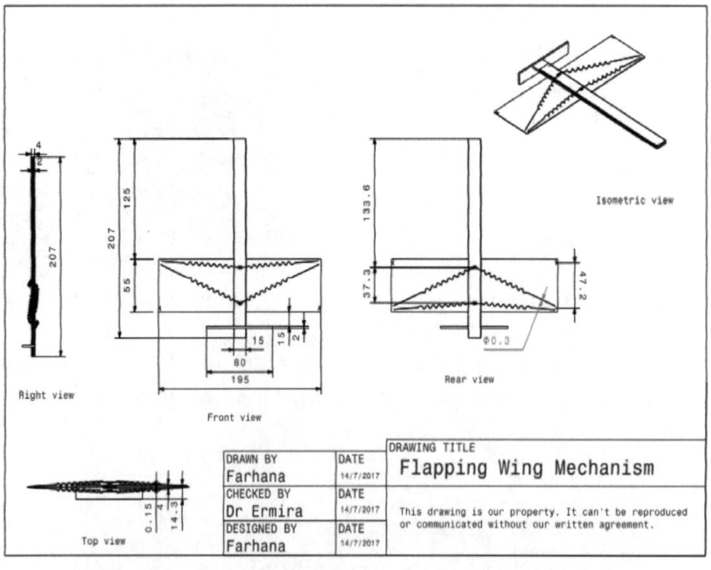

Figure 6: Detail description of the selected design

Table 3: Data obtained with power (upward flapping)

Power (W)	Deflection (deg °)
0	0
0.048	0
0.195	0.212
0.473	0.424
0.853	0.637
1.363	0.849
1.940	1.379
2.696	1.803
3.637	2.650
4.699	3.812
5.777	6.336
7.326	8.623
8.740	14.334

Table 4: Data obtained with power (downward flapping)

Power (W)	Deflection (deg °)
0	0
0.041	0
0.177	0.424
0.415	0.743
0.754	1.061
1.223	1.697
1.780	2.438
2.472	3.391
3.319	5.074
4.223	6.754
5.221	8.010
7.326	9.871
8.740	12.122

The rate of change of the flapping angle is nonlinear as illustrated in Figure 7 and Figure 8. The flapping of the wing becomes bigger and the flapping angle rises faster when the power is higher. It is also observed that the upward flapping produces a bigger deflection in comparison to the downward flapping angle of the wing. This is believed to be due to prototype fabrication that affects the flapping angle.

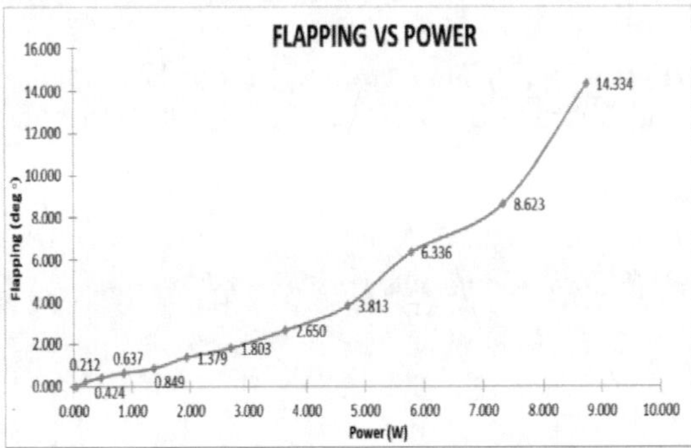

Figure 7: Power versus flapping (upwards)

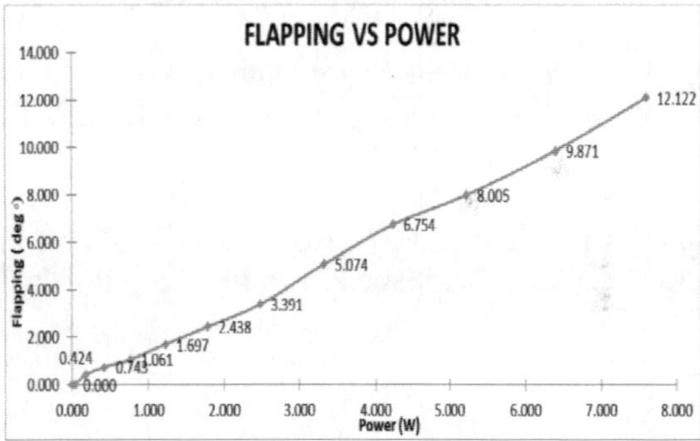

Figure 8: Power versus flapping (downwards)

5. Conclusions

This paper presents the design of smart actuator involving two types of SMA samples. The concept is then expanded to identify the best actuator concept to be imposed in this design, as the SMA spring and SMA wire chosen to be the main element. SMA wires are able to contract upon electrical heating and they have been used as artificial biceps that allow the wing to contract. From the results, it can be concluded that the design of actuator replacement and its integration into the flapping wing MAV is successful. Heat is applied to the wire using an input electric current. The cooling of the system is through convection with the surrounding area. In order to improve the proposed design in this paper, further studies have to be carried out like exploration of cooling techniques and investigation of wing morphing by means of SMA. With regards to the former, development of a viable means to increase the cooling rate of a SMA wire would prove advantageous with respect to the attainable stroke length and range of motion at increasing speed. The design requirement for designing the flapping wing such as amplitude and expected wing displacement also needs to be defined.

Acknowledgement
The authors would like to acknowledge that the research findings presented in this paper is funded by Research University Grant Scheme (RUGS): UPM/700-2/1/GPB/2017/9530800 from Universiti Putra Malaysia.

References
[1] Bolkcom C 2004 *Homeland Security: Unmanned Aerial Vehicles and Border Surveillance* CRS Report for Congress
[2] McCormack E D 2008 *Use of Small Unmanned Aircraft by the Washington State Department of Transportation* Washington State Transportation Center
[3] Ifju P G, Ettinger S, Jenkins D A and Martinez L 2001 *SAMPLE Journal* **37** 7-13
[4] Abdulrahim M, Garcia H and Lind R 2005 *Journal of Aircraft* **42** 131-7
[5] Markose S, Patange S, Raja S and Salice P 2014 *International Journal of Advanced Research in Electrical, Electronics and Instrumentation Engineering* **3** 444-51
[6] Michelson R C and Naqvi M A 2003 *RTO-AVT von Karman Institute for Fluid Dynamics Lecture Series*
[7] Shyy W, Aono H, Chimakurthi S K, Trizila P, Kang C K, Cesnik C E S and Liu H 2010 *Prog Aerosp Sci* **46** 284–327
[8] Lewin G C and Hajhariri H 2003 *J. Fluid Mech* **492** 339–62
[9] Sane S P and Dickinson M H 2002 *J. Exp.* **205** 1087–96
[10] Broering T and Lian Y 2012 *Acta Mech Sin* **28** 1557–71
[11] Lee J S, Kim J K, Han J H and Ellington C P 2012 *J Bion Eng* **9** 18–28
[12] Sun M and Xiong Y 2005 *J Exp Biol* **208** 447–59
[13] Lee J S, Han J H and Kim D K 2011 *8th International Conference on Ubiquitous Robots and Ambient Intelligence*
[14] Bashir S, Meng S Q, Mahmud A S and Wu Z 2017 *Journal of Material and Design* **124** 225-37
[15] Nespoli A, Besseghini S, Pittaccio S, Villa E and Viscuso S 2010 *Sensors and Actuators A: Physical* **158** 149–60
[16] Quan D and Hai X 2015 *Procedia Engineering* **99** 1241-46
[17] Chinese Patent No. 201080034831
[18] Calkins F T, Mabe J H and Ruggeri R T 2008 *ASME 2008 Conference on Smart Materials*

IOP Conf. Series: Materials Science and Engineering **270** (2017) 012018 doi:10.1088/1757-899X/270/1/012018

Concept design of a disaster response unmanned aerial vehicle for India

Y Vashi, U Jai, R Atluri, M Sunjii, Y Kashyap, V Ashok, S Khilari, K Jain and S Aravind Raj*

School of Mechanical Engineering, VIT University, Vellore – 632 014, Tamil Nadu, India

*aravindsakthivel@hotmail.com

Abstract. The Indian sub-continent experiences frequent flooding, earthquakes and landslides. During the times of peril, live surveillance of the disaster zone facilitates the disaster agencies in locating and aiding the affected people. For this reason, development of a micro unmanned aerial vehicle (UAV) can be an optimal solution. This article provides a conceptualization of a UAV model that meets the need of the country. A comparison of different aircraft components and their optimization and sensitivity analyses are presented. In the end, this research produces a preliminary design of UAV that can accomplish surveillance and payload dropping missions in disaster affected areas.

1. Introduction

New breakthroughs in the tele-networking and signalling technologies have propelled the research and development of unmanned aerial vehicles (UAVs) throughout the globe. Although previously they are specifically for military purposes, the UAVs have been accepted by many civilian agencies within the developed as well as the developing nations for various applications. From peace time surveillance to offshore oilrig patrols, these UAVs can perform many functions that have been previously done using a full size aircraft at a fraction of the operating cost. India, with the limited regional aviation facilities, has been a scene of many natural fallouts and this situation has adversely affected the economy of the involved regions. Due to climatic changes, the country has faced scores of erratic rainfalls in previous few years. The recent South Asian floods in 2017 have killed 1300 people and affected approximately 41 million people [1]. Similarly, past floods in India especially in the states of Tamil Nadu and Bihar have seen hundreds of fatalities and displacement of the entire districts in most cases. The high rate of fatalities is directly related to the delayed response by the authorities due to shortfall of rescue assets.

With no specialised UAV at their disposal, rescue authorities have to rely on the national Air Force assets for conducting search and rescue missions, which have to be relocated from the frontline bases to the affected areas. The UAVs are used as the command and control centres to direct rescue units. To mitigate the loss of human lives and damage to the property, this study aims to offer UAV prototype that can perform the surveillance and payload dropping missions while flying close to the ground. The cause of such endeavour is to reduce losses in disaster hit areas through quick and easy surveillance by a medium sized UAV that is capable of being deployed from any short strips or roads at minimal time delay. Table 1 lists the main characteristics of a surveillance UAV according to the military standard [2]. Though the requirements of battlefield do not directly correspond to that of disaster management, the strict limitations sanctioned by the army will facilitate in streamlining the mission statement and the meticulous design and execution.

Table 1: Characteristics of surveillance UAV

• Performance of efficient surveillance • Cost-effectiveness • Reliability • Usefulness and operational availability	• Maintainability • Day and night operations • Operating in a wide range of weather conditions • Beyond Line-of-Sight (BLOS) operation	• Real time operation • Multi-mission capability • Minimum turnaround time

2. UAV Conceptual Design Phase

The mission statement for this design and development study states the requirement of a medium sized UAV with short field capabilities and a loiter time of minimum 30 minutes with full electronics and payload setup. The typical aircraft design process often includes four major phases: conceptual design, preliminary design, detail design, and test and evaluation [3]. As the name implies, aircraft conceptual design phase is when the aircraft is still at its concept level. At this stage, general design requirements are entered into the process to generate a satisfactory configuration. The summary of activities within this conceptual design phase is shown in Figure 1.

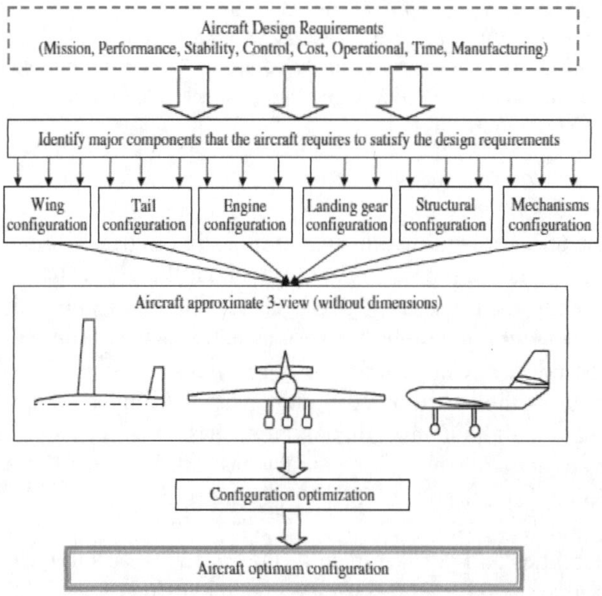

Figure 1: UAV conceptual design (reproduced from [3])

Meanwhile, Figure 2 depicts the decision making flow process that has been followed in designing the specified UAV in this study. The process involves step-by-step evaluation of the variables. During design parameter calculations such as for wing design, tail design, landing gear height, thrust required, etc., several iterations are performed with the help of C++ codes in order to eliminate the possibility of errors. The different sets of data obtained are individually evaluated with the help of aerodynamics software like XFLR, AAA and HyperWorks. Once the best set of values has been obtained, the overall aircraft parameters and planform is tested in XFLR and changes are made if necessary. Following the completion of the aircraft design concept, detailed part design of the UAV components is performed on Solid Works while the part stress analysis is performed using ANSYS. Once the required strength and integration parameters are met, the fabrication work is performed. The fabricated UAV undergoes a thorough dimensional and structural scrutiny. Once the airworthiness of the UAV is found to comply with all safety requirements, flight tests are performed and the obtained results are compared with the calculated ones.

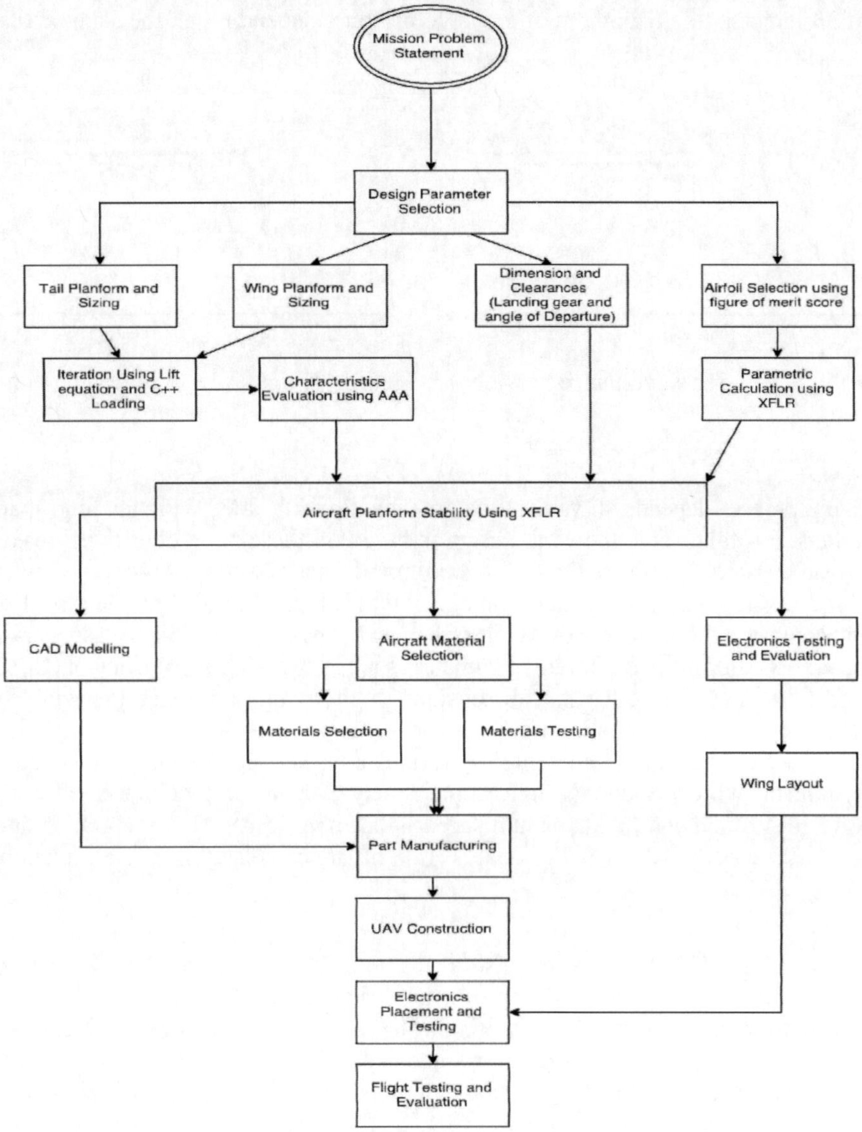

Figure 2: Methodology flowchart

The mission of the UAV is to perform surveillance and drop humanitarian packages during disaster relief. Both mission statements are represented as simple diagrams in Figure 3 and Figure 4. The static and dynamic payloads of the UAV can be varied depending on the user's requirements or the mission profile. The static payload of the UAV is generally the electronics and sensors that are carried onboard depending on the mission requirement. The sensor package may vary from a day time camera for day time surveillance to a night vision or thermal camera for night operations. The sensors may also vary from air density sensors to the sensors that can detect pollutant in the atmosphere at different altitudes. Dynamic payload of the UAV varies from medicine, food packages, location beacons for humanitarian and rescue missions to foliage density and seismic sensors for geological missions. The calculated payload capacity of the aircraft is 24lbs, which may be optimized depending on the mission statement. For a maximum payload delivery mission, the UAV carries minimum number of batteries and more amount of dynamic payloads, whereas for a maximum endurance mission, the payload is consisted of batteries only to maximize the time on station. Therefore, the endurance of the UAV may vary from a minimum of 30 minutes to a maximum of a couple of hours depending on the setup and atmospheric conditions.

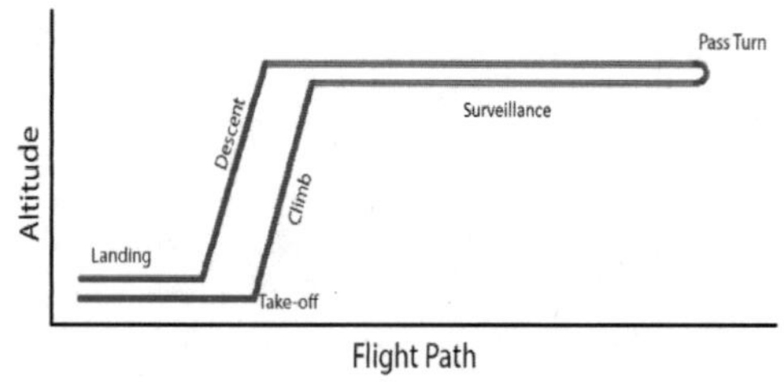

Figure 3: Flight path for surveillance mission

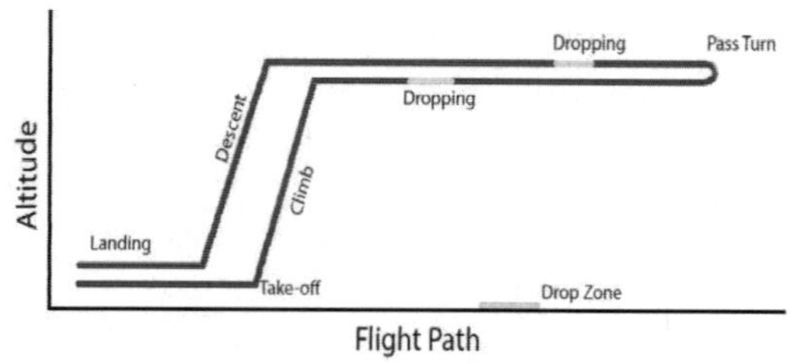

Figure 4: Flight path for dropping humanitarian aid

2.1. Wing Design

The wing sizing process depends on various factors including lift, drag, wing loading, span efficiency factor and others. For this study, the wing design is based on these main objectives: maximize lift to increase payload capacity, minimize drag for maximum efficiency and the UAV velocity must remain between 11-18 m/s for accurate payload dropping. Wing area, aspect ratio and the airfoil shape play a crucial role in achieving these objectives. Wing area must be maximized while the aspect ratio (AR) has to be optimized. AR determines the several parameters such as overall drag coefficient (C_D) of the wing, overall lift coefficient (C_L) of the wing, dimensions of the wing and hence the wing loading, and Oswald's efficiency of the wing profile (e).

Wing sizing is done based on optimizing the variables in the lift equation. The cruise condition is used to determine the wing area and its angle of incidence. Cruising altitude of the UAV is assumed to be 200ft above the ground and the maximum speed is taken as 15m/s. C_L is 1.14 at 7° and using this information, cruise velocity–12m/s and wing area–1.28 m^2 are calculated from the Equation 1 [4].

$$L = \frac{1}{2}\rho v^2 C_l \tag{1}$$

The AR is selected as 7.33 based on the historical data of UAVs with similar mission statements [5, 6]. A higher AR will decrease drag and improve the efficiency of the wing planform but it can also compromise the structural integrity. An iterative C++ code is used to determine the optimal AR based on C_D, e and cruise velocity at fixed lift (total payload is determined to be between 35-45 lbs), which is tabulated in Table 2. MAC is fixed at 16.5 inches. Here, the wing loading factor is not considered since the wing loading in any of the cases is very low (3-5 lbs/ft^2) and does not contribute significantly to the feasibility of the design.

Table 2: Wing sizing optimization

Wing Area (m^2)	**AR**	**Velocity (m/s)**	**Oswald's Efficiency Factor (e)**	**Drag Coefficient (C_D)**
1.20	6.83	14.60	0.870	1.112
1.24	7.06	14.35	0.880	1.076
1.28	7.33	14.12	0.886	1.018
1.32	7.51	13.91	0.881	0.998
1.36	7.74	13.70	0.860	0.954

IOP Conf. Series: Materials Science and Engineering **270** (2017) 012018 doi:10.1088/1757-899X/270/1/012018

Based on Table 2, the optimum configuration appears to be at AR = 7.33, which corresponds to the highest e = 0.886 and C_D = 1.018. The final dimensions of the wing are: wing area = 13.78ft^2, AR = 7.33, wing span = 121 inches and chord length = 1.375ft. This configuration has been tested in AVL software and the results are verified.

Wingtips are usually incorporated to change the spanwise lift distribution of the wing. In addition, it is found that wingtips also tend to reduce the induced drag produced by the wing due to the increased effective wing span that pushes the tip vortex away from the wingtip. The increase in effective wing span also means an increase in the aspect ratio of the wing, which is preferred as it offers a better lift-to-drag ratio. Often times, aspect ratio is limited by structural constraints. Therefore, wingtips provide a cost effective opportunity without necessarily straining the structural integrity of the airframe. For the proposed UAV design, winglets and Hoerner wingtips are considered. Winglets are small extensions of the wings that are usually set at an angle with respect to the surface of the wing. They can effectively reduce the magnitude of the wingtip vortices produced by the wing. However, although a well-designed winglet can reduce the induced drag, it also leads to an increase in skin friction drag at lower speeds (which covers the large proportion of the flight envelope for the UAV). This is because of its additional surface area and its form factor in general. On the other hand, Hoerner wingtips as depicted in Figure 5, do not increase the surface area significantly and therefore, do not result in the increase of skin friction drag. Moreover, the use of Hoerner wingtips result in a better span-wise lift distribution that alleviates the stress at the far ends of the wing. Based on this discussion, the Hoerner wingtips are selected and incorporated into the wings.

An important part for the wings is the airfoil shape. The airfoil selection is a crucial process, which determines the overall versatility of the UAV. Bearing in mind the subsonic flow regime around both sides of the wing and the relatively modest chord sizes, only airfoils that perform well in low Reynolds conditions are considered [8]. The data for several low Reynolds number airfoils are extracted from the open source website and compiled. By holding a minimum bar value on the lift curve slopes, the list of airfoils is sifted down to only five airfoils: CH10, S1223, FX63, GOE225 and NACA 4412 as tabulated in Table 3. The open source Vortex Lattice Code, XFLR5 [9] is applied to obtain the C_L and C_D graphs for an infinite span wing in a subsonic, incompressible flow regime. A figure of merit graph is plotted by weighing the factors of C_L, C_D and stall characteristics as shown in Figure 6. The weighing factors is decided by considering the goals aligned with the design objectives. The 2D characteristic values of the airfoils are then input into a VLM code, Tornado to obtain the ideal C_L and C_D values for a wing with a finite span and chord. The conversion process 2D to 3D values relies on the airfoil's lift curve slope and the wing planform [10]. In this case, the latter is assumed to be rectangular to maintain the consistency of the analysis. By analyzing the results, airfoil S1223 trumps the other airfoils by providing better lift characteristics. Besides providing better lift, its stall characteristics are more favourable during flight at the edge of flight envelope.

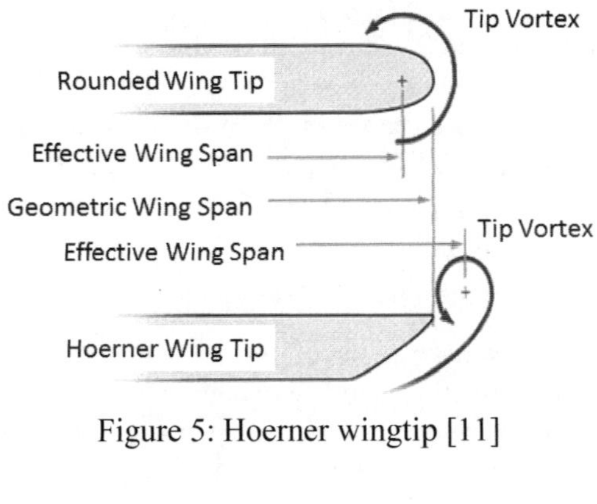

Figure 5: Hoerner wingtip [11]

Table 3: Stall angle and maximum lift coefficient for considered airfoils

Airfoil	**Stall Angle (Degrees)**	C_{Lmax}
S1223	11	2.15
CH-10	12	1.58
GOE425	10	1.36
NACA 4412	11.2	1.41
FX63-137	9	1.63

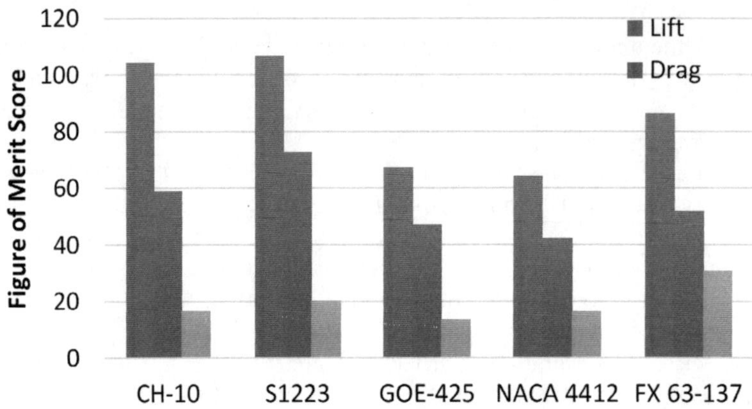

Figure 6: Airfoil comparison

2.2. Propulsion System Design

An aircraft requires a propulsion system to retain a sustained flight. The contribution of a power plant is to generate the propulsive force or thrust. An engine functional analysis is performed to select the right engine and determine the most suitable engine parameters. Apart from the selection, the arrangement of propulsion system also plays an important role, which can be arranged as tractor, pusher, double tractor or push-pull configuration. In the tractor type engine, engine and propeller are placed at the nose of the aircraft. It maintains the stability of the aircraft and reduces the weight of overall system. On the other hand, pusher type power plant offers safe handling characteristics, greater forward area for electronics, better field of vision for the cameras and also reduced fuselage wetted area. However, if the pusher configuration is to be used, engine cooling and propeller stall are the major factors of concern. On top of that, the pusher arrangement almost always operates in the turbulent wake of the engine and airframe that are ahead of it, which is one of the reasons its bigger noise than tractor configurations. In the push-pull type motor, propellers are placed individually on the nose and tail of the aircraft. This increases the weight of aircraft. After considering these factors, the decision of using a single tractor configuration is made.

The Jett SJ-46, a 0.46 cubic inch, two strokes, single cylinder, glow fuel engine, has been selected for the UAV. The engine is affected by the changes in climatic conditions and is tuned for the weather conditions of India [12]. Since this study is conducted in Tamil Nadu, the engine is tuned to have peak performance in regional conditions (i.e. average temperature of 32°C and humidity about 60-90% [13]. Based on the weather data obtained from Regional Meteorological Centre (RMC), Chennai, the engine is tested on multiple propellers and fuel mixtures to outperform the weather. Numerous iterations are carried out using different blends of ethanol, castor and synthetic oil blends and nitro methane [14]. A comparative study on engine temperature versus thrust generated using different fuel blends is done to understand the engine performance characteristics. For the design of the UAV, a static thrust of 7.8lbs is required for it to maintain level flight in ideal conditions and 9.86lbs of thrust in extreme weather conditions. The engine is further fine-tuned using a tuning pipe to generate additional 4lbs of thrust, considering relatively high FOS for superlative conditions. Hence, based on the mission requirements, high/low pitch propellers with different fuel blends can be used with pre-defined tuning pipe lengths to maximize the performance. An easy and user-friendly tuning-pipe length adjustment mechanism has also been made such that even a layman can operate it if required.

2.3. Tail Design

The overall stability of the UAV is decided by the nature of balancing forces or moments acting on it. The tail of the UAV plays a critical role in providing it with positive stability to counter and control the moment created at its wing. Adequately sized tail section provides positive handling characteristics such as pitching and yawing, and plays an important part in the recovery of the UAV. The tail design of the

UAV is based around several factors such as stability, aesthetic, structural integrity and weight. The basic function of the tail is to balance the forces, so that different tail configurations should have equal efficiency and approximately the same area. Several tail configurations like conventional tail, T-tail, H-tail and V-tail are evaluated.

Although T-tail provides the compensating phenomena and small surface sizing, the configuration requires a heavy integration for holding the configuration and the design characteristics of falling into the wing wake region at higher angle of attacks makes it unsuitable for the given task. Similarly, the H-tail provides an efficient area compensation and better stability but it proves to be a hindrance in the takeoff and landing performance of the UAV as the overhanging vertical surface reduced the departure and approach angles of the aircraft. In case of V-tail configuration, the excess weight of the integration and the surface assembly led to its rejection [4, 15]. Finally, a conventional three-piece tail design is selected. The final selected design consists of a single circular carbon fiber boom, which carries the single integration hub for both the vertical and horizon tails, thus reducing the overall tail weight of the UAV. Lightweight 6mm nylon screws and nuts are applied to fasten the structure on the circular boom, allowing for hassle free assembly and disassembly. The tail performance characteristics are quantified and compared in Figure 7, which demonstrates why conventional tail has been selected over other configurations.

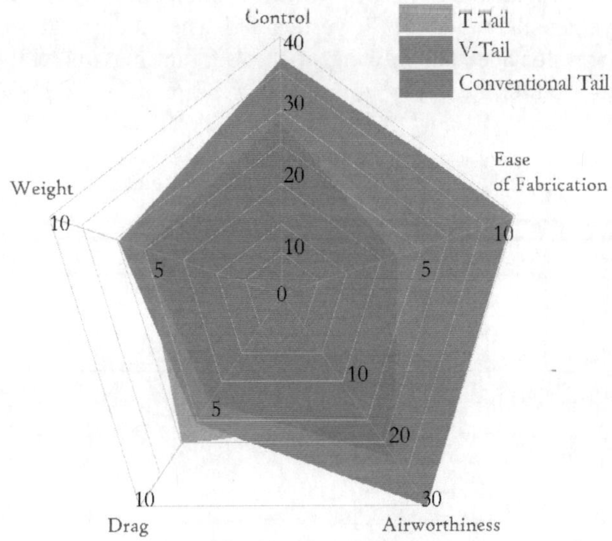

Figure 7: Tail selection

2.4. Control Surfaces Design

The directional changes such as roll, pitch and yaw are brought about by the control surfaces. Accurate sizing of the control surfaces ensures safe operation of the UAV within the aerodynamic parameters such as bank angle, pitch angle, angle of attack. The ailerons located on the back of the trailing end of each wing cause movement or roll of the aircraft around its longitudinal axis. The elevator located on the horizontal tail takes care of the pitching motion on its transverse axis. Finally, the rudder mounted on rear of the vertical tail causes a movement of the UAV about its vertical axis. Each of these control surfaces is actuated by a servo motor.

From graphical data obtained from XFLR and AVL, the values for pitch, roll and yaw coefficient are found out. These data are then utilized to determine the required roll, yaw and pitch rates, and the approximate required area of the control surfaces is decided to provide the UAV with appropriate roll, pitch and yaw authority. Finally, 10% of the excess area calculated is added to compensate for the empirical or manufacturing errors.

The control surface efficiencies are evaluated using Equation 2, where $C_l\delta_a$ is the control authority, *y1, y2* refer the control surface distance from the root and $C_{l_{aw}}$ is the lift curve slope of the control surfaces.

$$C_l\delta_a = 2C_{l_{aw}} \int_{y1}^{y2} cy\, dy \tag{2}$$

The span and cord of the ailerons are approximated at 16% and 30% of the wing of the UAV. The main factor governing the location and size of the ailerons is the desired maneuverability of the UAV. The location of ailerons also plays a major part in the structural aspect of the wing as outboard ailerons cause more twist to the wing, leading to fluttering on low aspect ratio wing. To avoid any fluttering in the wing and any potential weight increase, iterations are carried out to determine the control authority of the ailerons. Once the control authority, $C_l\delta_a$ is determined, the dimensional value is obtained by Equation 1. The moment arm of the aileron is measured to be 5.25ft and the aileron is shown in Figure 8. In the meantime, the elevator helps with pitch and level cruise of the aircraft. A similar procedure as for the ailerons is applied to the elevator sizing process to calculate its area, chord and span required for handling the aircraft. The pitch authority of the elevator is calculated and additional dimensional margin of 5% is considered to compensate for manufacturing defects. The elevator area is set to 27% of the horizontal spanning from end to end and the elevator is shown in Figure 9. Last but not least, the main function of the rudder is to provide the UAV with yawing motion in the z-axis. The same control surface equations are used to evaluate the rudder of the vertical tail. The rudder area is calculated to be 38% of the horizontal surface area for adequate yawing of the UAV at maximum flying weight. Figure 10 illustrates the rudder designed for the UAV.

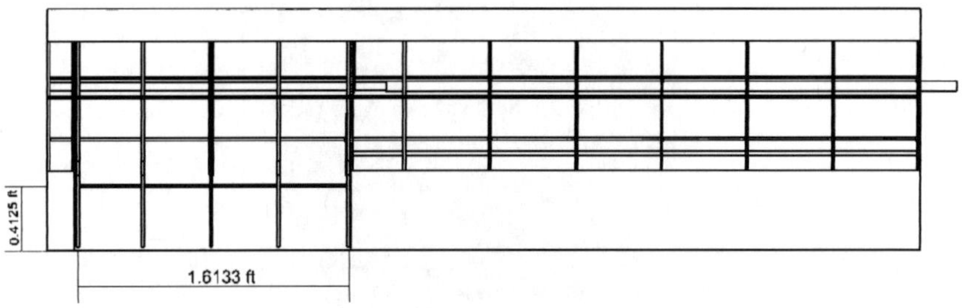

Figure 8: Aileron sizing

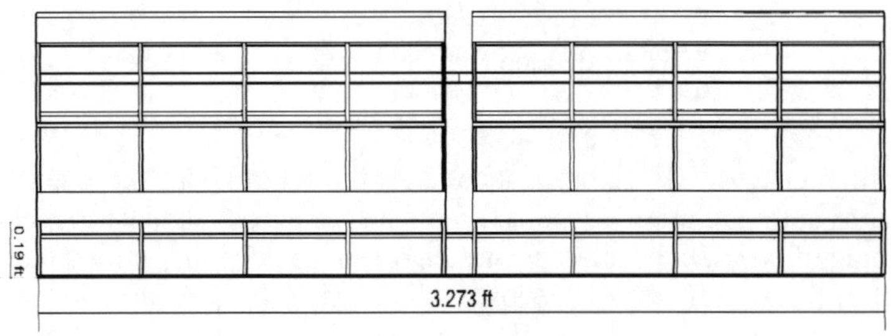

Figure 9: Elevator sizing

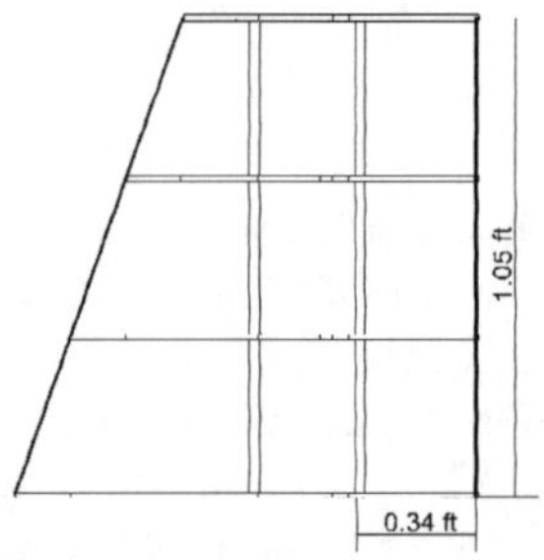

Figure 10: Rudder sizing

Selection of servos for the control surface is done by evaluating the torque required by the control surfaces using the moments acting on given surface area. Power requirements and distribution by the lithium-polymer battery is also taken into consideration. After calculating the residual power 16 kg.cm torque, turning servos are selected. To verify the authority of the aileron, a test jig simulating both the left and right ailerons is constructed. The parts simulating the ailerons are loaded with the maximum expected load and the control surface movements are performed for a predetermined duration of flight for several cycles. Power distributions are also verified during the testing. The positioning of the servo is shown in Figure 11.

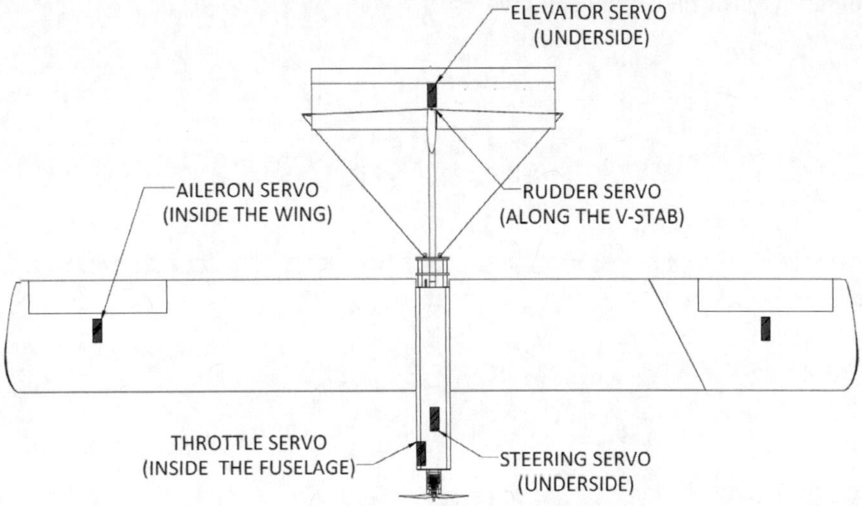

Figure 11: Servo positioning on the UAV

Stability of the UAV is of paramount importance when it comes to its maneuverability and overall endurance in rough weather conditions. Static stability is primarily influenced by tail sizing, tail arm and centre of gravity (CG) location of the aircraft. The volume coefficients are quantitative terms used to represent the effectiveness of the empennage, and they are factors at the designer's liberty and can be optimized accordingly. Volume coefficients are key values in determining the neutral point position [4]. The neutral point is always located at a distance from the CG of the UAV. This distance is termed as the static margin. It is nothing but the safe distance that the CG can shift without adversely affecting the overall the stability of the aircraft. However, having an extremely high static margin would not be feasible as it would compromise maneuverability of the vehicle. Therefore, an optimization process [16] is set up to calculate the suitable static margin that allows the control of the UAV to the pilot but at the same time will also augment the stability of the vehicle. The formula in Equation 3 gives the position of the neutral point [17], where x_n is the position of the neutral point, x_o is the position of the aerodynamic center, V_H is the volume coefficient of the horizontal stabilizer, $a_w$is the lift curve slope of the wing, a_t is the lift curve slope of the tail and $\frac{\delta\varepsilon}{\delta\alpha}$ is variation of downwash with angle of attack.

$$x_n = x_o + V_H\eta\frac{a_w}{a_t}\left(1-\frac{\delta\varepsilon}{\delta\alpha}\right) - \frac{C_{m_{af}}}{C_{L_{aw}}} \quad (3)$$

The value of static margin is then arrived by using the position of the neutral point of 10.5% and the CG location is calculated. The stability of the UAV is then put to test in XFLR5, which has yielded the C_m vs α plot. The trim angle of the aircraft (i.e. intercept of the x-axis) is established as 0.5°. At this angle, the UAV reaches a neutral state where all net moments acting on it would equal zero.

2.5. Payload Dropping Mechanism

Keeping the humanitarian nature of the UAV missions, it is provided with a dynamic payload capacity of 8lbs, which allows it to be used to drop medicines, food packs, canned containers and bottled water. In research configuration, the payload can consist of self-powered sensors to be dropped for geological readings. To keep the overall dimension and weight of the UAV to minimum, it is decided to construct an open load dropping mechanism where the dynamic payload is mounted outside the aircraft. The

IOP Conf. Series: Materials Science and Engineering **270** (2017) 012018 doi:10.1088/1757-899X/270/1/012018

dropping mechanism consists of two 16 kg-cm torque Turnigy servos that will release the drop line when triggered by the operator. To keep the drag profile of the loaded UAV low, a deflector screen is added ahead of the load and it is retracted back once the payload is released. The overall dropping is triggered with the help of the onboard camera, which relays real-time video back to the operator. The illustration of this mechanism is shown in Figure 12.

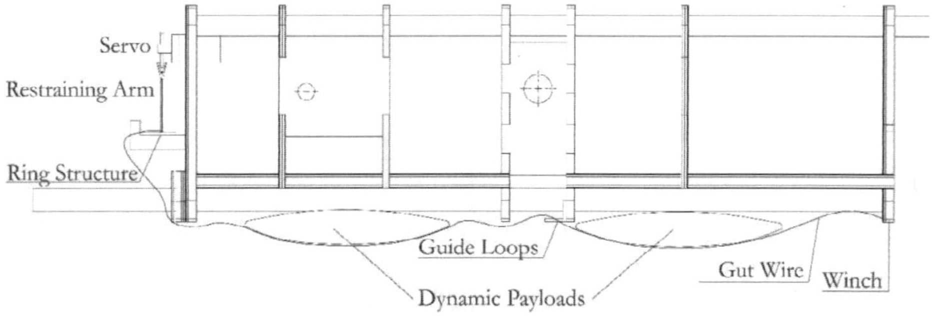

Figure 12: Payload dropping mechanism

2.6. Landing Gear Design

The landing gear is the principle support for the UAV when parked, taxiing, taking off and landing. It is essentially a one-dimensional structure that undergoes a large change in stress values, experiences thermal shocks and high friction, skidding and non-linear elongation from the start to end of landing or take off [18]. The landing gear configuration has several variations: bicycle, tricycle, tail dragger and multi-bogey. It can also be retractable or fixed [4]. A fixed tricycle gear is chosen as appropriate for the UAV in this study due to its supreme dynamic stability, horizontal floor angle, shorter wheel base, easiness in landing, faster acceleration during take-off due to lower angle of attack and absence of the ground looping phenomenon [3]. A fixed landing gear system increases the drag by nearly 10% but it is superior to retractable mechanism due to absence of heavy retraction mechanism. After fixing the type of gear, the overall design layout is decided by placing the main landing gear at 50-55% of the mean aerodynamic chord (MAC) of the wing. Thereafter, the distance between the aft and front CG of the UAV is calculated. The wheel base is calculated by applying 80-95% of the load on main landing gear and 5-20% on the nose landing gear [18]. The wheel track is determined such that the aircraft does not roll over when taxiing at sharp corners. The wheel and tire sizes are further calculated [4].

3. Conclusion

The prototype UAV is tested from both dirt and grass strips to validate the rough field short takeoff and landing characteristics. During testing, telemetric data obtained from the onboard data acquisition system has indicated a top velocity of 16m/s and a stall speed of 7m/s, which concur with calculated theoretical data. The cruise speed is obtained with a setup of static load of 16lbs and a dynamic load of 4lbs while being powered by two 5500 Mah 70C lipos. The flight endurance with the following set is obtained to be 19 minutes with the batteries being depleted to 43% charge. The designed UAV will serve the purpose of providing aerial surveillance and relief material delivery during the flood relief operations. The design parameters make it suitable for low speed cruising and also personnel detection missions. The onboard dynamic load capability allows the user to accurately deliver the relief material such as food packages or medicines and its simple and durable design will allow the relief forces to rapidly deploy this UAV from dirt strips or civil roads. The future development of the UAV will be based around a sturdier wing design to enhance its cruise characteristics by optimizing the wing aspect ratio. Lifting more dynamic cargo will be one of the key capabilities that will be maximized in order to improve mission performance of the UAV. Further enhancement will also be made on the autonomous guidance system and onboard electronics in order to incorporate several multiple UAVs under a single command structure.

Acknowledgement
The authors would like to thank VIT University (Vellore), School of Mechanical Engineering, Team Assailing Falcons for providing support and encouragement throughout the project. They would also like to thank Mr. Billton Vitus and Mr. Jatin Thaman for their help in the project.

References

[1] Dash J and Paul R 2017 *https://www.reuters.com/article/us-southasia-floods/worst-monsoon-floods-in-years-kill-more-than-1200-across-south-asia-idUSKCN1B510Z*
[2] Schuh C A and Nieh T G 2003 *Mater. Res. Soc.* **740** 27-31
[3] Sadrey M H 2012 *Aircraft Design: A Systems Engineering Approach* John Wiley & Sons, Ltd.
[4] Raymer D P 2012 *Aircraft Design: A Conceptual Approach* AIAA
[5] Tulapurkara E *http://nptel.ac.in/courses/101106035/026_Chapter%205%20_L19_(03-10-2013).pdf*
[6] Abbott I H and Von Doenhoff A E 2012 *Theory of Wing Sections: Including a Summary of Airfoil Data* Courier Corporation
[7] Spedding G R and McArthur J 2010 J. Aircr. **47** 120–8
[8] *http://web.mit.edu/drela/Public/web/xfoil/*
[9] Melin T 2000 *A Vortex Lattice MATLAB Implementation for Linear Aerodynamic Wing Applications* Thesis, Royal Institute of Technology (KTH)
[10] Roskam J 2003 *Airplane Flight Dynamics and Automatic Flight Controls* Roskam Aviation and Engineering Corporation
[11] Zenith Aircraft Company 2010 *http://www.zenithair.com/stolch801/design-wing.html*
[12] Jett Aerotech 2015 *http://www.dubjett.com/tech2015.html*
[13] Indian Meteorological Department 2017 *http://www.imdchennai.gov.in/obs_data.htm*
[14] *http://www.rccaraction.com/everything-you-need-to-know-about-nitro-fuel/*
[15] Gudmundsson S 2014 *General Aviation Aircraft Design: Applied Methods* Elsevier
[16] Bredehoft B, Briggs R, Chou A, Duelley R, Kovalic A, McNeilus J, Pesce P, Preus D, Prince M, Ricciardi A, Sherman M and Sunday E 2008 *Large Crowd Surveillance Unmanned Aerial Vehicle* Virginia Polytechnic Institute and State University
[17] Anderson J D 1984 *Fundamentals of Aerodynamics* McGraw Hill
[18] Currey N S 1988 *Aircraft Landing Gear Design : Principles and Practices* AIAA
[19] A. Jha 2009 *14th National Conference on Machines and Mechanisms*

IOP Conf. Series: Materials Science and Engineering **270** (2017) 012019 doi:10.1088/1757-899X/270/1/012019

System identification of an unmanned quadcopter system using MRAN neural

M F Pairan and S S Shamsudin*

Department of Aeronautical Engineering, Faculty of Mechanical and Manufacturing, Universiti Tun Hussein Onn Malaysia, 86400 Parit Raja, Johor, Malaysia.

*syafiq@uthm.edu.my

Abstract. This project presents the performance analysis of the radial basis function neural network (RBF) trained with Minimal Resource Allocating Network (MRAN) algorithm for real-time identification of quadcopter. MRAN's performance is compared with the RBF with Constant Trace algorithm for 2500 input-output pair data sampling. MRAN utilizes adding and pruning hidden neuron strategy to obtain optimum RBF structure, increase prediction accuracy and reduce training time. The results indicate that MRAN algorithm produces fast training time and more accurate prediction compared with standard RBF. The model proposed in this paper is capable of identifying and modelling a nonlinear representation of the quadcopter flight dynamics.

1. Introduction

Quadcopter is a type of rotorcraft that consists of four main fixed pitch rotors and does not require any complex mechanical control mechanism for its propellers. Because of this, the mechanical system of quadcopter is much easier to maintain compared to conventional helicopter. Quadcopter has a unique capability of taking off and landing vertically, besides hovering and cruising at lower speed compared to fixed wing unmanned aerial vehicle (UAV). Due to these advantages, quadcopter has attracted very strong interest worldwide and has been used in both military and civil applications such as in real-time reconnaissance surveillance, aerial photography, delivery services and also traffic monitoring. These applications require the quadcopter to fly in close proximity relative to the targets and therefore, need the application of a robust control system.

A robust controller is capable of adapting to change in dynamics and dealing with the quadcopter's uncertainty parameters. A robust control system is of utmost importance for autonomous flight and a robust controller is designed to provide accurate flight performance within the acceptable disturbance ranges [1]. It is a challenging process to identify an accurate dynamics model for a quadcopter since it is highly nonlinear, has multi-input multi-output (MIMO) characteristics, strongly coupled, under-actuated and has an unstable system [2,3]. Development of a dynamics model using the first principle of modelling based on Newton's law and Euler's angles are inaccurate, time-consuming and require deep mathematical and physical calculations about the system behaviour [4–8]. On the other hand, the system identification (SI) provides a method to derive dynamics models based on experimental data of the quadcopter control inputs and measured outputs. Neural networks (NN) based SI approach is an alternative for modelling techniques for quadcopter and it does not rely on any specific mathematical model. The NN approach is typically consisted of a flexible model structure and is characterized by a limited number of NN parameters. NN models are trained using a large number of efficient training methods with high prediction accuracy [8–11]. Nevertheless, this NN modelling method also has few disadvantages such as longer required training time, slow convergence rate and vulnerability to over-fitting problems.

Recursive training algorithms with a good selection of NN model structure can notably improve the prediction accuracy, reduce the NN training time and avoid data over-fitting [8,11–13]. The usage of recursive algorithms like Kalman Filter or recursive Gauss-Newton (rGN) with either Constant Trace (CT) or EFRA variation can be applied to reduce the computation complexity of the offline (batch) training method. However, the correct selection of hidden neurons remains to be a vital parameter to be considered since either excessive or insufficient neuron can lead to a poor prediction performance. Typical selection of hidden neurons is based on the trial and error method or rule of thumb approaches [11,14,15]. After several attempts or trials, the deduced number of hidden neurons obtained from the trial and error method is then used to estimate the weights using NN training algorithms [16–19]. In order to avoid the troublesome selection of hidden neurons to be used in NN model, automatic neurons selection, network growing and pruning strategy can also be employed to remove hidden neurons that consistently made little contributions to the network outputs. Fully-tuned radial basis function (RBF) with MRAN recursive training algorithm is proposed in this study to overcome the large numbers of hidden neurons and parameters selection dilemma. The MRAN algorithm introduces network growing and hidden neuron pruning approach to automatic tuning the neuron based error criteria thus reducing training time and over-fitting problem. In MRAN learning approach, new training sample is presented in recursive mode while all RBF network parameters will be continuously updated [20,21].

This paper presents the recursive system identification method for a quadcopter based on the RBF neural network structure. RBF is trained with Constant Trace (RBF-CT) and MRAN (RBF-MRAN) algorithm to model attitude dynamics of the quadcopter. Attitude output and pilot stick input during flight are recorded and used as training and validation data for the neural network. MRAN algorithm aims to reduce the training time, increase prediction accuracy and produce the minimal structure of the standard RBF. For this purpose, the addition and removal neuron strategy is incorporated into the RBF architecture. To assess the performance of the proposed model, RBF trained with MRAN algorithm is compared with the standard RBF structure trained with Constant Trace algorithm (CT) with regards to the percentage of root mean square error (RMSE), the coefficient of determination (R^2) criteria and the mean training time. The RBF-CT structure is selected for comparison because it has the same basic structure (three fixed layer), activation function (radial basis function) and also optimizing parameter method (CT algorithm).

2. Quadcopter Platform and Avionics System

The XuGong v2 Profoldable quadcopter shown in Figure 1, which is manufactured by ImmersionRC Ltd, has been chosen as the rotorcraft UAV platform for this project. The quadcopter's main frame is made from lightweight carbon fiber with a weight of 96g and it is capable of approximately producing 20 minutes flight time. It is also equipped with four AIR 2213 KV920 (rpm/V) brushless motors with T9545 plastic self-tightening propellers and is controlled by four AIR20A electronic speed controller (ESC). Radio transmitter controller used for this quadcopter is FrSkyTaranis X9D 2.4 GHz transmitter with FrSky X8R receiver. In this study, the DJI NAZA v2 is used as the main flight controller for the XuGong quadcopter. The XuGong quadcopter platform is primarily selected because of its main frame compatibility, open source and its capability of future upgrades. It is also equipped with crash-resistant arms that will minimize the damage during hard landing and crash. These arms contain a break-away pin that will take the crash impact by bending back into the original shape without breaking the arm.

The data acquisition system is developed to record the desired attitude data during flight to assess the performance of the proposed system identification method. The collected data is recorded on the flash drive as text file format. The overview of the data acquisition for the onboard system and ground station is depicted in Figure 2, which consists of a computer as the ground station, a VN-100 Inertia Measurement Unit (IMU) and National Instrument (NI) myRIO 1900, which is a compact and portable

IOP Conf. Series: Materials Science and Engineering **270** (2017) 012019 doi:10.1088/1757-899X/270/1/012019

reconfigurable I/O device with built-in wireless devices that is used as a data logger. MyRIO can be configured in NI LabVIEW software as the wireless host that provides 10 analogue inputs (AI), six analogue outputs (AO), 40 digital inputs and outputs (DIO) and power output (3.3V and 5V). The VN-100 IMU consists of 3-axis accelerometers, 3-axis gyros, 3-axis magnetometer, a barometric pressure sensor and a 32-bit processor. VN-100 is capable of computing data in a real-time continuously over a 360-degree range of motion with no limitation on the angles and it also has a built-in microcontroller that runs a quaternion based Extended Kalman Filter (EKF), which offers faster and smoother attitude estimates and angular rate measurements.

Figure 1: XuGong quadcopter platform

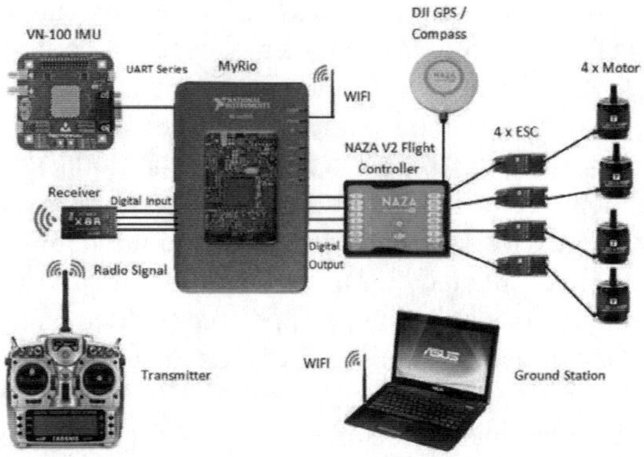

Figure 2: Quadcopter system identification setup

The task of the onboard system is to collect all necessary flight data from the IMU and receiver while the ground station is used to monitor the quadcopter's flight data during flight. Figure 3 shows a detailed overview of onboard SI system for the quadcopter under development. National Instrument myRIO 1900 board is used to collect flight information from the IMU through serial UART and store the data in a flash drive. The FPGA in the myRIO 1900 board is used to collect all control stick input signals from the receiver and these signals will be transferred to real-time processor for data saving.

3. System Identification based on MRAN Algorithm

Figure 4 illustrates the overall process of system identification based on neural network model. The RBF neural network model is located in parallel with the actual quadcopter plant where NN inputs are from actual inputs and outputs of the quadcopter. The NN model is trained with inputs data to predict the output of the quadcopter plant. The error between the predicted outputs of the RBF model and the outputs of the quadcopter is used in training algorithm to adjust the NN parameters. The overall neural network modelling approach consists of the flight test data gathering process, neural network model structure selection, and the model estimation and validation.

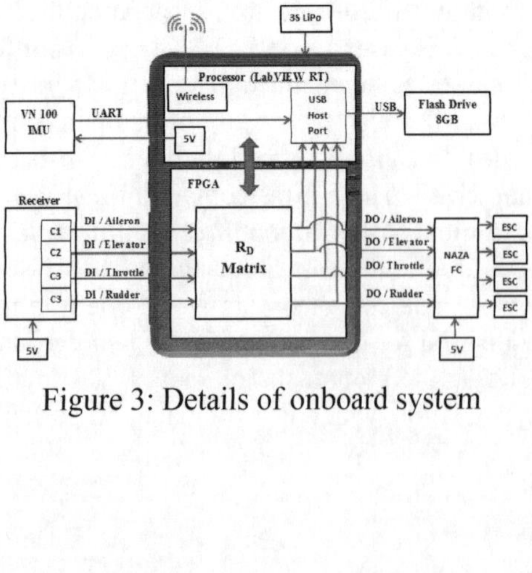

Figure 3: Details of onboard system

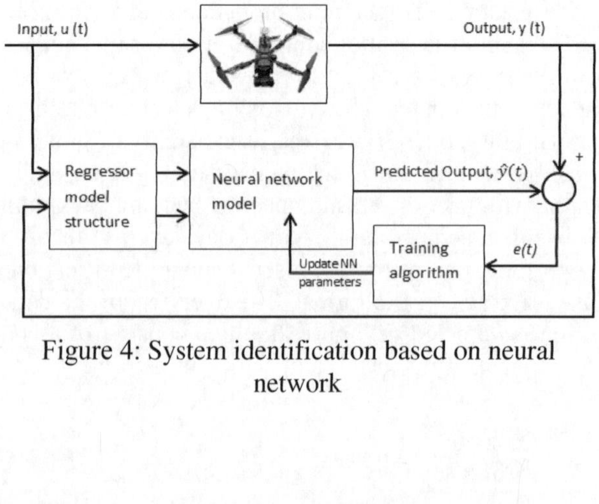

Figure 4: System identification based on neural network

3.1. Flight Test Data Collection

Flight test data collection is the experimental method where the input-output data pair is obtained from the quadcopter onboard system. Flight data collection is an important step to be conducted prior to the actual implementation of real-time system identification. For training and validation data collection, the frequency sweep excitation signals that have a quasi-sinusoidal shape of increasing frequency with variable amplitude are given to the quadcopter input [22,23]. The frequency sweeps signal issued by a pilot will start at low frequency and increases up to the higher frequency as desired by the user. Flight data for the quadcopter is collected for attitude information only. During the experiment, all control inputs and state measurements are recorded at a 50Hz sampling rate and the noises are filtered through a low-pass filter with 10Hz cut-off frequency. The data acquisition system on the quadcopter measures the Euler angles (i.e. roll, Φ; pitch, θ and yaw, ψ) and the angular rates (i.e. roll rate, p; pitch rate, q and yaw rate, r) as the output of the system. The control inputs associated with pilot commands are defined as elevator stick, δ_{lon}; rudder stick, δ_{pedal} and aileron stick, δ_{lat}. They are collected as input of the system and are stored in a flash drive. Table 1 tabulates the output responses based on frequency swept for each input channel for flight test data collection.

Table 1: Output responses from input channel

Input channel	Euler angles	Angular rates
δ_{lat}	ϕ	p
δ_{lon}	θ	q
δ_{ped}	ψ	r

3.2. Model Structure Selection

A model structure selection is a step where the model to represent the relationship between inputs and outputs of a system is chosen. The model output should replicate the expected value of the true system output. The most preferred model structure is the one with capability to predict accurately the actual flight dynamics from a set of test data. In this study, the neural network representation of nonlinear Auto Regressive eXogenous inputs (ARX) model structure is used to predict the relationship between input-output variables.

As indicated in Figure 5, RBF architecture with one hidden layer is chosen to learn the nonlinear relationship of NNARX model. Input-output relationship (i.e. coefficient of ARX model) and lagged input-output data are expressed in terms of parameter vector, θ and time regression vector, φ. For a single hidden layer case, the outputs formulation from the output layer of an RBF network is given by Equation 1 to Equation 4, where $w1_{kj}$ is the weights between hidden and output layers, μ_j and σ_j refer to centre and width of the Gaussian function of *j*th hidden layer, *h* denotes the number of neurons in the hidden layer while B_k is the bias elements for output layer. The number of inputs, hidden and outputs of neural network is presented by *i*, *j* and *k*, respectively. Parameter vector of RBF contains all RBF parameters (centres, weights, widths and bias) and regression vector consists of past input and output. The best regressor of RBF structure for system identification that produces low percentage of RMSE and short mean training time based on trial and error method are selected in this study [11].

$$\hat{y}(t) = B_k + \sum_{j=1}^{h} w1_{kj}\left(exp\left(-\frac{1}{\sigma_j{}^2}\left\|x-\mu_j\right\|^2\right)\right) \quad (1)$$

$$\hat{y}(t+k|t,\theta) = g(\varphi(t+k),\theta) \quad (2)$$

$$\theta = \left[w1_{kj}\mu_j\sigma_j B_k\right] \quad (3)$$

$$\varphi = [\varphi_1\varphi_2 \;\cdots\; \varphi_i] = \left[y(t-1) \cdots\; y\left(t-n_y\right)\;\; u(t-1) \;\cdots\; u(t-n_u)\right] \quad (4)$$

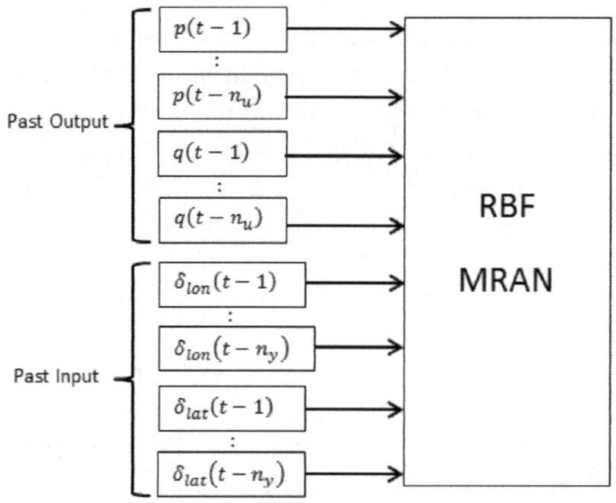

Figure 5: Regressor structure for RBF and MRAN

3.3. Model Estimation

The third step is to determine the NN parameters and minimize the error between the identified system and true model. The system collects the measurement data in each time step and, at the same time, a new model of the system is formed. In this study, RBF NN structure is trained with recursive training algorithm of Constant Trace (CT) and MRAN algorithm. Detailed methodology of MRAN algorithm is shown in Figure 6. The MRAN algorithm calculates the output of the RBF network with three error criteria, which are used to check if a new hidden unit should be added from the existing network. If all three criteria are satisfied, one neuron will be added and the algorithm will remove any hidden neurons that have the least contribution to the RBF output. If the error criteria do not satisfy any specification, the algorithm routine will proceed with Jacobian matrix and CT training calculation. After that, the MRAN method proceeds to prune step and continues with another set of input-output data pair.

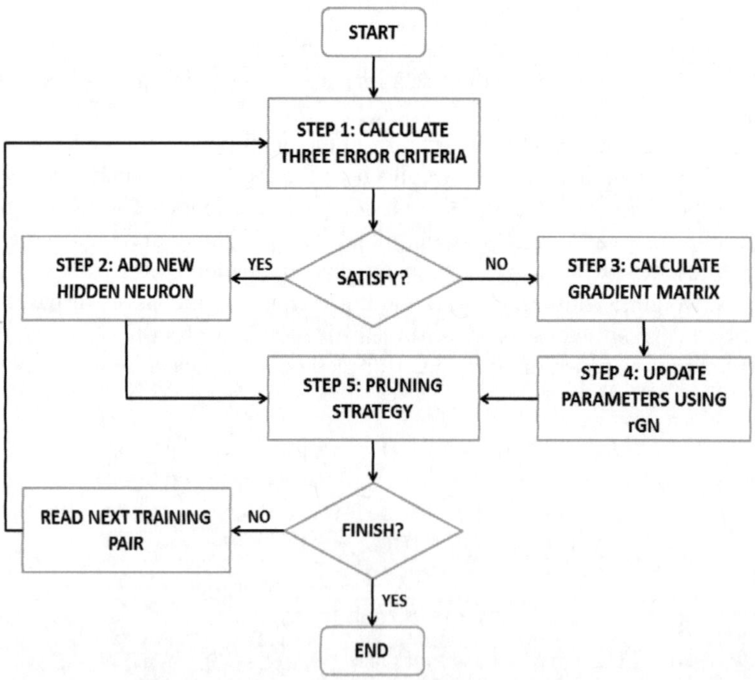

Figure 6: Summary of the MRAN algorithm flow

The first step of the algorithm is to check if the error criteria satisfy the condition for the addition of a new hidden unit. Error criteria E_1, E_2 and E_3 are user-defined thresholds that need to be satisfied to trigger the addition of a new hidden neuron condition. The calculations of the three error criteria are as given by Equation 5, Equation 6 and Equation 7, where e_t is error prediction, y_t is the output from measurement quadcopter output, $\hat{y_t}$ is the RBF prediction output, μ_{tr} is the centre of the hidden unit that is closest to the current input, γ is decay constant, i is number of iterations, eta_{max} and eta_{min} are respectively the maximum and minimum threshold to bound value of E_3, x_t is current input, N_w is sliding window size and e_j is current prediction error.

$$\|e_t\| = \|y_t - \hat{y_t}(x_t)\| > E_1 \tag{5}$$

$$e_{rmsi} = \sqrt{\sum_{j=t-(N_w-1)}^{t} \frac{\|e_j\|^2}{N_w}} > E_2 \tag{6}$$

$$d_t = \|x_t - \mu_{tr}\| > E_3 = max\{eta_{max}\gamma^i , eta_{min}\} \ 0 < \gamma < 1 \tag{7}$$

When all the criteria in the first step are met, a new hidden neuron is added with initial parameters associated with it in the second step. This is given by Equation 8, Equation 9 and Equation 10 where $\underline{w}_{h+1}^T$, μ_{h+1} , σ_{h+1} are respectively the initial parameters value of weight, center and width for new hidden neuron, e_t is current output error, x_t is the current input introduced into the network and k_p is the overlap factor that determines the overlap of the responses of the hidden units in the input space.

$$\underline{w}_{h+1}^T = e_t \tag{8}$$

$$\mu_{h+1} = x_t \tag{9}$$

$$\sigma_{h+1} = k_p\|x_t - \mu_{tr}\| \tag{10}$$

If any of the three criteria for adding a new hidden unit cannot be satisfied, the Jacobian matrix is calculated. The Jacobian matrix, Π is calculated by partial differentiating Equation 1 with respect to parameter vector, θ. The result from gradient vector calculation is shown by Equation 11.

$$\Pi = \begin{bmatrix} \mathrm{I}, \phi_1(x_t)\mathrm{I}, \phi_1(x_t)(2\, w_1^T/\sigma_1^2)(x_t - \mu_1)^T, \phi_1(x_t)(2\, w_1^T/\sigma_1^3)\|x_t - \mu_1\|^2, \cdots, \\ \phi_1(x_h)\mathrm{I}, \phi_h(x_t)(2\, w_h^T/\sigma_h^2)(x_t - \mu_h)^T, \phi_h(x_h)(2\, w_h^T/\sigma_h^3)\|x_h - \mu_h\|^2 \end{bmatrix}^T \tag{11}$$

Next, all RBF parameters $\theta = \left[w_0 , w_1 ,\mu_1^T , \sigma_1, \cdots , w_h ,\mu_h^T , \sigma_h\right]^T$ are updated using the CT algorithm. CT method is proposed to prevent covariance blow-up without loss of the tracking ability by limiting the eigenvalues value [13]. The CT training algorithm is shown by Equation 12 to Equation 15.

$$K(t) = P(t-1)\Pi(t)\left(1 + \Pi^T(t)P(t)\Pi(t)\right)^{-1} \tag{12}$$

$$\hat{\theta}(t) = \hat{\theta}(t-1) + K(t)e_t \tag{13}$$

$$\bar{P}(t) = P(t-1) - K(t)\Pi^T(t)P(t-1) \tag{14}$$

$$P(t) = \frac{\alpha_{max} - \alpha_{min}}{trace\left(\bar{P}(t)\right)}\bar{P}(t) - \alpha_{min}I \tag{15}$$

Next, in the pruning step, the MRAN algorithm will remove neurons that make little contribution to the RBF output for N_w consecutive observations. The output, $y_{nj}\,(j = 1 \ldots n_y)$ of the nth hidden neuron is normalized with the highest output obtained in order to reduce inconsistency in the output data. The calculation in this step is made using Equation 16 and Equation 17.

$$y_{nj} = w_{nj}\, exp\left(-\frac{1}{{\sigma_n}^2}\|x - \mu_n\|^2\right) \quad (16)$$

$$r_{nj} = \frac{y_{nj}}{max\,\{y_{1j}, y_{2j}, \cdots, y_{hj}\}} \quad (17)$$

The neuron is removed if r_{nj} value decreases below desired threshold value, δ for N_w consecutive inputs. After this pruning step, next input-output data pair is trained and the overall calculation routine is repeated.

3.4. Model Validation

One step ahead (OSA) prediction and mean square error (MSE) are used to validate the performance of the proposed network models. One step ahead prediction is a simple plot that compares the actual measurement data with the model prediction over a test data set. The overall accuracy of the prediction error is calculated according to the percentage of root mean square error (RMSE) and R^2 criterion as given by Equation 18 and Equation 19, respectively. The training performance of the NN models is gauged using the mean training time in each iteration.

$$\%RMSE = \sqrt{\frac{\Sigma(\hat{y}_i(t) - y_i(t))^2}{\Sigma(y_i(t) - \bar{y}_i(t))^2}} \quad (18)$$

$$R^2 = 1 - \frac{\Sigma(\hat{y}_i(t) - y_i(t))^2}{\Sigma(y_i(t) - \bar{y}_i(t))^2} \quad (19)$$

4. Results and Discussion

The performance of the proposed RBF architecture trained with MRAN algorithm is compared to the standard RBF to determine whether the proposed RBF model is capable of improving the prediction performance over the standard RBF model. The NN models are trained to predict the future response of the quadcopter attitude dynamics with two output variables i.e. roll and pitch rates. The best model structures for RBF and MRAN are selected for comparison by trial and error approach as tabulated in Table 2.

Table 2: Regression structure selection

NN model	**Regression structure**	**Roll rate (%RMSE)**	**Pitch rate (%RMSE)**	**Mean training time**
RBF-CT	1-1	20.39	20.59	6.71
	2-1	18.33	22.77	5.39
	2-2	17.04	23.05	12.42
	3-1	16.69	16.86	6.97
	3-2	18.92	21.23	16.96
	3-3	22.62	24.04	23.72
RBF-MRAN	1-1	12.61	16.42	2.34
	2-1	12.58	16.65	2.11
	2-2	12.73	15.55	6.35
	3-1	10.99	13.53	5.23
	3-2	10.31	14.86	6.18
	3-3	9.81	15.13	12.56

In this study, both RBF structures are selected with three past outputs and one past input lag structure for system identification because this structure type provides the lowest percentage of RMSE and the shortest mean training time. Table 3 displays all RBF parameters used for system identification of the quadcopter. The MRAN training parameters selected for this study using the trial and error method are shown in Table 4.

Table 3: RBF NN parameters

Network Specifications	**RBF-CT**	**RBF-MRAN**
Number of hidden neuron	5	1 (initial)
Number of past output	3	
Number of past input	1	
Activation function in hidden layer	Radial basis function	
Activation function in output layer	Linear function	
CT parameter	α_{max}= 10 α_{min} = 0.001	α_{max}= 20 α_{min} = 0.1
Number of regressors	8	

Table 4: MRAN training parameters

E_1	E_2	Eta_{max}	Eta_{min}	γ	*NW*	K_p	*Threshold*
0.04	0.2	4	2	0.96	15	0.3	0.001

Table 5 shows the comparison of the results of percentage RMSE, R^2, mean training time and final neuron of RBF and MRAN structure to identify roll and pitch rate of the quadcopter. From the table, it can be seen that MRAN algorithm produces better prediction accuracy with an RMSE of 10.99% and 13.53% for roll rate and pitch rate, respectively, compared to standard RBF. This table also indicates that the R^2 value of MRAN is closer to one and offers the best fit with the actual output compared to RBF trained with CT. Table 5 also highlights that MRAN structure offers faster system identification with 5.23 ms mean training time compared with RBF-CT (6.97 ms). This can be attributed to the fact that MRAN requires only four hidden neurons compared to five fixed RBF-CT that resulted in faster completion of the training process.

Table 5: Overall RBF and MRAN prediction performance

NN model	**Percentage of RMSE (%)**		**R^2**		**Mean training time(ms)**	**Final hidden neuron (unit)**
	Roll Rate (*q*)	**Pitch Rate (*p*)**	**Roll Rate (*q*)**	**Pitch Rate (*p*)**		
RBF	16.69	16.86	0.9696	0.9690	6.97	5
MRAN	10.99	13.53	0.9868	0.9801	5.23	4

Figure 7 presents the hidden neuron growth history along with MRAN training. Hidden neurons are added and pruned to obtain the best RBF structure throughout the sampling time. This result implies that the MRAN model adapts itself to the changes in flight dynamics very well. It is apparent that the proposed RBF model has achieved the maximum growth of hidden neurons at eight units before being reduced and stabilized at four hidden neurons.

Furthermore, the results obtained by one step ahead prediction responses from the RBF and MRAN model are respectively plotted against the actual attitude responses of the quadcopter in Figure 8 and

Figure 9. The red and green solid line represents the predicted output from the MRAN and RBF-CT, respectively, meanwhile the black line represents the measured rate from IMU. The predicted outputs from MRAN and RBF-CT are capable to track the actual attitude measurement of the quadcopter. As can be observed in Figure 8 and 9, MRAN outputs closely match the actual measured outputs of the quadcopter compared with that of the RBF-CT. Therefore, when there are sudden changes in attitude dynamics, MRAN is capable of following and tracking the change of the dynamics. Results from the graph tally with the RMSE and R^2 results that indicate MRAN is better than RBF-CT in predicting the attitude dynamics of the quadcopter.

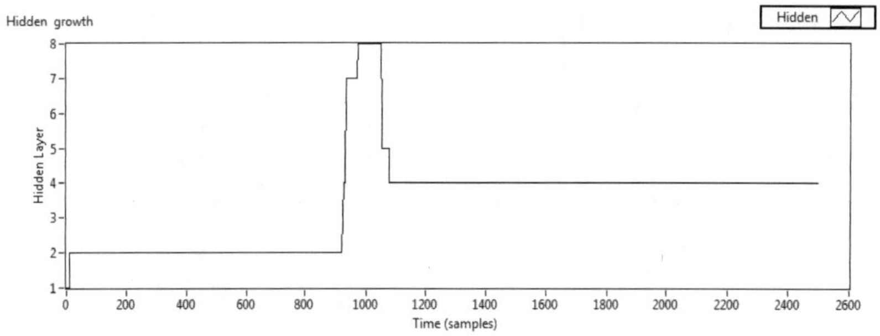

Figure 7: MRAN hidden neuron growth

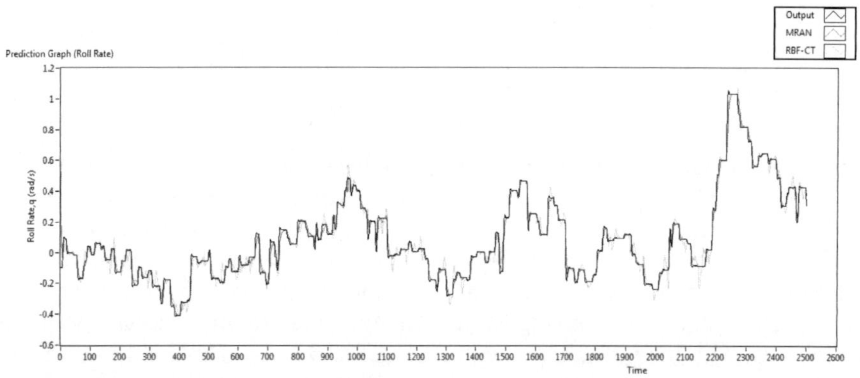

Figure 8: Prediction graph of roll rate for RBF-CT and MRAN

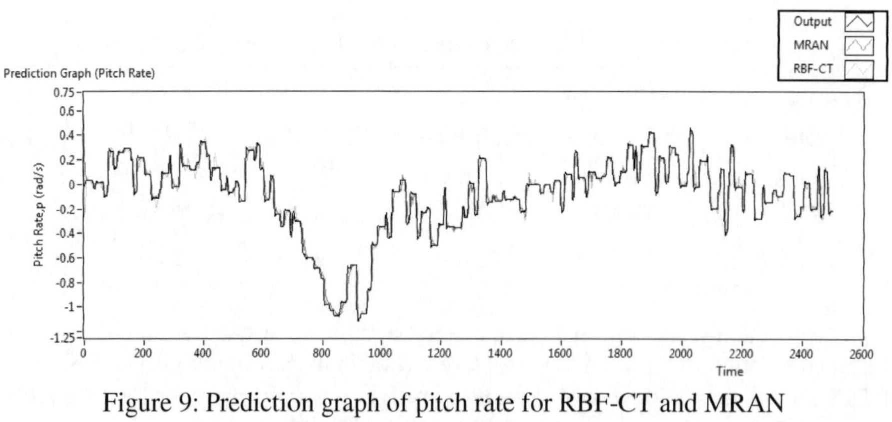

Figure 9: Prediction graph of pitch rate for RBF-CT and MRAN

5. Conclusion

In this study, nonlinear attitude dynamics of the Xugong quadcopter is identified from the flight data using recursive neural network approach. The RBF and MRAN models are proposed for the real-time system identification implementation. Fully-tuned RBF network is proposed in this project to improve prediction performance and training time of standard RBF model. The proposed MRAN model also offers advantages in the automatic tuning of RBF model parameters, which includes neurons selection as part of training updates. The real-time has predicted the outputs from the proposed neural network models and these outputs are compared with the actual output responses of the Xugong quadcopter for performance assessment. Validation of the results conducted in this study confirmed that the proposed MRAN model is suitable for modelling the quadcopter's attitude dynamics correctly with improvement in training time. The MRAN algorithm offers good attitude prediction with minimal NN structure (less neuron) and shorter training time compared to RBF-CT. Dynamics models obtained from the MRAN model can be further used for research on stability, control and design of the adaptive flight control for autonomous flight of a quadcopter.

Acknowledgement

This research is funded by research grant from the Ministry of Science, Technology and Innovation (MOSTI) and is technically supported by members of Aircraft System and Design Research (ASDR), UTHM.

References

[1] Zulu A and John S 2014 *Open J. Appl. Sci.* **4** 547–56
[2] Zhang X, Li X, Wang K and Lu Y 2014 *Abstr. Appl. Anal.* **2014** 320526
[3] McKerrow P 2004 *IEEE International Conference on Robotics and Automation*
[4] Abas N, Legowo A and Akmeliawati R 2011 *4th International Conference On Mechatronics*
[5] Putro I E, Budiyono A, Yoon K J and Kim D H 2008 *IEEE International Conference on Robotics and Biomimetics*
[6] Wei W 2015 *Development of an Effective System Identification and Control Capability for Quadcopter UAVs* PhD Thesis University of Cincinnati
[7] Tischler M and Remple R K 2006 *Aircraft and Rotorcraft System Identification: Engineering Methods with Flight Test Examples* American Institute of Aeronautics and Astronautic
[8] Shamsudin S S and Chen X 2014 *Int. J. Intell. Syst. Technol. Appl.* **13** 56–80
[9] Zurada J M 1996 *IEEE Aerospace Applications Conference*
[10] Lawryńczuk M 2014 *Computationally Efficient Model Predictive Control Algorithms* Springer International Publishing Switzerland
[11] Shamsudin S S and Chen X 2012 *Int. J. Model. Identif. Control* **17** 223–41
[12] Hunter D, Yu H, Member S, Pukish M S, Kolbusz J and Wilamowski B M 2012 *IEEE Trans. Ind. Informatics* **8** 228–40
[13] Shamsudin S S 2013 *The Development of Neural Network Based System Identification and Adaptive Flight Control for an Autonomous Helicopter System* PhD Thesis University of Canterbury
[14] Panchal F S and Panchal M 2014 *Int. J. Comput. Sci. Mob. Comput.* **3** 455–64
[15] Peyada N K and Ghosh A K 2009 *AIAA Atmospheric Flight Mechanics Conference*
[16] Bansal S, Jiang F J and Tomlin C J 2016 *IEEE Conference on Decision and Control*
[17] Dief T N and Yoshida S 2015 *IRACST - Int. J. Comput. Sci. Inf. Technol. Secur.* **5** 314–19
[18] Zhong H, Li S, Wang Y and Liu H 2016 *IEEE/ASME Int. Conf. Mechatron. Embed. Syst. Appl.*
[19] Wu J, Peng H, Chen Q and Peng X 2014 *ISA Trans.* **53** 173–85
[20] Suresh S, Sundararajan N and Saratchandran P 2008 *Neurocomputing* **71** 1345–58
[21] Sundararajan N, Saratchandran P and Li V 2002 *Fully Tuned Radial Basis Function Neural Networks For Flight Control* Kluwer Academic Publishers
[22] Cai G, Taha T, Dias J and Seneviratne L 2016 *Proc. Inst. Mech. Eng. Part G J. Aerosp. Eng.* **231** 30-46
[23] Adiprawita W, Ahmad A S and Sembiring J 2007 *J. Bionic Eng.* **4** 237–44

IOP Conf. Series: Materials Science and Engineering **270** (2017) 012020 doi:10.1088/1757-899X/270/1/012020

Implementation of high-gain observer on low-cost fused IR-OS sensor embedded in UAV system

E Mohd Nor[1,2], S B Mohd Noor[2], M R Bahiki[3] and S Azrad[3,*]

[1]Department of Electrical and Electronics Engineering, Universiti Pertahanan Nasional, Malaysia
[2]Department of Electrical and Electronics Engineering, Universiti Putra Malaysia.
[3]Department of Aerospace Engineering, Universiti Putra Malaysia

*syaril@upm.edu.my

Abstract. This paper presents discrete time implementation of a high gain observer (HGO) and extended term to estimate the state velocity and acceleration from the position measured by a low-cost sensor installed on-board the unmanned aerial vehicle (UAV). Owing to the low-cost sensor, the signal produced from fused IR–OS is noisy and therefore, additional filters are used to remove the noise. This study proposes an alternative to this standard and tedious procedure using HGO. The discrete time implementation of HGO and its extended term is presented and ground tests are conducted to verify the algorithm by inducing a dynamic motion on the UAV platform embedded with the fusion IR-OS onboard. A comparison study is conducted using standard numerical differentiation and ground truth measurement by OptiTrack. The results show that EHGO can produce a velocity signal with the same quality as that of differentiated signal from fused IR-OS using Kalman filter. The novelty of HGO lies in its simplicity and its minimal tuning of parameters.

1. Introduction

Unmanned aerial vehicles (UAVs) have been perceived to have good military and civilian potentials to serve various missions such as aerial surveillance, search and rescue, and monitoring applications. In the past few years, apart from controller development for attitude stabilization, the work in design and development of UAV platform is primarily focused on enhancing the autonomy level for autonomous navigation and collision avoidance control. Providing accurate state estimation is vital for successful and reliable implementation of the controller algorithm, which consequently results in stability control of the complete UAV. Position information of UAV can be directly obtained from Global Positioning System (GPS) for outdoor applications. Meanwhile, several studies on indoor flight have been directed to the use of low-cost sensors like infrared, sonar and ultrasonic [1-3]. However, sensor measurement for the velocity and integral position is not always available and approximation is usually required.

Directly differentiating the position signal using numerical calculation is the traditional method to obtain velocity. However, this method is highly susceptible to noise and errors [4]. A Kalman filter-based velocity estimator has been developed to get smooth velocity data from position measurement using IR and sonar sensors [1]. Similar procedure using Kalman filter to estimate the smooth velocity has been presented in another study using fused data from IR and ultrasonic sensor [2]. Kalman filter is widely used for smoothing noisy data and as an estimator due to its simplicity and easy application.

However, its effectiveness is limited only to the system that has been linearized. Other state estimation approaches reported are based on state transformation and therefore, the nonlinearities only depend on input and outputs [5], and the use of linear matrix inequality (LMI) techniques [6-7]. However, these methods involve complex gain formulas and require solving LMI or partial differential equations. The state estimation based on design of robust observers has been reported using high-gain [8] or sliding mode method [9]. The advantage of robust estimator is its capability to be directly applied to estimate rapid dynamics from complex, nonlinear system, in which linearization is not required.

This paper focuses on design of robust estimator using high-gain observer (HGO). Development of HGO over the last two decades has been reported in Ref. [8]. In an ideal differentiator, the noise in the signal will be differentiated together with the signal, thus generating large derivatives along with the differentiated signal. However, the basic concept of HGO is to simultaneously estimate derivative of a signal and quantify the noise in the signal through a few gain ratios [10]. An extended term from HGO has been proposed to estimate higher dynamics of nonlinear system, in addition to providing estimated velocity [11]. Extended high-gain observer (EHGO) has been demonstrated in an advance acquisition board with high-quality sensor of good precision. Therefore, the objective of this paper is to study the real time performance of HGO and its extended term in estimating velocity and acceleration from the position information obtained by fused measurement of low-cost infrared (IR) and ultrasonic sensor (OS).

2. Background Study

A nonlinear system is given in the normal form as shown in the following Equation 1, where$x \in R^2 = (x_1, x_2), x_1$ is the position and x_2 is velocity, y is measured output and u is control signal [12].

$$\begin{aligned} \dot{x}_1 &= x_2 \\ \dot{x}_2 &= f(x) + g(x)u \\ y &= x_1 \end{aligned} \tag{1}$$

In this system, only position state is measured. Given state feedback stabilizing controller, $u = \gamma(x_1, x_2)$, the velocity state x_2 has to be estimated before the controller can be implemented. The HGO algorithm is proposed in Equation 2, which is to obtain the first differentiation of position signal [13].

$$\begin{aligned} \dot{\hat{x}}_1 &= \hat{x}_2 + \frac{\alpha_1}{\varepsilon}(x_f - \hat{x}_f) \\ \dot{\hat{x}}_2 &= \hat{x}_a + \frac{\alpha_2}{\varepsilon^2}(x_f - \hat{x}_f) \end{aligned} \tag{2}$$

By adding one more integrator, the second differentiation term is estimated by Equation 3.

$$\dot{\hat{x}}_a = \frac{\alpha_3}{\varepsilon^3}(x_f - \hat{x}_f) \tag{3}$$

The complete expression of Equation 2 and Equation 3 is known as EHGO [11,14]. The term ε is the high-gain term and the value varies depending on the system. The values of $\alpha_1, \alpha_2, \alpha_3$ are called the set of polynomials, which is chosen such that the polynomials in Equation 4 are Hurwitz and the states $\hat{x}_2$ are estimates of velocity x_v, $\hat{x}_3$ are estimates of acceleration.

$$s^3 + \alpha_1 s^2 + \alpha_2 s + \alpha_3 \tag{4}$$

The estimator algorithms in Equation 2 and Equation 3 are in continuous form. The algorithm needs to be converted into a discreet form for the real-time implementation. Inspired by the work in Ref. [4], the digital form is obtained by the following procedures. First, new observer variables $q = [q_1, q_2, q_3]$ is defined as in Equation 5, where ε is the observer gain and $\hat{x} = [\hat{x}_1, \hat{x}_1, \hat{x}_1]$ are the estimated states from Equation 2 and Equation 3.

$$q_1 = \hat{x}_1 \, , q_2 = \epsilon \hat{x}_2 \, , q_3 = \epsilon^2 \hat{\sigma}_3 \tag{5}$$

Then, the dynamics of the new variables in Equation 5 are derived by Equation 6.

$$\begin{bmatrix} \dot{q}_1 \\ \dot{q}_2 \\ \dot{q}_3 \end{bmatrix} = \begin{bmatrix} -\alpha_1/\varepsilon & 1/\varepsilon & 0 \\ -\alpha_2/\varepsilon & 0 & 1/\varepsilon \\ -\alpha_3/\varepsilon & 0 & 0 \end{bmatrix} \begin{bmatrix} q_1 \\ q_2 \\ q_3 \end{bmatrix} + \begin{bmatrix} \alpha_1 \\ \alpha_1 \\ \alpha_1 \end{bmatrix} y$$

$$y = \begin{bmatrix} 1 & 0 & 0 \\ 0 & 1 & 0 \\ 0 & 0 & 1 \end{bmatrix} \begin{bmatrix} q_1 \\ q_2 \\ q_3 \end{bmatrix} + \begin{bmatrix} 0 \\ 0 \\ 0 \end{bmatrix} \tag{6}$$

Equation 6 is the state–space model that will be used for discretization to digital form. The state–space is converted into digital form in MATLAB environment. Sampling rate has to be defined at this stage. Upon conversion, new A, B, C and D matrices that are produced will be coded into C# in GUI-based terminal software developed by Microsoft Visual Basic as the integrated environment. Re-conversion from digital form to continuous form is as shown in Equation 7.

$$\begin{bmatrix} \hat{x}_1 \\ \hat{x}_2 \\ \hat{x}_3 \end{bmatrix} = \begin{bmatrix} 1 & 0 & 0 \\ 0 & 1/\varepsilon & 0 \\ 0 & 0 & 1/\varepsilon^2 \end{bmatrix} \begin{bmatrix} q_1 \\ q_2 \\ q_3 \end{bmatrix} \tag{7}$$

3. Low-Cost Collision Avoidance System

The hardware setup of motion tracking system is presented in this section. An experimental validation is performed on the low-cost collision avoidance system developed in the study presented in Ref. [2]. Data fusion from IR and OS sensors, combined with linear Kalman filter, is developed to estimate the relative position measurements between the UAV and the obstacles. The proposed collision avoidance algorithm requires the position and velocity states.

The IR–OS sensors are mounted in between quadrotor arms. The overall monitoring system shown in Figure 1 consists of quadrotor vehicle with mounted IR–OS sensors, pair of XBee wireless routers, ground PC station and OptiTrack motion tracking system based on IR cameras. The Optitrack will be the ground truth for validating the motion given by the onboard fused IR–OS sensors. Block diagram of the UAV monitoring system is illustrated in Figure 2.

Figure 1: Four sets of fused IR-OS sensors (shown in circled)

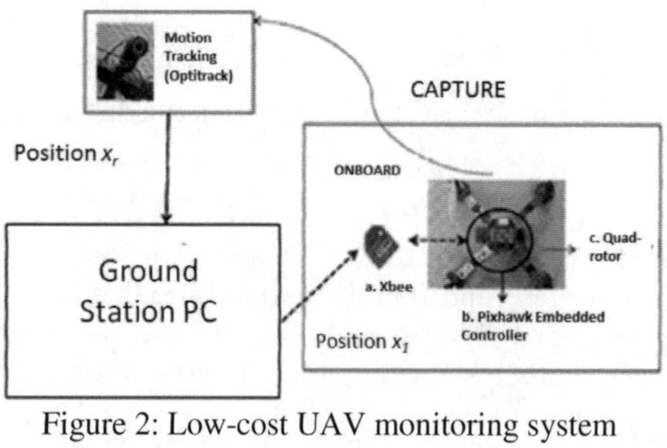

Figure 2: Low-cost UAV monitoring system

The ground station is a PC that runs on the Windows operating system using a GUI-based terminal software developed by the Microsoft Visual Studio as the integrated development environment. The quadrotor vehicle communicates with the ground station via Xbee wireless routers at a frequency of 20 Hz. The position x_1 is the signal obtained by the fused data using low-cost IR–OS sensor that has been filtered by a linear Kalman filter to obtain smooth position estimation, which is embedded onboard the quadrotor. The signal is wirelessly transmitted by Xbee to the ground PC. Meanwhile, the position x_r is the ground truth signal measured by OptiTrack indoor monitoring and transmitted to the ground PC station where the output feedback law is implemented.

In standard practice, the velocity of a vehicle is estimated from the rate of change of the position range data, and the data is filtered using linear Kalman filter to obtain smooth velocity estimation. The rate of change is based on basic numerical differentiation in Equation 8. Additionally, the acceleration based on numerical differentiation in Equation 9.

$$x_v = \frac{x_{1(n+1)} - x_{1(n)}}{t_{(n+1)} - t_{(n)}} \tag{8}$$

$$x_a = \frac{x_{v(n+1)} - x_{v(n)}}{t_{(n+1)} - t_{(n)}} \tag{9}$$

The overall structure of computing the velocity and acceleration from position x_1 by the fused IR-OS sensor using standard numerical differentiation procedure and the estimator is illustrated in Figure 3. The performance obtained from both methods is obtained.

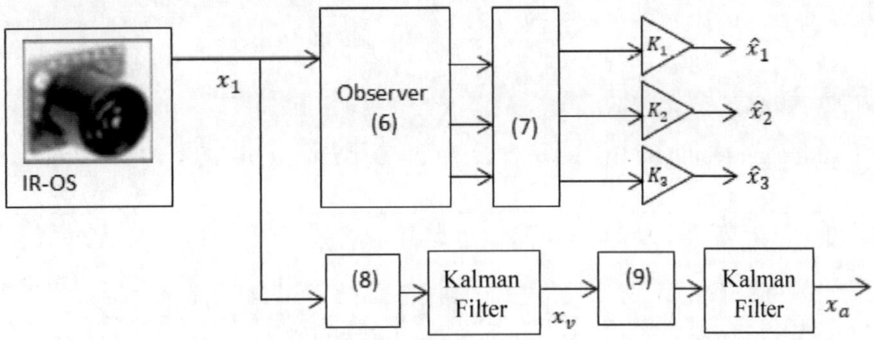

Figure 3: Structure to obtain velocity and acceleration from IR-OS position measurement

4. Results and Discussions

The observers sets of polynomial in Equation 4 are assigned as follows:$\alpha_1 = 9$, $\alpha_2 = 9$, $\alpha_3 = 3$. In the meantime, the sampling time $T_s = 0.05$, which is selected based on frequency of the motion tracking system that is 20 Hz. The performance of the position state estimated using Equation 7 and Equation 6 is first compared with two different observer gains, $\varepsilon = 0.05$ and 0.01 as shown in Figure 3. The results of this comparison are presented in Figure 4 and also Figure 5, which show the position of the UAV as measured using the proposed estimator, ground truth position as measured by OptiTrack and also the measurement using the fused IR–OS. The plots aim to highlight the closeness of the estimated position from EHGO with the position measured by IR–US. The results show that the estimated position with observer gain $\varepsilon = 0.01$ is nearly indistinguishable from the position x_1 measured by the fused IR–OS sensor, as shown in Figure 5. On the other hand, Figure 4 shows that the estimated position with the position from fused IR–OS. The amplitude is larger compared to the ground truth measurement. This result concludes that the proposed estimator is workable at a high gain. The aforementioned results indicate that the gain has to be 100. When the gain is reduced, the position error between the estimated and measured position will increase. Eventually, as gain is 50, the error becomes extremely large and unacceptable because the signal amplitude is larger than the ground truth measurement.

The real-time velocity generated from the proposed estimator is compared with the result obtained from the numerical differentiation using IR–OS and the ground truth using OptiTrack, and they are as shown in Figure 6. The result obtained from HGO uses gain $\varepsilon = 0.01$.

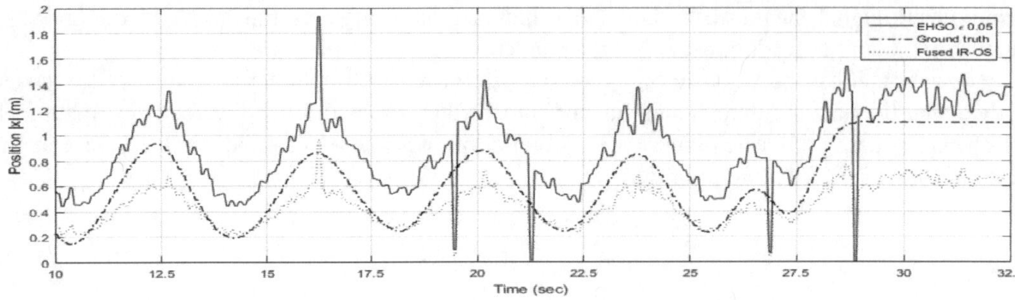

Figure 4 : Position generated by EHGO = $\hat{x}_1$ using $\varepsilon = 0.05$, fused IR-OS = x_1 and ground truth = x_r

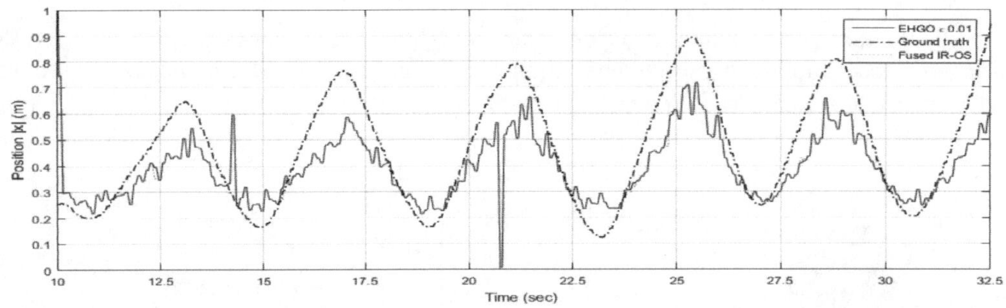

Figure 5: Position generated by EHGO = $\hat{x}_1$ using $\varepsilon = 0.01$, fused IR-OS = x_1 and ground truth = x_r

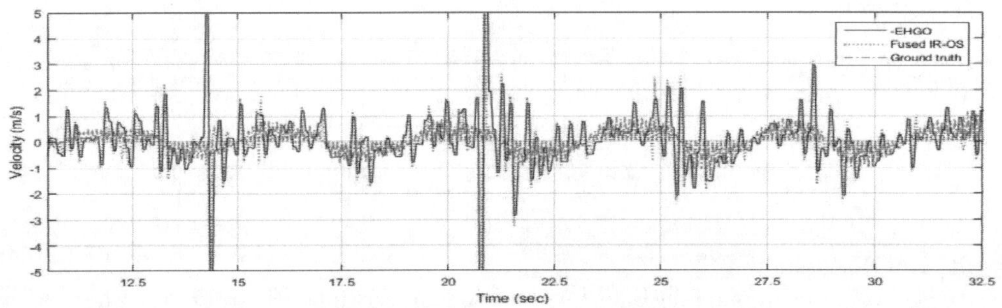

Figure 6: Velocity generated by EHGO = $\hat{x}_2$ using $\varepsilon = 0.01$, fused IR-OS = x_v and ground truth

Based on Figure 6, the estimated velocity by EHGO has a similar performance to the velocity from fused IR-OS. The velocity from fused IR-OS is obtained as illustrated by x_v in Figure 3. The results show that EHGO is evidently smoother in comparison to the sharp spiking observed in fused IR–OS. However, the quality of velocity signal from fused IR-OS is not as good as the quality obtained from the ground truth using OptiTrack. Although both signals are actually differentiated using the same rate of change algorithm in Equation 8, the fused IR–OS needs further filtering to obtain smoother velocity that is comparable to the ground truth.

Finally, the results in the acceleration signal generated by HGO and filtered differentiation signal are as in Figure 7. The obtained performance is similar to previously discussed velocity performance. The EHGO can estimate acceleration signal with nearly similar performance to the result in fused-IR, which is obtained by the numerical differentiation in Equation 9 and filtered noise using linear Kalman filter. The estimated signal from HGO is smooth and shows less amplitude spiking. However, similar to the velocity signal, the quality of acceleration signal obtained from the fused IR–OS is not as good as the quality generated by the ground truth. A filter is required to filter the noise of velocity signal prior the differentiation to obtain the acceleration signal.

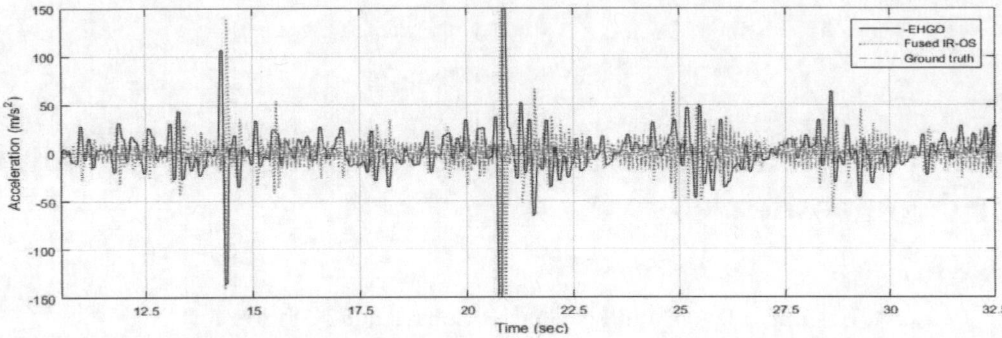

Figure 7: Acceleration generated by EHGO = $\hat{x}_3$ using $\varepsilon = 0.01$, fused IR-OS = x_a and ground truth

5. Conclusion

The real-time performance of EHGO has been presented to estimate velocity and acceleration states from position signal measured by on-board fused IR–OS motion tracking sensors. In standard practice, the signal needs to pass through two Kalman filters to remove the noise amplitude before and after numerical differentiation. However, by using the proposed estimator with the correct gain values, the performance quality of the velocity and acceleration signals is made slightly better. In particular, the signal obtained is smooth and has no sharp spiking. Therefore, the estimator based on HGO algorithm can be used as an alternative method to obtain the velocity and acceleration signals to replace the long computational procedure using filters. The HGO algorithm is robust such that it is able to provide smooth velocity and acceleration signals are provided without the need to use additional Kalman filter to remove noise.

References

[1] Sobers D M, Chowdhary G and Johnson E N 2009 *AIAA Guidance, Navigation and Control Conference*
[2] Rambabu R, Bahiki M R and Md Ali S A 2015 *ARPN J. Eng. Appl. Sci.* **10** 10012–7
[3] Chee K Y and Zhong Z W 2013 *Sensors Actuators A. Phys.* **190** 66–76
[4] Dabroom A M and Khalil H K 1999 *International Journal of Control* **72** 1523-37
[5] Ding Z 2005 *IEE Proceedings - Control Theory and Applications* **152** 460–4
[6] Fan X and Arcak M 2003 *Syst. Control Lett.* **50** 319–30
[7] Açikmeşe B and Corless M 2011 *Automatica* **47** 1339–48
[8] Khalil H K and Praly L 2014 *International Journal of Robust and Nonlinear Control* **24** 991–2
[9] Drakunov S and Utkin V 1995 *34th IEEE Conference on Decision and Control*
[10] Vasiljevic L K and Khalil H K 2006 *45th IEEE Conference on Decision and Control*
[11] Freidovich L B and Khalil H K 2008 *IEEE Trans. Automat. Contr.* **53** 2324–34
[12] Isidori A, Marconi L and Serrani A 2003 *IEE Trans. Automtic Control* **48** 413–26
[13] Khalil H K 2015 *Nonlinear Control* Pearson

IOP Conf. Series: Materials Science and Engineering **270** (2017) 012021 doi:10.1088/1757-899X/270/1/012021

Vacuum infusion method for woven carbon/Kevlar reinforced hybrid composite

N Hashim[1,*], D L Majid[1], N Uda[2], R Zahari[1] and N Yidris[1]

[1]Department of Aerospace Engineering, Faculty of Engineering, Universiti Putra Malaysia, 43400 Serdang, Selangor , Malaysia.
[2]Department of Aeronautics and Astronautics, Graduate School of Engineering, Kyushu University, 744 Motooka, Nishi-ku, Fukuoka, Japan.

*nurain.binti.hashim@gmail.com

Abstract. The vacuum assisted resin transfer moulding (VaRTM) or Vacuum Infusion (VI) is one of the fabrication methods used for composite materials. Compared to other methods, this process costs lower than using prepregs because it does not need to use the autoclave to cure. Moreover, composites fabricated using this VI method exhibit superior mechanical properties than those made through hand layup process. In this study, the VI method is used in fabricating woven carbon/Kevlar fibre cloth with epoxy matrix. This paper reports the detailed methods on fabricating the hybrid composite using VI process and several precautions that need to be taken to avoid any damage to the properties of the composite material. The result highlights that the successfully fabricated composite has approximately 60% of fibres weight fraction. Since the composites produced by the VI process have a higher fibre percentage, this process should be considered for composites used in applications that are susceptible to the conditions where the fibres need to be the dominant element such as in tension loading.

1. Introduction

Composite material, which has been used domestically since 1960s, is one of the superior materials that can replace metals and wood in various application fields. Due to its lighter weight, anti-corrosion and also good capability in sustaining high loads, composite material is often preferred as alternative for steel, aluminium and wood in some applications like bridge structures, wind turbine blades, aircraft structures and automobile parts. Since composite material consists of two different constituents (fibre and matrix), they can be chosen accordingly to design various composite materials with characteristics based on the required application. Polymer matrix composite material is one of composite materials that have been used widely in the manufacturing industries. This material generally consists of fibres and polymer matrix, and also called as Fibre Reinforced Polymer (FRP) composite. The fibres used in these materials commonly have higher strength and modulus than the matrix, hence making it the main component to sustain the load [1].

There are different processes invented for fabricating the FRP composites, which are hand lay-up, prepregs and Vacuum-assisted Resin Transfer Moulding (VaRTM). Hand lay-up method is one of the easiest methods but it can be rather inconvenient as the resin has to be distributed evenly by hand. Air bubbles that appear during the process are also difficult to be removed using human power, especially in a large scale fabrication. Using prepregs is one of the expensive processes that use the autoclave to cure. This method is widely used for fabricating the primary aircraft structures [2–4]. VaRTM or also known as Vacuum Infusion (VI) process is invented to reduce the manufacturing cost of the composite using prepregs [5]. This process has been developed such that composite materials can be fabricated and cured without using the autoclave. It has also been proven that the composites fabricated using the VI process are not only low in cost [4] but also of high quality that makes them adequate to be used in primary structures of an aircraft [3–4]. Previous works have compared the performance of composites made by hand lay-up and VI processes, and it is found that the composites made from the VI process have better performance in mechanical properties [6–8].

Although VI process has been proven to be able producing high quality composite materials that are suitable to be used in aircraft structures, this fabricating process has always been overlooked since using prepregs is much more popular in recent days. Most of previous studies have investigated on the flow rate, permeability and compaction pressure effect on the composite laminates' properties [9–11]. However, a detailed discussion on the preparation technique and its effects on the properties of the composite materials are rarely found. To address this gap, this work discusses the steps followed and the precautions taken during fabrication of woven carbon/Kevlar reinforced epoxy hybrid composite laminates. The results are presented by weight fraction of fibres obtained from the finishing products and several steps that affect the properties of the final product are also discussed.

2. Methodology

The fabrication process of hybrid composite laminate starts with the preparation of the materials and tools. Moreover, the right dimension of the composite and required thickness are very important to be determined first to ensure that it is consistent with the requirements for mechanical testing. The fibres used in this work are the 3K carbon fibres and Kevlar-29. These two fibres are weaved together in one cloth, where carbon fibres as the warp and Kevlar fibres as the weft. The area of one cell of the carbon fibre is approximately 3mm^2 and 3.75mm^2 for the Kevlar fibre (Figure 1). The thickness of the woven cloth is approximately 0.26mm. The matrix of the composite was made of EpoxAmite 100 base epoxy resin, which can be cured in room temperature. To avoid the resin starts hardening too early during the process, 103 slow hardener is used with mix ratio by weight 100:24 (resin:hardener). The specification of the resin can be found in Table 1 as provided by the products' manufacturer.

The schematic diagram of the vacuum infusion method is shown in Figure 2. Ten plies of woven fibre cloths are stacked together on a glass laminate that has been coated with mould release wax. Two spiral tubes are placed at both ends of the fibre cloths. One of them is connected to outlet tube while the other is connected to inlet tube. The outlet tube is connected to the vacuum pump and this is where the vacuum pressure is applied during the fabrication process.

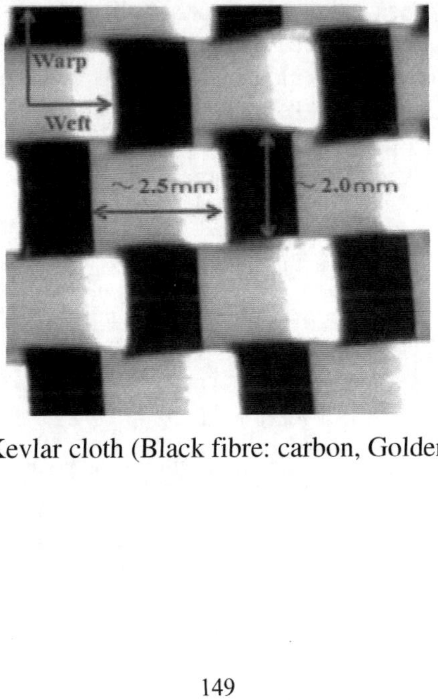

Figure 1: Woven carbon/Kevlar cloth (Black fibre: carbon, Golden yellow fibre: Kevlar 29)

IOP Conf. Series: Materials Science and Engineering **270** (2017) 012021 doi:10.1088/1757-899X/270/1/012021

Table 1: Specification of EpoxAmite 100 epoxy base resin with 103 slow hardener

Handling properties		Physical Properties	
Specific gravity	1.108 g/cm^3	Ultimate tensile strength	54.54 Mpa
Mixed viscosity	650	Tensile modulus	3.12 Gpa
Specific volume	9.1 x 10^{-4} m^3/kg	Flexural strength	84.81 Mpa
		Flexural modulus	2.96 Gpa
		Compressive strength	72.39 Mpa
		Heat deflection	53 °C

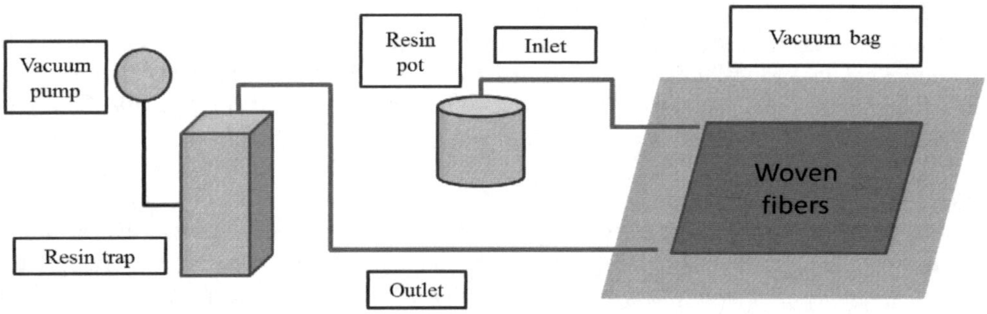

Figure 2: Schematic diagram of vacuum infusion process

A peel ply that works as a barrier between the fibre cloth and the vacuum bag is laid up first on the fibre cloth. It is very important to make sure that the peel ply used is big enough to cover the fibre cloths and spiral tubes. A mesh sheet, which works to accelerate the resin distribution inside the fibre cloths, is then placed on the peel ply. Finally, the fibre cloth is covered and sealed tightly with the vacuum bag. The arrangement diagram is shown in Figure 3. After the vacuum bag is tightly sealed, it is turned on to check for any occurrence of leak before the resin is infused. The leak is checked using the leak detector. The resin infusion process is started by setting the pump with low pressure, around -30kPa for several minutes. This step is done to observe any leaks around the vacuum bag and it is also claimed that this can minimize the air trap between the fibre layers and ensure that they are compacted steadily before the resin infusion [12]. The pressure is then increased to -60kPa and the resin started to be infused from the inlet tube. When the resin has already been distributed into half of the area of the fibre cloths, the pressure is once again increased to -80kPa and the resin is let to flow until it reaches the spiral tube at the other end of the fibre cloths. During this time, the flow rate of the resin starts to decrease and the pressure increases to -100kPa. When all resin has been completely infused into the fibre clothes, both inlet and outlet tubes of the laminates are closed tightly and they are set to cure in room temperature for 24 hours in vacuumed condition.

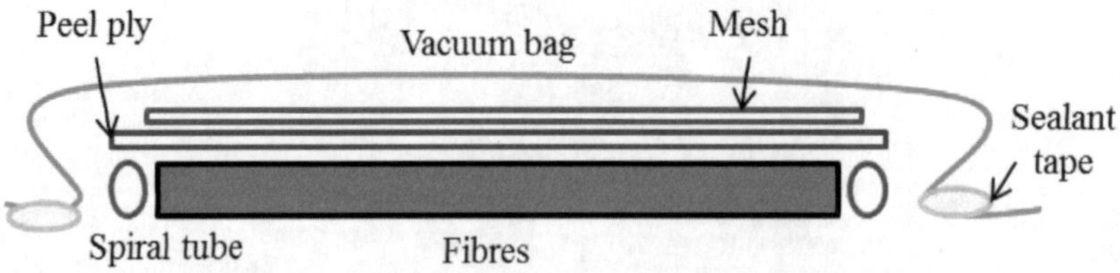

Figure 3: Arrangement of fibre cloths and fabrication tools for fabrication process

It is very important to identify several precautions during the process to avoid any damage to the composite laminate. Minor leaks sometimes can occur during resin infusion process, which is caused by tiny tear at the vacuum bag and leaks between the vacuum bag and the sealant tape. These leaks usually can be detected by observing any substantial air bubbles inside the laminate. When air bubbles appear during the resin infusion process, leak spots have to be closed immediately before closing the outlet tube. Air bubbles also can be removed by adding extra resin, in which the resin flow will take all air bubbles to the outlet tube. Failure in removing the air bubbles will possibly cause the occurrence of voids and the composite laminate cures in non-vacuum condition, which happens when the vacuum pressure stopped. The air trapped inside the laminate is eventually accumulated on the surface of the laminate and causes the vacuum bag to be lifted. An example of failed fabrication caused by leaks and air trap can be seen in Figure 4.

To measure the weight fraction and volume fraction of fibre and matrix in the fabricated composite laminates, density of the material have to be determined first. Density test is carried out following the ASTM D792 procedures. The composite laminates are cut into small pieces and tied with a string and immersed in distilled water (Figure 5). The amount of distilled water has to be enough for the material to be fully immersed and the material must not touch the bottom of the water container. Assuming that the weight of the string is negligible, density of the composite material can be obtained. Meanwhile, after the laminates are completely cured, its final weights are measured to determine the fibre weight fraction W_f and fibre volume fraction V_f. Weight fraction of fibres calculation is done using Equation 1, where w_f and w_c are the mass of fibres used and the mass of the composite laminate, respectively. The weight fraction of fibres obtained is then used to calculate the volume fraction of fibres, V_f by using Equation 2 and Equation 3. V_m is the volume fraction of the matrix in the composite laminates.

$$W_f = \frac{w_f}{w_c} \tag{1}$$

$$V_m = (1 - W_f)\,\frac{\rho_c}{\rho_m} \tag{2}$$

$$V_f = 1 - V_m \tag{3}$$

Figure 4: Air bubbles inside the laminate that cause the vacuum bag to be lifted up after resin infusion process is finished

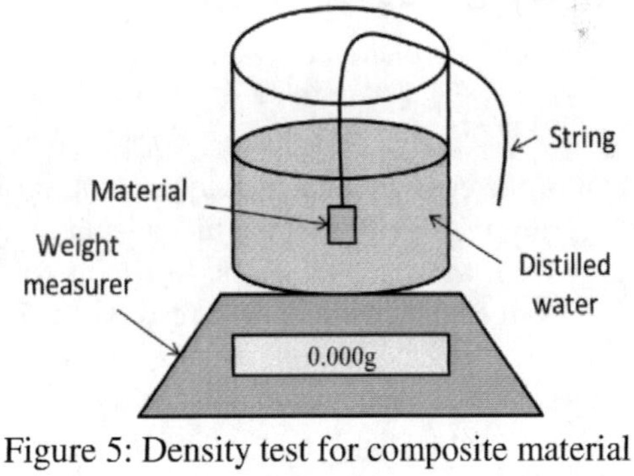

Figure 5: Density test for composite material

3. Results and Discussions

Eight laminates are fabricated using the VI process in this work. Three laminates have the dimension of 300mm x 500mm while the rest are fabricated with the dimension of 350mm x 500mm. The results from the density test are tabulated in Table 2. From the laminates' weight obtained, the weight fraction of fibre, W_f and fibre volume fraction, V_f are calculated and compiled in Table 3. In addition, several random samples have been cut into smaller rectangular shape (i.e. 250mm x 25mm) from laminate B. They are scanned under ultrasonic damage detector machine and the results are indicated in Figure 6. The yellow coloured area indicates that the area is not fully covered with resin. The thickness of these samples is also measured and the average value is shown in Table 4.

Table 2: Density of composite material obtained from the density test

Sample	Weight in air (g)	Weight immersed in water (g)	Composite material's density (g/cm²)
1	2.286	1.793	1.275
2	2.132	1.701	1.253
3	2.147	1.677	1.280
	Average and standard deviation:		**1.269 ± 0.01**

Table 3: Composites' weight fraction and volume fraction of each composite laminates

Laminate	Size (mm x mm)	Fibre's mass (g)	Composites' mass (g)	Fibre weight fraction, W_f	Fibre volume fraction, V_f
A*	500 x 300	280.61	537.82	0.522	0.447
B	500 x 300	282.26	448.89	0.629	0.571
C*	500 x 300	286.30	528.0	0.542	0.470
D	500 x 350	327.56	514.66	0.637	0.565
E	500 x 350	326.90	527.01	0.621	0.561
F	500 x 350	331.58	530.8	0.625	0.566
G	500 x 350	328.59	541.09	0.607	0.546
H	500 x 350	329.80	517.84	0.637	0.580

*Air bubbles occurred during fabrication and extra resin was infused to remove it. However, the leaks still occurred and the vacuum bag lifted up after the vacuum process finished.

Figure 6: Distribution of resin inside composite laminate B

Table 4: Thickness of composite laminates

Laminate	Thickness average (mm)
500mm x 300mm	2.2 ± 0.06
500mm x 350mm	2.39 ± 0.06

Out of the eight laminates of hybrid composite fabricated, six of them have been successfully cured without any major voids. From the calculation, it is found that the weight fraction of each laminate is ranging from 60% to 64% of the total composites' mass while the volume fraction of fibres obtained ranges from 55% to 58%. These results show that the complete fabricated composites are dominated by fibres in weight and volume composition, which also similar to the results obtained in [7, 8, 13]. However, for the failed composite laminates (i.e. laminate A and laminate C), adding the extra resin as a method of removing the air bubbles might be ineffective as it is found that it has many voids and lower fibre weight fraction compared to the other laminates. During the fabrication process, it is very crucial to ensure that the vacuum bag is sealed tightly to prevent any leaks that can bring damage to the composite laminate. Failure of removing the air bubbles can cause the appearance of voids.

From the ultrasonic scan results, resin can be seen to be distributed evenly at the centre area of the composite laminates. The resin is lacking at both ends of the specimen where the inlet and outlet tube are located. The uneven distribution of the resin is caused by the high pressure used during infusion process. It has been reported in a previous study that higher speed and pressure applied during vacuum infusion process for woven reinforcements will draw the resin out of the laminates faster and at the same time, the remaining resin in the inlet spiral tube would have partly cured and resulted in reduced flow [11]. This phenomenon will eventually cause insufficient wetting at the laminate area close to the inlet. Furthermore, lubrication effect can cause the fibre cloth's thickness to decrease once its wetted with resin [12, 14]. Lubrication effect is where the resin causes the fibre to rearrange and nested more closely, and decreases the thickness of the wetted area. Thus, the changes of fibre thickness are also causing the flow of resin towards the outlet tube becomes more difficult, resulting in insufficient and uneven resin distribution in the outlet area. Based on the results shown, if the sample cut from the laminate needs to have even distribution of resin, it is recommended to consider fabricating a bigger size of laminate with 10% excess area from the required sample size at both ends as shown in Figure 7.

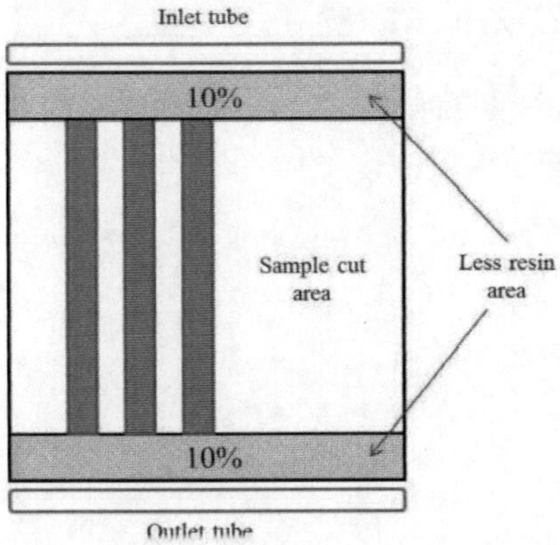

Figure 7: Recommended extra area needed during the vacuum infusion process

The results shown in previous Table 3 highlight that, even when each laminate has almost the same volume fraction of matrix, thickness of smaller samples cut from the laminates is slightly different. As the bigger plate has longer distance for the resin to flow, less resin can flow successfully to the outer area of the laminate. The findings indicate that different flow rates are achieved during the infusion process. Thus, different pressure level is needed to be applied according to the size of the fibre cloth.

4. Conclusion

All in all, woven carbon/Kevlar hybrid composite laminates are successfully fabricated using vacuum infusion method in this work. However, as the results show, even a small leak can damage the whole composite laminates. This process should be improved to minimize the occurrence of leaks by giving extra cautions during the leak checking process. The results also show that the amount of resin infuse, infusing pressure and fibre cloth size have significant influence to the thickness and resin distribution of composite laminate. The composite laminates made from the VI process have higher weight ratio of fibres and might be the most suitable to be applied on structures that are prone to tension loading.

Acknowledgement

This research work is financially supported by Universiti Putra Malaysia (Grant number: 9463100). The authors also wish to acknowledge the support and guidance received from Aerospace Structures & Structural Dynamics Lab, Graduate School of Engineering, Kyushu University.

References

[1] Wang R-M, Zheng S-R and Zheng Y-P 2011 *Polymer Matrix Composite and Technology* Woodhead Publishing
[2] Light Structures B. V. *http://www.lightweight-structures.com*
[3] Kamae T, Kochi S, Wadahara E and Shinoda T 2009 *17th International Conference on Composite Materials*
[4] Takeda F, Nishiyama S, Hayashi K, Komori Y, Suga Y and Asahara N 2005 *Mitsubishi Heavy Ind. Ltd. Tech. Rev.* **42** 1–6
[5] Vacmobiles 2012 *http://www.vacmobiles.com*

IOP Conf. Series: Materials Science and Engineering **270** (2017) 012021 doi:10.1088/1757-899X/270/1/012021

[6] Yuhazri M and Phongsakorn P 2010 *Int. J. Basic & Appl.* **10** 63–6
[7] Kim S Y, Shim C S, Sturtevant C, Kim D D W and Song H C 2014 *Int. J. Nav. Archit. Ocean Eng.* **6** 723–36
[8] Sevkat E, Brahimi M and Berri S 2012 *Polym. Polym. Compos.* **20** 321–32
[9] Ren P, Zhang Z, Li X, Ren F, Jin Y, Di Y and Fang Q 2010 *Polym. Compos.* **31** 2131–7
[10] Hammami A and Gebart B R 2000 *Polym. Compos.* **21** 28–40
[11] Hammami A 2001 *Polym. Compos.* **22** 337–48
[12] Yenilmez B, Senan M and Sozer E M 2009 *Compos. Sci. Technol.* **69** 1710–9
[13] Manjunatha C M, Taylor A C, Kinloch A J and Strasse C 2010 *Compos. Sci. Technol.* **70** 193–9
[14] Song X 2003 *Vacuum Assisted Resin Transfer Molding (VARTM): Model Development and Verification*

IOP Conf. Series: Materials Science and Engineering **270** (2017) 012022 doi:10.1088/1757-899X/270/1/012022

Buckling of thin walled composite cylindrical shell filled with solid propellant

A P Dash[1], R Velmurugan[1,*] and M S R Prasad[2]

[1]Department of Aerospace Engineering, Indian Institute of Technology, Madras, Chennai, India

[2]Defence Research and Development Organization, India

* ramanv@iitm.ac.in

Abstract. This paper investigates the buckling of thin walled composite cylindrical tubes that are partially filled with solid propellant equivalent elastic filler. Experimental investigation is conducted on thin composite tubes made out of S2-glass epoxy, which is made by using filament winding technique. The composite tubes are filled with elastic filler having similar mechanical properties as that of a typical solid propellant used in rocket motors. The tubes are tested for their buckling strength against the external pressure in the presence of the filler. Experimental data confirms the enhancement of external pressure carrying capacity of the composite tubes by up to three times as that of empty tubes for a volumetric loading fraction (VLF) of 0.9. Furthermore, the finite element based geometric nonlinearity analysis predicts the buckling behaviour of the partially filled composite tubes close to the experimental results.

1. Introduction

Composites have established itself as a strong alternative to metals in the field of rocketry. Composite rocket motor casings (CRMC) have become more popular in rocketry due to its inherent advantage in terms of specific strength. CRMC are filled with solid propellant at few different volumetric loading to meet the various mission requirements. In order to reduce the size of the rocket, volumetric loadings of 90% and above have become common practice in many cases. Solid propellant is a polymeric material that is poured into CRMC and cured at a specified temperature environment. This polymer is similar to a soft elastomeric material with regards to its mechanical behavior. At the time of structural design of the CRMC, one considers all possible load cases that the rocket undergoes throughout its mission. There are occasions when the rocket motor casings experience external pressure loading, which gives rise to a critical load case for which one has to design CRMC for buckling.

It has been a general practice to design CRMC alone to withstand buckling loads and the presence of solid propellant inside it is ignored. In this process, design of CRMC ends up with a conservative estimate. Efficient use of filament wound fiber reinforced CRMC requires a better knowledge of the structural response of the composite structure in the presence of solid propellant. Buckling behavior of thin shells with inner soft elastic filler, both with axial and external pressure loading, was investigated by many early researchers [1-4]. The closed form analytical approach to handle such cases has been discussed for isotropic cylindrical shells. The shell supported by an elastic foundation is the closest approach used by the authors to replicate the behavior of the shells in the presence of solid propellant. Karam and Gibson (1995) have reported experimental results of cylindrical silicon rubber shells with

IOP Conf. Series: Materials Science and Engineering **270** (2017) 012022 doi:10.1088/1757-899X/270/1/012022

and without complaint core under uniaxial load [5]. In their study, the thickness to radius ratio (t/r) of shells were varied from 0.02 – 0.23. The thickness variation of the shells for the test was indicated to be of the order of 5% of nominal thickness and the elastic modulus ratio of the shell to core (E_s / E_c) was of the order of 100. An increase of axial load carrying capacity of filled shells to an extent of 10 times with respect to empty tube has been reported in this study. Furthermore, Li (1998) has reported improvement of stability based on the results of finite element based theoretical investigation on metal and composite shells in the presence of elastic fillers [6]. In this study, composite shell was modeled using Sander's non-linear shell theory and filler propellant was modeled as Winkler-type foundation. This study reports extensively on the effect of initial geometric imperfections. A drop in buckling load carrying capacity to the tune of 11% has been indicated for isotropic shell with 1-20% of thickness imperfection. However, the effect of geometric imperfection was reported to be negligible in the case of the composite cylindrical shells.

It is a well-known fact that the buckling behavior of cylindrical structures is very much influenced by initial imperfections. Tsouvalis et al (2003) have studied the effect of geometric imperfections on buckling behavior of laminated composite cylindrical shells under external hydrostatic pressure [7]. In the study, both experimental and theoretical results are reported on carbon epoxy composite cylinders. The research concluded that it is essential to have a non-liner analysis to capture the behavior and it is enough to model the cylinder as a perfect mid-surface with average thickness that gives a knockdown factor of 0.6 to 0.7 on the buckling load in comparison to a perfect cylinder. This present study focuses on evaluating the effect of the presence of solid propellant (soft elastic filler) with various volumetric loading in a composite shell. The enhancement in the buckling load carrying capacity of the CRMC under external pressure loading at various volumetric loading has been experimentally investigated for quantification. Finite element based theoretical estimates are also presented for comparative purposes.

2. Composite Tubes

Depending upon the application, sizes of CRMC vary in their geometric parameters. In the present day rocketry, many of the CRMC sizes fall in the bracket of L/D between 1 to 5 and r/t between 50 to 150, where L, D, r and t are length, diameter, radius and thickness of the casing, respectively. The emphasis of this study is on behavior of thin walled composite tubes that are partially filled with solid propellant equivalent material. The cylindrical shells with L/D = 3 and r/t = 100 are chosen in the investigation, which is a fair representation of the class of CRMC in practical use. Based on the application, various types of composite material such as carbon-epoxy, glass-epoxy and Kevlar-epoxy have been used in the manufacturing of CRMC. In this study, glass epoxy is chosen as the material for the tubes. Figure 1 shows basic dimensions of the finished composite tubes. The selected tubes have an internal nominal diameter of 150mm and thickness of 2mm.

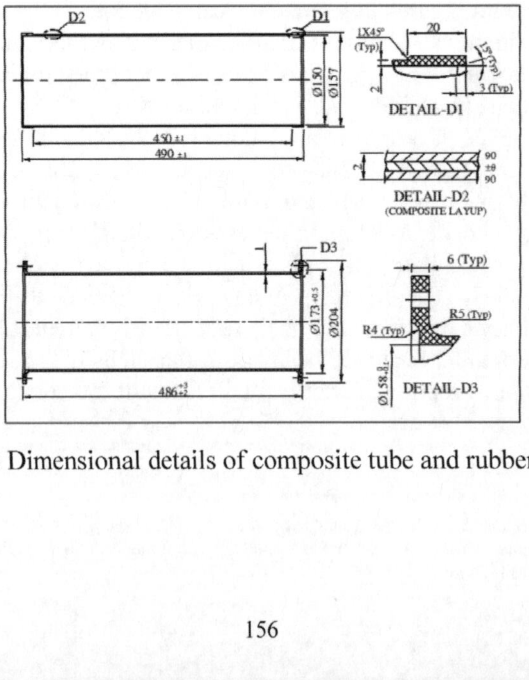

Figure 1: Dimensional details of composite tube and rubber bladder

S2-glass based tow pregs with epoxy resin has been chosen for the fabrication of composite tubes using filament winding technique. Figure 2 shows the steel mandrel on which CNC winding program is used to ensure a good quality product with acceptable repeatability in winding while Figure 3 shows the glass-epoxy tubes under winding. Infrared lamps are employed to ensure the proper resin softening during winding of tubes using tow pregs.

Figure 2: Steel mandrel with POP end domes ready for winding

Figure 3: Glass epoxy tube under winding using S2 glass tow preg

The nominal winding thickness of 2mm was achieved with individual layer thickness of 0.5mm. On the components, fibre volume fraction (V_f) of 0.6 and density of 2.09 g/CC were obtained. After the winding and extraction of the composite tubes, they were inspected using ultrasonic technique to ensure there was no delamination of layers or any other damage to the tubes during processing. Two different winding patterns were chosen to make two types of tubes. They were designated as 'Series-1' with 90°, $\pm$ 45°, 90° winding and 'Series -2' with 90°, $\pm$ 30°, 90° winding pattern. Eighteen tubes from each of the series were fabricated for conducting experimental research. It is a well known fact that buckling load is a strong function of the initial geometry imperfection of the structure. All of these 36 tubes prepared for the experiments were dimensionally inspected to check for any geometric variation in diameters. Based on this measurement, initial radial geometric imperfection of 0.1mm was noticed on the tubes, which is 5% of the nominal thickness. This data was further utilized for finite element analysis later. Prior to the manufacturing of the tubes, the flat laminate level testing was conducted on S2 glass epoxy tow preg material. The achieved composite material properties are reported in Table 1. During the characterization of the laminates, suitable strain gauges were utilized for the determination of the Young's modulus and Poison ratio. In case of the Young's modulus, single axis gauge was used while for the Poisons ratio, a 90° rosette was employed.

Table 1: Laminate level engineering properties of S2-glass towpreg

Sl. No.	Property	ASTM standard used for testing	Average Value
1	Longitudinal tensile strength	D3039	1648 MPa
2	Longitudinal tensile modules	D3039	68 GPa
3	Major poison ratio (ν_{12})	D3039	0.26
4	Transverse tensile strength	D3039	20 MPa
5	Transverse tensile modulus	D3039	21 MPa
6	Longitudinal compressive strength	D3410	969 MPa
7	Transverse compressive strength	D3410	124 MPa
8	Inplane shear strength	D3518	40 MPa
9	Inplane shear modulus	D3518	5 GPa
10	Density	D792	2.09 g/cc
11	Glass transition temp	E1356	184°C

The finished composite tubes were then internally lined with standard ROCASIN rubber of 0.5mm radial thickness and cured. This rubber lining is vital to ensure a good bonding between the composite tube and solid propellant fill. The rubber-lined tubes were filled with inert solid propellant filler that has equivalent mechanical properties as that of standard HTPB based solid rocket motor propellant composition with 18% Aluminum. Flat samples of the inert propellant were tested to determine their mechanical properties using ASTM standard D638. The set up used for testing the samples and the test specimens are shown in Figure 4. A typical force–extension curve obtained from this uni-axial testing of the above inert propellant material is presented in Figure 5. Since use of strain gauge is not possible for propellant samples, extensometer is utilized to record the extension history. Data of this curve is utilized to derive the Mooney Rivlin constants, which are applied in finite element analysis. The filling of composite tubes was done for five different volumetric loading fractions (VLF), which is defined as the ratio of volume of propellant to the inner volume of the empty tube. Empty and filled composite tube samples prepared for testing are depicted in Figure 6.

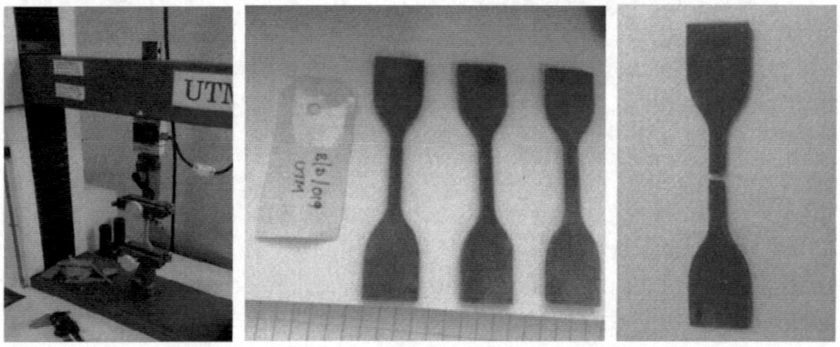

Figure 4: Inert propellant sample under uni-axial testing as per ASTM D638

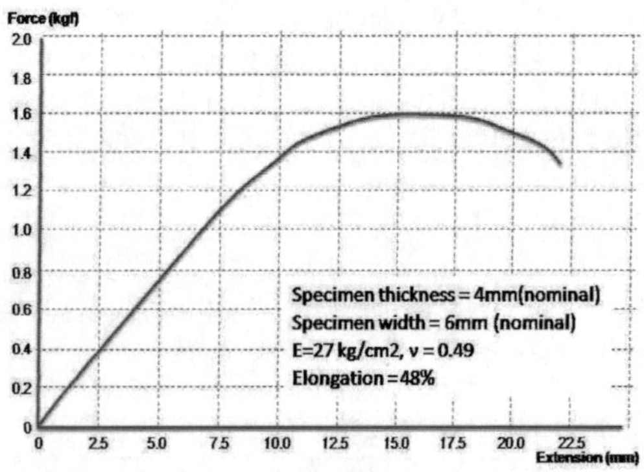

Figure 5: Experimental stress-strain curve of filler propellant material

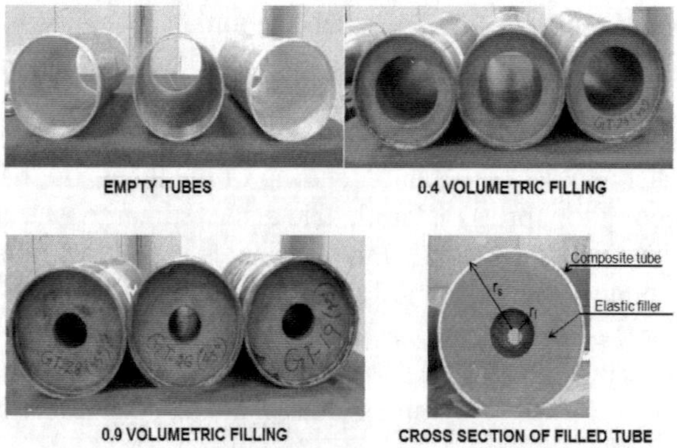

Figure 6: Typical photographs of empty and filled composite tubes

3. Experimental Test Setup

Since this investigation focuses on buckling of composite tubes under external pressure environment, a suitable tailor made test setup is realized to conduct these tests. Figure 7 presents the external pressure buckling setup. In short, a pressure chamber was fabricated out of structural steel with two end plates. Provision was made to support the shell under testing at either ends. The end conditions ensure that there is no translation and rotation of the shells at the point of support. The propellant filled composite tubes are assembled into the pressure chamber with two end plates supporting either end of the tube. The filament wound composite tubes are quite porous in nature and they need a mechanism to stop the entry of water through the thickness of the filament wound tubes. This has been achieved by sliding a thin

IOP Conf. Series: Materials Science and Engineering **270** (2017) 012022 doi:10.1088/1757-899X/270/1/012022

neoprene-based rubber bladder of 1mm thickness over the tube with the suitable dimension. To summarize, Figure 8 presents the photographs of the shell assembly at different stages.

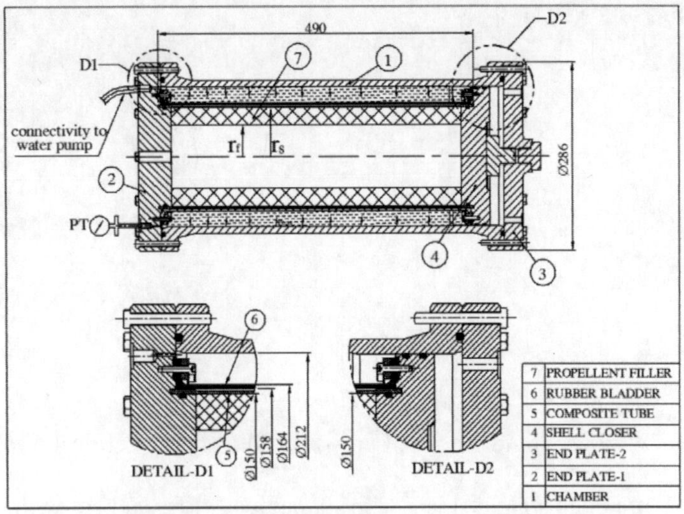

Figure 7: Schematic of test setup for external pressure buckling

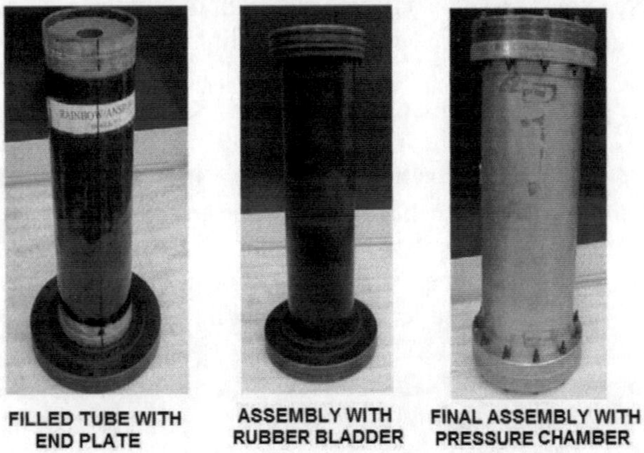

Figure 8: Stages of assembly of hardware for testing

The setup has a provision to fill water and pressurize the annular gap between the shell and the pressure chamber using pneumatic driven pump to apply external pressure to the specimen tube. Two pressure transducers are connected to the pressure chamber to record time history of pressurization. A suitable computerized data acquisition system is used to acquire the pressure data at the rate of 100 samples per second. Typical pressure time history record of the annular volume between the chamber and the tube is presented in Figure 9. The first pressure peak in the curve, after which annular pressure drops, is considered as the initiation of buckling of the composite tube. This pressure is referred to as experimental critical pressure, P_{cr}. Once the shell starts to buckle, the annular volume increases and the pressure starts dropping after this point.

4. Finite Element Analysis

Finite element (FE) analysis was done in this study to predict the behavior of empty and filled tubes. The composite shell was modeled using shell-181 while the solid elastic propellant was modeled using solid-186 element of commercially available ANSYS software.

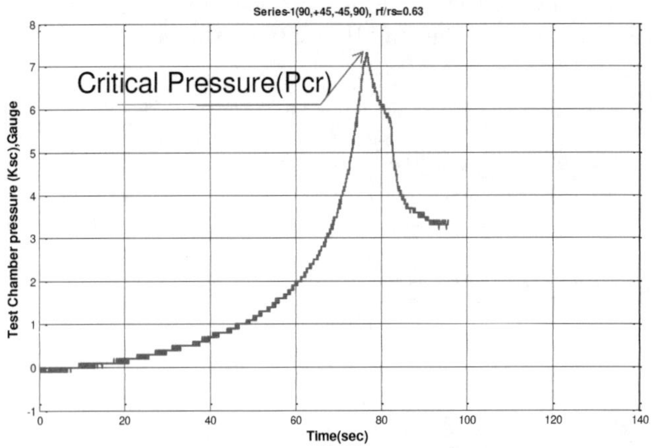

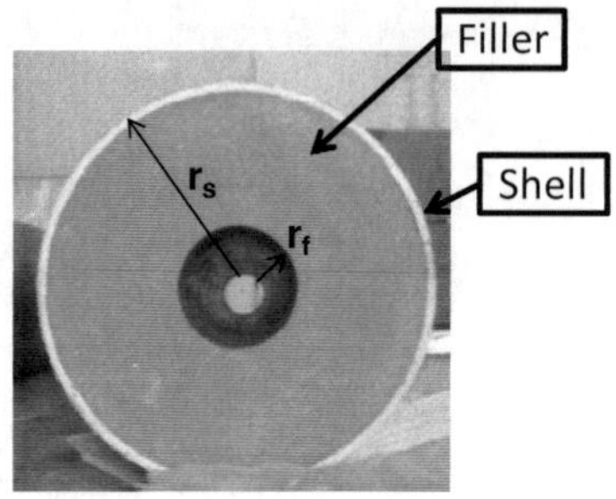

$$Radius\ ratio = \frac{r_f}{r_s} = \frac{Filler\ Radius}{Shell\ Radius}$$

$$Volumetric\ loading = \frac{Volume\ of\ filler}{Volume\ of\ shell}$$

Figure 9: Typical time history of pressure indicating the critical pressure

Solid-186 is a 3-D 20-noded solid element with six degrees of freedom at each node whereas shell-181 is a 4-noded element with six degrees of freedom at each node. Shell element caters for modeling the filament wound structure as a layered element. The shell elements available for modeling layered composites in ANSYS are shell-181 and shell-281, and both of them are suitable for analyzing thin to moderately-thick shell structures. Both are Rreissner-Mindlin shell theory elements and are well suited for large strain nonlinear applications. Nevertheless, an advantage of shell-181 over shell-281 is the reduced computation time required for solving the problem.

The accuracy of results by shell-181 to a particular problem is checked by mesh refinement studies. In this case, four layers of 0.5mm each with appropriate layer property were defined in shell element of the FE model. Figure 10 shows typical mesh adopted for the structure. Mesh convergence studies were taken up to select the size of elements. Convergence is taken with respect to the size of elements in terms of the curvature normal angle (Φ) and linear size (h) as indicated in Figure 11. Based on the results of this study, $\Phi = 3^o$ and h = 4mm is selected for all FE modeling.

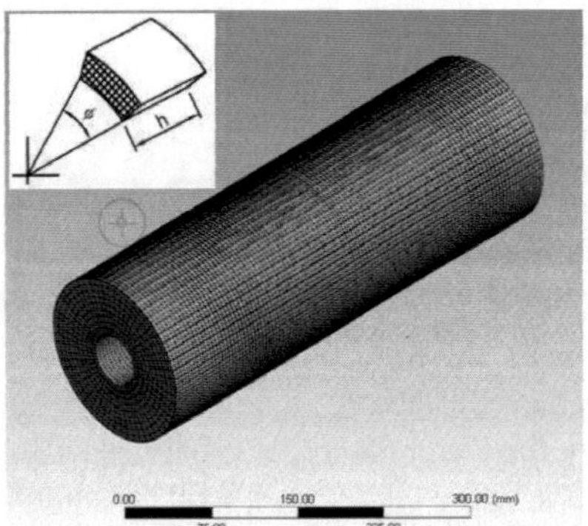

Figure 10: Typical mesh adopted for FE modeling

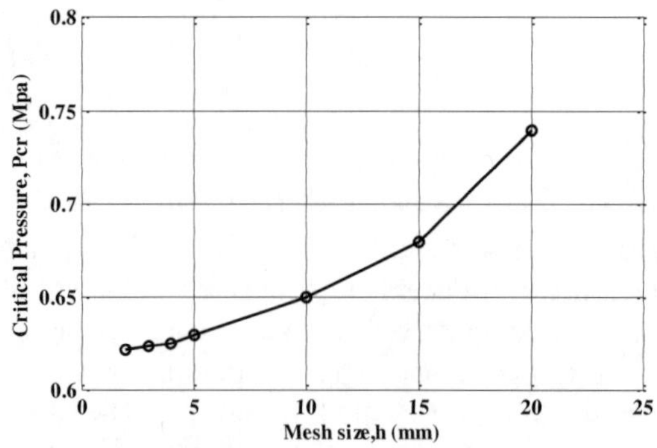

Figure 11: Mesh size convergence study results

The physical element size of 4mm x 4mm is reasonably good for the modeling. Clamped boundary condition was applied to either ends of the shell but the filler ends were not constrained. The external pressure was applied in steps to complete the non-linear analysis. At every step of the load application, deformed geometry of the tube due to previous load step was updated. Initial geometric imperfection of magnitude 0.1mm in the radius was considered for the non-linear buckling analysis. Using uni-axial test data reported in previous Figure 5, two-parameter Mooney-Rivlin model was extracted and used to represent the engineering property of the elastic filler propellant. Convergence of the solution was monitored to ensure convenience of each load stop. Pressure load versus radial displacement history of the tube structure was extracted from the FE analysis. Typical convergence graph reported in ANSYS is presented in Figure 12. The instability instance is indicated as the bisection occurrence point, which is the instance marked as the critical load.

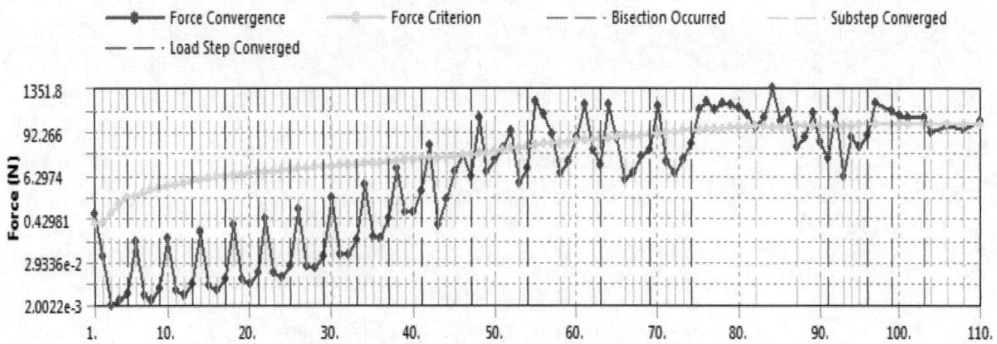

Figure 12: Convergence plot in ANSYS

Plots of load-displacement curves and critical pressure for Series-1 and Series-2 glass tubes are as presented in Figure 13 and Figure 14, respectively, for various volumetric loading factions obtained from FE non-linear analysis. The maximum pressure before initiation of instability is picked up as the Pcr from respective curve. It is observed that there is a small zone of linearity in the load-displacement curves generated in FE simulation. The radial deformation in these linear zones is less than 0.25mm, which is very small in magnitude. On the other hand, nonlinearity is observed early in the response. Since the solution in each load step has converged, the prediction of Pcr is accepted to be accurate. However, the overlapping trend of slope for the curves in the linear zones needs further investigation. In this present study, the focus is to reach the critical pressure zone. Hence 25 load steps were selected for the solution, which is assumed to be adequate.

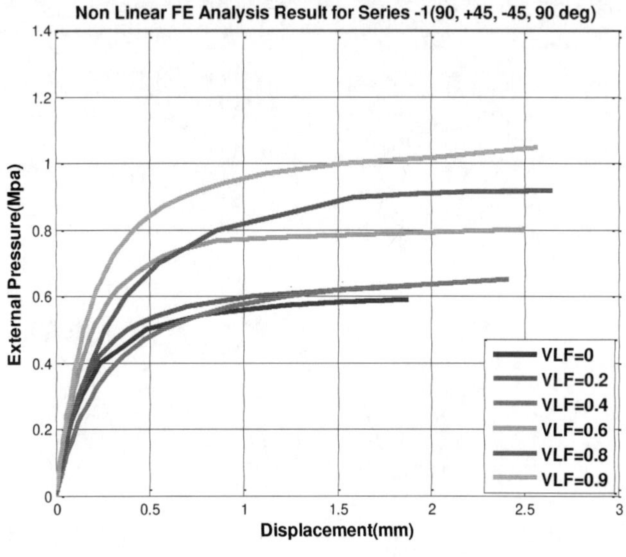

Figure 13: FE based load deflection curve for series-1 tubes

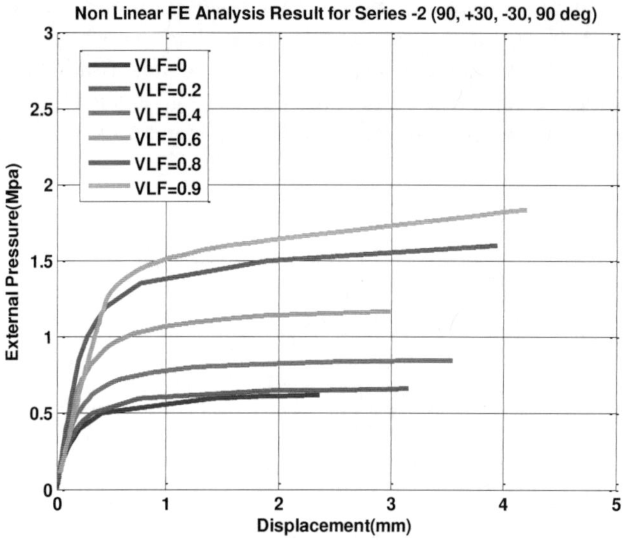

Figure 14: FE based load deflection curve for series-2 tubes

The first mode shapes of the composite tubes that are obtained in the linear buckling analysis with different VLFs are presented in Figure 15. It can be seen that the tubes under external pressure exhibit four full lobes circumferentially and one half lobe in longitudinal direction. No mode shape pattern change is observed between different filling ratios for the given dimensions of the tube combinations and the mechanical properties of tube and elastic filler. Therefore, the increase in VLF is not sensitive in terms of changing the mode shape for the tube under investigation.

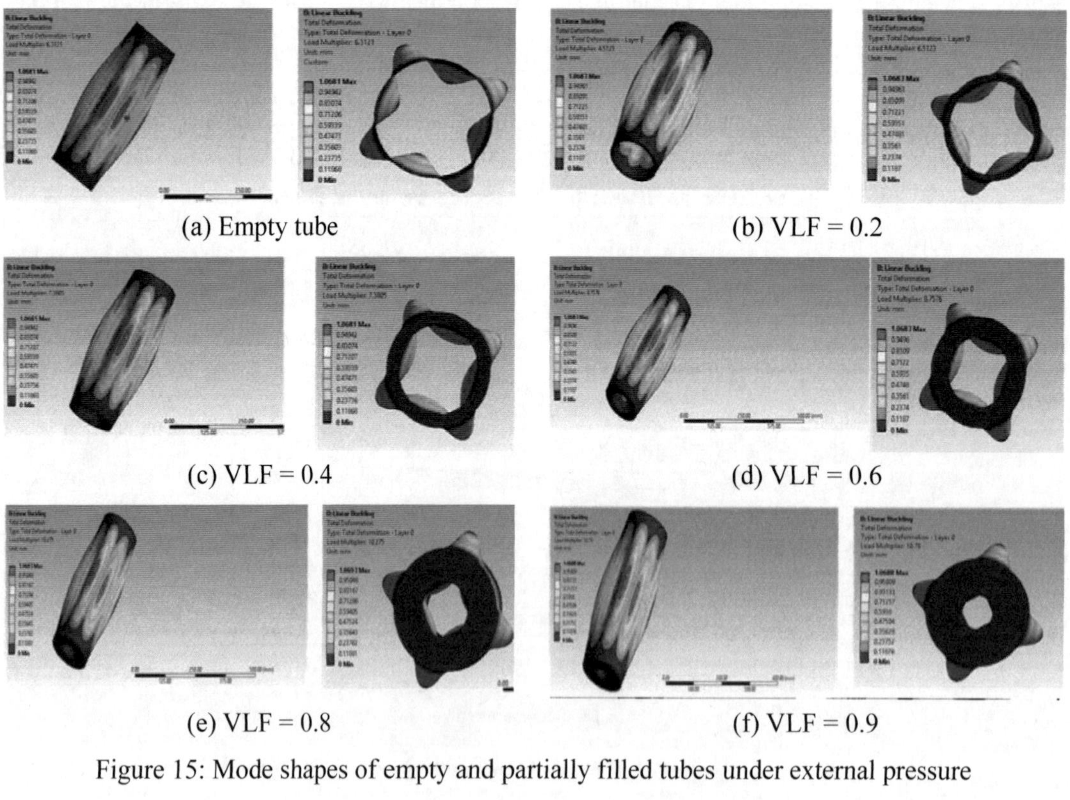

(a) Empty tube (b) VLF = 0.2

(c) VLF = 0.4 (d) VLF = 0.6

(e) VLF = 0.8 (f) VLF = 0.9

Figure 15: Mode shapes of empty and partially filled tubes under external pressure

5. Results and Discussion

Examples of pressure time curves recorded during experiments for Series-1 and Series-2 composite tubes for empty tube and filled tube are presented in Figure 16 and Figure 17. It can be observed that time of rise of the history is different for the three tests due to the manual mode of the presentation. Repetitiveness of test data for the same volumetric loading can be seen in these two figures as well. Similar test data record was obtained for all volumetric loading fractions for experimental results of both Series-1 and Series-2 tubes. Critical pressure data was extracted from each of these experimental curves. Table 2 and Table 3 summarize the experimental data as recorded in the tests and tabulate the corresponding computed mean of Pcr for Series 1 and Series 2 tubes, respectively. The spread of the experimental data over the mean is also indicated in the tables. Finite element based theoretical critical pressure for Series-1 and Series-2 tubes, derived from previous Figure 13 and Figure 14, are tabulated in Table 2 and Table 3 for comparison.

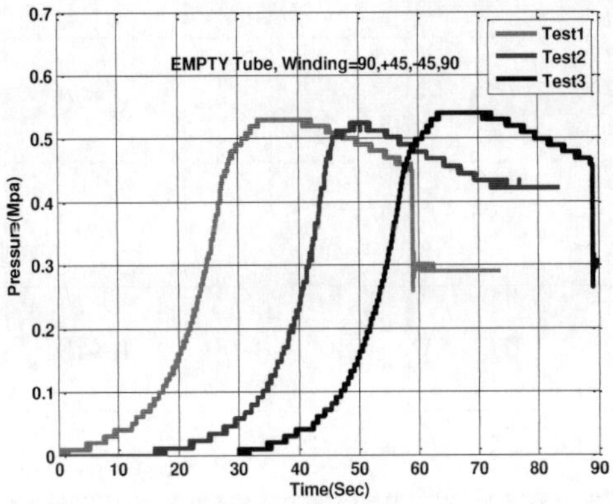

Figure 16: Typical experimental pressure record for Series 1 empty tubes

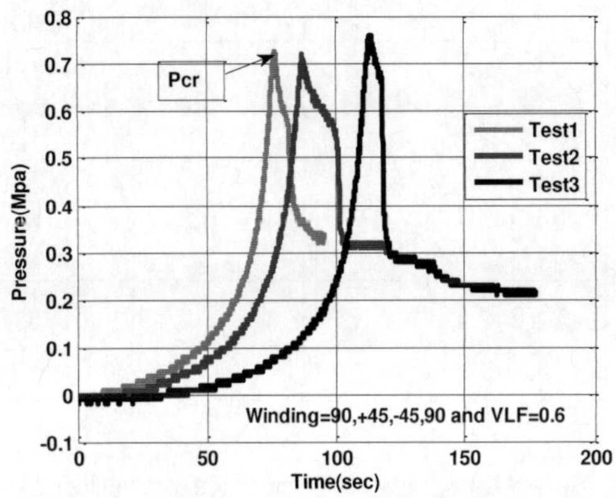

Figure 17: Typical experimental pressure record for Series 1 filled (0.6 VLF) tubes

Table 2: Critical pressure for partially filled tubes of Series-1

Sample No.	Ratio of radius (r_f/r_s)	Volumetric loading fraction	Experimental Data All (MPa)	Mean (MPa)	Spread Over Mean (%)	FEM Result (MPa)
1 2 3	1.00	0.0	0.540 0.530 0.525	0.532	+1.5 -0.38	0.590
4 5 6	0.89	0.2	0.570 0.574 0.571	0.572	+0.35 -0.35	0.630
7 8 9	0.77	0.4	0.594 0.600 0.590	0.595	+0.85 -0.84	0.660
10 11 12	0.63	0.6	0.730 0.740 0.770	0.747	+3.1 -2.3	0.800
13 14 15	0.45	0.8	0.980 1.020 1.050	1.016	+3.3 -3.7	0.920
16 17 18	0.31	0.9	1.170 1.170 1.120	1.155	+1.3 -3.1	1.050

Table 3: Critical pressure for partially filled tubes of Series-2

Sample No.	Ratio of radius (r_f/r_s)	Volumetric loading fraction	Experimental Data			FEM Result (MPa)
			All (MPa)	Mean (MPa)	Spread Over Mean (%)	
1 2 3	1.00	0.0	0.595 0.641 0.595	0.610	+5.1 -2.5	0.620
4 5 6	0.89	0.2	0.640 0.720 0.660	0.673	+6.9 -5.1	0.662
7 8 9	0.77	0.4	0.880 0.920 0.890	0.897	+2.5 -1.4	0.849
10 11 12	0.63	0.6	1.140 1.180 1.200	1.173	+0.6 -2.9	1.170
13 14 15	0.45	0.8	1.430 1.540 1.460	1.477	+4.3 -3.2	1.600
16 17 18	0.31	0.9	1.950 2.220 2.000	2.023	+9.7 -3.7	1.840

Figure 18 provides the graphical variation of the data spread, which shows a standard deviation of 3.7% over the mean. Comparative analysis graphs of the aforesaid results are presented in Figure 19 and Figure 20 for Series-1 and Series-2 composite tubes, respectively. Results from both experimental and theoretical results using FE analysis highlight the buckling load carrying capacity of the composite tubes has been enhanced up to 1.8 to 3 times as that of an empty tube for a volumetric loading fraction of 0.9. The spread in the experimental data for each set ranges between 2% to 12%, which can be contributed to the variation in cylindricity of the filament wound tubes, resulting in different initial imperfection.

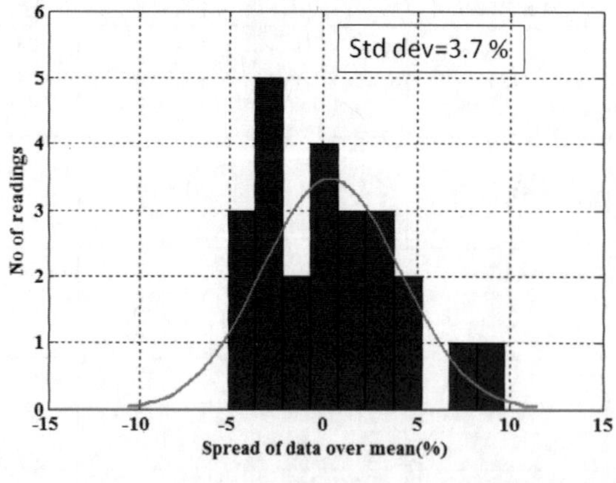

Figure 18: Spread of experimental data obtained for both Series-1 and Series-2 tubes

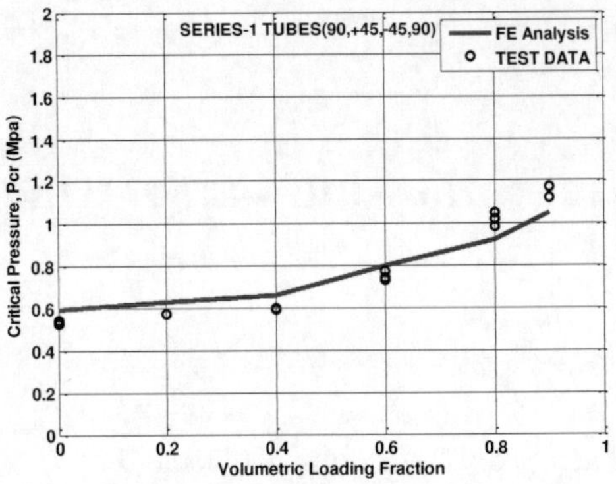

Figure 19: Comparison of experimental data and FE results for Series-1 tubes

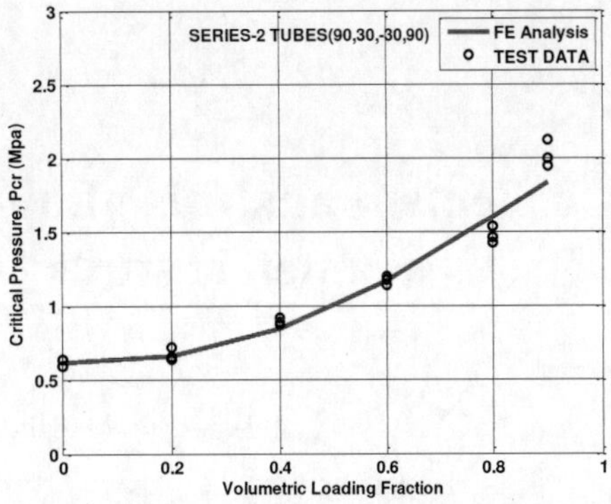

Figure 20: Comparison of experimental data and FE results for Series-2 tubes

6. Conclusion

This experimental investigation, which is supported by the theoretical simulation, concludes that there is an improvement in the buckling strength of thin glass composite shells filled with propellants under external pressure. The non-linear FE analysis captures the behavior of the filled tube very closely to the trend established from the experiments (i.e. within 10% accuracy in terms of predication of Pcr). Volumetric loading fraction of 0.9 with solid propellant equivalent elastic filler increases the strength of the tube in comparison to the empty one by 1.8 to 3 times. This improvement in buckling strength can be utilized for structural optimization of CRMC in rocketry application.

References

[1] Seide P 1962 *J AEROSPACE SCI* **29** 851-62
[2] Seide P and Weingarton V I 1962 *American Rocket Society Journal* **32** 680-8
[3] Kerr A D and Myint T U 1965 *Int J Mech Science* **7** 373-81
[4] Goree W S and Nash W A 1962 *Experimental Mechanics* **2** 142-9
[5] Karam N G and Gibson L J 1995 *INT J SOLIDS STRUCT* **32** 1285-306
[6] Li H P 1998 *Investigation of the stability of metallic/composite cased propellant rocket motors under external pressure* Virginia Polytechnic Institute and State University
[7] Tsouvalis N G, Zatiraton A A and Paazolou V J 2003 *COMPOS PART B - ENG* **34** 217-26

IOP Conf. Series: Materials Science and Engineering **270** (2017) 012023 doi:10.1088/1757-899X/270/1/012023

Finite element analysis of high aspect ratio wind tunnel wing model: A parametric study

N A Rosly and M Y Harmin*

Department of Aerospace Engineering, Faculty of Engineering, Universiti Putra Malaysia, Serdang, Selangor, Malaysia

*myazdi@upm.edu.my

Abstract. Procedure for designing the wind tunnel model of a high aspect ratio (HAR) wing containing geometric nonlinearities is described in this paper. The design process begins with identification of basic features of the HAR wing as well as its design constraints. This enables the design space to be narrowed down and consequently, brings ease of convergence towards the design solution. Parametric studies in terms of the spar thickness, the span length and the store diameter are performed using finite element analysis for both undeformed and deformed cases, which respectively demonstrate the linear and nonlinear conditions. Two main criteria are accounted for in the selection of the wing design: the static deflections due to gravitational loading should be within the allowable margin of the size of the wind tunnel test section and the flutter speed of the wing should be much below the maximum speed of the wind tunnel. The findings show that the wing experiences a stiffness hardening effect under the nonlinear static solution and the presence of the store enables significant reduction in linear flutter speed.

1. Introduction

Recent advancement of new aircraft designs leads to more flexible and lighter aircraft, thus improving its operational range and also operating cost per mile. Nonetheless, an unfortunate consequence of this development is the substantial increase in nonlinear effects, which gives rise to the dramatic changes in terms of the static and dynamic characteristics of the aircraft. One of the characteristics of a highly flexible aircraft is the nonlinear behaviour of their structural stiffness due to either geometric stiffening or softening. Another characteristic is the significant changes in air loads direction under a follower force, where the force direction is dependent on the wing surface deformation. These circumstances defy the validity of conventional linear approach, hence consideration of the above effects is crucial in order to correctly model the system.

A study has been conducted on nonlinear aeroelasticity and flight dynamics for High Altitude Long Endurance (HALE) aircraft, which discovered that the HALE wing does not necessarily have to be in extreme flight state to be highly curved and this condition might occur at relatively low aerodynamic loading too [1]. The study implies that the curved wing exhibits differences in modal characteristics as compared to the straight wing especially in terms of torsional and edgewise mode coupling, causing the aeroelastic behaviour to act differently. Hence, the inclusion of geometric nonlinearity is of great importance and cannot be neglected. A great deal of researches have been devoted to study the effect of nonlinearities on the aeroelastic behaviour. However, most of these studies are solely focussing on developing numerical formulation and there is a lack of research in validating the reliability of their numerical model [2-10]. Tang and Dowell [11-14] are among the researchers that have made a great effort in performing experimental studies on the effects of the structural geometrical nonlinearity with respect to flutter boundary, limit cycle oscillation (LCO) and gust responses. In their studies, a wing store of slender body is introduced at the wing tip of a high aspect ratio (HAR) wing in order to reduce the frequency gap between flutter modes, resulting in the reduction of flutter speed within operational envelope of the wind tunnel. This is in line with the results of another study that show the influence of vibration and flutter properties is greater by varying the position of the store along the span instead of having the mass equally distributed along it [15]. This finding suggests the flutter frequency reduces as the store is moved towards the tip of the wing. This work has been validated through an experimental investigation [16].

In this study, procedure for designing a HAR wing for the wind tunnel is developed, taking into account the geometric nonlinearities and the tip store effects with respect to the static deflection and flutter characteristics. The solutions are selected such that all defined design limits and constraints are satisfied.

2. Methodology

In this study, the experimental model is divided into two main parts: HAR wing model with a tip store and a wing root mechanism. These components are designed such that the wing root mechanism could rigidly hold the root of the wing as well as able to rotate in order to control the root angle of attack of the wing system. On top of that, it is crucial to ensure the basic features of the experimental model to have a simple manufacturing and assembly process. This enables easy modification and repairs when required.

The wing model is designed with a reasonably simple layout and loosely based on the wing model by Tang and Dowel (2001) [11]. It consists of spar, ribs, fairings and tip store as illustrated in Figure 1 and Figure 2(a). The wing spar is made of spring steel, which is chosen due to its high yield strength and durability in spite of significant deformation. One end of the spar is held tightly by the wing root mechanism while the other end is used for store attachment. The wing ribs are made of Acrylonitrile Butadiene Styrene (ABS) with a cross-sectional of NACA0012 airfoil. Each rib has a slot with the size of the spar cross section. The spars are slid through the slots and they are distributed equally along the span with 50mm interval between them. The gap between each rib is covered by fairings that are made of Styrofoam, which completes the aerodynamic contour of the wing. In the mean time, the store is a symmetrical cylinder aluminium bar that is consisted of lower and upper parts in order to enable the clamping mechanism to the tip of wing spar. Figure 2(b) shows the wing root mechanism that consists of turning disc and clamp component, which are made out of model foam and block of aluminium, respectively. They are attached together via fasteners and are inserted to the slot that has been created at the middle of one of the sidewalls of the test section. Once the desired angle is set, the wing root mechanism is then locked to the test section.

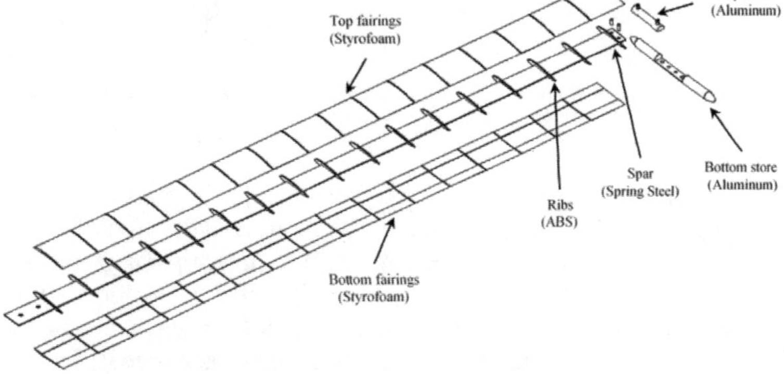

Figure 1: Exploded wing for experimental wing model

IOP Conf. Series: Materials Science and Engineering **270** (2017) 012023 doi:10.1088/1757-899X/270/1/012023

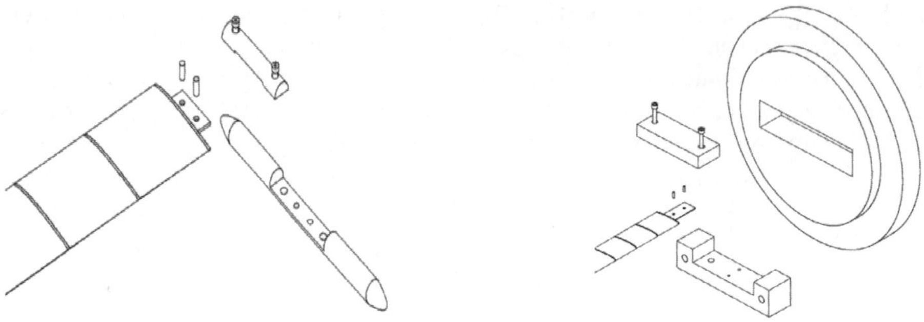

(a) (b)

Figure 2: Exploded view of (a) tip store, (b) wing root mechanism

The wind tunnel is of a closed loop type, with the test section having a cross sectional dimension of 1m×1m and maximum operating speed of 40m/s. Since the wing model is attached to the middle part of one side of the test section walls, this leaves a margin of ±0.5m between the upper and lower part of the test section. Therefore, the maximum allowable tip deflection is constrained to be within a range of ±0.3m in order to reduce the effects of the wall interference in the wind tunnel. On the other hand, the flutter speed has to be as low as possible to ensure the safety of the testing envelope. In this case, it is constrained to be lower than 40m/s. To assess the HAR wing type, the aspect ratio of the wing design has to be more than 10. The finite element representation of the studied wing consisting of spar, ribs, fairing and store is modelled with MSC NASTRAN shell elements of CQUAD4, CTRIA3, CHEXA and CPENTA as shown in Figure 3. Besides that, the lumped mass of CONM2 element is introduced to represent the weight of fasteners that attach the store to the spar. Fully clamped boundary condition is applied at the wing root to replicate a cantilever-like condition. The doublet lattice aerodynamic panel illustrated in Figure 4 is employed to represent the aerodynamic model and it is coupled with the structural model in order to establish a complete system of an aeroelastic system of the studied wing.

In this work, parametric study in terms of spar thickness, span length and store diameter is coded using MATLAB and the system is integrated with the finite element solver of MSC NASTRAN. The wing is examined under linear and nonlinear analyses to demonstrate undeformed and deformed wing cases.

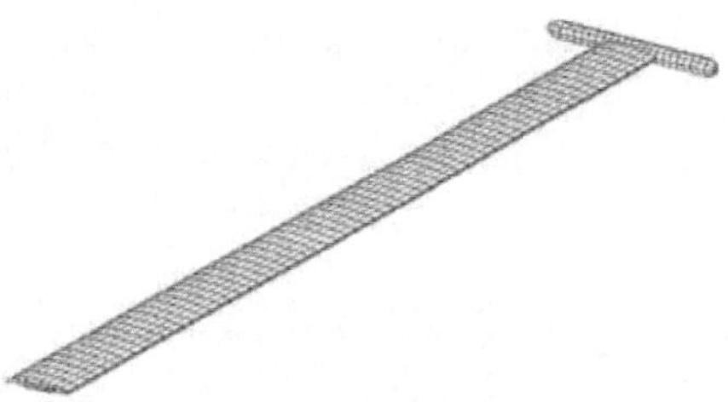

Figure 3: FE Model

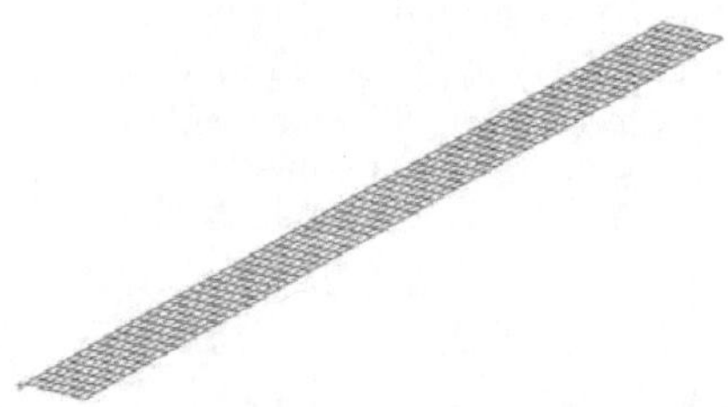

Figure 4: Aerodynamic Model

3. Results and Discussions

The parametric study is performed by conducting linear and nonlinear analyses under the gravitational loading defined at the sea level in order to obtain the maximum tip deflection of the wing for the wind off condition. The nonlinear analysis is carried out due to the characteristic of the HAR wing that can undergo high deflection beyond the linear region even when it is under gravitational loading. The nonlinear solution is solved in iterative manner where the stiffness is updated upon convergence of the nonlinear problem for each of considered load steps. The design space is specified by the variables of spar thickness ranging from 1.0mm to 1.5mm and span length from 500mm to 800mm. The effect of store is also studied by varying its diameter from 10mm to 14mm. It should be noted that the wing chord is kept constant at 50mm, which is twice the length of spar width, and the length of the store is fixed at 150mm. Since the system can be represented as a cantilever beam like model, the variation in the width parameter is insignificant for the bending calculation under uniform loading.

Figure 5 provides tip deflection results for both linear and nonlinear cases for all the considered design variables. It can be observed that the variation in spar thickness, span length and store diameter provide significant impact in terms of tip deflection. The tip deflection increases with increasing span length, decreasing spar thickness and increasing in store diameter. It is noted that the tip deflection for linear and nonlinear cases are significantly different as the deflection for nonlinear case is lower than the linear case. This demonstrates that the wing has experienced a stiffness hardening effect, where the effect is visible at higher magnitude of deflection. This finding suggests the importance of considering the geometrical nonlinearity in the analysis. Since the linear system is only valid for a small deflection, it may overestimate the actual bending deflection for highly deflected system. Overall, the nonlinear results show that all wing design options are acceptable as the wing tip deflection under gravitational loading is lesser than the allowable margin of ±0.3m.

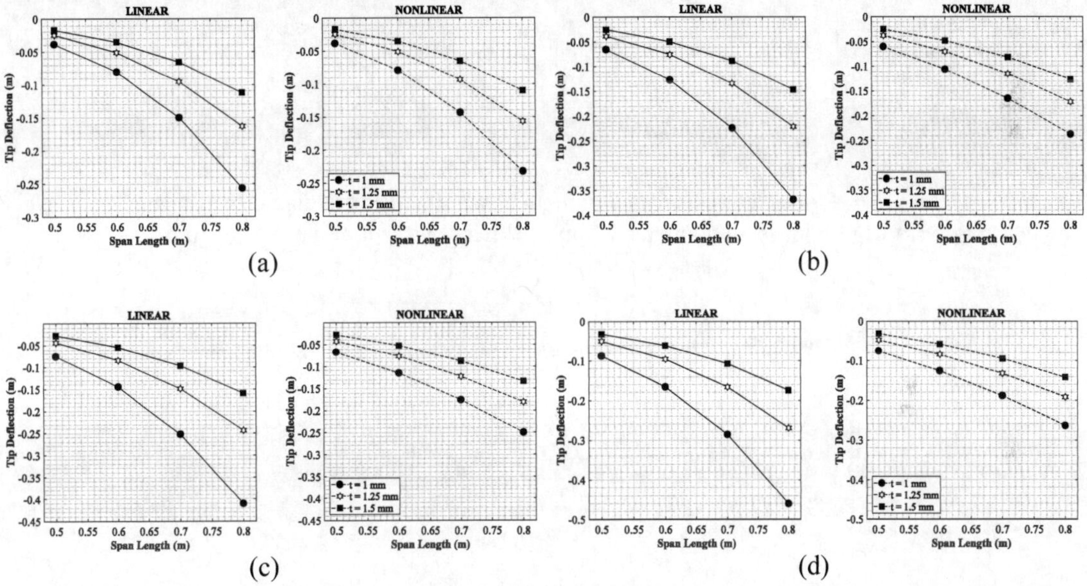

Figure 5: Linear and nonlinear static analyses for (a) wing, (b) wing with store diameter 10mm, (c) wing with store diameter 12mm, (d) wing with store diameter 14mm

Subsequently, with similar consideration of design variables and design space, the flutter analysis using p-k method is performed for a range of airspeeds at sea level condition. Figure 6 depicts the considered parametric study with respect to flutter speed and flutter frequency. From the plots, the shaded region provides an acceptable design options that is below than the flutter speed constraint of 40m/s. It can be seen clearly that, as the store diameter increases (which is reflected in increment of tip mass), the flutter speed reduces significantly as well as flutter frequency.

To understand more on this characteristic, the graphical representation of V-f and V-ζ is plotted between the flutter modes of the clean wing and the wing with the biggest store diameter as illustrated in Figure 7. The flutter speed and frequency are identified when the damping ratio changes in sign, which indicates that the oscillation cannot be damped out at this particular condition. For both cases, flutter occurs due to coupling between second bending and first torsional modes, where the frequency gap between them decreases with increasing of airspeed. It should be noted that, with the presence of the store, the frequency gap between the flutter modes has reduced significantly compared to a clean

IOP Conf. Series: Materials Science and Engineering **270** (2017) 012023 doi:10.1088/1757-899X/270/1/012023

wing, which in turn brings down the flutter speed from 52m/s to 33m/s. Hence, this enables the flutter testing to be performed at a much lower speed and lowering the risk level at the same time.

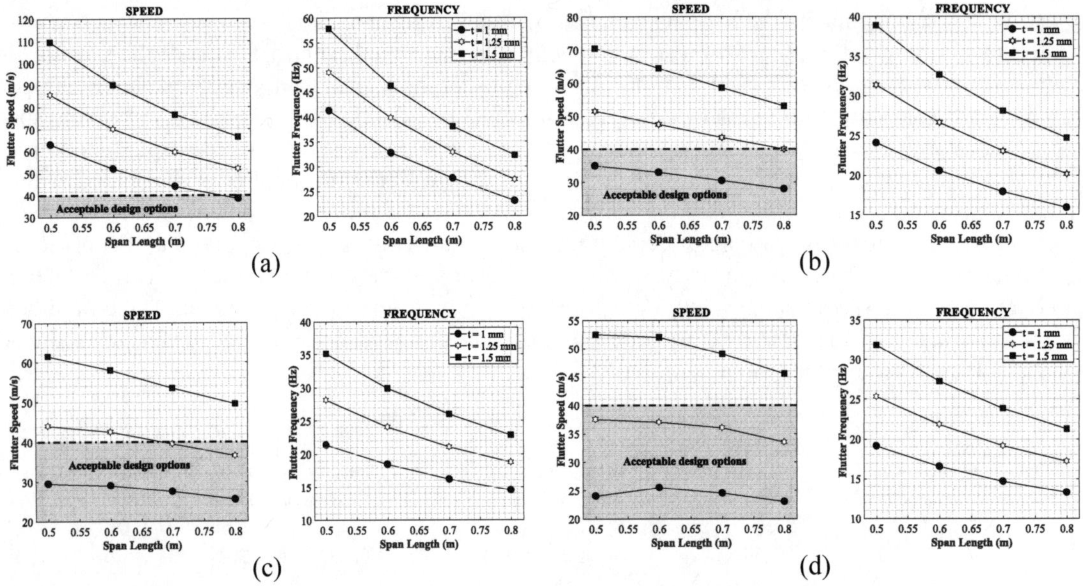

Figure 6: Flutter analysis for (a) wing, (b) wing with store diameter 10mm, (c) wing with store diameter 12mm, (d) wing with store diameter 14mm

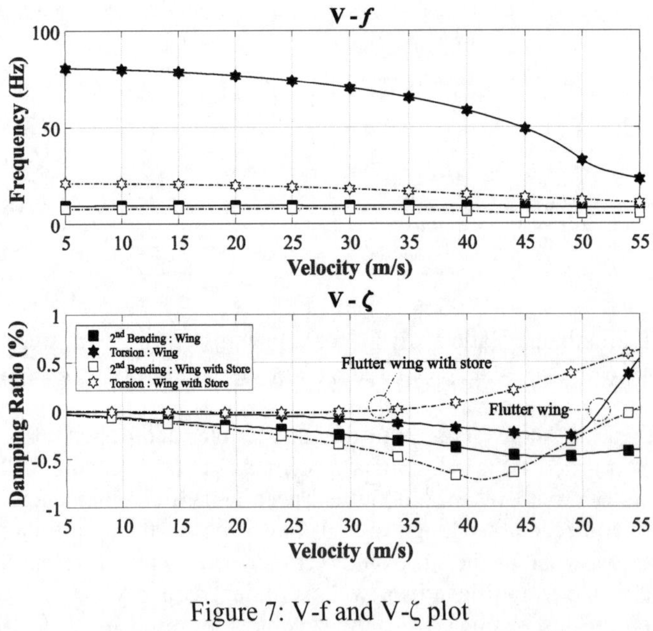

Figure 7: V-f and V-ζ plot

Table 1 provides the linear and pre-stressed normal mode analyses for first seven modes of both clean wing and wing with the biggest store. The pre-stressed analysis is first carried out by performing nonlinear static analysis of the considered wing under gravitational loading and the solution is then restarted in the linear normal mode analysis upon the updated stiffness about its nonlinear statically deformed state. From the finding, the out-of-plane bending modes are not affected much as the wing is deflected. Nonetheless, the chordwise bending and torsional modes of the undeformed condition have changed into 'chordwise bending - torsion' and 'torsion - chordwise bending' modes when the wing is in deformed state. The 'chordwise bending - torsion' implies that the mode is more prominent in the chordwise direction rather than torsional while the 'torsion - chordwise bending' mode indicates that the mode is more prominent in torsional direction than chordwise. From the results, it is clearly shown that the torsional-related mode of the deformed wing appears at a much lower frequency compared to the undeformed condition. Hypothetically, this finding suggests that the flutter speed can be reduced further as the torsional-related mode is getting closer to flutter mode of second bending in comparison to the fully linear system.

Table 1: Natural frequencies for linear and pre-stressed normal mode analysis

WING				WING WITH STORE			
Undeformed		Deformed		Undeformed		Deformed	
Mode	**Freq**	**Mode**	**Freq**	**Mode**	**Freq**	**Mode**	**Freq**
1st Bending	1.53	1st Bending	1.59	1st Bending	1.09	1st Bending	1.49
2nd Bending	9.62	2nd Bending	9.61	2nd Bending	7.85	2nd Bending	8.69
3rd Bending	26.95	**Chordwise-Torsion**	15.37	**Torsion**	21.18	**Chordwise-Torsion**	9.63
Chordwise Bending	30.17	3rd Bending	26.72	**Chordwise Bending**	21.48	3rd Bending	23.68
4th Bending	52.88	4th Bending	52.58	3rd Bending	23.31	**Torsion-Chordwise**	28.24
Torsion	81.15	5th Bending	87.19	4th Bending	47.29	4th Bending	47.60
5th Bending	87.55	**Torsion-Chordwise**	98.02	5th Bending	79.96	5th Bending	80.07

4. Conclusion

The study is set out to determine the design options for the wind tunnel model of the HAR wing. One of the findings shows that the flutter speed can be reduced significantly with the presence of a store. In the mean time, another finding provides a possibility of the flutter speed to be further reduced when the wing is in the deformed state compared to the undeformed state. This highlights the importance of including geometric nonlinearity in the analysis to correctly model the system. Further research should be undertaken to validate the finite element HAR wing through experimental modal analysis and wind tunnel testing.

Acknowledgement

The authors acknowledge the support from the Ministry of Higher Education (MOHE) for this study through research grant 03-01-15-1625FR.

References

[1] Patil M J, Hodges D H and Cesnikz C E S 1999 *International Forum on Aeroelasticity and Structural Dynamics*
[2] Peters D A, Karunamoorthy S and Cao W 1995 *Journal of Aircraft* **32** 313-22
[3] Tang D M and Dowell E H 2004 *Journal of Fluids and Structures* **19** 291-306
[4] Su W and Cesnik C E S 2011 *AIAA Journal* **49** 324–39
[5] Xie C, Wang L, Yang C and Liu Y 2013 *Chinese Journal of Aeronautics* **26** 514-21
[6] Hesse H and Palacios R 2014 *AIAA Journal* **52** 1717-32
[7] Catellani M, Cooper J E and Lemmens Y 2016 *15th Dynamics Specialists Conference*

[8] McEwan M I, Wright J R, Cooper J E and Leung A Y T 2001 *Journal of Sound and Vibration* **243** 601–24
[9] Przekop A and Rizzi S A 2006 *AIAA Journal* **44** 1767–78
[10] Tiso P, Jansen E and Abdalla M 2011 *AIAA Journal* **49** 2295-304
[11] Tang D M and Dowell E H 2001 *AIAA Journal* **39** 1430-41
[12] Tang D M and Dowell E H 2002 *AIAA Journal* **40** 419-29
[13] Tang D M and Dowell E H 2002 *Journal of Aircraft* **39** 885-8
[14] Tang D M and Dowell E H 2010 *AIAA Journal* **48** 427-42
[15] Gern F H and Librescu L 1998 *Aerospace Science and Technology* **5** 321-33
[16] Moradi M, Sadeghi M H and Dowell E H 2017 *Journal of Aircraft*

IOP Conf. Series: Materials Science and Engineering **270** (2017) 012024 doi:10.1088/1757-899X/270/1/012024

Parametric instability of shaft with discs

A M Abdul Wahab, Z A Rasid*, A Abu and N F Mohd Noor Rudin

Malaysia-Japan International Institute of Technology, Kuala Lumpur, Malaysia.

*arzainudin.kl@utm.my

Abstract. The occurrence of resonance is a major criterion to be considered in the design of shaft. While force resonance occurs merely when the natural frequency of the rotor system equals speed of the shaft, parametric resonance or parametric instability can occur at excitation speed that is integral or sub-multiple of the frequency of the rotor. This makes the study on parametric resonance crucial. Parametric instability of a shaft system consisting of a shaft and disks has been investigated in this study. The finite element formulation of the Mathieu-Hill equation that represents the parametric instability problem of the shaft is developed based on Timoshenko's beam theory and Nelson's finite element method (FEM) model that considers the effect of torsional motion on such problem. The Bolotin's method is used to determine the regions of instability and the Strut-Ince diagram. The validation works show that the results of this study are in close agreement to past results. It is found that a larger radius of disk will cause the shaft to become more unstable compared to smaller radius although both weights are similar. Furthermore, the effect of torsional motion on the parametric instability of the shaft is significant at higher rotating speed.

1. Introduction

There are two types of resonance: force resonance and parametric resonance. In rotor-dynamic field of study, the force resonance has been widely researched [1-4]. It can result in catastrophic failure of the rotor system if the external excitation such as the speed of the shaft equals to the natural frequency of the rotor system. Parametric resonance or parametric instability of a rotor system, on the other hand, occurs due to external excitation that modifies the stiffness, inertia or other parameters of the rotor system with time. In comparison to force resonance, parametric instability can occur more often at the excitation speeds that are integral or sub-multiple of the frequency of the rotor. This highlights the importance of the study on the parametric instability of the rotor system. Furthermore, the parametric instability of the rotor system can lead to its fatigue failure [5]. Many research have been devoted on improving parametric instability of rotor systems subjected to periodic changes of inertia or stiffness that could be due to the anisotropic bearing support [6], asymmetry of disk and shaft [7], periodic axial compressive load [8] and the transverse crack propagation [9].

Studies on dynamic instability analysis of a rotating beam/shaft subjected to periodic axial forces have been conducted using the discrete singular convolution method [10-11]. While considering the external viscous damping and internal material damping, it has been shown that the instability region increases with the increase of spinning speed for shaft with damping, and the results correlated well with those obtained using the Floquet's method. In addition, the instability region can also be modified by shear deformation, rotatory inertia, distributed axial loads, elastic soil and damping [12]. In term of damping, it has been suggested that internal damping of the Kelvin-Voigt has stabilizing effect on the primary parametric resonance dynamic instability of Timoshenko beam under the periodic axial loads [13]. However, it depends on the size of the damped segment and its location. The system is stable if the damped segment is located near the fixed end where it enlarged from the fixed end. On the other hand, static load component has destabilizing effect to the shaft system as it is increased. Recently, a study has shown that the location of disk has significant effect on dynamic instability of a shaft system [14]. Using the transfer matrix method, the critical axial force is found to decrease rapidly at first, then remains almost constant, and finally decreases linearly as the disk is located closer to the support at the right end when shear deformation is not considered.

Nonetheless, the above studies do not focus on shafts that may rotate at very high speed. Machines that operate at high shaft speed are well sought nowadays since high overall efficiency of the rotating equipment can be achieved when the speed of the rotating equipment lies in between 10000 RPM and 100000 RPM [15]. Furthermore, while focusing on the effects such as translational and rotary inertia, gyroscopic moments, bending and shear deformation, and damping, an important effect of axial torque of the shaft has been neglected. Though the effect of axial torque has been applied by Nelson (1980) [16] and Rao (2011) [17], their studies are limited to the whirling frequency and critical speed of the shaft. In this study, apart from focusing on the effect of attaching discs to the shaft, the parametric instability of high speed rotor system subjected to periodic axial load is investigated while considering the effect of the axial torque. The shaft speed is up to 40000 RPM. The disc parameters under study here are the disk location with respect to shaft, disk mass and the number of disc. The finite element method (FEM) form of the Mathieu-Hill equation is developed considering the translational and rotary inertias, gyroscopic moments, bending and shear deformations, and the axial torque. The results are validated and compared with those from the formulation that does not consider the axial torque effect.

2. Material and Methods

The rotor system investigated in this study consists of a shaft, one or more discs and two bearings, represented by the supports as shown in Figure 1. L and *l* are the length of the shaft and the distance of the disk from the left bearing, respectively, while d and t are the diameter of the shaft and thickness of the disk, respectively. Table 1 shows the material properties and the dimensions of components of the rotor system being investigated.

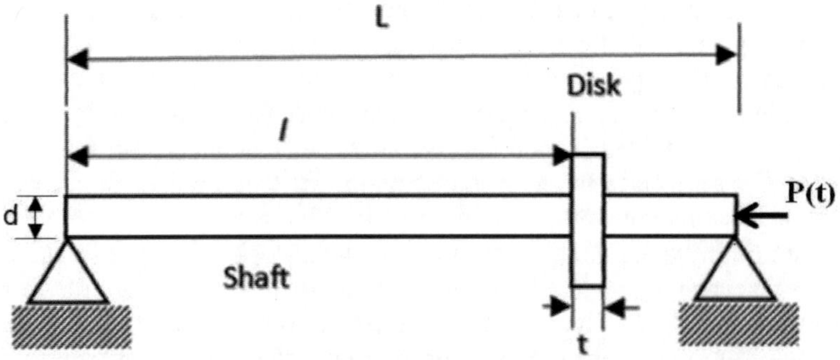

Figure 1: The rotor system being investigated

This study employs FEM to estimate the whirling natural frequency and the parametric instability region of the rotating shaft system. The developed FE formulation is based on Ref. [15], where the Timoshenko beam theory and the Nelson's FEM approach that considers the torsional effect are used. The effect of torsional motion is added to the formulation as the degree of freedom such that its effect on resonance can be determined. In this study, beam model having elements with 5 degrees of freedom per node including the axial twist of the shaft is called the 5DOF model. The results will be compared to those of the beam model having element with 4 degrees of freedom per node that is 4DOF model. Figure 2 shows the degrees of freedom of the shaft element used in this study. The model involves two translation components and three rotary inertia components that are denoted by v, ω, θ_x, θ_y and θ_z, respectively. The translation components are due to bending, transverse shear deformation and also torsional motions. Meanwhile, the rotary inertia components are due to bending and torsional motions.

Table 1: Dimensions and material properties of the shaft system in the study on the effect of the location of disc (Study 1), the mass of the disc (Study 2) and the number of discs attached (Study 3)

Shaft	
Young's modulus, E	207 GPa
Modulus of rigidity, G	79.6 GPa
Poisson's ratio, v	0.303
Density, ρ	7833 kg/m^3
Radius, r	round tube with r_{inner} = 0 m, r_{outer} =0.0508 m
Length, L	1.27 m
Shear factor, κ	0.9
Bearing	
Direct stiffness coefficient, K_{by}, K_{bz}	$7x10^7$N/m
Direct damping coefficient, C_{by},C_{bz}	0
Disk in Study 1	
Young Modulus, *E*	207 GPa
Density, ρ	7833 kg/m^3
Thickness, t	0.03 m
Diameter, d	0.3 m
Disk in Study 2	
Young Modulus, *E*	207 GPa
Density, ρ	7833 kg/m^3
Diameter, d_1 and d_2	0.3 m
Thickness, nominal weight, t_1	0.03 m
Thickness, heavy weight, t_2	0.07 m
Diameter, d_3	0.16 m
Thickness, nominal weight, t_3	0.1055 m
Disk in Study 3	
Young Modulus, *E*	207 GPa
Density, ρ	7833 kg/m^3
Diameter, d_1, d_2	0.4 m
Thickness, t_1, t_2	0.005 m
Diameter, d_3	0.4 m
Thickness, t_3	0.006 m

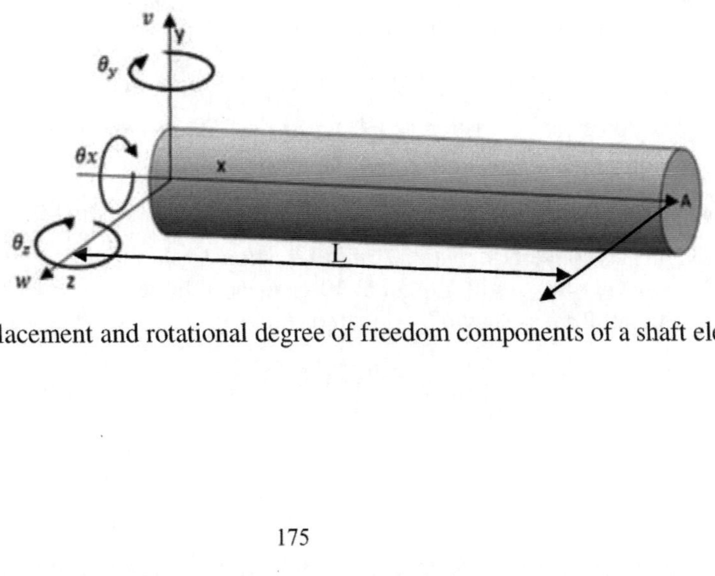

Figure 2: The displacement and rotational degree of freedom components of a shaft element

In Figure 1, the rotating shaft is loaded with a pulsating axial force, $P(t)$ (given by Equation 1) that acts along the un-deformed axis of the shaft. Component ϕ denotes the frequency axial excitation of the dynamic force component. P_s and P_t are the static and time dependent components of the force, respectively.

$$P(t) = P_s + P_t \cos \phi t \tag{1}$$

The total potential energy stored in the beam element that is due to bending and torsional motions can be expressed as in Equation 2, where E is the Young's modulus of the shaft, G is the shear modulus of the shaft, I_p is the total second moment of inertia, J is the polar moment of inertia, A is cross-sectional area of the shaft, k is the shear correction factor, P is the axial force and θ_x, θ_y and θ_z are the rotary displacement in x-, y- and z-axis, respectively.

$$U = \frac{1}{2}\int_{-a}^{+a} EI_p(\acute{\theta}_z^2 + \acute{\theta}_y^2)\, dx + \frac{1}{2}\int_{-a}^{+a} \kappa GA(\acute{v} - \theta_z)^2 + \kappa GA(\acute{w} + \theta_y)^2\, dx + \frac{1}{2}\int_{-a}^{+a} GJ\acute{\theta}_x^2\, dx - \frac{1}{2}\int_{-a}^{+a} P[(\acute{v})^2 + (\acute{w})^2]dx \tag{2}$$

$$I_p = I_z + I_y = \int z^2\, dA + \int y^2\, dA \tag{3}$$

$$J = \int_{-a}^{+a} \left(\frac{\partial\psi}{\partial y} - z\right)^2 + \left(\frac{\partial\psi}{\partial y} + y\right)^2\, dA \tag{4}$$

$$\acute{\theta} = \frac{\partial\theta}{\partial x} \tag{5}$$

$$\acute{v} = \frac{\partial v}{\partial x} \tag{6}$$

On the other hand, kinetic energy of the rotating shaft is given by Equation 7, where the first, second, third and fourth terms correspond to translation energy term, rotary inertia term, gyroscopic element and torsional element terms, and ρ is the density of the shaft.

$$T = \frac{1}{2}\int_{-a}^{+a} \rho A\left(\left(\frac{\partial v}{\partial t}\right)^2 + \left(\frac{\partial w}{\partial t}\right)^2\right) dx + \frac{1}{2}\int_{-a}^{+a} \rho I_y\left(\frac{\partial\theta_y}{\partial t}\right)^2 + \rho I_z\left(\frac{\partial\theta_z}{\partial t}\right)^2 dx + \frac{1}{2}\int_{-a}^{+a} \rho I_x\left(\frac{\partial\theta_x}{\partial t}\right)^2 dx \tag{7}$$

The shaft under study is discretized into several elements and each element consists of two nodes. Following the standard FEM procedure and using Lagrange's formula, the standard FEM governing equation is given by Equation 8, where $[M]$, $[G]$, $[K]$ and $[K_g]$ are the elemental mass, gyroscopic, stiffness and geometric stiffness matrices, respectively, Ω is the spin speed, $P(t)$ is the periodic axial force and $\{q\}$ is the nodal displacement vector.

$$[M]\{\ddot{q}\} + \Omega[G]\{\dot{q}\} + ([K] - P(t)[K_g])\{q\} = 0 \tag{8}$$

The periodic axial force can be written as $P(t) = \alpha P^* + \beta P^* \cos \phi T$ where α and β are the static and dynamic load factors, respectively. Then, the Mathieu-Hill type of equation is given as Equation 9.

$$[M]\{\ddot{q}\} + \Omega[G]\{\dot{q}\} + ([K] - [K_g](\alpha P^* + \beta P^* \cos \phi T\,(t))\{q\} = 0 \tag{9}$$

The Bolotin's method converts Equation 9 into Equation 10 that represents the parametric instability problem to be solved in this study. Equation 10 gives an infinite eigenvalue problem as indicated by Equation 11, or in full form as shown by Equation 12.

$$(\phi^2[M_E] + \phi[G_E] + [K_E])q = 0 \tag{10}$$

IOP Conf. Series: Materials Science and Engineering **270** (2017) 012024 doi:10.1088/1757-899X/270/1/012024

$$|(\phi^2[M_E] + \phi[G_E] + [K_E])| = 0 \tag{11}$$

$$\left\| \begin{bmatrix} [K] + \left(\alpha - \frac{\beta}{2}\right)[K_g] & 0 \\ 0 & [K] + \left(\alpha + \frac{\beta}{2}\right)[K_g] \end{bmatrix} + \phi \begin{bmatrix} 0 & -\frac{1}{2}\Omega[G] \\ \frac{1}{2}\Omega[G] & 0 \end{bmatrix} + \phi^2 \begin{bmatrix} -\frac{1}{4}[M] & 0 \\ 0 & -\frac{1}{4}[M] \end{bmatrix} \right\| = 0 \tag{12}$$

3. Results and Discussion

Dynamic instability analysis is conducted on the static and rotating simply supported shafts, subjected to axial compressive load. The results are then compared to the work of Chen and Ku (1990) [18] that applied Timoshenko beam element, which requires four degrees of freedom per node. Figure 3 shows the plots of dynamic load factor, β against the frequency ratio that represents the parametric instability regions for the shafts, where regions in between the V-shape lines represent the instability region (**US**). For other regions (**S**), the system goes toward stable equilibrium position as disturbance occurs. Figure 3 also shows that the instability regions of current works are in good agreement with Ref. [18].

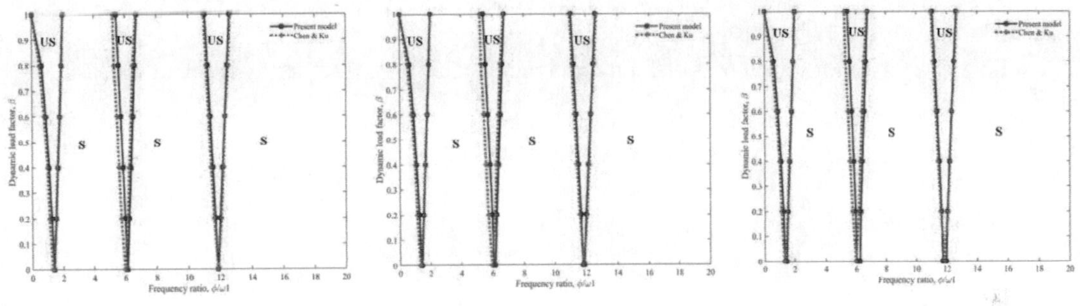

(a) (b) (c)

Figure 3: Validation of the dynamic instability charts for simply supported shaft at: (a) 0 RPM, (b) 8000 RPM (c) 20000 RPM

In reference to Table 1, three cases are compared in this study: Case 1 - shaft without any disk (no disk), Case 2 – shaft with a disk located at the middle of the shaft (disk @ 50%) and Case 3 – shaft with a disk positioned near to the end of the shaft (disk @ 75%). Disk with the same properties is used throughout the analysis. Case 1 (no disk) is used as the benchmark for any difference that might occur in the dynamic behaviour of the shaft as the disk is added. Rigid bearings are used in this study. Figure 4 to Figure 6 show the plots of instability regions of the shaft at 0 RPM, 9000 RPM and 40000 RPM, respectively, for the three cases of disk location based on 4DOF and 5DOF models. In all cases, as the position of the disc is moved to the right, the instability charts can be seen to be moved to the left. This shows that having the disk location near to the end of shaft results in more unstable shaft as compared to having the disk placed at the center of shaft, which is in line with the work in Ref. [14]. Comparing the plots corresponding to 4DOF and 5DOF models, the effect of torsion considered in 5DOF model is significant at high speed. At high speed, the plots have no centre of instability whereas the instability region for 5DOF model can be seen to be wider, which indicates that the torsional effect has increased the instability of the shaft. The difference in frequency ratio between that of 4DOF model and 5DOF model reaches 60% at the spin speed of 40000 RPM.

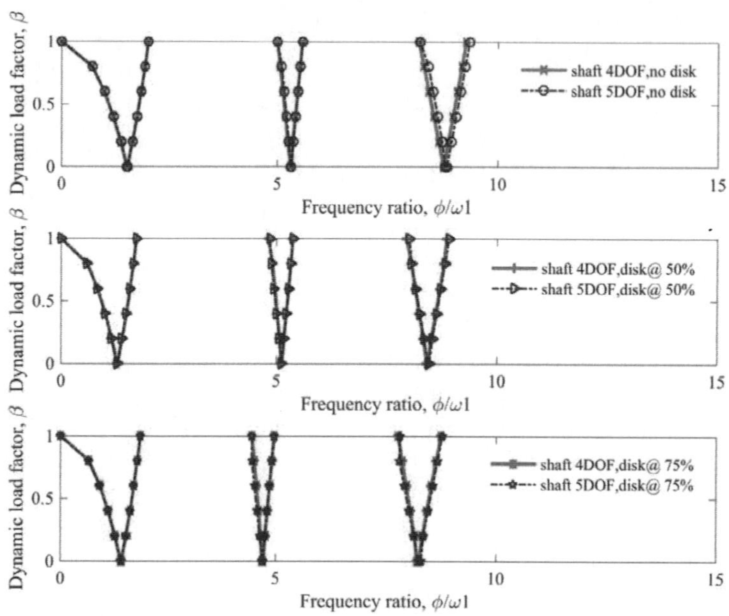

Figure 4: Effect of location of disk on dynamic instability of the shaft rotating at 0 RPM

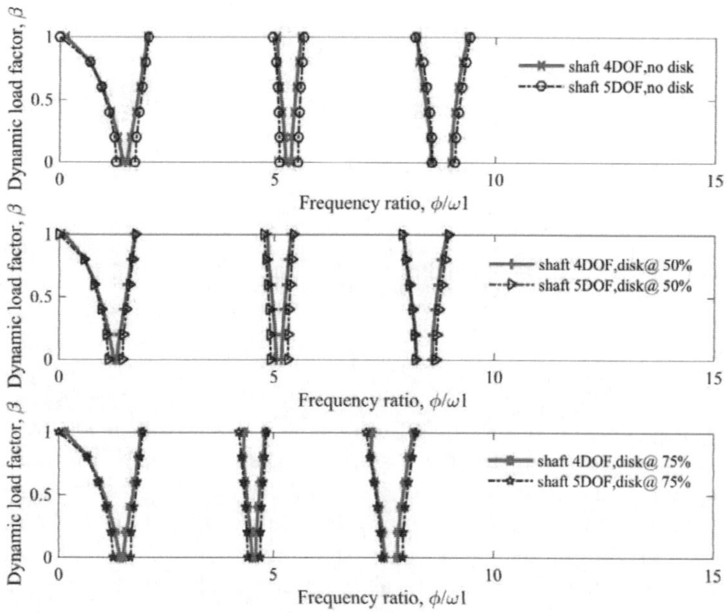

Figure 5: Effect of location of disk on dynamic instability of the shaft rotating at 9000 RPM

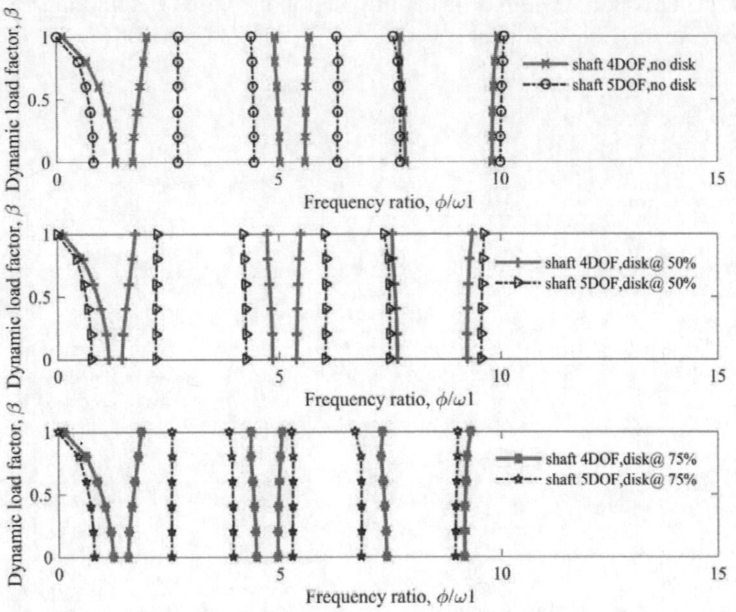

Figure 6: Effect of location of disk on dynamic instability of the shaft rotating at 40000 RPM

Moreover, the effect of disk mass on the dynamic behaviour of shaft is also studied. In reference to Table 1, for Study 2, a rotor with rigid bearings and centred single disk is used. The comparison is between nominal disk, heavy disk and nominal thick disk to be attached to the shaft. The nominal disk and the nominal thick disk differ in their radius and thickness but their masses are the same. Figure 7 to Figure 9 depict the plots of dynamic load factor, β against frequency ratio that show the instability regions of the shaft at 0 RPM, 9000 RPM and 40000 RPM, respectively, modelled using 4DOF and 5DOF models. The advantage of using 5DOF model can be seen at higher rotational shaft speed where the dynamic instability chart is wider due to torsional degree of freedom effect that is considered in 5DOF model. The wider instability chart of the 5DOF model as compared to that of the 4DOF model can be seen at 9000 RPM and even much clearer at 40000 RPM. Figure 7 shows that the parametric instability regions of shafts consisting of the three disk types are quite similar at static condition. In Figure 9, at rotational shaft speed of 40000 RPM, parametric instability corresponding to the nominal disk is more stable compared to that of the heavy disk according to the 5DOF model. However, the instability region for nominal disk can be seen to slightly shift towards the higher frequency level as compared to that of the nominal thick disk. This makes the shaft becomes more stable in the case of disk with smaller radius. The width of disk centre orbit of the shaft system is proportional to the radius of the disk while at the same time, the instability condition depends on the width of disk centre [19]. Therefore, for disks with equal weight, as the axial force fluctuates, the one with larger radius is more unstable compared to that of smaller radius.

In the meantime, the comparison made in Study 3 from Table 1 is on the dynamic behaviours of shafts having one, two and three disks. In all cases, rigid bearings are used to support the shaft at both its ends. Figure 10 to Figure 12 show the plots of dynamic load factor, β against the frequency ratio corresponding to shafts having one, two and three number of disks at a speed of 0 RPM, 9000 RPM and 40000 RPM, respectively, where 4DOF and 5DOF models are applied. In all cases, the increase in the number of disk will shift the instability region inward, showing that the shaft becomes unstable as the number of disk increases although the width of the instability chart remains equal. This condition is expected since natural frequency is inversely proportional with mass. The effect of torsional degree of freedom considered in the 5DOF can be seen in the widening of the parametric instability chart as the speed is increased. In general, the width of instability region increases as rotating speed increases, thus making the system more unstable at higher speed.

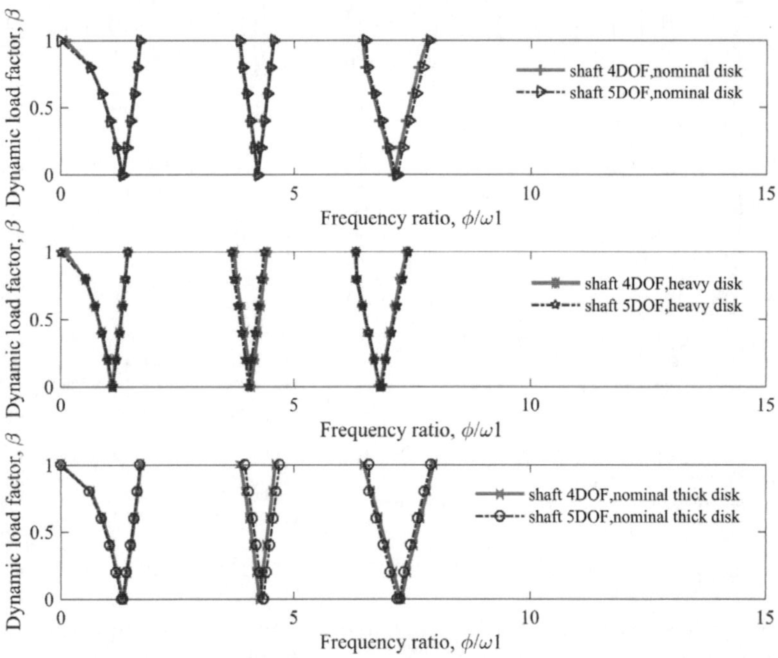

Figure 7: Effect of disk mass on parametric instability of shaft at 0 RPM

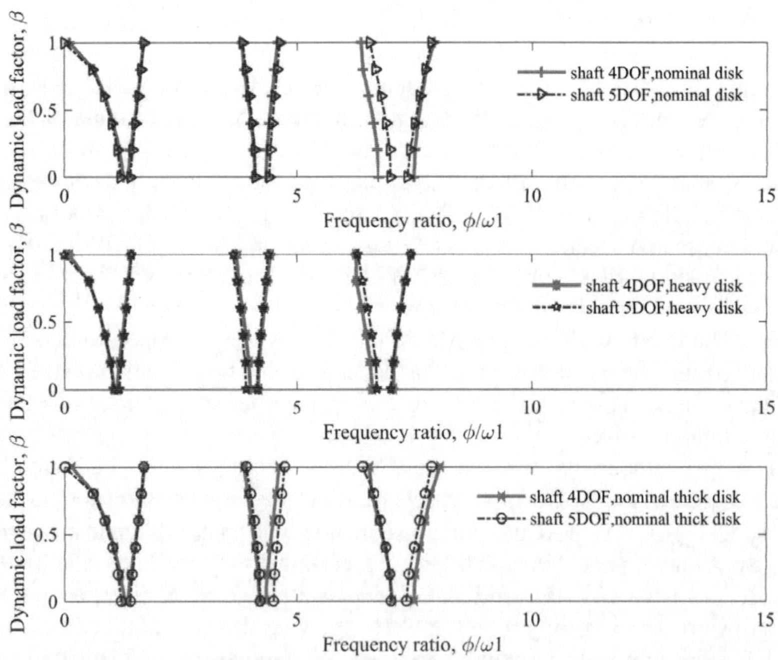

Figure 8: Effect of disk mass on parametric instability of shaft at 9000 RPM

IOP Conf. Series: Materials Science and Engineering **270** (2017) 012024 doi:10.1088/1757-899X/270/1/012024

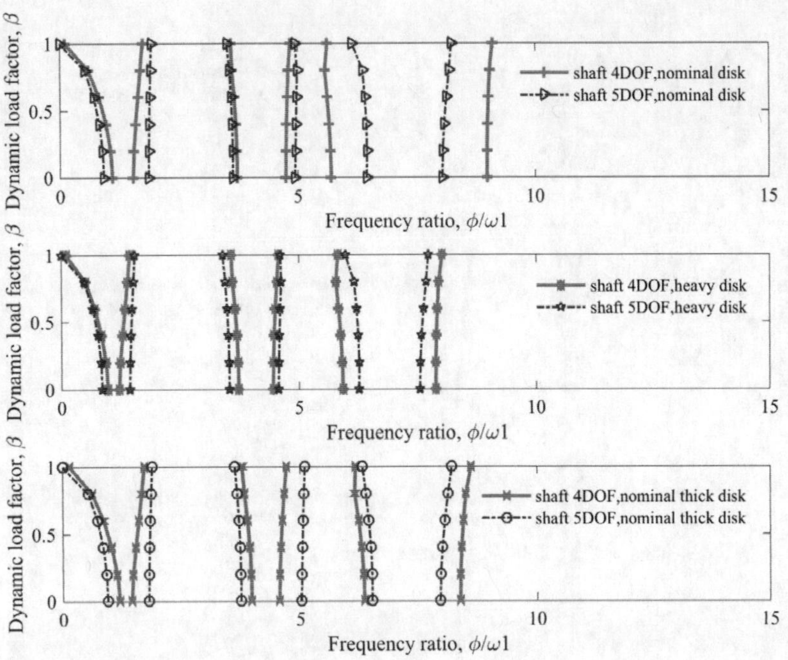

Figure 9: Effect of disk mass on parametric instability of shaft at 40000 RPM

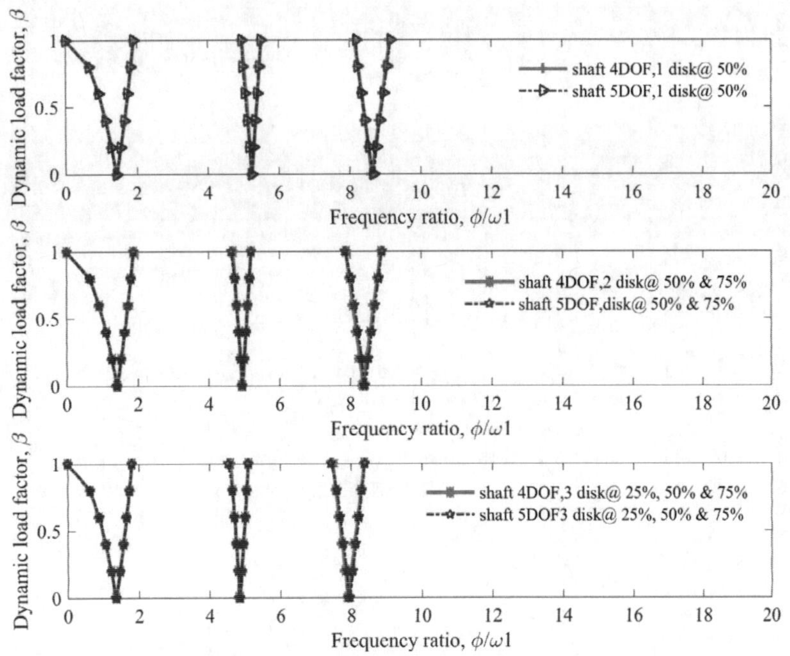

Figure 10: Effect of number of disk on parametric instability of shaft at 0 RPM

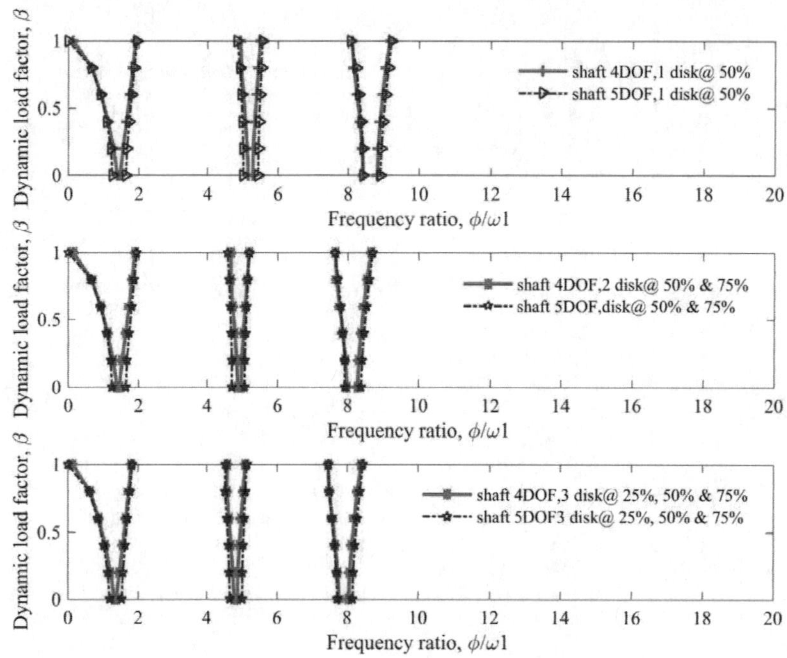

Figure 11: Effect of number of disk on parametric instability of shaft at 9000 RPM

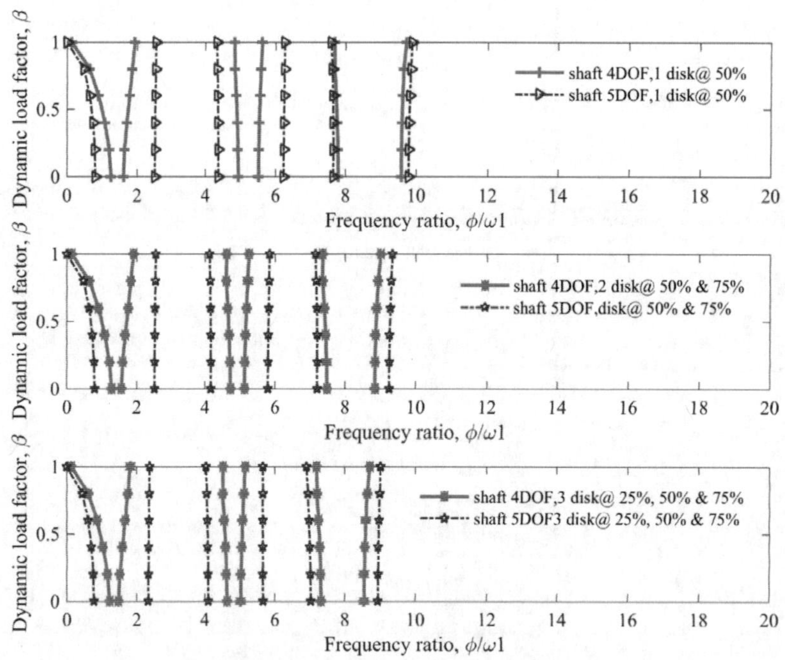

Figure 12: Effect of number of disk on parametric instability of shaft at 40000 RPM

4. Conclusion

A study has been conducted on the effect of attaching disks on the parametric instability of high speed shaft while considering the torsional effect of the shaft. Two FEM formulations (i.e. 4DOF and 5DOF models) are developed for parametric instability study of shaft, where the latter model considers the torsional effect of the shaft. In general, the width of instability region of a shaft increases as rotating speed increases, thus the system becomes more unstable at higher speed. Furthermore, the instability charts corresponding to the 5DOF model show wider instability region as compared to those of 4DOF model at high speed. This shows that the torsional effect has caused the rotor system to become more unstable at high speed and the difference in the width of the instability chart can reach as high as 60% at the spin speed of 40000 RPM for the case of the effect of disk location in study 1. As such, it can be concluded that, for the shaft-disk system, the effect of torsional motion of the shaft on the parametric instability of the shaft is significant at higher rotating speed. Moreover, a rotor system becomes more unstable at lower frequency as the number of disk increases and, for static shafts with disks that differ in weight and radius, their parametric instability regions are quite similar. Last but not least, attaching disk with larger radius will cause the shaft to become more unstable.

Acknowledgement

The authors acknowledge the support in this study from Malaysia-Japan Institute of Technology and Universiti Teknologi Malaysia through the provision of the GUP Tier 2 Grant – PY/2017/00468.

References

[1] Gayen D, Chakraborty D and Tiwari R 2017 *European Journal of Mechanics A/Solids* **61** 47-58
[2] Bai B, Zhang L, Guo T and Liu C 2012 *Procedia Engineering* **31** 654–8
[3] Hu Z, Tang J, Zhong J, Chen S and Yan H 2016 *Mechanical Systems and Signal Processing* **76-77** 294–318
[4] Montagnier O and Hochard C 2014 *Journal of Sound and Vibration* **333** 470–84
[5] Bolotin V V 1965 *Am. J. Physics* **33** 752
[6] Wang S, Wang Y, Zi Y and He Z 2015 *Journal of Sound and Vibration* **359** 116-35
[7] Mailybaev A A and Spelsberg-Korspeter G 2015 *Journal of Sound and Vibration* **336** 227-39
[8] Chen W R and Chen C S 2014 *International Journal of Structural Stability and Dynamics* **14** 1-24
[9] Han Q and Chu F 2012 *Communication in Nonlinear Science Numerical Simulation* **17** 5189-200
[10] Li W, Song Z and Chai Y 2015 *Journal of Engineering Mechanics* **141** 04015033
[11] Song Z, Chen Z, Li W and Chai Y 2017 *Meccanica* **52** 1159-73
[12] Dakel M, Baguet S and Dufour R 2014 *Journal of Sound and Vibration* **333** 2774-99
[13] Chen W R and Chen C S 2014 *International Journal of Structural Stability and Dynamics* **14** 1450014
[14] Yim K B and Yim J T 2013 *J. Mech. Sci. Technology* **27** 359–66
[15] Boglietti A, Gerada C and Cavagnino A 2014 *IEEE Trans. Ind. Electron* **61** 2943-5
[16] Nelson H D 1980 *Journal of Mechanical Design* **102** 793-803
[17] Rao J S 2011 *Finite Element Methods for Rotor Dynamics* in *History of Rotating Machinery Dynamics* Springer Netherlands
[18] Chen L W and Ku D M 1990 *J. Sound Vibration* **143** 143–51
[19] Pei Y C, Lu H and Chatwin C 2010 *Proc. Inst. Mech. Eng. Part K J. Multi-Body Dynamics* **224** 211–9

IOP Conf. Series: Materials Science and Engineering **270** (2017) 012025 doi:10.1088/1757-899X/270/1/012025

Effect of low velocity impact damage on the natural frequency of composite plates

E Y L Chok, D L A A Majid* and M Y Harmin

Department of Aerospace, Faculty of Engineering, Universiti Putra Malaysia, 43400 Serdang, Selangor, Malaysia

*dlaila@upm.edu.my

Abstract. Biodegradable natural fibers have been suggested to replace the hazardous synthetic fibers in many aerospace applications. However, this notion has been limited due to their low mechanical properties, which leads to the idea of hybridizing the two materials. Many aircraft components such as radome, aft body and wing are highly susceptible to low velocity impact damage while in-service. The damages degrade the structural integrity of the components and change their dynamic characteristics. In worst case scenario, the changes can lead to resonance, which is an excessive vibration. This research is conducted to study the dynamic characteristic changes of low velocity impact damaged hybrid composites that is designed for aircraft radome applications. Three materials, which are glass fiber, kenaf fiber and kenaf/glass fiber hybrid composites, have been impacted with 3J, 6J and 9J of energy. Cantilevered and also vertically clamped boundary conditions are used and the natural frequencies are extracted for each of the specimens. The obtained results show that natural frequency decreases with increasing impact level. Cantilevered condition is found to induce lower modes due to the gravitational pull. To eliminate mass and geometrical effects, normalized modes are computed. Among the three materials considered, glass fiber composites have displayed the highest normalized frequency that reflects on its higher stiffness compared to the other two materials. As the damage level is increased, glass fiber composites have shown the highest frequency reduction to a maximum of 35% while kenaf composites have the least frequency reduction in the range of 1 – 18%. Thus, kenaf fiber is taken to be helpful in stalling the damage progression and reducing the effect of damage. This has been proven when the percentage frequency decrement shown by kenaf/glass fiber composite lies between glass fiber and kenaf fiber composites.

1. Introduction

Owing to their comparably high mechanical properties, natural fibers especially kenaf fibers have been suggested to replace many highly hazardous synthetic fibers such as glass fibers that are commonly in used for structural loads. However, as established in several studies, kenaf fibers have been unable to fully replaced glass fibers because the differences in mechanical properties between the two fibers are very high. This is why a lot of research works nowadays are focusing more on the hybridization of the natural and synthetic fibers [1–4]. Hybridization process can maintain the individual advantages of the two fibers and also alleviate some of their disadvantages. Hence the idea of hybridizing kenaf fiber with glass fiber is believed to be helpful in reducing the hazards from glass fiber and also enhancing the mechanical properties of kenaf fibers through the aid of glass fiber. Since then, a lot of researches have been carried out to investigate the mechanical properties of this hybrid material.

According to these studies, while non-hybrid composites exhibit the highest or lowest mechanical properties, the hybrid composites lies in between depending on the reinforcement fibers ratio [5]. This is generally referred to as the rule of mixture. Although hybrid composite is less likely to exhibit the highest material properties, it is often considered in order to tailor the material to the exact need of the structure under design. The needs typically revolve around the cost and weight factors, and also the biodegradability of the material. However, an impact damage can act as a limitation to the composites' widespread application in the field. During operation, an aircraft is exposed to a number of events that can cause impact damage, mostly involving low velocity impact damages. For instance, dropped tools during maintenance routine and in-service impacts by foreign objects or debris can introduce an out of plane impact on the aircraft structures, which can be of low and high velocity impact. High velocity impact damage usually involves instant damage and failure on the impacted subject as it is dominated by wave propagation [6]. However, since fiber reinforced composites are highly sensitive to the low velocity impacts [7], the effect of this type of impact is studied in this research work. The effect of low velocity impact damage on the dynamic characteristic of a structure can be studied through a vibration testing and there are a number of studies that have been carried out on this matter. For instance, carbon fiber has been utilized as the reinforcement of a composite material and from the obtained results, it is concluded that natural frequency tends to decrease with increasing impact level [7–9]. Similar results have also been observed and reported in other experimental studies [10–12]. Therefore, it is crucial to assess the changes in dynamic characteristic at the early stage through accurate, time and cost effective tool such that service life of the structure can be extended and coincidence with resonant frequencies can be avoided.

The dynamic characteristics with impact damage have been widely researched for synthetic fibers and the induced impact has been shown to vary with low velocity, medium velocity and/or in ballistic form. However, there is an absence in post impact dynamic characteristics studies that involve natural fibers. There is also no studies found on the effects of hybridizing natural fiber with synthetic fibers to the modal properties after being damaged in a low velocity impact. Hence, this study is dedicated to study the effect of a low velocity impact damage on the modal properties of a natural fiber, synthetic fiber and hybrid natural/synthetic fiber composites, and investigate the resultant difference in dynamic response between the materials. Moreover, comparison of modal parameters between these materials without damage is also of interest in this study. The relationship between impact damage level and natural frequency has been highlighted in several previous studies. The effect of impact damage on the natural frequency of the materials is discussed but significant reviews on the vibration characteristics of natural fibers and its hybrid with synthetic fibers after impact damage can hardly be found, resulting in the lack of understanding when comparing the modal parameters change between the two materials.

2. Research Methodology

The flow of experimental work for this study is shown in Figure 1. The main materials used are kenaf fiber and glass fiber, together with the resins to produce the kenaf, glass and kenaf/glass fiber hybrid laminates. To obtain uniform thickness, the composites are designed according to their own physical properties as described below:

- Glass fiber composite – six layers of chopped strand mat fiber (random oriented)
- Kenaf fiber composite – two layers of woven kenaf fiber mats
- Hybrid kenaf/glass fiber composite – two layers of glass fiber mats with a layer of kenaf fiber mat sandwiched in between.

Typical setup for modal testing is depicted in Figure 2. Specimens for both damaged and undamaged cases are impacted with the roving hammer.

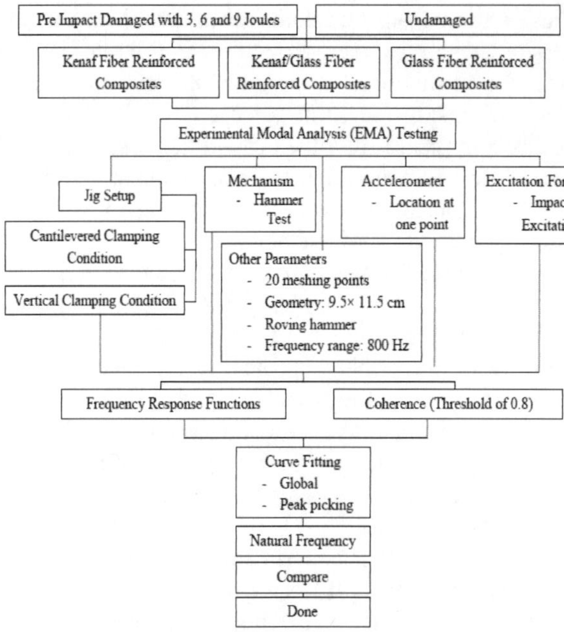

Figure 1: Flowchart of experimental work

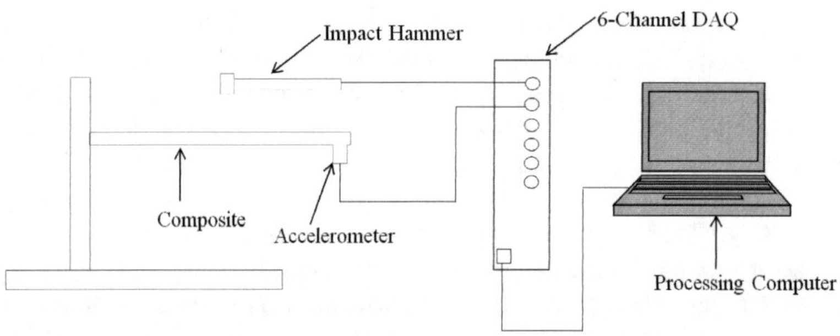

Figure 2: Setup for modal testing

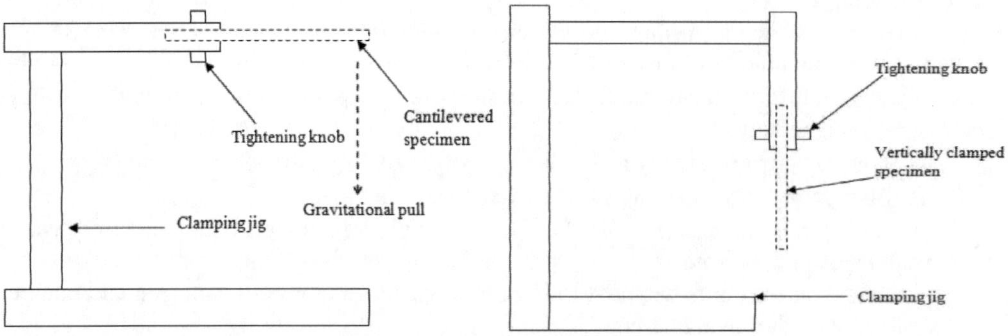

Figure 3: Clamping jig for (a) cantilevered structure, (b) vertically clamped structure

The experiment is conducted by varying two different clamping conditions: vertically clamped and cantilevered as shown in Figure 3. It is very common for EMA test to be conducted with cantilevered

IOP Conf. Series: Materials Science and Engineering **270** (2017) 012025 doi:10.1088/1757-899X/270/1/012025

condition as it represents the actual situation of most applications in the various fields such as the wing of an aircraft. However, in order to eliminate the gravitational force of a specimen under cantilevered condition, this study has also utilized the vertical clamping technique to investigate the behaviour of a structure without the interference of gravitational pull.

Modal analysis is conducted on the three types of materials: kenaf fiber reinforced composite, glass fiber reinforced composite and kenaf/glass fiber hybrid reinforced composite. This is the continuation work from previous study in [13]. All specimens are modelled to the dimension of 9.5cm × 14cm with the average weight is about 57g for prior impact test [13]. For this study, the dimension is taken to be 9.5cm × 14cm, with 2cm of the length is used for clamping. Each category of specimen for impact damaged (3J, 6J and 9J) and undamaged conditions is tested three times to obtain the average reading.

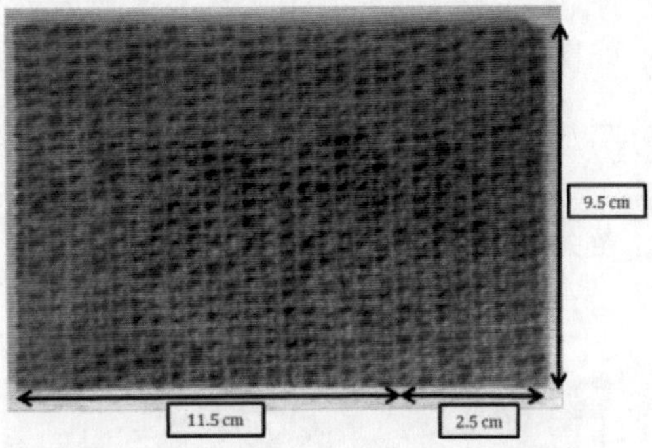

Figure 4: Geometry of the specimen tested

3. Results and Discussion

From frequency response function (FRF) graph comparing all three undamaged materials in vertical clamping condition in Figure 5, it can be seen that the pristine kenaf/glass fiber composite exhibits the highest natural frequency in all four modes. It is followed by the glass and kenaf fiber composites. The difference in frequency is not obvious in the first mode (i.e. first peak) among all three. Nonetheless, the variance gets higher while progressing to higher modes. On the other hand, for the cantilevered boundary condition shown in Figure 6, a similar trend to that of vertical clamping condition can be seen. However, the material with the lowest frequency in this case is the kenaf fiber composite and the highest is the hybrid kenaf/glass fiber composite. The main difference between the two FRF graphs for both cantilevered and vertically clamped is the variance of peaks in higher frequency. In this boundary condition, the shifting in higher modes or frequency is less significant compared to the previous graph.

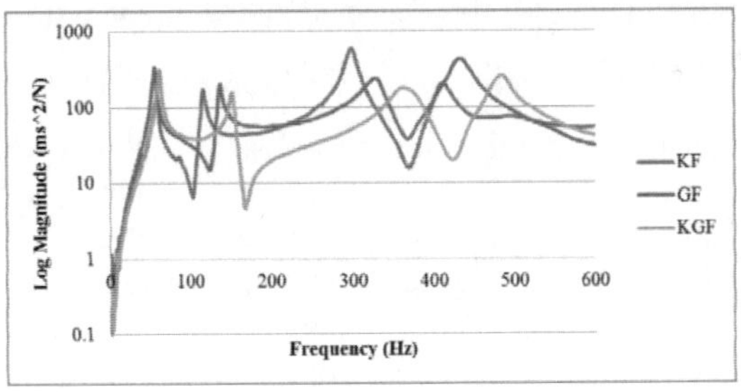

Figure 5: FRF overlays of undamaged materials for vertical clamping

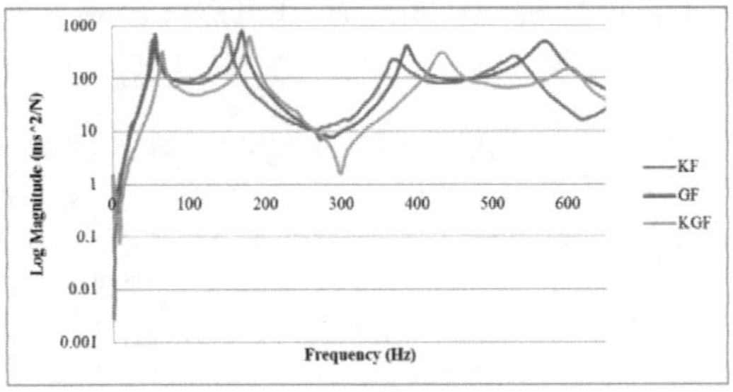

Figure 6: FRF overlay of undamaged materials for cantilevered

The finding for all three types of materials is the same when comparing between cantilevered and vertical clamping conditions. Cantilevered is found to exhibit a lower frequency than vertical clamping condition. Although cantilevered is the boundary condition normally found in actual applications such as the aircraft wing, vertical clamping is preferred in order to maximize the isolation of gravitational pull for the purpose of designing. At the same time, vertical clamping is also used in some parts of the aircraft structures like the stiffeners. The conclusion that can be drawn from the overlaid FRF graphs is that any shift in frequency as an effect of material or clamping conditions will take place in higher frequencies (higher modes).

The extent of damage effects on natural frequencies of the three configurations is discussed based on the graph representations for both clamping conditions shown in Figure 7. It can be observed that regardless of materials and clamping direction, the natural frequency will decrease with the increasing impact damage level and the natural frequency obtained under cantilevered condition is lower than the other clamping condition. This is because cantilevered structures suffered gravitational pull that tends to deform the structure. In other words, cantilevered boundary condition tends to reduce the stiffness of structure. The reduction trend has been expected because the introduction of damage will reduce the strength of a material.

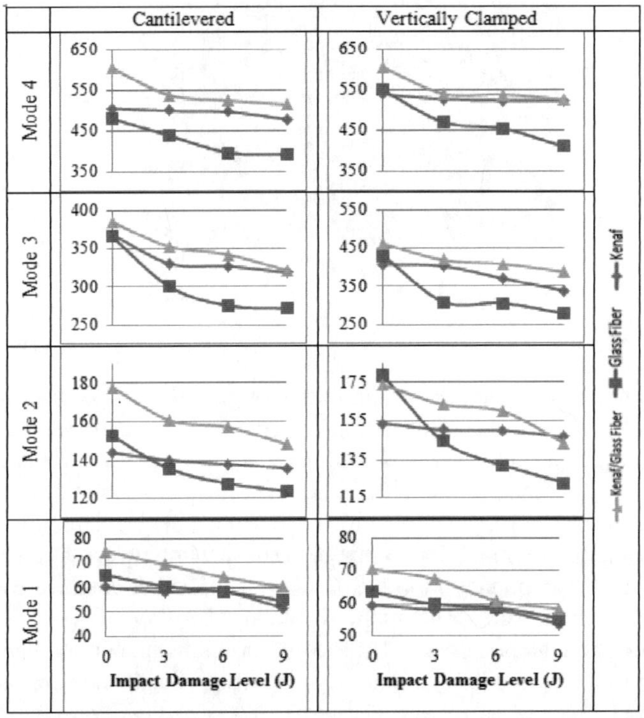

Figure 7: Effect of impact damage to natural frequency for vertically clamped and cantilevered conditions

An important thing to highlight in this study is the change in the dynamic response of each material with increasing level of damage. This can be discussed by analyzing the modes shift and percentage difference (shown in Figures 8 – 13). The frequency shift from undamaged to damaged states (3J, 6J and 9 J) for all three types of materials is observed to be high in the higher frequency range or higher modes. Based on this observation, it can be concluded that the effect of impact damage on a structure is much more apparent in higher frequency. The frequency shift in kenaf fiber composites is less when compared with that of sole glass fiber composite and hybrid kenaf/glass fiber composite. This shows that the impact resistance of the kenaf fiber composite is better than the other two materials as it can absorb the impact energy.

The trends obtained for cantilevered condition is similar to the previous boundary condition. The frequency shift can be noticed in all categories of materials for both clamping conditions and the shift is the lowest in kenaf fiber composite. As mentioned before, the frequency shift is more pronounced in higher frequency. Kenaf composite has the lowest natural frequency when it is undamaged. Once the damage is introduced, its natural frequency starts to decrease but the decrement (i.e. minimum 1.1% and maximum of 18.03%) is not as significant as the other two materials. Natural frequency decrement of kenaf/glass fiber composite is in the range of 4.3-25.7% whereas glass fiber composites exhibit the highest decrement up to 35.4%. From the obtained results, two significant statements can be drawn: natural frequency of kenaf fiber composite decreases the least when impact damage in induced and the frequency decrement when the glass fiber composite is damaged is significant since the magnitude of decrement is high.

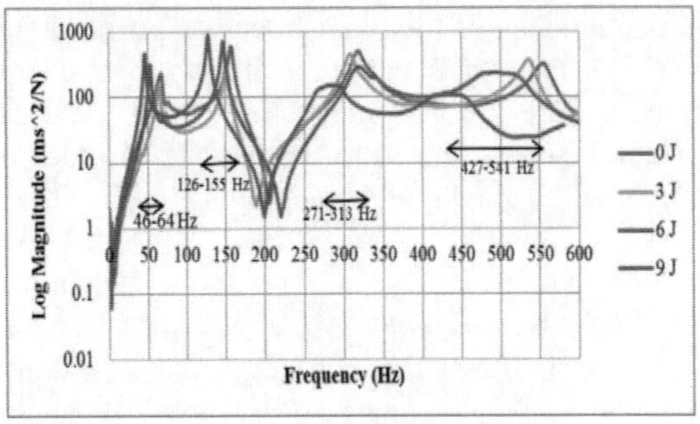

Figure 8: FRF overlay of kenaf composite for vertically clamped condition with different impact damage level

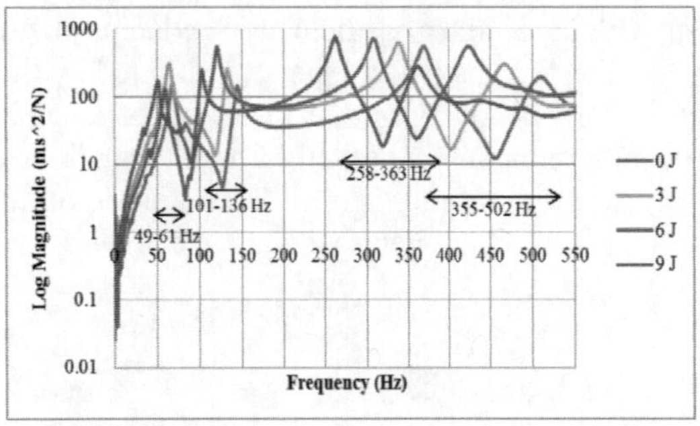

Figure 9: FRF overlay of glass composite for vertically clamped condition with different impact damage level

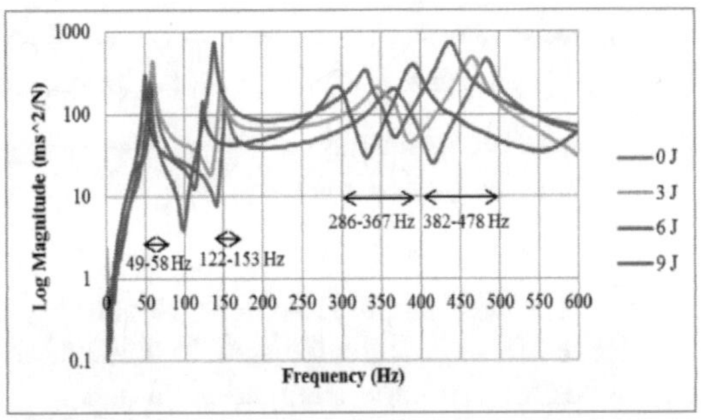

Figure 10: FRF overlay of kenaf/glass fiber composite for vertically clamped condition with different impact damage level

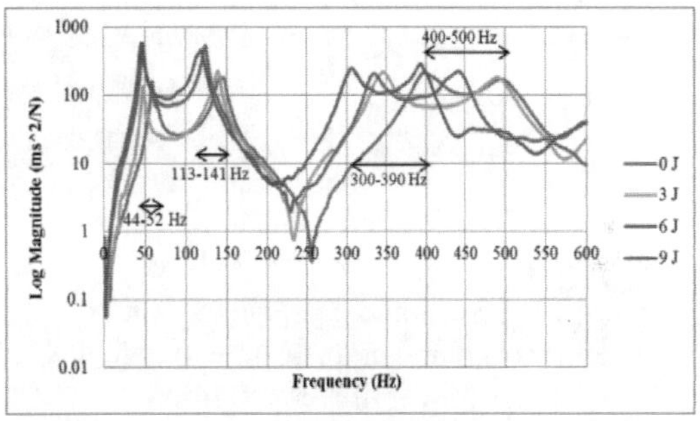

Figure 11: FRF overlay of kenaf composite for cantilevered condition with different impact damage level

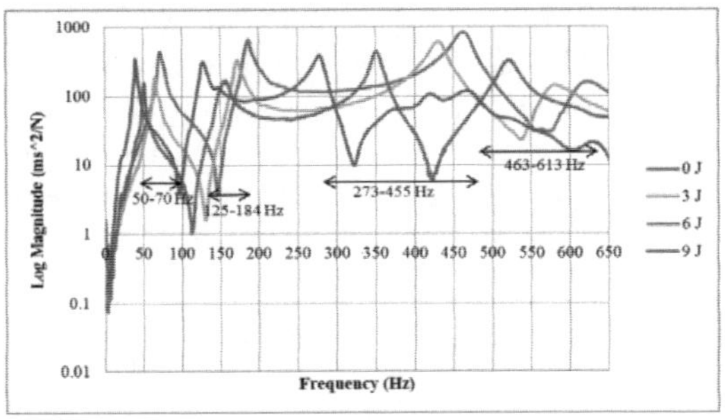

Figure 12: FRF overlay of glass fiber composite for cantilevered condition with different impact damage level

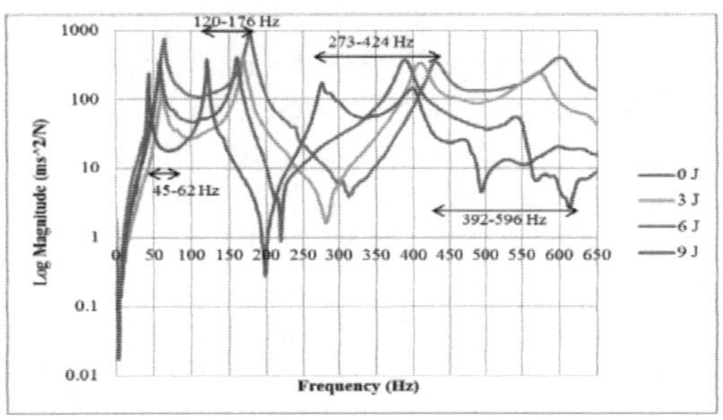

Figure 13: FRF overlay of kenaf/glass fiber composite for cantilevered condition with different impact damage level

IOP Conf. Series: Materials Science and Engineering **270** (2017) 012025 doi:10.1088/1757-899X/270/1/012025

Figure 14 and Figure 15 show the percentage decrement from undamaged to damaged states. From the figures, the effect of impact damage on glass fiber composite on both clamping conditions is very high even when the impact induced is of low velocity. This may be caused by the brittle nature of the composite. It is also determined in [13] that glass fiber composite has higher stiffness as compared to the other two. This means that the composite has less ability to absorb the impact energy when it is impacted, which explains the big reduction of its natural frequency after being damaged. From all the observations, two statements can be drawn as conclusions. The first being the hybridization of kenaf with synthetic fiber is shown to improve the natural frequencies. In general, the act of hybridizing glass fiber with kenaf fiber has helped in reducing the natural frequency of the glass fiber composite by up 19.4%. The high natural frequency exhibited by the hybrid kenaf/glass fiber composite is due to several factors namely the structural mass, stiffness and damping. The second conclusion is that the least frequency reduction exhibited by the kenaf fiber composite can be related to the better resistance of woven fiber composites against the initiation of delamination, hence the high residual strength. This is also supported in previous studies done [14–15].

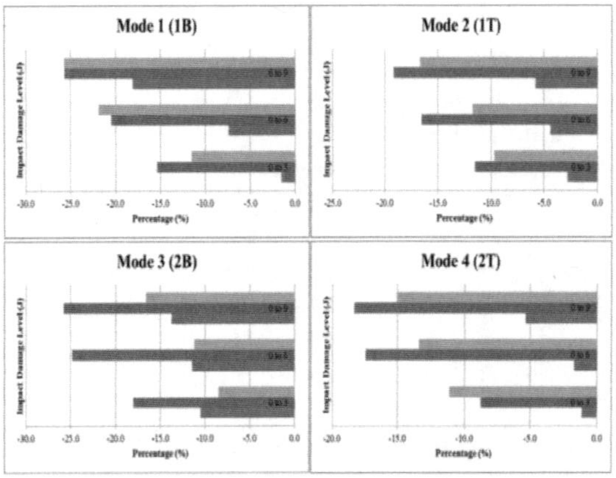

Figure 14: Percentage decrements of natural frequency from undamaged state to damaged states for cantilever condition

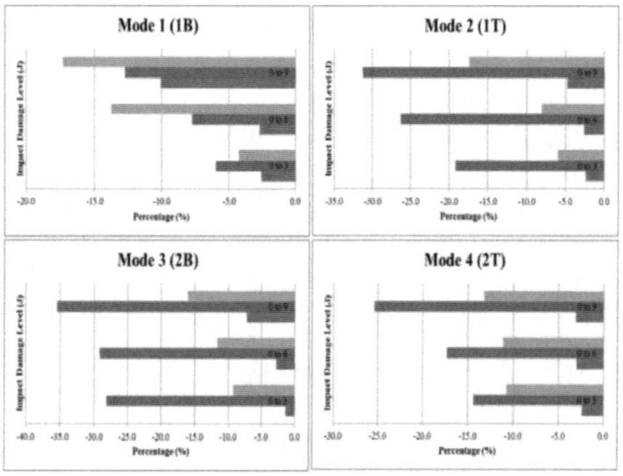

Figure 15: Percentage decrements of natural frequency from undamaged state to damaged states for vertically clamped condition

Glass fiber composite consistently shows the lowest natural frequency in all damaged cases among the three materials. Meanwhile, kenaf/glass fiber hybrid composites generally have the highest natural frequency in almost all cases. As this research work is the continuation of previous study in [13], glass fiber composite has been predicted to produce higher frequencies due to its high stiffness. However, the lower frequencies obtained by this composite material in this study can be contributed to several factors such as mass and geometrical variations. In order to eliminate these contributions, Figure 16 shows normalized frequencies in which glass fiber composite has portrayed the highest frequencies. It can also be concluded that the hybridization of kenaf with synthetic fiber has positive effects in term of natural frequencies decrement with post impact damage. This statement is made by referring to the slopes in Figure 16 where the glass fiber composite has higher steepness than the other two materials. Hence, the effect of kenaf in kenaf/glass fiber hybrid composite is believed to be capable in reducing the effect of damage on the composite as the slope difference between the glass fiber composites and kenaf/glass fiber composites is in the range of 20% to up to 64%.

4. Conclusion

In general, natural frequency decreases as the level of impact damage is increased. Furthermore, the frequency obtained under a cantilevered condition is lower than that in a vertically clamped condition. This is due to the gravitational pull in the cantilevered condition. Besides, the natural frequencies of undamaged kenaf/glass fiber composites prior to normalization are higher than glass fiber composites, with kenaf fiber composites exhibiting the lowest natural frequency for both cantilevered and also vertically clamped boundary conditions. From the results obtained, it is shown that in kenaf/glass fiber hybrid composite, glass fiber is responsible in the strength attribute in which higher natural modes can be acquired while kenaf functions as the restrainer of the damage propagation. This is apparent from the following results:

- After normalized frequency is computed, glass fiber composite has the highest natural frequency, followed by kenaf/glass fiber and then kenaf fiber composites.
- Natural frequency of glass fiber decreases the most as it marks a maximum of 35.4% decrement in natural frequency decrement when damage is induced.
- Kenaf composite suffered the least natural frequency decrement at the range of 1 – 18% while the reduction of kenaf/glass fiber composites lie between kenaf and glass fiber composites.
- Kenaf fiber is then inferred as a useful material to slow down and reduce the effect of damage on a structure.
- The effect of damage on a glass fiber composite can be reduced (lower percentage decrement of natural frequency) through the hybridization with kenaf fiber.

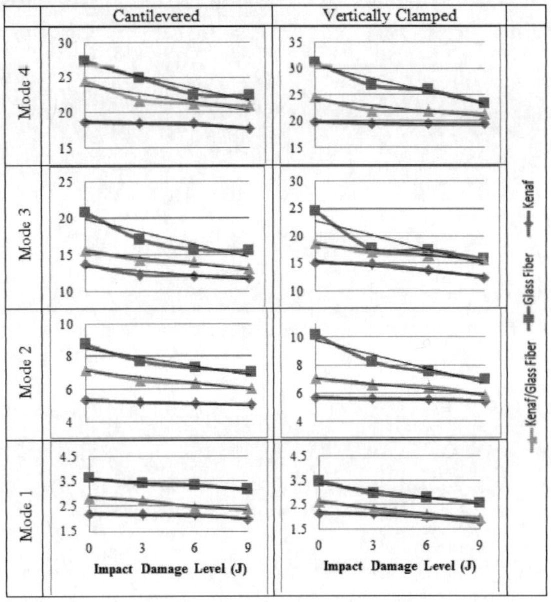

Figure 16: Normalized frequencies for all materials under cantilevered and vertically clamped boundary condition

Acknowledgements
The authors acknowledge support from Universiti Putra Malaysia through the research grant number 9463100.

References

[1] Jawaid M, Abdul Khalil H P S, Bhat A H and Abu Baker A 2011 *Advanced Materials Research* **264** 688-93
[2] Berhan M N, Yunus S, Salleh Z, Masdek N R N, Taib Y M and Ghazali F A 2015 *Jurnal Teknologi* **76** 37-41
[3] Ghani M A A, Salleh Z, Hyie K M, Berhan M N, Taib Y M D and Bakri M A I 2012 *Procedia Engineering* **41** 1654-9

[4] Sharba M J, Leman Z, Sultan M T, Ishak M R and Hanim M A A 2015 *BioResources* **11** 1448-65
[5] Faruk O, Bledzki A K, Fink H P and Sain M 2012 *Progress in polymer science* **37** 1552-96
[6] Razali N, Sultan M T H, Mustapha F, Yidris N and Ishak M R 2014 *The International Journal of Engineering and Science* **3** 8-20
[7] Shahdin A, Mezeix L, Bouvet C, Morlier J and Gourinat Y 2009 *Engineering Structures* **31** 2833-41
[8] Pérez M A, Gil L and Oller S 2014 *Composite Structures* **108** 267-76
[9] Nossol P, Czech A and Kroll L 2013 *Composites Science and Technology* 13 128-34
[10] Grady J E and Meyn E H 1989 *NASA Technical Memorandum 4115*
[11] Consuegra F and Irfanoglu A 2008 *14th World Conference on Earthquake Engineering*
[12] Lakhdar M, Mohammed D, Boudjemâa L, Rabiâ A and Bachir M 2013 *Energy Procedia* **36** 888-97
[13] Jamal Q M 2016 *Validation of low velocity impact on biocomposite flat plate laminates* Master Thesis, Universiti Putra Malaysia, Malaysia
[14] Kim J K, Leung L M, Lee S W R and Hirai Y 1996 *Polymers & polymer composites* **4** 549-61
[15] Strait L H, Karasek M L and Amateau M F *Journal of Composite Materials* **26** 1725-40

IOP Conf. Series: Materials Science and Engineering **270** (2017) 012026 doi:10.1088/1757-899X/270/1/012026

Ballistic impact velocity response of carbon fibre reinforced aluminium alloy laminates for aero-engine

I Mohammed[1,2], A R Abu Talib[1,3*], M T H Sultan[1,3] and S Saadon[1]

[1]Department of Aerospace Engineering, Faculty of Engineering, Universiti Putra Malaysia, Serdang, Selangor Malaysia.
[2]Department of Mechanical Engineering, College of Engineering, Hassan Usman Katsina Polytechnic, Katsina, Nigeria.
[3]Aerospace Manufacturing Research Centre, Faculty of Engineering, Universiti Putra Malaysia, Serdang, Selangor Malaysia.

*abdrahim@upm.edu.my

Abstract. Aerospace and other industries use fibre metal laminate composites extensively due to their high specific strength, stiffness and fire resistance, in addition to their capability to be tailored into different forms for specific purposes. The behaviours of such composites under impact loading is another factor to be considered due to the impacts that occur in take-off, landing, during maintenance and operations. The aim of the study is to determine the specific perforation energy and impact strength of the fibre metal laminates of different layering pattern of carbon fibre reinforced aluminium alloy and hybrid laminate composites of carbon fibre and natural fibres (kenaf and flax). The composites are fabricated using the hand lay-up method in a mould with high bonding polymer matrix and compressed by a compression machine, cured at room temperature for one day and post cure in an oven for three hours. The impact tests are conducted using a gun tunnel system with a flat cylindrical bullet fired using a helium gas at a distance of 14 inches to the target. Impact and residual velocity of the projectile are recorded by high speed video camera. Specific perforation energy of carbon fibre reinforced aluminium alloy (CF+AA) for both before and after fire test are higher than the specific perforation energy of the other composites considered before and after fire test respectively. CF +AA before fire test is 55.18% greater than after. The same thing applies to impact strength of the composites where CF +AA before the fire test has the highest percentage of 11.7%, 50.0% and 32.98% as respectively compared to carbon fibre reinforced aluminium alloy (CARALL), carbon fibre reinforced flax aluminium alloy (CAFRALL) and carbon fibre reinforced kenaf aluminium alloy (CAKRALL), and likewise for the composites after fire test. The considered composites in this test can be used in the designated fire zone of an aircraft engine to protect external debris from penetrating the engine shield due to higher values of impact strength and specific perforation energy as highlighted by the test results.

1. Introduction

Due to demands for lightweight body in aircraft construction, the need for improved fibre reinforced laminate composites has been progressively increasing. Fibre metal laminates play an important role in protecting the aircraft components against foreign objects and providing an excellent ballistic impact penetration resistance on them. Improvement in the performance of impact properties, lightweight and toughening of aircraft engine nacelle materials has become necessary to protect the components from cracking after collision with debris, birds, rain and/or hail. Fibre reinforced composite is among the materials with excellent properties of impact loading. In aircraft industry, this material type has been produced with different fibre types that uses different types of matrix, formed using different methods, and can have various shapes with different thickness [1]. Carbon fibre reinforced plastic (CFRP) has superior properties of high strength, stiffness and also resistance to impact loading [2]. However, the most frequently identified constraints of fibre-reinforced plastics are their poor resistance to localized impact loading and their load bearing properties [3, 4]. The study of hybrid laminate composites has become necessary due to limitation of each constituent. For instance, when they are considered alone, carbon fibre reinforced composite has the disadvantages of poor impact and residual strength whereas poor fatigue strength affects the performance of aluminium alloy. Combination of these two materials can overcome these problems and increase the cost effectiveness of the composite [5].

There are several factors to be considered in the hybridization of natural fibre with synthetic fibres, which include availability, lightweight, cost and environmental factors. Bast fibre has been reported to be the most suitable hybridized natural fibre [6, 7]. Furthermore, some advantages of Kenaf hybridized with synthesis fibre have also been highlighted [8]. An investigation on ballistic impact of synthetic fibre hybridized with natural fibre laminates has also been conducted. Different layers of aramid and kenaf are hybridized, and the resultant laminate composites that are used for protective helmet showed the same performance in high impact penetration as that of aramid only [9]. In addition, it is found that improved thickness and epoxy ductility of the composites will produce excellent impact properties and improve energy absorption of the material (in this study, kenaf is hybridized with glass fibre) [10]. In the meantime, it has been found out in a study on the effect of stacking a sequence of Kevlar fibres on failure mechanism and residual velocity, where the ballistic limit of coir yarn and Kevlar/epoxy hybrid composite are determined, that the main failure on a high velocity impact test is based on fibre fracture and delamination of the metal part of the composite [11].

Many researchers have studied the impact velocity of composite used in aircraft components. It is reported that the energy absorption of Kevlar-29/epoxy and Kevlar 29/epoxy-Al_2O_3/Al with different stacking sequences using a cylindrical steel projectile increases with increase of composite thickness [12]. The elastic, petaling and dishing work are all ingested in perforation energy [13]. Moreover, the relationship that exists between thickness of the composites and ballistic impact velocity of cylindrical bullet has been established, whereby the total loss of kinetic energy of the cylindrical bullet projectile is assumed to be equal to total work done in the deformation of the plate 14]. The energy dissipation of the component depends on the types of polymer used and the high impact strength of the composites is dependent on bonding adhesions that exists between the fibre and matrix, and the delamination of the composites, whereas the brittle matrix dissipation energy is suppressed [15]. The hybrid composite of synthetic fibre (Kevlar, glass fibre and carbon fibre) for constructing armour are studied and the result indicates an increase in ballistic impact performance, likewise on Kevlar and polypropylene (PP) [16, 17]. Fibre is the material responsible for crack in fibre reinforced composite. It controls the elongation of the composite during breakage by reducing impact energy absorption. Agglomeration also occurs on it at high fibre loading, causing fibre-to-fibre contact that reduces the stress transfer between the fibre and the matrix, resulting in composite failure [18-20].

Different modes are used in determining the level of damages in the composites during impact. There are various internal damages after the impact on the composites such as delamination, matrix cracking, fibre breakage and others, especially when they are impacted with low velocity. Ultrasonic image or radiography and backlighting are some of the equipment used to detect the internal damage made from impact [21-27], matrix crack for each ply of fibre are determined by Bar-Cohen and Crane [28] and also detect some of the defects that affect the fibres such as misalignment. In this study, four types of laminate composites are fabricated to withstand a high temperature in fire designated zone of an aircraft engine using fibre metal laminate of aluminium alloy 2024-T3, carbon fibre, natural fibres (kenaf and flax) and epoxy resin/hardener. Two of the composites are consisted only of carbon fibre and aluminium alloy: one with aluminium alloy at front and rear face of the composite only, and the other composite consists of alternating layer of aluminium alloy. Two other composites consist of hybrid synthetic and natural fibres with aluminium alloy at the front and rear face of the composites: one consists of carbon fibre and kenaf at the centre whereas the other consists of carbon fibre and flax.

The main purpose of this study is to investigate the impact strength resistance and the specific perforation energy of the fibre metal laminate composites. The test is performed in a fire designated zone of an aircraft engine, using propane-air burner according to ISO 2685 standard. The temperature range and heat flux range of the standard are 1100±80^{0}C and 116±10kW/m^2. The materials used in fire test are affected by the impact velocity based on the external forces that affect the aircraft engine nacelle during take-off and landing. Hybrid composites of synthetic/natural fibre and synthetic fibre with aluminium alloy 2024-T3 have been developed with varying proportion of materials layers, so that different effect of hybridization of the materials used in the composites are evaluated. Ballistic impact velocity of each composite is determined using gun tunnel system machine, and the impact and residual velocities each composite before and after fire test are evaluated using the high-speed video camera. Four main results are presented: the impact and residual velocity, potential energy, impact strength and the specific perforation energy before and after fire test of the composites considered.

2. Experimental Setup

Four composite samples have been prepared for the ballistic impact velocity target using fibre metal laminates of aluminium alloy 2024-T3, woven carbon fibre and natural fibres with the epoxy resin/ hardener of different numbers of layers, with 0^0-90^0 orientation. The plate area used for the impact is 100mm x 100mm and the target area is at the centre of the plate of around 15±2mm with the thickness is 3.5±0.2mm. Table 1 gives the summary and stacking configuration of the composites.

Table 1: Composites and their stacking configuration

S/No.	Composite Type	No. of layers	Stacking Configuration	Density (kg/m^3)
1	Carbon fibre kenaf reinforced aluminium alloy with aluminium alloy at the front and rear face of the composite (CAKRALL)	7	Al/2CF/KF/2CF/Al	1483
2	Carbon fibre flax reinforced aluminium alloy with aluminium alloy at the front and rear face of the composite (CAFRALL)	9	Al/2CF/FF/CF/FF/2CF/Al	1506
3	Carbon fibre reinforced aluminium alloy with alternating aluminium alloy (CARALL)	11	Al/2CF/Al/CF/Al/CF/Al/2CF/Al	1414
4	Carbon fibre reinforced aluminium alloy with aluminium alloy at the front and rear face of the composite (CF+AA)	13	Al/11CF/Al	1240

The composite of fibre metal laminates are fabricated by stacking the proper number of plies of the fibre materials and aluminium alloy sheets in a mould using the appropriate volume of epoxy resin/ hardener (of ratio 2:1) by hand lay-up method and then placing the stack into a compression machine. The laminates composites are then cured at room temperature for 24 hours and post cured in an oven for three hours at 80^{0}C. Figure 1 illustrates the laminate composite of the plate in a fixed position and

Figure 2 depicts the picture of the sample after impact. A plate of square edge of length 100 mm is clamped in a steel support and it is bolted to the steel block. Two sets of composites are tested: first set are composites before the fire test while the second set of composites are those after the fire test.

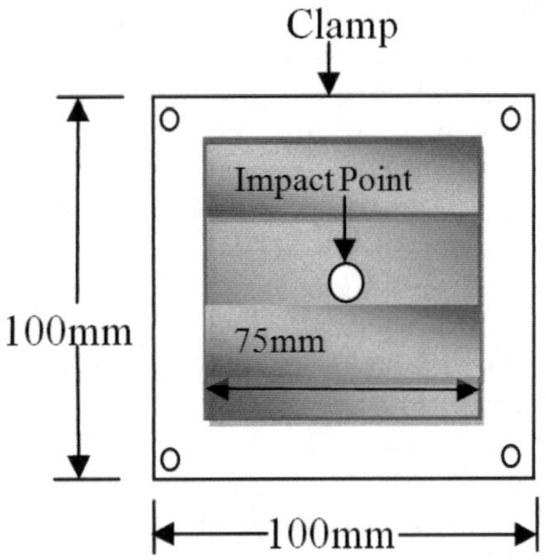

Figure 1: Composite laminate plate

Figure 2: Picture of a sample after impact

A flat cylindrical mild steel bullet of mass 5g, diameter 8.5mm and a length of 13.5mm are used in the test. In addition, a gun tunnel system, high speed video camera and a spot light are also used. The set up picture of the test is as shown in Figure 3, where the gun reservoir is pressurised at 20bars with helium gas and is fired by fast active valves. The flat cylindrical bullet accelerated horizontally in the barrel to impact chamber where the composite samples are fixed. The amount of the air in the barrel is ejected prior to the firing test by evacuating 100 mbar from the barrel.

The experimental impact velocity tests of the four composites are carried out using the test rig. The rig is 365mm length with nominal bore of 20mm. The composites to be tested are clamped between two steel constraining plates and are firmly tightened. During the test, the velocity before hitting the target and the residual velocity after hitting the target are measured by high speed video camera, which is set to record the incident impact at 30,000 frames per second with an image size of 277.3 pixels per image. The velocities of the flat cylindrical bullet before and after penetration of the composites, and the horizontal distance are recorded and the travel projectile time is calculated.

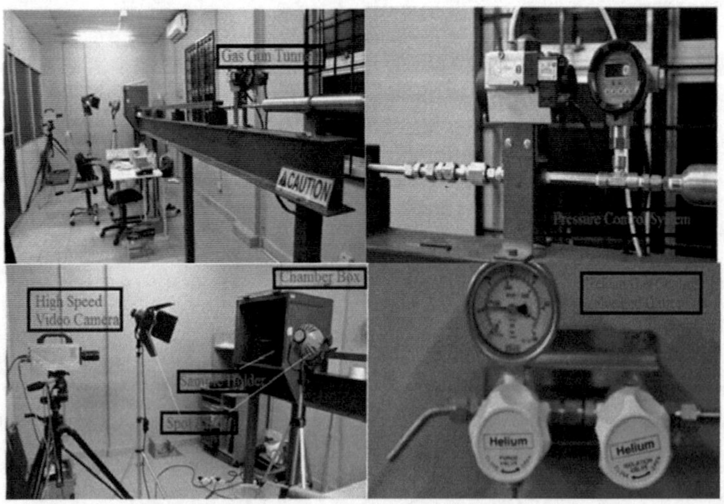

Figure 3: Ballistic impact test facility setup

Impact and residual velocities are obtained from different types of laminate composites before and after hitting the target, and also before and after the fire test. The ballistic impact test is performed on the composites for a range of impact energies until the complete perforation of the composite.

3. Results and Discussion

During the test, it has been observed that penetration occurs on all composite samples under study that are subjected to a velocity of 214±2m/s. The debris is pulled out from the front and rear faces of the composites, consisting of the matrix and fibre components, and delamination of the aluminium alloy. A bulge is observed on the rear face of the composites with much debris, while the front face takes the shape of a flat cylindrical projectile. Meanwhile, perforation at the front face has the same diameter with the projectile as shown in Figure 4. The bulge at the rear face shows that the projectile ceased at the exit of the composites with continuous crack development. As the fibre is fractured due to impact, the polymer matrix of that area will no longer hold the fibre (no stress acts on the area), therefore the matrix undergoes shear stress which in turn leads to failure of the panel. This situation agrees with a previous work in Ref. [29]. The flat cylindrical bullet has penetrated all composites during the impact, with increasing hole diameter from the front face to the rear face. Due to different materials used in each composite, the immensity of frictional force affects the impact energy absorption of the samples owing to much delamination that occur on the composite as reported by Ref. [30].

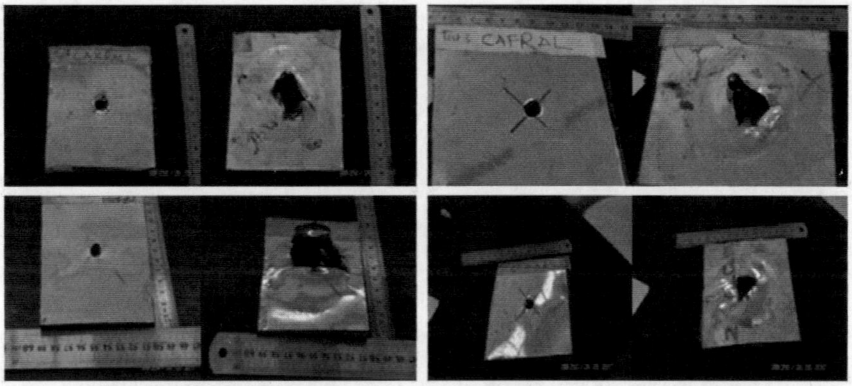
Figure 4: Front and rear faces of the samples after penetration

The lay-up of the samples are unidirectional and the impact damage is very extensive, especially in the delamination on the fibre axis leading to fragmentation of the fibre on the rear face. This is shown by the different shapes on different composites, which looks like rhombo-hedral and a cone shape, but the front face shows the same circular shape. Similar observation has been made in Ref. [31, 32], with a little bit higher deflection of the composites as reported in Ref. [33]. Before the fibre is fractured, the crack grows through the matrix inside the plies of the fibre up to the inter-laminar region where the composites are weakened and then failed as seen in Ref. [1]. The composites absorbed and dissipated the energy under the impact condition due to nature of composite constituents, as also observed in Ref. [34]. As the applied force has exceeded the fibre-matrix interfacial bond, de-bond occurs and the fibre breaks. The ductile tearing of the fibres and the delamination of the aluminium alloy of FML are in agreement with Ref. [35] and the mode of failure comes before the delamination as those observed in Ref. [36], which leads to plate deformation.

The velocities of each composite before and after the impact are presented in Figure 5. The CF+AA composites shows more impact resistance than the other composites. CAKRALL burns during the fire test and therefore, the velocity before the impact equals to the velocity after impact since no fibre is left on the centre of the composite during the impact test. From the results obtained, the velocity for CF+AA before the fire test is 6.7%, 26.6% and 17.6% greater than that for CARALL, CAFRALL and CAKRALL, respectively. Meanwhile, after the fire test, the velocity for CF+AA is 4.9% and 10.0%

greater than that for CARALL and CAFRALL, respectively. As the fibres are pulled out from the matrix, energy is dissipated. Impact energy is calculated from the difference of the potential energy of impact before and after hitting the samples. Moreover, impact strength of each composite is estimated by dividing the impact energy with their cross-sectional area as indicated in Figure 6.

The result shows that CF+AA has the highest impact strength for both before and after fire test. It has 32.98%, 50.0% and 11.7% greater impact strength (i.e. total energy absorbed before crack begins in the composite) than CAKRALL, CAFRALL and CARALL, respectively, for samples before fire test. On the other hand, for samples after the fire test, CF+AA has 75.0% and 36.6% greater impact strength than CAFRALL and CARALL, respectively. The result for CARALL shows that alternating layers of the composites causes delamination between the inner layers, as reflected by less travelling distance of the projectile in the sample that indicates less surface energy dissipation [37, 38]. Similar result has also been observed in Ref. [39], where the impact strength of different load percentages of Roselle fibre was investigated. There are some factors that affect the impact strength of the composites after fire test, which are poor adhesion, poor interfacial bonding between the matrix and fibre, and also stress concentration at the end of fibre. These factors are in agreement with the factors identified in Ref. [40] in the study of change in impact strength with filler content of rice husk with polypropylene composites.

Figure 7, meanwhile, shows the specific perforation energy of both composites before and after fire test. The CF+AA composite before fire test has the highest value of 23.40 Jm^2/kg and the least value of 3.38 Jm^2/kg is observed on CAFRALL after fire test.

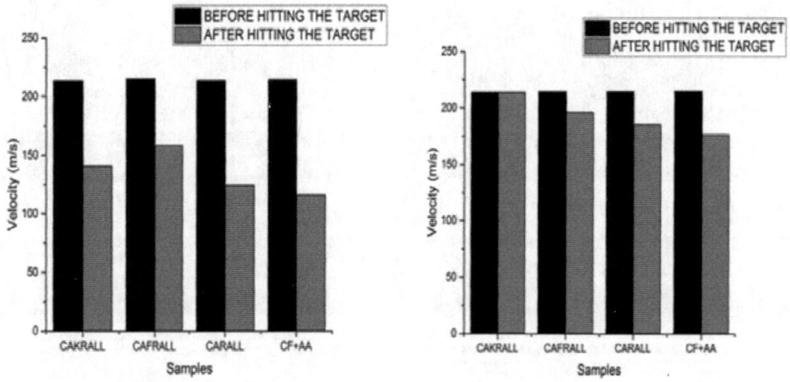

(a). Before Fire Test (b). After Fire Test

Figure 5: Velocity of the composites: (a) before impact and (b) after impact

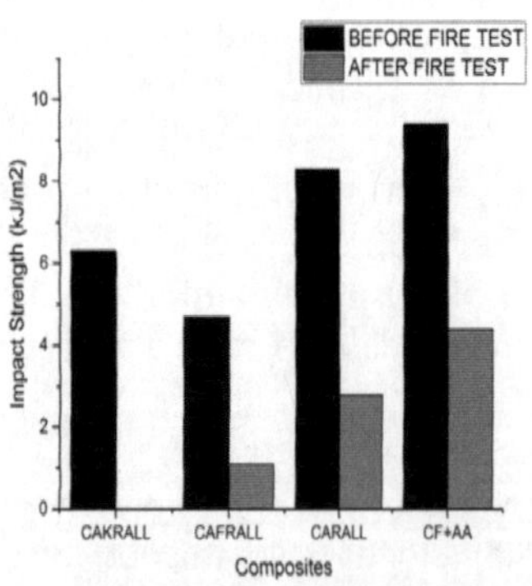

Figure 6: Impact strength of the composites before and after impact

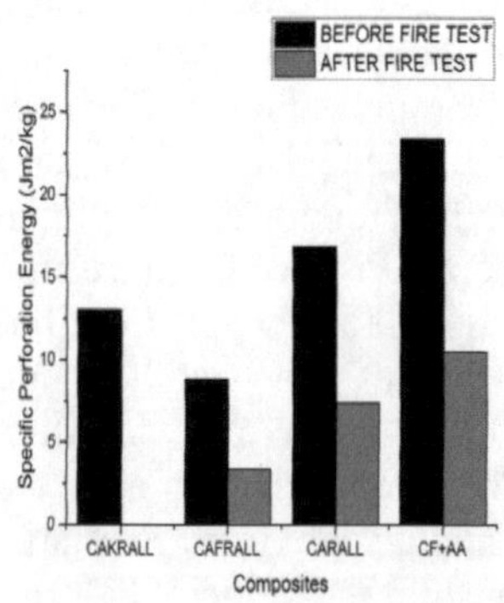

Figure 7: Specific perforation energy of the composites before and after the fire test

Specific perforation energy of the fibre metal laminates based on the type of fabrication materials increases with the increase of their tensile strength and laminate thickness. This result is in agreement with the result reported in Ref. [41]. In all the composites considered, the specific perforation energy values of the samples before the fire test are higher than those after the test. This can be contributed to the higher tensile and thickness values in the samples before the fire test. Furthermore, after the fire test, the composites lost most of adhesive bond properties between the fibre and matrix, and in some composites, there is total delamination of aluminium alloy. The specific perforation energy of CF+AA is higher than that of CARALL despite the fact that they have the same thickness. This implies that the carbon fibre plies are able to withstand more strength than a sheet of aluminium alloy and they are more bonded to epoxy resin/hardener. Therefore, the increase of carbon fibre corresponds to increase in perforation resistance in the composite. In addition, synthetic fibre (carbon fibre) is more resistant to perforation than the natural fibre (kenaf and flax). The results obtained in the specific perforation energy suggested that the fibre metal laminates that use thin sheets of aluminium alloy and reinforced fibres offer significant advantages to be utilized in the fire designated zones of an aircraft engine as a protective shield against external forces on the engine, including debris flying on the run way during landing or take-off or dropping any tools during an engine maintenance.

4. Conclusion

Moderate velocity impact performance of different types of fibre metal laminates based on reinforced synthetic and natural/synthetic fibres with aluminium alloy 2024-T3 and epoxy resin/hardener, before and after the fire test, has been studied. The results indicate that the hybrid composite laminates are able to be applied as lightweight energy-absorbing structures in the fire designated zone of an aircraft engine. The tests show that carbon fibre reinforced aluminium alloy, with aluminium alloy at the front and rear face of the composites, presents an excellent impact strength and specific perforation energy than the other considered composites due to good energy absorption behaviour of the carbon fibre. A crack growth is observed from front to rear faces of the samples, especially on the natural/synthetic hybrid fibre metal laminates, due to the de-bonding of the fibre and matrix. The lower modulus of the plies in composite makes aluminium alloy layers deform permanently by exciting remarkable energy through confined membrane stretching. The higher strain needed to break down natural/synthetic fibre with the laminate composites permits a greater amount of energy to be soaked up in the break down techniques, thereby increases perforation resistance of the composites. In terms of specific perforation energy and impact strength, the composites before the fire test have 223% and 214% higher than after the fire test, respectively. An examination on the samples shows that the composites can withstand a lower velocity impact on an external object either during take-off or landing.

Acknowledgement

The authors like to thank Universiti Putra Malaysia for providing the financial support through the IPS Grant Scheme (Vot Number: 9538500). Moreover, the authors also wish to thank Mr. Mohamad Nasir Johari and Mr. Saffairus Salih for their given technical support in this study.

References

[1] Hammond R, Proud W, Goldrein H and Field J 2004 *International Journal of Impact Engineering* **30** 69-86
[2] Matthews F and Rawlings R 1999 *Composites Structures and Materials* CRC Press
[3] Cantwell W, Curtis P and Morton J 1986 *Composites Science and Technology* **25** 133-48
[4] Zhou G 1995 *Composite Structures* **31** 185-93
[5] Vogelesang L and Vlot A 2000 *Journal of Materials Processing Technology* **103** 1-5
[6] Salman S D, Leman Z, Sultan M T, Ishak M R and Cardona F 2015 *BioResources* **11** 1176-88
[7] Yahaya R, Sapuan S, Leman Z and Zainudin E 2014 *Applied Mechanics and Materials* **564** 400-5
[8] Salman S D, Leman Z, Sultan M T, Ishak M R and Cardona F 2015 *BioResources* **10** 8580-603

[9] Salman S D, Leman Z, Sultan M T, Ishak M R and Cardona F 2016 *BioResources* **11** 7282-95
[10] Davoodi M, Sapuan S, Ahmad D, Aidy A, Khalina A and Jonoobi M 2012 *Materials Letters* **67** 5-7
[11] Azrin Hani Abdul R, Roslan A, Jaafar M, Roslan M N and Ariffin S 2011 *Advanced Materials Research* **277** 36-42
[12] Ramadhan A, Talib A R A, Rafie A S M and Zahari R 2012 *Applied Mechanics and Materials* **225** 213-8
[13] Abbud A T and Ali M 2011 *International Journal of Engineering and Technology* **8** 61-6
[14] Abu Talib A, Abbud L, Ali A and Mustapha F 2012 *Materials & Design* **35** 12-9
[15] Larsson F and Svensson L 2002 *Composites Part A: Applied Science and Manufacturing* **33** 221-31
[16] Bandaru A K, Chavan V V, Ahmad S, Alagirusamy R and Bhatnagar N 2016 *International Journal of Impact Engineering* **89** 1-13
[17] Bandaru A K, Vetiyatil L and Ahmad S 2015 *Composites Part B: Engineering* **76** 300-19
[18] Rojo E, Alonso M V, Oliet M, Del Saz-Orozco B and Rodriguez F 2015 *Composites Part B: Engineering* **68** 185-92
[19] Sreenivasan V, Ravindran D, Manikandan V and Narayanasamy R 2011 *Materials & Design* **32** 2444-55
[20] Manshor M, Anuar H, Aimi M N, Fitrie M A, Nazri W W, Sapuan S, El-Shekeil Y and Wahit M 2014 *Materials & Design* **59** 279-86
[21] Abrate S 1991 *Appl Mech Rev* **44** 155-90
[22] Malvern L, Sun C and Liu D 1987 *Damage in Composite Laminates from Central Impacts at Subperforation Speeds* in *Recent Trends in Aeroelasticity, Structures and Structural Dynamics* University Press of Florida
[23] Ramkumar R and Chen P 1983 *AIAA J* **21** 1448-52
[24] Frock B, Martin R, Moran T and Shimmin K 1988 *Imaging of Impact Damage in Composite Materials* in *Review of Progress in Quantitative Nondestructive Evaluation* Plenum Press
[25] Chen J and Sun C 1985 *Composite Structures* **3** 97-118
[26] Buynak C, Moran T and Donaldson S 1988 *SAMPE Journal* **24** 35-9
[27] Boll D, Bascom W, Weidner J and Murri W 1986 *Journal of Materials Science* **21** 2667-77
[28] Bar-Cohen Y and Crane R 1982 *Materials Evaluation* **40** 970-5
[29] Day R and Rodrigez J C 1998 *Composites Science and Technology* **58** 907-14
[30] Babu M G, Velmurugan R and Gupta N 2006 *Latin American Journal of Solids and Structures* **3** 21-39
[31] Rogers K, Sidey G and Kingston-Lee D 1971 *Composites* **2** 237-41
[32] Gorham D and Field J 1976 *Journal of Physics D: Applied Physics* **9** 1529
[33] Choi I *Composite Structures* **142** 7-14
[34] Santhosh J, Balanarasimman N, Chandrasekar R and Raja S 2014 *International Journal of Research in Engineering and Technology* **3** 144-50
[35] Abdullah M and Cantwell W 2006 *Composites Science and Technology* **66** 1682-93
[36] Hagenbeek M 2001 *Impact Properties* in *Fibre–Metal Laminates: An Introduction* Springer Netherlands
[37] Zhang D, Sun Y, Chen L, Zhang S and Pan N 2014 *Materials & Design* **54** 315-22
[38] Sabet A R, Beheshty M H and Rahimi H 2009 *Polymer Composites* **30** 1497-509
[39] Razali N, Sapuan S, Jawaid M, Ishak M R and Lazim Y 2016 *BioResources* **11** 9325-39
[40] Yang H-S, Kim H-J, Son J, Park H-J, Lee B-J and Hwang T-S 2004 Composite Structures **63** 305-12
[41] Cantwell W, Wade G, Guilien J, Jones G-V and Compston P 2001 *ASME Applied Mechanics* **250** 1-12

IOP Conf. Series: Materials Science and Engineering **270** (2017) 012027 doi:10.1088/1757-899X/270/1/012027

Carbon nanotube: A review on its mechanical properties and application in aerospace industry

Raunika A[1], S Aravind Raj[1,*], K Jayakrishna[1] and M T H Sultan[2]

[1]School of Mechanical Engineering, VIT University, Vellore, Tamil Nadu, India
[2]Department of Aerospace Engineering, Universiti Putra Malaysia, Selangor, Malaysia

*aravindsakthivel@hotmail.com

Abstract. Carbon nanotube (CNT) is a prominent material that has good potential to be used in numerous aerospace applications. This paper presents a review about CNT on its various aspects such as fabrication methods, mechanical properties and applications in aerospace. The evolution of CNT is discussed to its recent applications. The aim of this review article is to highlight the recent advancements in CNT and its possible applications in aerospace.

1. Introduction

Carbon is the 15th most abundant element in the crust of the earth. Having a tetravalency and hence a hybridization of sp3, carbon can form a vast variety of crystalline structures. The allotropic forms of carbon include graphite, diamond, Buckminster fullerenes, carbon nanotubes, graphene and about 500 hypothetical allotropes as illustrated in Figure 1. The carbon nanotubes (CNTs) are thin and long tubes that are basically graphene sheets rolled into cylinders (graphene is a hexagonal carbon structure). The CNTs have been known for their extraordinary mechanical, thermal and electrical properties, and they are members of fullerene structural family, with length-to-diameter ratio greater than 10^6.

CNTs have found applications in a lot of fields including nanotechnology, material science, optics and electricity due to their astonishing properties, and they can also be useful for aerospace industry on multiple grounds. CNTs are categorized into two major types: multi-walled carbon nanotubes (MWNTs) and single-walled carbon nanotubes (SWNTs) [1]. The first discovery of MWNT is made by occurrence of hard deposits on the cathode using electric arc experiments to form fullerenes [2]. On the other hand, the formation of SWNT is first confirmed through two separately published studies in 1993 [3]. SWNT is formed by rolling one sheet of graphene in the form of a cylinder and forming a lattice represented by the vector (m, n). Depending on m and n values, SWNT can be further classified into three different categories: armchair ($m = n$), zigzag ($m = 0$) and chiral (other values of m and n) as depicted in Figure 2. These naming conventions are used to classify SWNT according to their different wrapping style. m and n are basically integers that are used for representing the mapping.

2. CNT Synthesis Methods

Three main methods for CNT synthesis are arc discharge method, chemical vapour deposition method and laser ablation method. First, the apparatus used in the arc discharge method is a vacuum chamber containing two graphite electrodes with distance of 1-2mm between them, which helps in attaining a steady discharge. The chamber is filled with low-pressure gas before a direct current arc discharge (i.e. electrical breakdown of a gas to generate plasma) is introduced between the electrodes. Gas inside the chamber ionizes into electron and ions, which in turn result in the hot plasma formation between the electrodes. Arc discharge generates a temperature as high as 4000-6000K that makes the carbon pre-cursor at anode sublime. The carbon vapours produced gets deposited on the cathode and cools down there. When this is observed under the Transmission Electron Microscopy (TEM), a columnar texture with thickness of about 6mm is seen to be formed, which is categorized as MWNTs [4]. Meanwhile, SWNTs are produced through arc discharge method using catalyst [Iron(Fe) or Cobalt (Co)]. SWNTs are produced in the soot and not on the cathode. For bulk production of SWNTs through arc discharge method, metals like Nickel and Yttrium are used [5].

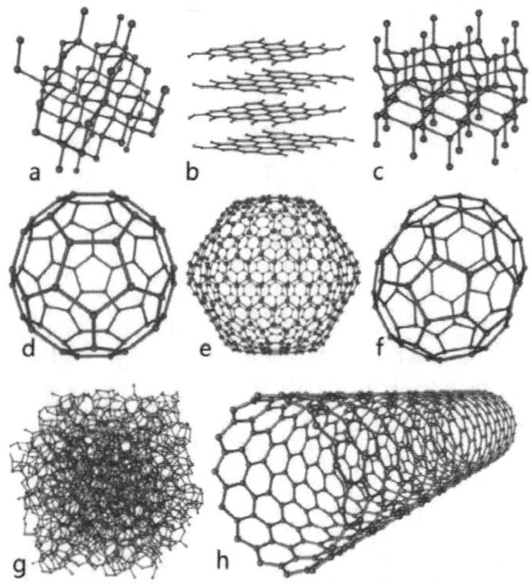

Figure 1: Allotropes of carbon: (a) diamond, (b) graphite, (c) lonsdaleite, (d) C_{60} Buckminster fullerene, (e) C_{540} fullerite, (f) C_{70}, (g) amorphous carbon, and (h) SWNT fullerenes [6]

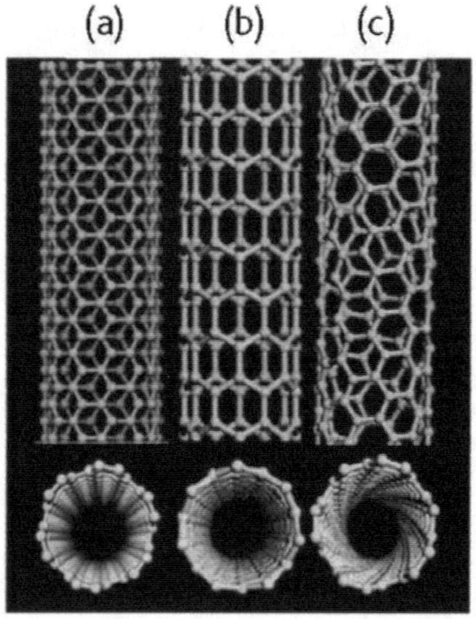

Figure 2: Categories of SWNT: (a) armchair, (b) zigzag, and (c) chiral [1]

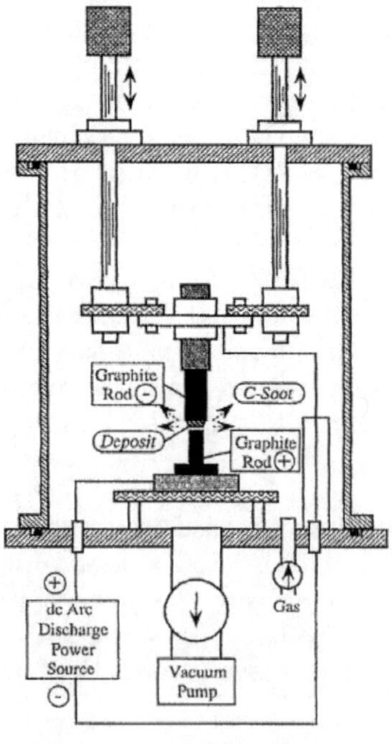

Figure 3: Working process of arc discharge device [4]

The second method, which is chemical vapour deposition (CVD) method, is currently used a lot as it is simpler and costs less in comparison to both arc discharge and laser ablation methods. In the CVD method, pre-treatment of the substrate surface can be done and this helps in controlling multi-walled and single-walled architectures [7]. The other two methods: laser ablation and arc discharge are less effective as they produce nanotubes with a higher amount of impurities. The product through CVD is likely to be more pure and therefore, the desired type of CNT can be obtained [8]. However, although CVD method gives a high yield, the products are more structurally defective. The schematic diagram of the CVD method is shown in Figure 4.

One type of CVD is thermal catalytic. In this method, hydrocarbons or other carbon feed stocks are decomposed at elevated temperature (i.e. ranging between 500-1200°C) by diluting them in continuous flow of noble gas over transition metal catalysts (i.e. iron, cobalt, nickel, etc.) in a furnace [9]. Metal catalysts applied in this method are obtained from sources like organometallocenes (i.e. cobaltocene, nickelocene, ferrocene) [10]. The decomposed carbon will dissolve the metal particles until a point of super saturation is achieved where carbon precipitate in the form of fullerene dorms gets formed. This CVD method is useful in producing CNT in variety of forms. Low temperatures (i.e. ranging between 600-900°C) will yield MWNTs whereas high temperatures (i.e. ranging between 900-1200°C) help in producing SWNTs [11]. The nanotube's diameter and the catalyst particles are correlated. Hence, the diameter of CNT obtained can be controlled by using catalyst of controlled sized [12]. Furthermore, another type of this method is plasma enhanced CVD (PECVD). In PECVD, the thermal energy of the system is substituted by energy sources that can provide energy for hydrocarbon decomposition and support the production of CNT at a comparatively lower temperature. In PECVD, hydrocarbon gas is in ionized state and it is used over transition metal. The sources, which are used as the plasma energy source, are hot filament PECVD, direct current PECVD, radio frequency PECVD and also microwave PECVD [8].

Last but not least, laser ablation or laser vaporization method for CNT synthesis uses a laser beam (Yttrium Aluminum Garnet or CO_2 laser), which is directed on a graphite rod placed inside the reactor. Graphite rod gets vaporized due to the elevated temperature of argon buffer gas that is further cooled and collected to copper collector, as illustrated in Figure 5. This method is advantageous as it helps in production of high quality SWNT (i.e. with minimal defect). This method also offers diameter control, which can be done by changing the temperature, flow rate and catalytic metal [8].

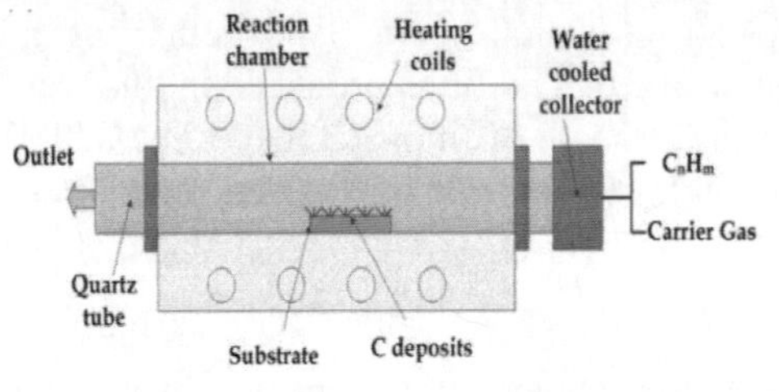

Figure 4: Schematic diagram for CVD [13]

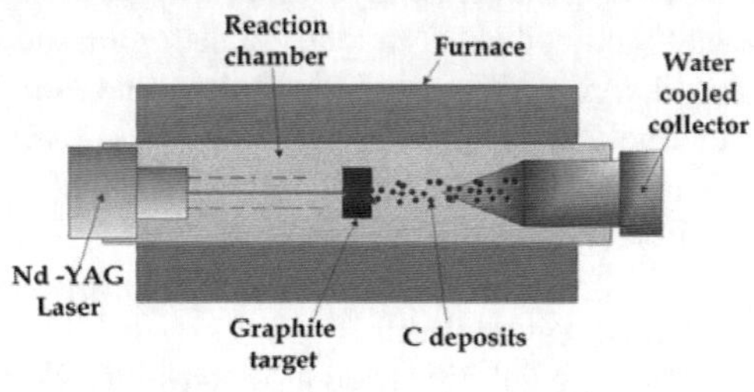

Figure 5: Schematic diagram for laser ablation [13]

3. CNT Mechanical Properties and Composites

Tensile strength of a material is the measurement of the force required to pull something such as rope, wire or structural beam to the point where it breaks. Tensile strength of a multi-walled CNT is found to be 63GPa [14]. CNT shells have been found to have a strength of 100GPa. Theoretically, Young's modulus of CNTs should be similar to Graphene sheet if their diameter is not so small to misshape the bonds. Young's modulus of CNT is found to have an order of 1TPa. This strength of CNT is due to the covalent sp^2 bonds between the carbon atoms [15]. These results are being observed mainly through Atomic Force Microscopy (AFM) and High-Resolution Transmission Electron Microscopy (HRTEM). The tensile strength and elastic modulus of CNTs make them the strongest and most inelastic materials ever discovered [16]. Young's modulus of MWNTs is found using Transmission Electron Microscopy

(TEM). TEM can measure the mean square vibration amplitudes of arc grown MWNTs over a wide temperature range (i.e from the room temperature to 800°C). The average value obtained for 11 tubes is 1.8TPa, ranging between 0.40TPa to 4.15TPa [17]. For SWNTs, the average for 27 tubes yielded a value of 1.3TPa. However, this result is found to have two systematic errors: one due to temperature measurement and other due to the nanotube length [18].

The real strength of materials is often lower than the theoretical value since the actual strength of a material cannot solely be determined with the help of its modulus. Strength of a material depends on several factors like its history, atmosphere, pressure and temperature. This situation applies to CNTs as well. Temperature is a key factor for checking the strength of a material because dislocation takes place due to temperature. CNTs, like other covalent materials, are very brittle at low temperature. At room temperature, flexibility of CNT is due to its strength and unique capability of hexagonal network to distort for relaxing stress [18].

In the meantime, due to low weight, high aspect ratio and high Young's modulus, CNTs are also ideal for making composites. A common method of producing nanotube composites is by combining the nanotubes dispersion with polymer solution and the solvent is then evaporated afterwards. Various methods are being used to produce CNT composites for mechanical purposes [19]. Nevertheless, there are some difficulties while fusing CNT in the polymer matrix since the CNT does not easily dispersed in it [20]. CNT composites are mostly prepared to get composites that have stiffness and high amount of strength for mechanical use. CNTs effectively enhance the capability of a polymer but some of the initial results were not as successful. For instance, polyvinyl alcohol (PVA)/nanotube composites are made by blending aqueous nanotube dispersion along with aqueous solution of polymer. The mixture obtained is then casted as films and the water is evaporated. The composites are obtained at the room temperature and they have stiffness way lower than that of isolated nanotubes. At room temperature, the modulus of the composites is 150MPa. Composites from MWNTs have been produced by the arc method and then submerged in a copolymer MPC-DEA. The nanotubes and polymers are mixed in the acidified solution, then applied to TEM matrix and dried. To perform the deformation studies, electron beams from TEM at a low accelerating voltage (75Kv) are focused on the film to produce the thermal stress. In turn, this initiates the forming of cracks inside the composites. The cracks grew wider but the tubes did not break. This can be due to better nanotubes that are produced by arc discharge and not by catalytic method [2]. Another effective way of enhancing the mechanical properties of composites is by deposing SWNTs layer by layer and polymer on the substrate, and then they are cross-linked. The layers are stuck together due to Vander Waals forces and electrostatic attraction among SWNTs that possesses negative charge with the polyelectrolyte that is positively charged. The films produced in this method have relatively better mechanical properties. Average tensile strength of films is found to be 220MPa, which is greater than most strong plastics. Thus, this layer by layer approach of making composites can be very promising in future.

4. Aerospace Applications

Being a high strength lightweight material, CNTs are an obvious choice for aerospace industry as they help in reducing the weight. Aerospace organizations and companies such as National Aeronautics and Space Administration (NASA) Ames Research Center, NASA Johnson Space Center and also NoPo Nanotechnologies Private Ltd. have been working on finding the potential roles of these materials. For example, Carbon Nanotube Reinforced Polymer (CNRP) can help serve as radiation protection, heat dissipation on coatings and static discharge. Moreover, CNTs can also be used in electric wirings to help in increasing the conductivity of current.

Space elevator is a revolutionary idea to carry payloads to space, which is conceptually illustrated in Figure 6. The elevators are shown to be using cables to carry the payloads to space. This means that high tensile strength cables are required. If a cable of low carbon density of 1300kg/m^3 is assumed with a fixed cross section, then the maximum stress to carry payloads from the earth's surface to the Geosynchronous Earth Orbit is found to be equal to 63GPa [21]. Such high strength can be achieved by CNTs. Nonetheless, even a few vacancies in the SWNTs can lead to a sudden change as suggested by

Quantized Fracture Mechanics criteria (i.e. vacancy defects in CNTs are produced after synthesis and this disorder in the structure makes the material weak) [21]. In such a huge cable structure, the vacancies are inevitable due to statistical reasons [22]. Statistical and different deterministic models to estimate the strength of mega CNT cable have suggested a strength reduction by a factor of at least 70% as compared with theoretical strength of CNT and an increase in mass by a factor of larger than 300% [21]. These results are confirmed by the first observations of long meter nanotube cable strength [23]. This reduction poses a doubt in effective realization of use of such long cables in construction of a space elevator.

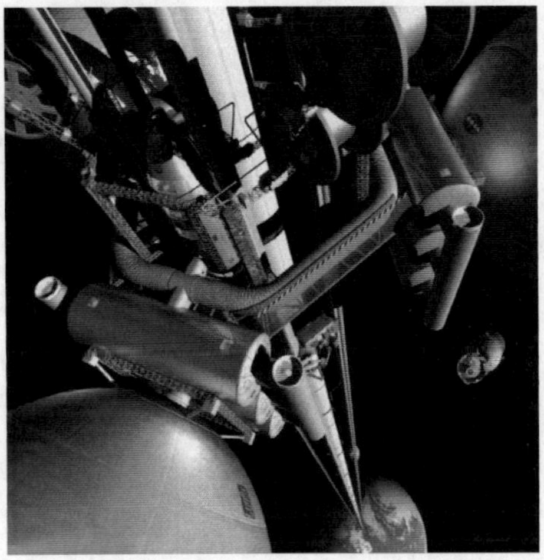

Figure 6: NASA's description of artwork by Pat Rawlings [24]

In addition, CNTs can also be used in large satellites and spacecrafts to reduce their weight. Using CNT in satellite can reduce the weight to one third of its current weight [25] The use of silane grafted MWNTs can result in better mechanical properties and thermal stability compared to neat resin with normal MWNTs [26]. Thermal conductivity of MWNTs is of higher value than diamond [27] and thus they offer an immense potential for thermal protection systems. To reduce the payload, aluminium is replaced with CNRP on the Reusable Launch Vehicle, which has lead to reduction in weight by 82%. Furthermore, CNTs can be used as propellant additives for chemical propulsion systems. An electric motor is applied in electric aircraft and in such motors, CNT can replace the iron and copper materials to fulfil the requirements of next generation motors by increasing efficiency and reducing weight. In space exploration, usage of electric propulsion is catching eyes over the usage of chemical propulsion. Electric propulsion has been efficiently utilized in different space missions [28]. In the nano-particle Field Extraction Thruster (nanoFET), thrust is produced by emitting charged CNT particles.

Reduction in weight can also be achieved by replacing the current copper wiring in aircraft by CNT based wires [29], where use of CNT cable known as 1553B can result in 69% weight reduction [30]. The weight of an aircraft is increased many folds due to wiring. For instance, a commercial plane like Boeing 747 has about 217261m of copper wiring that contributes 181.437kg to its weight. Moreover, CNT can be used in International Space Station and other future space settlements for the purification of water since nano materials possess a number of properties that enable them to purify water. The nanotechnology can serve as recyclable ligands for toxic metal ions, organic/inorganic solutes and also radio nuclides in aqueous medium [31].

5. Conclusion

CNT research field has come far from where it started in 1991. New methods of CNT synthesis have been achieved with CVD being the most promising one due to its economic advantages and simple methods, even though CNT obtained through this method is structurally ineffective. In its initial stages due to its high tensile strength and elastic modulus, CNT seems to be a promising material due to its amazing mechanical properties. However, even after years of efforts, there is still work that need to be done further in the field of CNT composites to increase their practical usages. The application of CNT in reducing the payload of aircraft and space shuttles appears to be really promising as of now. Many researchers have shown that the overall weight of aircraft and space shuttles can be effectively reduced by using CNT based composites or CNT for electric wiring.

References

[1] Baughman R H, Zakhidov A A and de Heer W A 2002 *Science* **297** 787-92
[2] Harris P J F 1999 *Carbon Nanotubes and Related Structures* Cambridge University Press
[3] Ajayan P M and Lijima S 1993 *Nature* **361** 333-4
[4] Ando Y and Zhao X 2006 *New Diamond and Frontier Carbon Technology* **16** 123-37
[5] Saito Y, Inagaki M, Shinohara H, Nagashima H, Ohkohchi M and Ando Y 1992 *Chemical Physics Letters* **200** 643-8
[6] Karthik P S, Himaja A L and Singh S P 2014 *Carbon Letters* **15** 219-37
[7] Ebbesen T W and Ajayan P M 1992 *Nature* **358** 220-2
[8] Koziol K, Boskovic B O and Yahya N 2010 *Synthesis of Carbon Nanostructures by CVD Method* in *Carbon and Oxide Nanostructures* Springer
[9] Popov V N 2004 *Materials Science and Engineering: R: Reports* **43** 61-102
[10] Sen R, Satishkumar B C, Govindaraj A, Harikumar K R, Rehnganathan M K and Rao C 1997 *J. Mater. Chem.* **7** 2335-7
[11] Dai H, Hafner J H, Rinzler A G, Colbert D T and Smalley R E 1996 *Nature* **384** 147-50
[12] Ago H, Shaffer M S P, Ginger D S, Windle A H and Friend R H 2000 *Physical Review B* **61** 2286-90
[13] Gore J P and Sane A 2011 *Flame Synthesis of Carbon Nanotubes* in *Carbon Nanotubes - Synthesis, Characterization, Applications* InTechOpen
[14] Yu M, Lourie O, Dyer M J, Moloni K, Kelly T F and Ruoff R S 2000 *Science* **287** 637-40
[15] Shailaja K, Sameena T, Sethy S P, Patil P and Ashraf M O *http://www.pharmatutor.org*
[16] Harris B and Bunsell A R 1977 *Structure and Properties of Engineering Materials* Prentice Hall Press
[17] Krishnan A, Dujardin E, Ebbesen T W, Yianilos P N and Treacy M M J 1998 *Physical Review B* **58** 1401-3
[18] Salvetat J P, Bonard J, Thomson N H, Kulik A J, Forro L, Benoit W and Zuppiroli L 1999 *Applied Physics A* **69** 255-60
[19] Thostenson E T, Li C and Chou T 2005 *Composites Science and Technology* **65** 191-516
[20] Li Y L, Shen M Y, Su H S, Chiang C L and Yip M C 2012 *Journal of Nanomaterials* **2012** 262694
[21] Pugno N 2006 *Int J Fract* **141** 313-23
[22] Carpinteri A and Pugno N 2005 *Nature Materials* **4** 421-3
[23] Zhang M, Fang S, Zakhido A, Lee S B, Aliev A E, Williams C D, Atkinson K R and Baughman R H 2005 *Science* **309** 1215-9
[24] *https://upload.wikimedia.org/wikipedia/commons/f/f0/Nasa_space_elev.jpg*
[25] Meador M A, Files B, Li J, Manohara H, Powell D and Siochi E J 2012 *Nanotechnology Roadmap: Technology Area 10* NASA
[26] Jin S B, Son G S, Kim Y H and Kim C G 2013 *Composites Science and Technology* **87** 224-31
[27] Kim P, Shi L, Majumdar A and McEuen P L 2001 *Physical Review Letters* **87** 215502
[28] Rayman M D, Varghese P, Lehman D H and Livesay L L 2000 *Acta Astronautica* **47** 475-87
[29] Roco M C, Mirkin C A and Hersam M C 2011 *Journal of Nanoparticle Research* **13** 897-919
[30] McKenna E *http://www.aviationtoday.com*
[31] Savage N and Diallo M S 2005 *Journal of Nanoparticle Research* **7** 331-42

Microstructural characterization of fly ash particulate reinforced AA6063 aluminium alloy for aerospace applications

A M Razzaq[1,5], D L Abang Abdul Majid[2,*], M R Ishak[2] and M B Uday[3,4]

[1]Department of Mechanical and Manufacturing Engineering, Faculty of Engineering, Universiti Putra Malaysia, 43400 UPM Serdang, Selangor, Malaysia.
[2]Department of Aerospace Engineering, Faculty of Engineering, Universiti Putra Malaysia, 43400 UPM Serdang, Selangor, Malaysia.
[3]Faculty of Mechanical Engineering, Universiti Teknologi Malaysia, Skudai, Johor Bahru, Malaysia.
[4]UTM Centre for Low Carbon Transport in Cooperation with Imperial College London, Institute for Vehicle Systems and Engineering, Universiti Teknologi Malaysia, Skudai, Johor Bahru, Malaysia.
[5]Oil Products Distribution Company, Ministry of Oil, Iraq.

*dlaila@upm.edu.my

Abstract. Aluminium-fly ash (FA) particulate reinforced composites (AA6063-FA) have been used in automotive and aerospace industries because of their low density and good mechanical properties. Three different weight fraction of FA: 2%, 4% and 6% are added to AA6063 alloy using compocasting method. The effect of FA particulates on microstructure, density and compression strength of AA6063- FA composites are investigated. Field Emission Scanning Electron Microscope (FESEM) micrographs reveal that the FA particulates are uniformly distributed in AA6063 alloy. The results also show that density, compression strength and microstructure of the AA6063-FA composites are significantly influenced by the FA amount. The increase in the weight fraction of FA will improve the microstructure and enhance the compression strength. The density of AA6063-FA composites decreases as the incorporation of FA increases.

1. Introduction

Metal matrix composites (MMCs) are the macroscopic combination of two or more different materials (i.e. one of them is a metal and the other a non-metal) in which tailored properties are determined [1]. In recent years, MMCs have received considerable attention due to their low density and high strength and stiffness as compared to those of conventional materials [2]. The reinforcement is commonly non-metals, in the likes of conventional ceramic materials such as SiC, Al_2O_3, fly ash, etc. The composites can be used to fabricate many important components for automotive and aerospace. The design of the process is an important part of synthesis for MMCs. The processes employed usually depend on the type of reinforcement and matrix used for the composite. Stir casting is the simplest and cheapest one, but a problem associated with this process is the non-uniform distribution of the particulate due to poor wettability and also gravity regulated segregation [3]. Achieving the uniform distribution of the reinforcement within the matrix is still a challenge in the stir casting process, which directly affects the properties and quality of the produced composites. The interfacial strength between the matrix and the reinforcement plays a significant role in determining the properties of MMCs [4]. In order to achieve the desirable properties of the metal matrix composite, several factors that have to be considered that include achieving a uniform distribution of the reinforcement material in molten matrix and improving wettability or bonding between the matrix and reinforcement [5, 6]. Improvement of the wettability can be achieved by increasing surface energy, decreasing surface tension of the molten alloy and also decreasing liquid – solid interfacial energy at the interface of the reinforcement matrix [7-9].

The need for high performance and lightweight materials for a lot of applications has led to wide research efforts in the development of aluminium matrix composites. Researches on aluminium fly ash composite have shown a great promise in various applications. Fly ash is one of the materials that has low density and cheap, and it is abundantly available as a solid waste by-product during combustion of coal in thermal power plants. The chemical composition of fly ash consists of Al_2O_3, SiO_2 and Fe_2O_3 as major constituents and the oxides of Mg, K, Ca, K and Na as minor constituents [10-12]. It has been observed that there is slight decrease in density and strength of aluminium composites with increasing weight percent of fly ash while the hardness is increased slightly up to 10 wt % fly ash and decreased afterwards [13]. An increase in the volume percentages of fly ash is shown to increase the hardness, impact strength and compression strength values in aluminium – fly ash composites, but the density is decreased with content increase of fly ash [2]. Beyond 20 wt.% of fly ash, both compressive strength and hardness of the composites begin to decrease due to increase of particle clustering and decrease in solid solution strengthening [10].

AA6061-fly ash aluminium matrix composites (AMCs) have been successfully fabricated using the compocasting method [14]. It is observed there is no formation of any inter-metallic compound due to interfacial reaction. The fly ash particles are thermodynamically stable at compocasting temperature and they are dispersed homogeneously in the AMCs. The interface between the aluminium matrix and fly ash particle is clear and the bonding is proper. The incorporation of fly ash particles into semi-solid aluminium alloy improved the wettability [14]. Industrial wastes such as fly ash and granulated blast furnace (GBF) slag are utilized successfully for the production of aluminium based MMCs. Al-fly ash (ALFA) and Al-GBF slag composites have been successfully produced by stir casting method. There is a uniform distribution of reinforcement particles in the matrix phase and a good bonding between the matrix and the reinforcements is obtained. The hardness of the composites increases whereas their density decreases with the presence of reinforcement as compared to the base alloy. Higher hardness value has been reported for ALFA composite than Al-GBF slag composite, but enhanced mechanical properties are observed for both composites under compression as compared to alloy [15]. It is proven that the fly ash released the Si in the melt in reduction reactions. A sufficient quantity of Si forms an Mg_2Si phase with Mg of the melt for precipitation strengthening. Microstructural observations reveal that, after the wall of fly ash particles cracks, the molten metal will infiltrate the pores and cracks in the non-solid fly ash particles. Increasing the reaction time will reduce the porosity in the aluminium matrix composites whereas the density and hardness are increased [16]. Addition of fly ash particles enhances the hardness and tensile strength of the fabricated AMCs. The fabricated AMCs with 6wt% of fly ash reinforcement has been shown to exhibit almost 50% higher hardness and 30.4% higher tensile strength as compared to un-reinforced LM6 alloy [17].

This work is motivated by the limited information available on AA603-FA composites. The aim of this study is to investigate the effect of fly ash addition on the microstructure, density and compression of AA6063 alloy, and to develop a benchmark of property data that can be used in future to evaluate the materials. Furthermore, it is of great importance to carry out a research based on the AA6063 alloy to improve understanding on aluminium fly ash composites.

2. Materials and Experiments

In this study, AA6063 alloy is used as the matrix alloy and fly ash as the reinforcement material to produce AA6063-FA composites. The fly ash particulates are collected from a thermal power station in Johor, Malaysia and they are sieved to 44μm to get smaller particle size. The chemical composition of AA6063 alloy and fly ash are presented in Table 1 and Table 2, respectively.

Table 1: Chemical composition of tested AA6063 in wt. %

Elements	Cu	Fe	Mg	Mn	Si	Ti	Zn	Cr	Al
wt.%	< 0.01	0.19	0.71	0.04	0.86	0.008	<0.01	<0.01	Balance

Table 2: Chemical composition of tested fly ash in wt. %

Elements	SiO_2	Al_2O_3	Fe_2O_3	TiO2	K_2O	CaO	MgO	Na_2O	CuO
wt.%	59.98	19.09	2.78	1.14	1.09	0.63	0.38	0.34	0.01

The fly ash particulates are initially preheated at temperature of 900°C for two hours to remove the impurities and water contents, and improve the wettability of the reinforcement particles with AA6063 alloy. Weight fraction of the fly ash particulates to synthesise the composites is set between 0-6 wt.%, in steps of 2%. The AA6063 rod is cut and charged into crucible inside the induction furnace, heated to temperature of 750°C ± 50°C above the liquidus temperature of the alloy and then the temperature in the furnace is reduced to 600°C to get semi-solid state of the alloy. At this temperature, the preheated fly ash is added to the molten AA6063 alloy and the mix is manually stirred for about 15 minutes. The composite slurry is then superheated to 750°C and stirred using the mechanical stirrer at a speed of 300 rpm to improve the fly ash particles distribution in the molten AA6063 alloy. The slushy composites are cast into preheated steel moulds. AA6063 alloy is prepared by casting for experimentation control. The cast test samples are then heat-treated for two hours at 560°C, where they are quenched in water at room temperature and then aged for three hours at 160°C.

The specimens prepared are polished and etched as per the standard metallographic procedure. The microstructures of these etched specimens are observed using the Field Emission Scanning Electron Microscope FESEM. The etchant that is applied to reveal the microstructure in this study is 10g NaOH and 90mL HNO_3 in 190ml distilled water (caustic etch). These specimens are immersed at the room temperature. Density of the composite specimens is obtained experimentally by Archimedes principle. Archimedes' principle states that, when a body is immersed in a fluid, there is a buoyant force acting upward on the body equals to the weight of the displaced fluid. The weight of displaced fluid equals its volume when water is used (density of water = 1 g/cm^3), hence the volume of the water displaced is equal to the volume of the body immersed. The density of the samples (with dimensions of 12mm x 12mm x 100mm) is evaluated by weighing them using electronic densimeter MD-300S with resolution of 0.001g/cm^3 and 300g capacity.

Compression tests are carried out on the rectangular specimens of pure aluminium and Al-fly ash composites with dimensions 12mm x 12mm x 24mm and a height/width (aspect ratio) of 2 with the parallel ends. These specimens of standard dimensions are prepared using wire cutting and their edges are chamfered to minimize folding. The compressive strength is measured in accordance with standard ASTM E 9. The samples are compressed using the Instron Machine (Model 3382) at crosshead speed of 0.8mm/min. Plotting of load versus displacement is done continuously through the data acquisition system. The compression tests are carried out until 50% reduction in height. Each test is repeated three times and the average value is taken to calculate the compression strength according to Equation 1 and Equation 2, where σ_c is the compression strength in N/mm^2, F is a force in N, A is cross sectional area of the specimen in mm^2, ε is the strain of specimen, Δl is length change of the specimen and lo is the original length of specimen.

$$\sigma_c = \frac{F}{A} \quad (1)$$

$$\varepsilon = \Delta l / l_0 \quad (2)$$

3. Results and Discussion

The particle size analysis is measured using a Malvern particle Mastersizer. The average particle size of the fly ash powder is 21.96μm and its density is found to be 2.33g/cm^3. Figure 1 depicts the FESEM micrograph that shows the shape of fly ash as a solid sphere structure. On the other hand, Figure 2 and Figure 3 are FESEM micrographs of AA6063 alloy composites reinforced with fly ash, which show a uniform distribution of FA reinforcement in the AA6063 alloy matrix. There is no evidence of crack, voids and defects in the casting. There is a very low agglomeration of fly ash in the low percentage content. Figure 3 reveals a clear interface between AA6063 alloy matrix and FA. The compocasting enhances the wettability between FA and AA6063 alloy matrix that results in a good bonding between FA and AA6063 alloy matrix. However, the problems of fly ash particles agglomeration and defects are increasing with increment of the fly ash content. The FA particulates are distributed around grain boundary as shown in Figure 2.

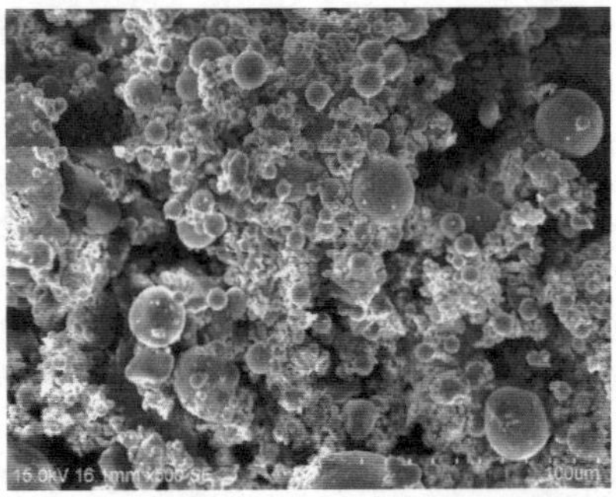

Figure 1: FESEM micrograph of FA

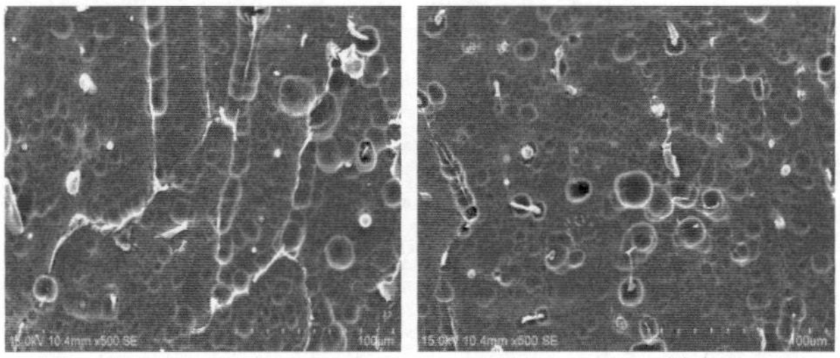

(a) (b)

Figure 2: Microstructure of AA6063-FA composite with different wt% fly ash (a) 4 wt. %, (b) 6 wt. %

Figure 3: FESEM micrograph for a good interfacial bonding microstructure

The bulk density and apparent porosity measurements of the cast AA6063-FA composite samples with different weight percentage of fly ash are determined by Archimedes' method with water as the immersion medium. The variation of the density of composites with different fly ash content samples is shown in Figure 4. It is observed that the bulk density values decrease with increasing reinforced fly ash particles. Since the density of fly ash (2.33g/cm^3) is less than that of AA6063 alloy (2.6871g/cm3), the overall density of the AA6063-FA composites is reduced by increasing amount of porosity in the

IOP Conf. Series: Materials Science and Engineering **270** (2017) 012028 doi:10.1088/1757-899X/270/1/012028

casting conditions. AA6063-FA composites containing 0 wt. %, 2 wt. %, 4 wt. % and 6 wt. % of FA particulates exhibit densities 2.6871g/cm^3, 2.6588 g/cm^3, 2.6468 g/cm^3 and 2.6396 g/cm^3, respectively.

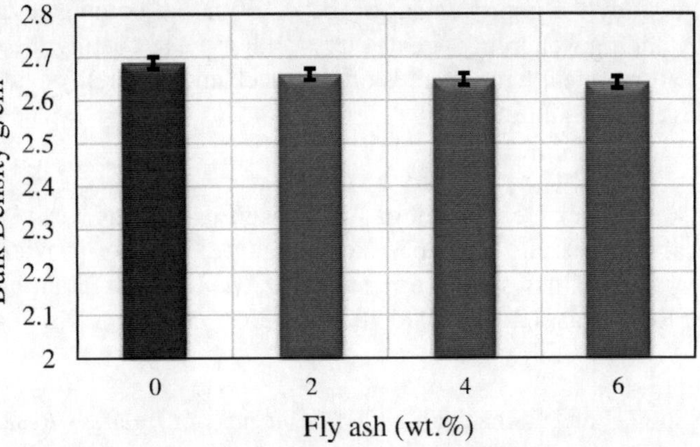

Figure 4: Density variations for different percentage compositions of AA6063–FA composites

Figure 5 reveals the typical stress strain diagrams for AA6063 matrix and AA6063–FA composites under compression test. The average ultimate compression strength (UCS) of AA6063–FA composites with various wt. % of FA is presented in Figure 6. It can be observed that the compressive strength is significantly increased with an increase in the weight percentage of fly ash particles. The maximum ultimate compression strength is found at 6 wt. % with 510MPa, which the improvement percentage is 100.1%. This is due to the hardening of the base alloy by fly ash particles [2, 18]. As the deformation increases, the hardness is also increasing. For the pure AA6063-fly ash composites, the rate of increase is greater for higher deformation with larger fly ash content. This might be due to the existence of the strain hardening effects from matrix alloy and mixture rule of composites strengthening [19].

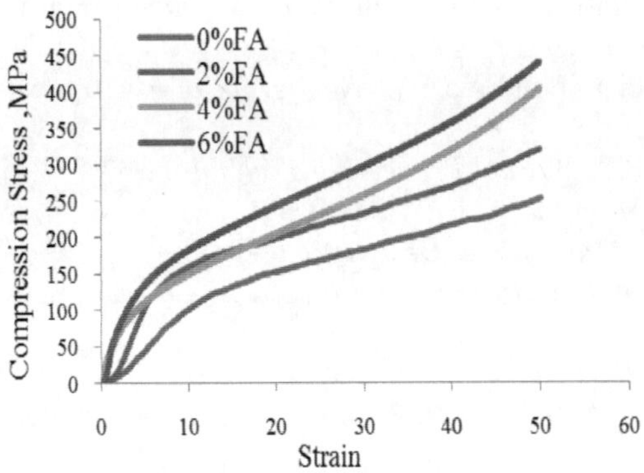

Figure 5: Stress vs. strain in compression for AA6063 alloy and AA6063–FA composites

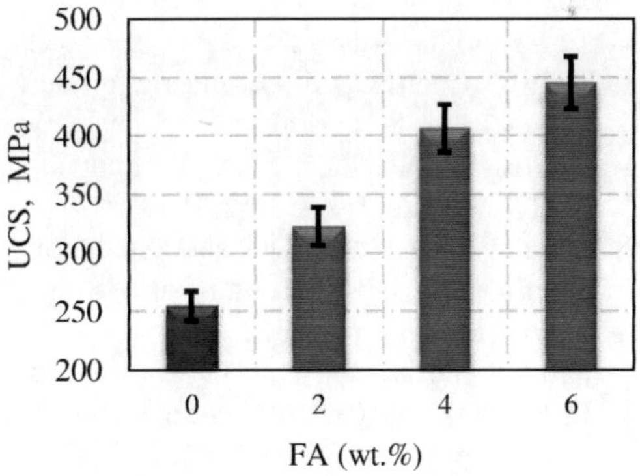

Figure 6: Effect of weight percentage of fly ash on ultimate tensile strength

4. Conclusion

The effects of weight fraction of FA on AA6063 aluminium alloy with regards to the properties of the AA6063-FA composites are investigated. Fabrication of the AA6063-FA particulate composites have been successfully done by compocasting technique and a good wettability is obtained. Based on the FESEM micrographs, the FA particulates are homogenously dispersed in the AA6063 matrix alloy. In addition, density of the AA6063-FA composites specimen decreases with an increase in FA particulate weight fraction. The increase in the weight fraction of FA particulates also results in an increase in the strength compression of the AA6063- FA composites, where the fabricated composite with 6 wt. % of

FA reinforced exhibits almost 100.1% higher compression strength as compared to the un-reinforced AA6063 aluminium alloy. Moreover, preheating the reinforcement particles and using compocasting technique without agents' addition are the engineering methods used to increase the wettability in this work. It is possible to study the improvement of the AA6063-FA composites' properties and increase the fly ash content by adding wettability agent, fluxes addition and coating the reinforcement particles. Effect of different heat treatment temperatures on the mechanical and physical properties of AA6063-fly ash can also be investigated in future.

Acknowledgement
The authors would like to thank Department of Aerospace Engineering, Universiti Putra Malaysia and Faculty of Mechanical Engineering, Universiti Teknologi Malaysia for providing research facilities in this study. This research work has been supported by UPM Research Grant (Vot. Number: 9463100) and UTM Research FRGS Grant (R.J1300000.7824.4F723).

References

[1] Prasad D S, Shoba C and Ramanaiah N 2014 *Journal of Materials Research and Technology* **3** 79-85
[2] Mahendra K. and Radhakrishna K 2007 *Materials Science-Poland* **25** 57-68
[3] Sarkar S, Sen S, Mishra S C, Kudelwar M K and Mohan S 2010 *Journal of Reinforced Plastics and Composites* **29** 144-8
[4] Akbari M K, Baharvandi H and Mirzaee O 2013 *Composites Part B: Engineering* **52** 262-8
[5] Hashim J, Looney L and Hashmi M 1999 *Journal of Materials Processing Technology* **92** 1-7
[6] Tjong S C and Ma Z 2000 *Materials Science and Engineering: R: Reports* **29** 49-113
[7] Singh V K, Chauhan S, Gope P C and Chaudhary A K 2015 *High Temperature Materials and Processes* **34** 163-70
[8] Laha T, Kuchibhatla S, Seal S, Li W and Agarwal A 2007 *Acta Materialia* **55** 1059-66
[9] Razzaq A M, Majid D L, Ishak M R and Uday M B 2017 *IOP Conference Series: Materials Science and Engineering* **203** 012002
[10] Shanmughasundaram P, Subramanian R and Prabhu G 2011 *European Journal of Scientific Research* **63** 204-18
[11] Tang Z, Chen X, Liu D, Zhuang Y, Ye M, Sheng H and Xu S 2016 *Journal of Environmental Sciences* **48** 169-78
[12] Barroso J, Ballester J, Ferrer L M and Jimenez S 2006 *Fuel Processing Technology* **87** 737-52
[13] Guo R, Rohatgi P and Nath D 1997 *Journal of Materials Science* **32** 3971-4
[14] Selvam J D R, Smart D R and Dinaharan I 2013 *Materials & Design* **49** 28-34
[15] Murthy I N, Babu N A and Rao J B 2014 *Journal of Minerals and Materials Characterization and Engineering* **2014**
[16] Fan L-J and Juang S H 2016 *Materials & Design* **89** 941-9
[17] Efzan M E, Syazwani N S and Abdullah M M A B 2016 *IOP Conference Series: Materials Science and Engineering* **133** 012046
[18] Anilkumar H, Hebbar H and Ravishankar K 2011 *International Journal of Mechanical and Materials Engineering* **6** 41-5
[19] Rao J B, Rao D V and Bhargava N 2010 *International Journal of Engineering, Science and Technology* **2** 50-9

IOP Conf. Series: Materials Science and Engineering **270** (2017) 012029 doi:10.1088/1757-899X/270/1/012029

Aeroelastic tailoring via ribs orientation of NASA Common Research Model

Y N Chan, M Y Harmin* and A S M Rafie

Department of Aerospace Engineering, Faculty of Engineering, Universiti Putra Malaysia, Malaysia

*myazdi@upm.edu.my

Abstract. A baseline model in reference to the NASA Common Research Model is selected to illustrate the concept of aeroelastic tailoring of wing ribs structure by varying the orientation of the ribs. This enables the torsional-bending modes characteristic to be altered, which results in a possibility of improvement in the aeroelastic performance without having to compromise its overall weight. Two strategies are implanted in this work: the first strategy considers the whole ribs to be oriented in parallel while the second one is by dividing the wing into six parts where the ribs are allowed to be oriented in parallel within their part only. In many cases, there are a significant improvement with up to 5.5% of flutter increment via orientating the ribs section, hence leading to a possibility in significant structural weight reduction.

1. Introduction

A massive growth of air transportation is seen in the past decades since it is the only sensible way to travel across long distance. The demand for air transport has increased significantly and able to act as a catalyst in economic and social growth. However, this rapid growth also consequently increases the impact on the environment through the aircraft carbon emissions. Various emission reduction options have been proposed in which the key agenda is to improve the aircraft overall efficiency while at the same time maintaining its overall performance.

Aeroelastic instabilities, namely divergence and flutter, are two most common design parameters that dictate the flight envelope of the aircraft [1]. Numerous studies have been done in delaying these aeroelastic instabilities through advancement in aircraft materials, aerodynamics, structures and also control technologies. Over the past decades, the composite technology has increasingly been employed in aerospace industry due to its attractive property of higher strength-to-weight ratio in comparison to the conventional metal alloys. This solution has been widely used in the current design of aircraft. For instance, AIRBUS aircraft has employed about 25% and 50% of composite materials on its A380 and A350XWB variants, respectively [2,3]. On the other hand, the advancement in aerodynamics is mostly directed in reducing the total drag with relative improvement in lift-to-drag ratio. One way to idealize this is through the usage of boundary layer control to delay the flow separation. In addition, morphing wing technology has also gained much attention with a focus to investigate a shape-changing ability in acquiring the best possible aerodynamic shape given the flight condition [4,5]. Apart from these, the approach of aeroelastic tailoring on aircraft design has also become a possible solution in improving the aircraft overall efficiency. This can be realized through active and/or passive control of aeroelastic wing system [6]. It is intended that the aeroelastic instabilities such as the flutter, divergence and gust loads response to be delayed and influenced in a beneficial way without any compromise in weight.

Several researches have been performed in finding the optimum fibre orientation for a composite wing box with aeroelastic instability parameters as a maximizing cost function. One of the researches studies the effect of laminate lay-up of composite wing box with respect to the flutter speed [7]. It is found that the torsional rigidity properties can be altered by varying the composite stacking sequence, which in turn changes the coupling behaviour of the flutter modes. This offers an improvement in the flutter instability without affecting the overall structural weight. Another study performed aeroelastic tailoring on a composite forward-swept wing with a parametric study involving composite thickness and orientation parameters [8]. It is found that the overall structure weight of initial wing design could be reduced and the wing aeroelastic performance in terms of flutter and divergence speeds could also be improved simultaneously. Recently, there has been an interest in understanding the effect of wing structure arrangement to its aeroelastic performance. The concept presented by Harmin et al. (2011) is based upon the modification of wing ribs arrangements of a metallic wing by rotating each rib in the same orientation [9]. Meanwhile, Othman et al. (2016) has also adopted a similar aeroelastic tailoring technique on a Goland rectangular wing box by varying each of the ribs individually [10]. Both of the studies have resulted in an increment of frequency gap between the bending-torsional modes, which led to encouraging improvement in aeroelastic performance without any weight penalty. In addition, a few aeroelastic tailoring methods including the use of curvilinear rib and spar with varying orientation, tow steering composite laminates and also material and thickness grading have been implemented on a Common Research Model wing box [11]. In this study, each method has shown significant outcomes with varying degrees of success in weight reduction and improvement of flutter performance. A recent research has included also the effect of varying wing sweep angle and ribs arrangement on an un-tapered wing in terms of its static and dynamic aeroelastic behaviour [12]. With this motivation, further investigation is carried out using a much realistic wing representation nowadays.

2. Methodology

The parameterized model of the NASA Common Research Model (CRM) wing box is considered in this study. This model describes a single-aisle subsonic transport aircraft configuration with full-scaled cantilever wing that is generated by NASA for the purpose of research work within the aerodynamic community [13]. The wing has a semi span of 26.18m, root chord of 7m, taper ratio of 0.229, sweep angle of 35° and an aspect ratio of 9. It consists of 51 ribs and 3 spars where each gap between them are covered by the skins. In the baseline setup, the ribs are aligned perpendicular to the leading edge and they are equally distributed along the span. It should be noted that the height and thickness of the spars and ribs gradually decrease from the root to the tip of the wing. The studied wing is modelled with MSC NASTRAN shell elements of CQUAD4 and CTRIA3, and its material properties have been defined as anisotropic aluminium based [14]. The baseline wing model is illustrated in Figure 1 and a fully constrained boundary conditions are defined at the root of the wing. On the other hand, Figure 2 shows the finite element (FE) representation of the wing box section that is coupled with the doublet lattice aerodynamic panel in order to establish a complete system of aeroelastic model.

Figure 1: Baseline wing FE model

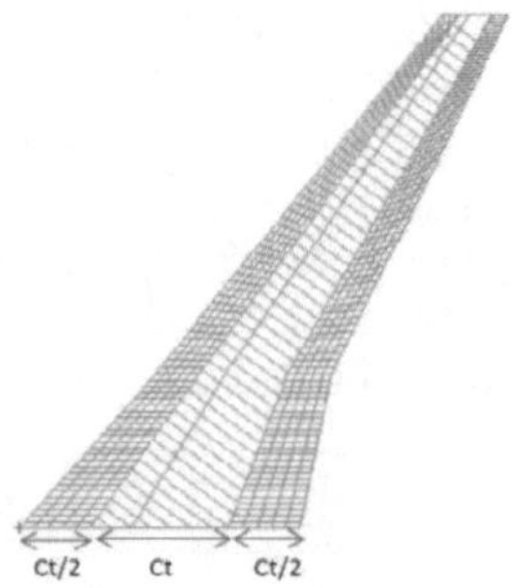

Figure 2: Aeroelastic wing model

The ribs have been numbered accordingly from root to tip in increasing order from 1 to 51. Only the ribs that are connected to all three spars are considered for varying orientations. In this case, they are the 8th rib and those onwards. The centre of rotation is set in the middle of each rib of the baseline configuration and nine different degrees of rotation are defined between -10° to +10°. The negative angle of orientation represents clockwise rotation while the positive angle of orientation refers to anti-clockwise rotation about its baseline position. Each of these ribs is allowed to vary with the constraint that they do not overlap each other. In addition, as the rib is orientated, its thickness is accordingly adjusted due to the change in its length. This enables the total weight of the wing to remain the same in all cases. Figure 3 shows the schematic of possible rib's orientation as explained in this paragraph. If each of the ribs is allowed to be orientated via the defined setup, this leads to a total of 9^{44} possible orientations, which makes it extremely challenging to cover the whole design space. For ease of the computational effort and initial understanding, the wing is divided into six parts as shown in Figure 4. Table 1 provides the list of case settings that are considered in this study. All solutions involved in this work are coded in MATLAB and the system is integrated with MSC NASTRAN solver under solution 145 of the p-k method flutter analysis [15].

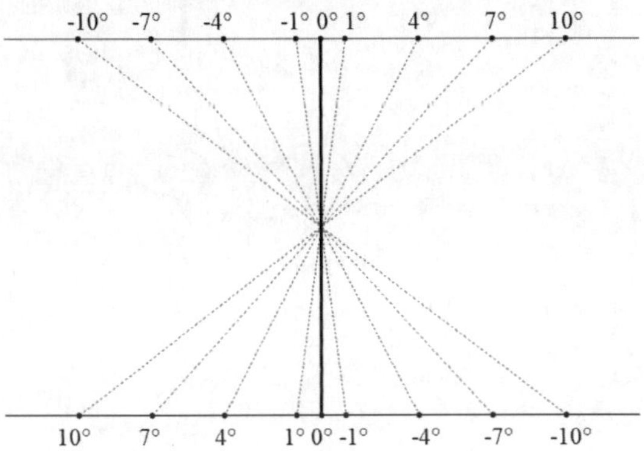

Figure 3: Possible rib's orientation, θ

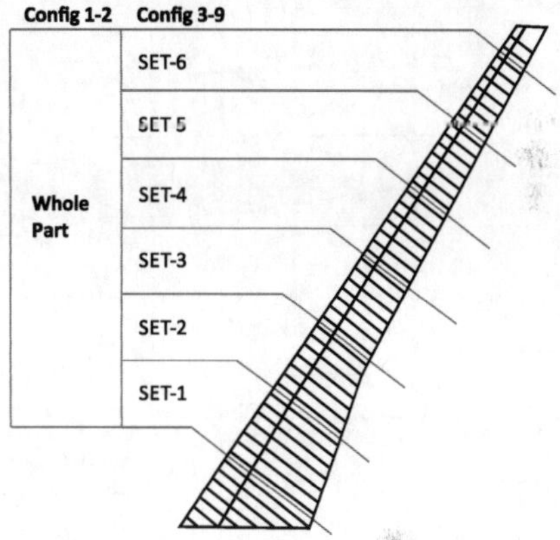

Figure 4: Division of the wing into 6 sets of ribs

Table 1: List of considered case studies

Case	Description
1	All ribs are rotated together in parallel with rotation angle θ
2	Odd numbered ribs are rotated in defined angle θ, while even numbered ribs rotated in opposite angle θ
3	Only ribs in Part 1 (8th to 14th rib) are rotated in parallel with angle θ
4	Only ribs in Part 2 (15th to 21st rib) are rotated in parallel with angle θ
5	Only ribs in Part 3(22nd to 28th rib) are rotated in parallel with angle θ
6	Only ribs in Part 4 (29th to 35th rib) are rotated in parallel with angle θ
7	Only ribs in Part 5(36th to 42nd rib) are rotated in parallel with angle θ
8	Only ribs in Part 6 (43rd to 49th rib) are rotated in parallel with angle θ
9	Ribs in odd numbered parts are rotated in defined angle θ, while ribs in even numbered parts are rotated in opposite angle θ

3. Results and Discussions

The combinations of ribs orientations that maximized the air speed at which the flutter occurred are of interest in this study. It should be noted that the divergence instability is not taken into account since it is found from the findings that divergence occurs later than flutter, thus making flutter instability as a maximum limit of the wing system. A series of flutter analysis using p-k method has been conducted for the nine considered cases to assess the bending-torsional characteristics with respect to aeroelastic performance [15]. The flutter result is represented in terms of ***V-f*** and ***V-ζ*** plots, which describe the trends of frequency and damping ratio against the air speeds. Meanwhile, flutter speed and frequency parameters are identified at the condition where the damping ratio changes in sign, which implies that the oscillation cannot be damped out at the particular state.

Figure 5 shows graphical representation of ***V-f*** and ***V-ζ*** plots for the first five modes of the baseline system that consist of three bending and two torsional modes. It can be clearly observed that there is a strong coupling between second bending and first torsional modes as the airspeed increases. Note that the flutter occurs when the two modes are about to coalesce and move close enough for them to couple unfavourably. On the other hand, Figure 6 depicts the comparison between first torsional and second bending modes for both baseline configuration and Case-1 of positive 10° rib rotation cases. From the figure, an increment in the frequency gap is observed that results in the delay of flutter speed. For the considered oriented of Case-1, the flutter occurs right after the mode switch between the first torsion and second bending modes, which demonstrates that flutter is not merely due to the coupling between these two modes also the coupling effect from the third bending mode as well.

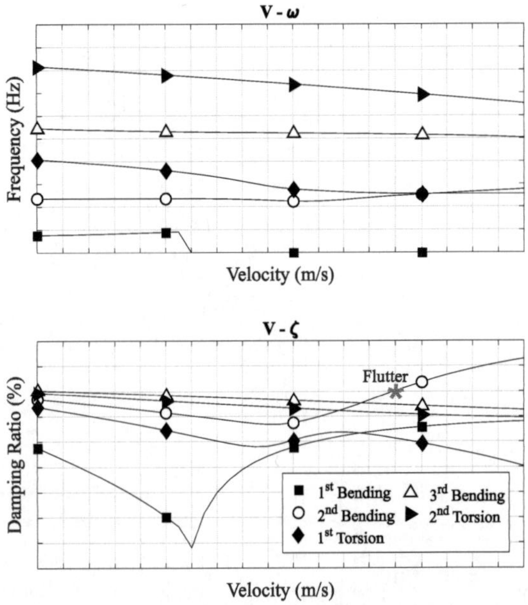

Figure 5: V-ζ and V-ω plots of baseline system

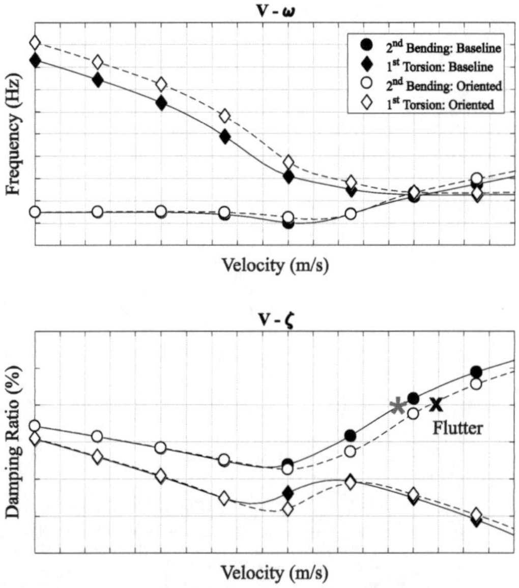

Figure 6: Comparison of V-ζ and V-ω plots between baseline system and Case-1

The influence upon bending-torsional characteristics is generally caused by changes in wing mass and stiffness distribution in both spanwise and chord wise directions, which result in different flexural axis. The torsional modes are observed to have greater impact compared to the bending modes, hence the frequency gap of flutter modes can only be increased with a possibility of an increment in torsional frequency. This finding is in agreement with the earlier study made by Othman et al. (2016) [10]. In addition, this also demonstrates that varying ribs orientation offers a greater torsional rigidity while the bending rigidity is not much affected.

To highlight the improvement and degradation of flutter performance, flutter results are normalized with respect to the flutter speed of the baseline system. Figure 7 shows the normalized flutter speed of Case-3 to Case-8, where only a set of ribs are oriented at a time. From the figure, it can be observed that the flutter results are improved as the ribs are oriented in positive angle and reached the maximum at the furthest possible orientation. Meanwhile, the opposite trend with degradation in flutter is seen as the ribs are oriented in the negative angle orientation. Design-1 of Case-5 provides the greatest flutter improvement, followed by Design-2 of Case-6 and Design-3 of Case-4. It should be noted that all the three designs are comprising a set of oriented ribs located in the middle of the wing, which indicate the impact is greater at this location. Figure 8 shows the normalized flutter speed of Case-1, Case-2 and Case-9, where the ribs are varied in parallel and zig-zag arrangements. It is found that Design-4 of the parallel ribs orientation provides an overall best solution with the highest increment of flutter speed by 5.5% compared to the baseline system. On the other hand, the zigzag arrangements offer improvement in flutter speed for both positive and negative angles of orientation with Design-5 of set ribs zigzag arrangement outperformed the Design-6 of individual zigzag arrangement.

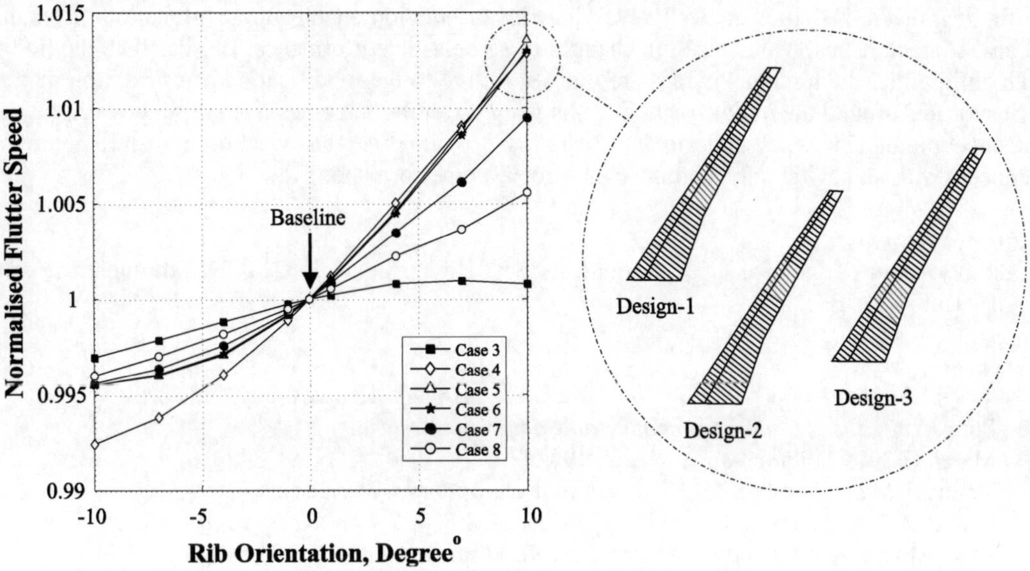

Figure 7: Normalized flutter speed for Case 3-8

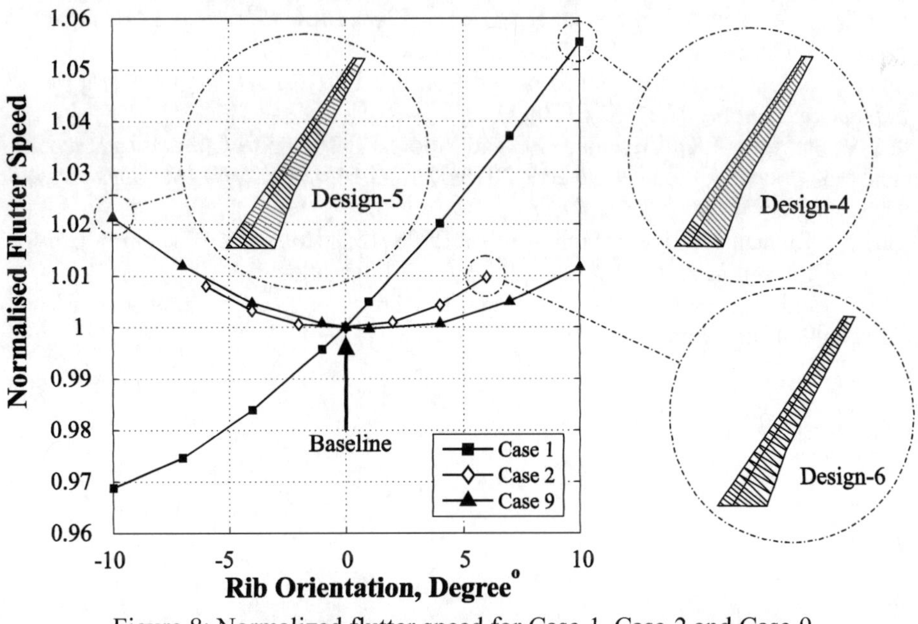

Figure 8: Normalized flutter speed for Case-1, Case-2 and Case-9

The optimum solutions for all the cases are still at their furthest possible angle of orientation, hence hypothetically the flutter speed may be further delayed if spacing between the ribs can be extended. One way to do this is by reducing the number of ribs and another way is by distributing the ribs with increasing spacing along the spars. A major drawback that might arise from this is the proneness of the wing to skin wrinkling. Therefore, further investigation is required to assess the proposed wing design in further details. All in all, the findings from this study are encouraging and suggests that there is a possibility of weight reduction on the wing via the oriented rib concept without having to compromise its aeroelastic performance.

4. Conclusion

The findings from this study show that varying ribs orientation enables the exploitation of bending-torsional characteristics, which result in changes of aeroelastic performance. Besides that, the findings also highlight that the torsional modes are more sensitive to the modification of orientation made on the ribs located around the middle section of the wing. Overall, outcomes from this study can be taken to be encouraging since they indicate the possibility of wing's weight reduction by using the oriented rib concept without having to compromise its aeroelastic performance.

Acknowledgements

The authors acknowledge the support from Ministry of Higher Education (MOHE) through research grant 03-01-15-1625FR.

References

[1] Wright J R and Cooper J E 2007 *Introduction to Aircraft Aeroelasticity and Loads* Wiley
[2] Pora J 2001 *International Committee on Composite Materials*
[3] March G 2015 *Reinforced Plastics* **59** 190-3
[4] Young T M and Hirst M 2012 *Innovation in Aeronautics* Woodhead
[5] Leal P B, Petterson R and Hartl D J 2017 *25th AIAA/AHS Adaptive Structures Conference*
[6] Shirk M H, Hertz T J and Weisshaar T A 1986 *Journal of Aircraft* **23** 6-18
[7] Guo S J, Banerjee J R and Cheung C W 2003 *Journal of Aerospace Engineering* **217** 115-22
[8] Tian W, Yang Z, Gu Y and Ouyang Y 2016 *Journal of Aerospace Engineering* **29** 04016056
[9] Harmin M Y, Ahmed A T and Cooper J E 2011 *52nd AIAA/ASME/ASCE/AHS/ASC Structures, Structural Dynamics and Materials Conference.*
[10] Othman M S, Chun O T, Harmin M Y and Romli F I 2016 *IOP Conference Series: Materials Science and Engineering* **152** 012009
[11] Jutte C V, Stanford B K, Wieseman C D and Moore J B 2014 *52nd AerospaceSciences Meeting*
[12] Francios G, Cooper J E and Weaver P 2016 *57th AIAA/ASCE/AHS/ASC Structures, Structural Dynamics, and Materials Conference.*
[13] Vassberg J, Dehaan M, Rivers S and Wahls R 2008 *AIAA Aplied Aerodynamics Conference*
[14] Bauccio M 1993 *ASM Metal Reference Book* ASM International
[15] Rodden W and Johnson E 1994 *MSC/NASTRAN Aeroelastic Analysis User Guide* The Macneal-Schwendler Corporation

IOP Conf. Series: Materials Science and Engineering **270** (2017) 012030 doi:10.1088/1757-899X/270/1/012030

Shape memory alloy resistance behaviour at high altitude for feedback control

W T Ng, M F Sedan, E J Abdullah*, S Azrad and A S M Harithuddin
Department of Aerospace Engineering, Faculty of Engineering, Universiti Putra Malaysia, Serdang, Selangor, Malaysia

*ermira@upm.edu.my

Abstract. Many recent aerospace technologies are using smart actuators to reduce the system's complexity and increase its reliability. One such actuator is shape memory alloy (SMA) actuator, which is lightweight, produces high force and large deflection. However, some disadvantages in using SMA actuators have been identified and they include nonlinear response of the strain to input current, hysteresis characteristic that results in inaccurate control and less than optimum system performance, high operating temperatures, slow response and also high requirement of electrical power to obtain the desired actuation forces. It is still unknown if the SMA actuators can perform effectively at high altitude with low surrounding temperature. The work presented here covers the preliminary process of verifying the feasibility of using resistance as feedback control at high altitude for aerospace applications. Temperature and resistance of SMA actuator at high altitude is investigated by conducting an experiment onboard a high altitude balloon. The results from the high altitude experiment indicate that the resistance or voltage drop of the SMA wire is not significantly affected by the low surrounding temperature at high altitude as compared to the temperature of SMA. Resistance feedback control for SMA actuators may be suitable for aerospace applications.

1. Introduction

The shape memory alloy (SMA) actuators have been proposed for numerous commercial applications such as automotive, aerospace, robotic and biomedical [1]. In the aerospace research, SMA actuator has found its way in many systems including aircraft, helicopters and satellites [2]. SMA actuator can also be considered as the enabling technology for morphing wing concept, which allows the lift coefficient to be varied accordingly to produce more efficient flight [3-6]. Furthermore, engine nozzle sections can also be varied with application of SMA to reduce noise during take-off [7], and SMA wires have been used in the release mechanisms in small satellites and vibration counter mechanisms [2]. Moreover, the helicopter blades can be folded or retracted using SMAs to ease storage in the designated bay [8], which is an innovation as most helicopters utilize manual folding that requires more manpower and significant amount of time to accomplish it.

In a recent development, it is highlighted that limitations have been found in using SMA actuators for a morphing wing on a real aircraft. These include high operating temperatures, slow response and high need in electrical power to obtain the desired actuation forces [3]. This present study is carried out to understand the behaviour of SMA and its feasibility to be utilized as an actuator at high altitude for aerospace application. Resistance of SMA is chosen as an alternative to a few common feedback control variables such as position and temperature feedback control. Resistance feedback control is expected to be less affected by the significant changes in temperature as compared to other variables like temperature and strain feedback control. High altitude experiments are conducted to verify this notion.

2. SMA Actuator

SMA wires are able to change their length when a change in temperature occurs. Two types of heating that can introduce temperature changes are conventional and Joule heating. The conventional heating involves placing the SMA wire near a heat source. On the other hand, Joule heating can be achieved by connecting the two ends of the SMA material to an electrical power source. Many works that are carried out by researchers used Joule heating over conventional heating [9-12]. SMA undergoes phase transformation in two types of conditional change: temperature and loading. The phases are austenite, twinned martensite and de-twinned martensite [13]. The common initial phase is twinned martensite. Upon loading, the SMA changes into de-twinned martensite phase. If the de-twinned martensite SMA is heated, it changes into austenite. Allowing the SMA to cool down will return it to the initial phase of twinned martensite. SMA has been labelled as high energy density, large energy storage and high force produced-to-weight ratio material. Force is produced by means of strain, particularly when the SMA undergoes a change in its dimension, most commonly, length. The small size of SMAs enables them to replace the conventional actuators that are mostly much larger in size and to be used in small workspaces or areas.

The shape memory effect in SMA can be divided into two categories: one-way and two-way [14-15]. SMA with one-way shape memory effect operates in a way that it can return to its original state after deformation. For example, after a load is applied to deform the SMA, heat can be used to revert it to its original state. On the other hand, SMA with two-way shape memory effect can remember two states: one at high temperature and another at low temperature. SMAs require training for this purpose or effect. Several types of variables can be used to control the SMA actuator that include temperature, resistance, strain and position. Temperature and resistance are associated with the SMA actuator itself whereas strain and position are usually related to the system that is being actuated. Using temperature as a feedback control requires some compensations to be carried out. As the current is allowed to flow through the SMA wire, the temperature of the wire increases. However, when the current is stopped or removed, the temperature curve does not go back down through the same line. In general, two types of approaches can be taken to remove the hysteresis effects. The first one is the open loop compensation, which can be done by applying the inverse of the generalized Prandtl-Ishlinksii model that is used to model the hysteresis [16]. On the other hand, the second one is the closed loop feedback, whereby a corrective heat input is generated based on the error [17].

Furthermore, resistance is studied as an option to overcome hysteresis. Resistance of a material can be measured using a two-wire or four-wire configuration [18]. Both concepts require the presence of a small power supply to the device under test (DUT). The voltage drop across the DUT and the current in the circuit are used to compute the resistance. For a four-wire setup, additional resistors are placed in the circuit to reduce the measurement error. This setup is recommended for resistance measurement under 100 ohms. In previous work by Dhanalakshmi (2014), differential resistance feedback control is applied to control two antagonistic SMA actuators [12]. The values of resistance are computed from voltage drop measurements of both SMA wires and the difference between them is used to determine rotational movement of the shaft-disc. The angular change of shaft-disc is also shown in a manipulator arm. Moreover, the capability of resistance feedback control to control the position of a slider has also been demonstrated [19]. A Linear Variable Differential Transformer (LVDT) position sensor is placed against the slider to measure the displacement.

In certain applications of SMA actuators, the actuation is indirectly controlled. For example, strain feedback uses strain gauges to monitor the strain of the system being actuated. For the morphing wing, the change in strain experienced by the wing structure due to the SMA wire exerting a force is used as a feedback signal to obtain the desired shape. A certain shape of the wing corresponds to a specific lift coefficient [5]. In another work, the strain is measured and controlled to study the performance of the proportional-integral (PI) controller on a flexible aluminium beam [20]. To track position or deflection of an actuated system, a laser range sensor is required for the position feedback. The tip position of the composite beam with embedded SMA wires can be controlled accurately by position feedback using a laser range sensor, with the aid of feed-forward action, PD controller and a robust compensator. In this work, the feasibility of using resistance as feedback control at high altitude for aerospace applications is explored. A composite plate, actuated by a Ni-Ti shape memory alloy wire, is used for this study.

3. Methodology

Prior to this experimental work, finite element method is used to gauge the structural response of the composite plate actuated by SMA actuators. Different configurations have been analyzed by changing the properties of the composite material, position of SMA actuators within the structure and the forces exerted on the structure by the actuators. The placement of the actuators is critical in obtaining the desired plate displacement. The design objective that has been set then is to produce a plate that can be displaced at least 10mm along the z-axis at the tip. A different combination of applied forces by the SMA actuators is analyzed. The analysis is repeated for different configurations of the SMA actuators. It has been found that the number of SMA actuators used and their orientation will produce significant variation in plate displacement [21].

Figure 1 illustrates the schematic diagram of the SMA actuators on the composite plate used in this study. The dimension of the composite plate is 52cm in length and 20cm in width. Three SMA wires with a length of 35.5cm are instilled on the surface of the plate at a distance of 2.5cm between each other. One strain gauge is placed on the surface of the plate and a thermocouple has been attached to the SMA actuator. The experiments are conducted using single cantilever plate, where one end is fixed while another is free to move. The laminated woven composite model is composed of six fiberglass/ epoxy layers with equal 0^0 layup. The SMA actuator selected is Ni-Ti wires provided by Dynalloy, in which sold under the trade name FLEXINOL.

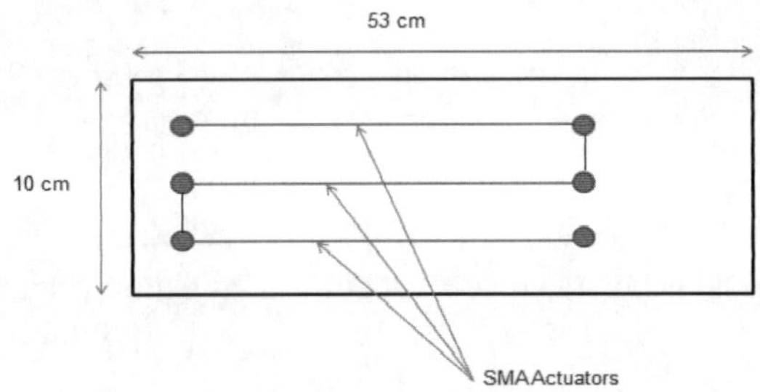

Figure 1: Schematic diagram of composite plate with SMA actuators

An experimental test bench has been designed to analyze the performance of strain feedback smart structure system composed of the composite laminate plate with SMA actuators placed on the surface of the plate. A proportional-integral-derivative (PID) controller has been designed using mathematical model of the SMA actuator and its performance has been verified to be effective through a hardware-in-the-loop experiment [22]. However, using the strain of the composite plate as feedback can be quite complicated as the input parameter to the control system does not come directly from SMA actuator. Hence the objective of this current work is to determine the feasibility of using resistance of the SMA actuator as input variable to the control system. It is believed that this will produce accurate response for the deflection of the composite plate in order to reduce overshoot and steady state error.

For the preliminary stages, basic experiments are carried out to determine the relationship between several variables. The manipulated variable is voltage of power supply and the responding variables are temperature of SMA, voltage drop of SMA, force produced by SMA and also resistance of SMA as shown in Figure 2. Meanwhile, the measurement process for the experiments is depicted in Figure 3. For the temperature of SMA, a voltage is supplied from 0V up to 2.5V to activate the SMA, and the temperature readings are recorded using thermocouple. As for the voltage drop, similar voltage supply is used and the voltage drop is recorded using two wires. Both responding variables are measured by their respective measurement device, alongside data acquisition cards from NI and LabView. As for the force produced, a digital hanging scale is aligned horizontally and hooked on to the SMA wire to measure the pulling force. The resistance of the SMA is computed by dividing the voltage drop by the current. The conducted measurement process for this is depicted in Figure 4 and Figure 5.

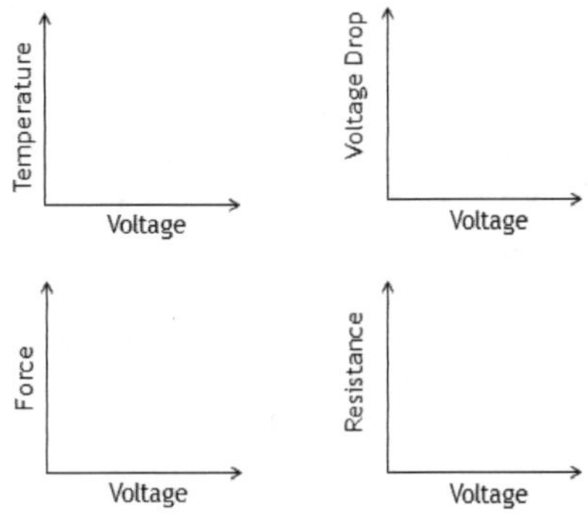

Figure 2: Graphs of outputs against voltage

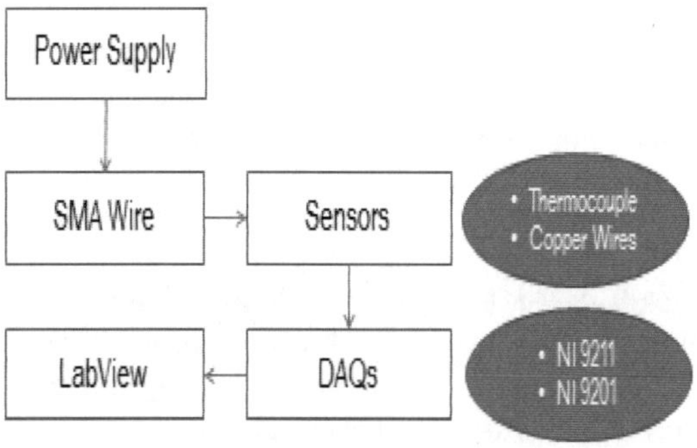

Figure 3: Measurement process of temperature and voltage drop

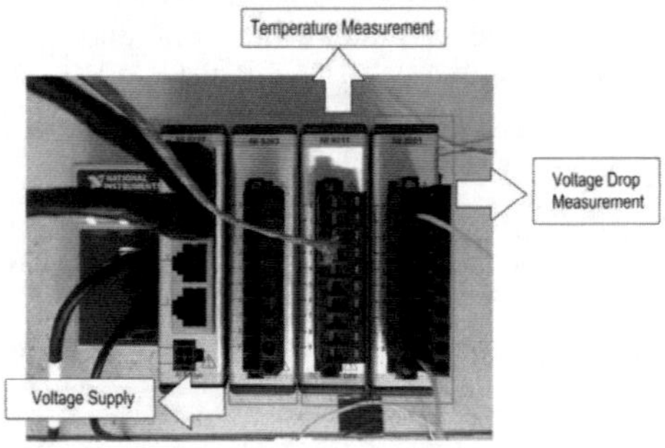

Figure 4: DAQ to acquire and generate signals

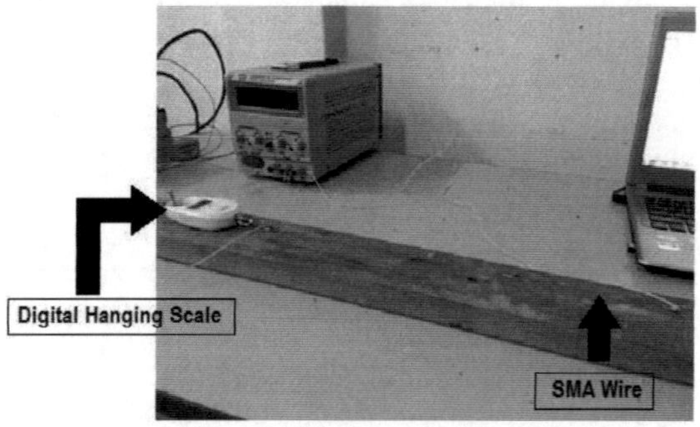

Figure 5: Digital hanging scale to measure force produced by SMA

For the high altitude testing, resistance of the SMA wire is measured with a different power source. The adjustable power supply is replaced with a battery, which will introduce a constant power supply. This alternative is suggested since there was very limited onboard space when the high altitude testing is carried out. A composite plate comprising of fiberglass and epoxy hardener is drilled at both of its ends, with the diameter of a small screw's thread. One end of the plate is screwed onto a small wooden block glued in the glider while the other end is left to remain a free end. The fixed end has two screws, diagonal to each other, to prevent the plate from rotating. The free end of the composite plate has been supported by another piece of wooden block. Figure 6 shows the composite plate with SMA placed in the High Altitude Testing platform. The sensor onboard is a thermocouple as shown in Figure 7, where the SMA temperature and the ambient temperature in the glider are measured. A 3.7V rating battery is connected to the SMA wire to complete the circuit. Figure 8 shows the Arduino Uno board that is used to control the actuation that sent signals to connect or disconnect the circuit. The board measured the voltage drop of the SMA during actuation. The SMA is actuated at five different altitudes during the ascent of the balloon: 4km, 8km, 12km, 16km and 20km.

The High Altitude Fixed Wing Experimental Platform is flown with a 1600g balloon and 2.5kg payload (fixed wing). It is launched from the paddy field in Langkap, Perak, Malaysia (4°2'31.62"N 101°10'4.45"E) at 12:00 pm, carrying the experimental test bed onboard. The flight is fully controlled by the control module system. When the balloon reached 108268ft, it burst and the fixed wing payload

IOP Conf. Series: Materials Science and Engineering **270** (2017) 012030 doi:10.1088/1757-899X/270/1/012030

began its free fall, which is then slowed down by a parachute at 88582.7 ft. During descent, the release mechanism is activated to release the parachute and the fixed wing began its autonomous mission back to the desired landing site, which has been designated to be at Seberang Perak, Malaysia (4°3'31.51"N 100°51'54.00"E). After arriving at the landing coordinate, the fixed wing began to loiter autonomously to descend slowly before the ground station controller took over control from the autopilot to make a smooth landing. Figure 9 shows the flight path profile of the glider and Figure 10 shows the change in altitude of the balloon. After 2 hours 48 minutes, the fixed wing safely landed by manual control at 50m from the ground station. All experiments and system hardware are recovered with very minimal damage to the fixed wing body.

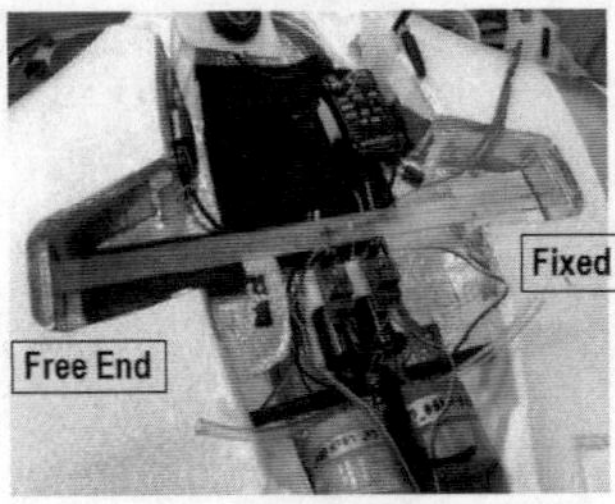

Figure 6: SMA fixed to composite plate for high altitude testing

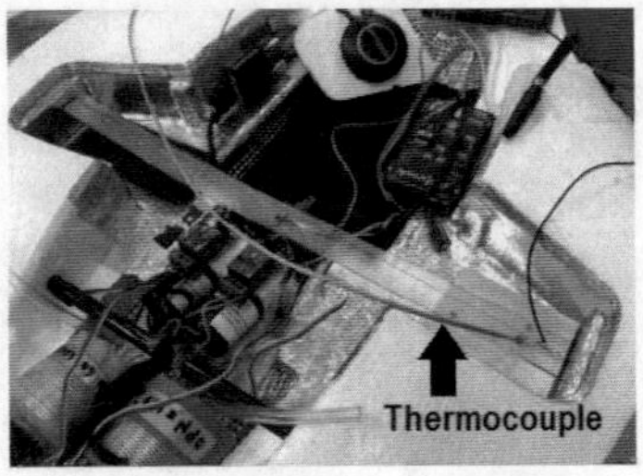

Figure 7: Thermocouple used to measure temperature

Figure 8: Arduino UNO for automated on-off power supply

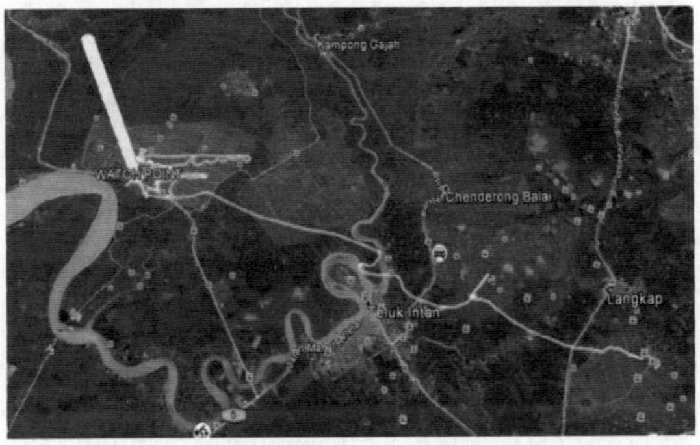

Figure 9: Flight path profile

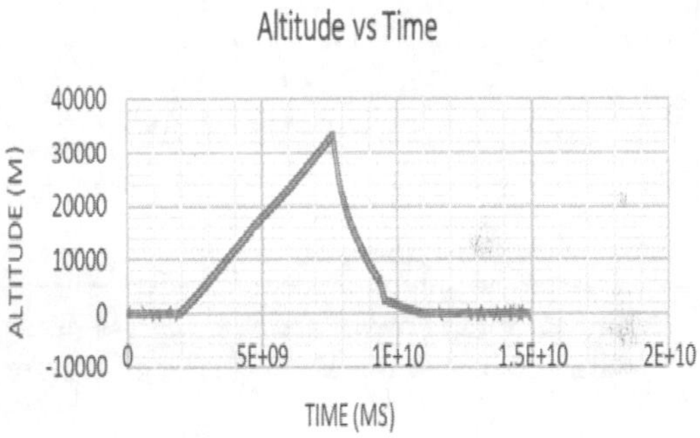

Figure 10: Altitude variation of high altitude platform with time

For the next stage, resistance is used as a variable for feedback control to actuate a composite plate. Three SMA wires are arranged in a series electrical connection on the surface of a composite plate as shown in previous Figure 1. For the feedback control process, the voltage drop is measured across the SMA wires that are arranged in series. This value is obtained on the LabView interface using NI 9201 and then compared to a user input value. The input is keyed in numerically in the visual interface. The PID controller ensured appropriate amount of current is supplied to allow the measured voltage drop to match the input value. Figure 11 shows the schematic diagram of the feedback control system.

4. Results and Discussion

The graphs of voltage drop and temperature against voltage supply indicates that the SMA temperature contains more hysteresis compared to the voltage drop when it is being heated and cooled as shown in Figure 12 and Figure 13. Therefore, based on the results, the option of using voltage drop in feedback control reduces the need for compensation to be applied. The voltage drop response is more linear than

IOP Conf. Series: Materials Science and Engineering **270** (2017) 012030 doi:10.1088/1757-899X/270/1/012030

temperature response, which can be contributed to the independence of voltage drop towards external factors. On contrary, temperature is easily affected by the ventilation or the external heat source. The electrical resistivity of SMA is dependent on the temperature of the SMA that causes the modification of electronic structure of the material [23]. Figure 14 shows the force produced by the SMA wire with the change in voltage, where the force increases with the voltage supply but underwent a sudden spike after approximately 2.25V. This could be due to the phase transformation of the SMA wire. The large magnitude of force can be applied to do some heavier work whereas the linear part of the relationship is useful for controlling a range of force.

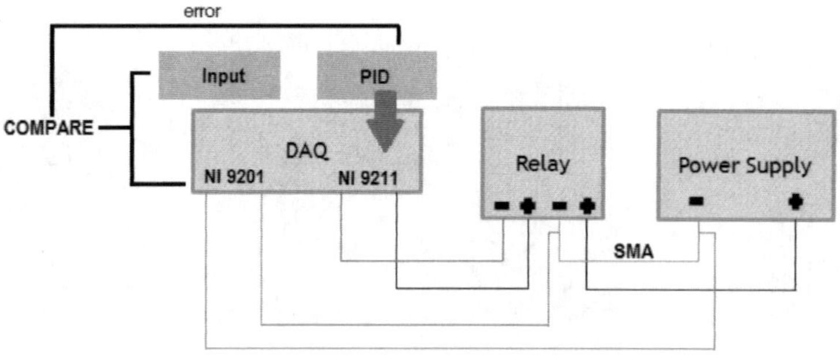

Figure 11: Schematic diagram of feedback control system

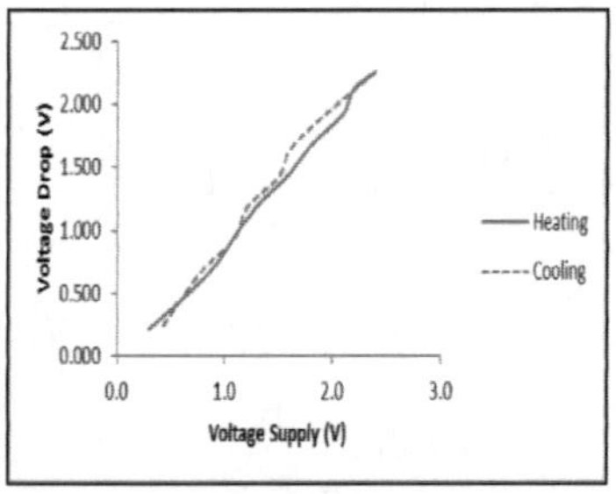

Figure 12: Graph of voltage drop against voltage

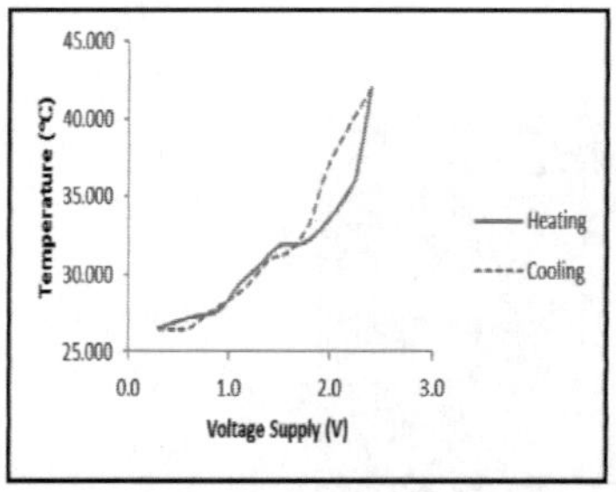

Figure 13: Graph of temperature against voltage

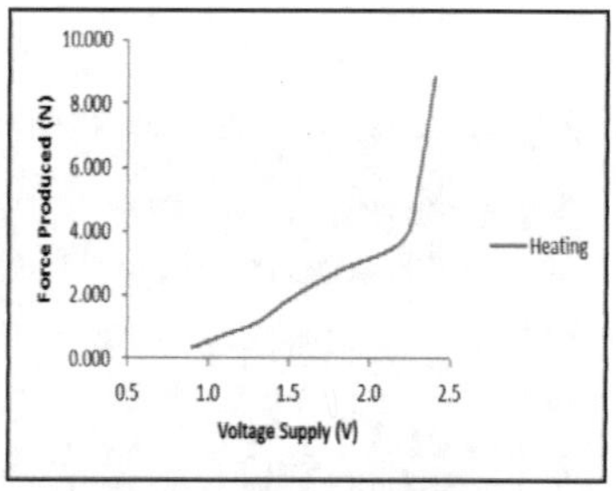

Figure 14: Graph of force produced against voltage

The SMA actuator is activated by providing PWM at different altitudes as shown in Figure 15. A few parameters have been observed including voltage drop, resistance, SMA temperature and ambient temperature in the fixed wing glider. The battery of 3.7V has managed to supply voltage above 3.5V consistently as shown in Figure 16. From Figure 17, it can be seen that both the ambient temperature and SMA temperature decreases when the altitude increases. The power supply is activated at altitudes of 4km, 8km, 12km, 16km and 20km, and it is deactivated after 2km in all occasions. The ambient temperature reading in the fixed wing glider can be observed before each SMA peak temperature, with the first phase recording was 30°C, second was 26°C, third was 17°C, fourth was 8°C and the fifth was 0°C. From the readings for the temperature and resistance measurement, it can be clearly seen that the temperature of SMA is highly affected by the change in altitude and ambient temperature. However, the resistance of SMA is only slightly affected by the change in the altitude and ambient temperature as shown in Figure 18. This shows that resistance feedback control of SMA actuator is more suitable for aerospace applications. For the resistance feedback control at ambient or room temperature, a step input of 8V has been selected. The result in Figure 19 shows that the system is able to provide the desired response. However, the steady state error is large, which is approximately 22.5%. In view of this, further improvement on the controller design may improve the transient and steady state response of the resistance feedback system for the SMA actuated composite plate.

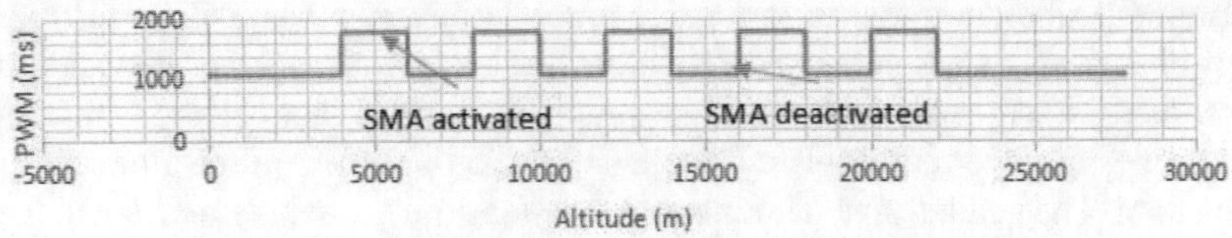

Figure 15: Graph of PIXHAWK PWM against altitude

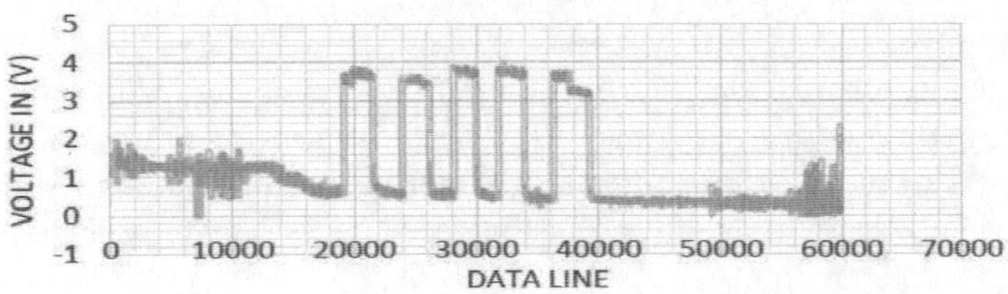

Figure 16: Graph of voltage supply against data line

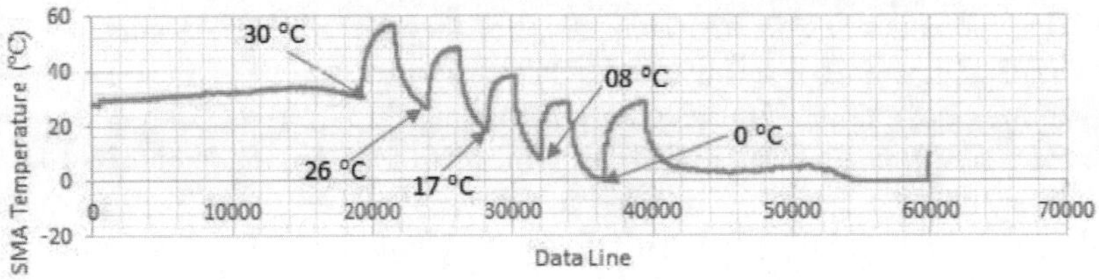

Figure 17: Graph of SMA temperature against data line

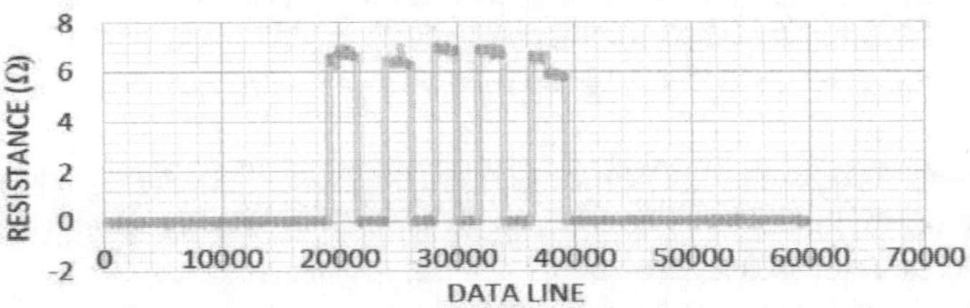

Figure 18: Graph of SMA resistance against data line

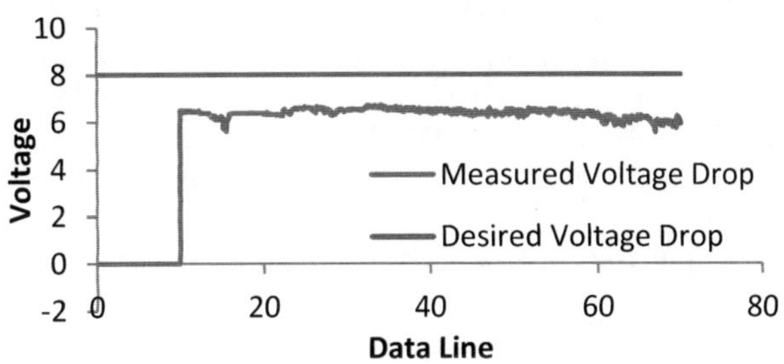

Figure 19: Graph of voltage against data line

5. Conclusion

The experimental data obtained in this study has indicated that temperature contains a higher level of hysteresis than voltage drop. This comparison results suggest that voltage drop provides better control without requiring further compensation. The results from the high altitude experiment also have shown that the resistance of SMA wire is not notably affected by low surrounding temperature. Independence towards surrounding temperature makes resistance a good option as the variable for feedback control. Feedback control has been successfully carried out using voltage drop as the feedback variable for the SMA actuated composite plate. An output voltage supply response to actuate the composite plate is also achieved. The composite plate represents a simple simulation or portrayal of a morphing structure for aerospace application. The results have demonstrated the feasibility of using resistance feedback control for SMA actuating system. By employing this method, the external sensor can be eliminated since SMA actuator itself acts as the sensor. This will help to further reduce the complexity of the system.

Acknowledgement

The authors would like to acknowledge the AeroPutra Space team from Universiti Putra Malaysia for their significant contribution in this research work. Their participation in the GSBC 2016 competition has allowed the authors to successfully conduct the experimental work at high altitude. Moreover, the authors also like to acknowledge the funding for this research by Universiti Putra Malaysia under the GP-IPS/2016/9513400 research grant.

References

[1] Jani J M, Leary M, Subic A and Gibson M A 2014 *Materials & Design* **56** 1078-113
[2] Hartl D J and Lagoudas D C 2007 *Proceedings of the Institution of Mechanical Engineers, Part G: Journal of Aerospace Engineering* **221** 535-52
[3] Kammegne M, Grigorie T L, Botez R M and Koreanschi A 2016 *Proceedings of the Institution of Mechanical Engineers, Part G: Journal of Aerospace Engineering* **230** 132-45
[4] Grigorie T L, Botez R M and Popov A V 2015 *Journal of Aerospace Engineering* **28**
[5] Bil C, Massey K and Abdullah E J 2013 *Journal of Intelligent Material Systems and Structures* **24** 878-98
[6] Grigorie T L, Popov A V, Botez R M, Mamou M and Mébarki Y 2010 *The International Conference on Informatics in Control, Automation and Robotics*
[7] Quan D and Hai X 2015 *Procedia Engineering* **99** 1241-6
[8] Ali M I F and Abdullah E J 2016 *IOP Conference Series: Materials Science and Engineering* **152** 012016
[9] Dong Y, Boming Z and Jun L 2008 *Materials Science and Engineering: A* **485** 243-50
[10] Lara-Quintanilla A, Hulskamp A and Bersee H E 2013 *Journal of Intelligent Material Systems and Structures* **24** 1834-45
[11] Song G, Kelly B and Agrawal B N 2000 *Smart Materials and Structures* **9** 711
[12] Dhanalakshmi K 2014 *ISA Transactions* **53** 289-97
[13] Liang C and Rogers C A 1990 *Journal of Intelligent Materials and Structures* **1** 207–34
[14] Anadon J R S 2002 *Large Force Shape Memory Alloy Linear Actuator* PhD Thesis, University of Florida
[15] Zanaboni E 2008 *One-Way and Two-Way Shape Memory Effect: Thermo-mechanical Characterization of Ni-Ti Wires* PhD Thesis, Universitá Degli Studi Di Pavia
[16] Feng Y, Rabbath C A, Hong H, Su C Y and Lin W 2010 *49th IEEE Conference on Decision and Control*
[17] Dickinson C A and Wen J T 1998 *Journal of Intelligent Material Systems and Structures* **9** 242-50
[18] Tong A 2001 *Sensor Review* **21** 193-8
[19] Ma N, Song G and Lee H J 2004 *Smart Materials and Structures* **13** 777
[20] Lima W M, Araújo C J D, Valenzuela W A V and Rocha Neto J S D 2012 *Journal of the Brazilian Society of Mechanical Sciences and Engineering* **34** 413-22
[21] Majid D L, Abdullah E J, Harun N F, Lim G Y 2012 *Procedia Engineering* **50** 445-52
[22] Abdullah E J, Majid D L, Romli F I, Gaikwad P S, Yuan L G and Harun N F 2015 *International Journal of Mechanics and Materials in Design* **11** 25-39
[23] Airoldi G, Lodi D and Pozzi M 1997 *Journal de Physique IV Colloque* **7** C5-507-C5-512

The effect of changing disk parameters on whirling frequency of high speed rotor system

A M Abdul Wahab, Z A Rasid*, A Abu, N F Mohd Noor Rudin and F Yakub

Malaysia-Japan International Institute of Technology, Universiti Teknologi Malaysia, Kuala Lumpur, Malaysia.

*arzainudin.kl@utm.my

Abstract. The requirement for efficiency improvement of machines has caused machine rotor to be designed to rotate at high speeds. It is known that whirling natural frequency of a shaft changes with the change of shaft speed and the design needs to avoid points of resonance where the whirling frequency equals the shaft speed. At high speeds, a shaft may have to carry a huge torque along and this torsional effect has been neglected in past shaft analyses. Whirling behaviour of high speed rotating shaft is investigated in this study with consideration of the torsional effect of the shaft. The shaft system under study consists of a shaft, discs and two bearings, and the focus is on the effect of the disc parameters. A finite element formulation is developed based on Nelson's 5 degrees of freedom (DOF) per node element that includes the torsional degree of freedom. Bolotin's method is applied to the derived Mathieu-Hill type of equation to get quadratic eigenvalues problem that gives the forward and backward frequencies of the shaft. Campbell's diagrams are drawn in studying the effect of discs on the whirling behaviour of the shaft. It is found that the addition of disks on the shaft decreases the whirling frequency of the shaft and the frequency is lower for mass located at the centre of the shaft compared to the one located near to the end. The effect of torsional motion is found to be significant where the difference between critical speed of 4DOF and 5DOF models can be as high as 15%.

1. Introduction

Vibration control is one of the main concerns in the design of rotating shaft. The demand of the market today for more powerful and efficient rotating machines has led to the design of machines that need to be operated at high speeds [1]. Classification of high speed rotating equipment lies in between 10000 RPM and 100000 RPM [2]. Such rotating equipment that operates at high speed has been used in wide ranges of industrial applications like textile [3], automotive [4] and electrical spindle [5]. An important type of vibration that needs to be controlled is the lateral vibration, which is called whirling. Whirling is a phenomenon where the plane of the enclosed bent up shaft–axis and the bearing centre line rotates about the bearing centre line, causing the shaft to bow out as illustrated in Figure 1. This condition can occur due to the mass imbalance of the shaft or disks and it occurs at resonance where whirling natural frequency is equal to the spin speed. Due to the gyroscopic effect of the shaft, whirling frequency is known to not only split into two but also changes with the spin speed. The frequency might increase with speed as the shaft whirls in the rotational direction of the shaft and this frequency is called the forward frequency (FF). In opposite manner, the shaft whirls in direction opposite to the shaft rotation while this backward frequency (BF) decreases with the spin speed. This behavior is usually captured in the Campbell diagram as shown in Figure 2. Resonance occurs at point R where the FF is equal to the spin speed. The spin speed at resonance is called the critical speed, Ω_{cr}. To avoid this resonance condition that may occur when this changing frequency of the shaft equals the shaft speed, it is very important for accurate modal analysis to be conducted in the rotor system.

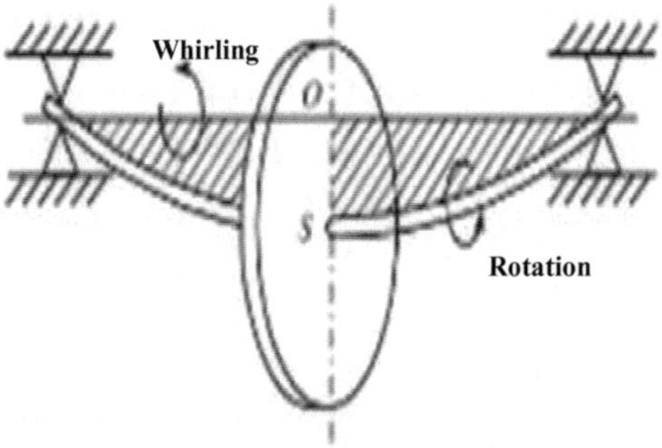

Figure 1: The whirling of a shaft [6]

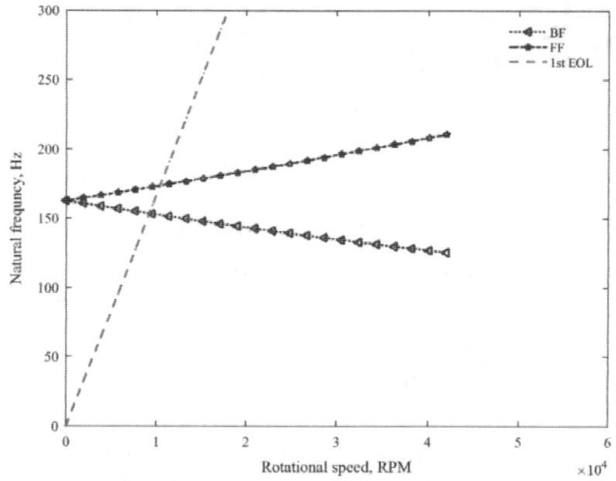

Figure 2: The Campbell diagram of a shaft

Lien-Wen and Der-Ming (1991) [7] applied the three-nodal, C^0 Timoshenko beam finite element model to investigate natural whirl speeds of rotating shaft with various end conditions and slenderness ratios. Effects of translational and rotary inertia, gyroscopic moments, bending and shear deformation are included in the mathematical model. The numerical results of the study indicated that the natural whirl speeds are considerably affected by the end conditions of the rotating shafts, shear deformation and the rotary inertia. In a different method of analysis, free vibrations of shafts supported on resilient bearings have been analyzed by Karunendiran and Zu (1999) [8]. Using Timoshenko beam theory, the exact frequency equation in the complex compact form is derived. The rotary inertia and gyroscopic effects are included while shear deformation effect is neglected. Meanwhile, a study on free vibration of a simply supported rotating shaft with stretching nonlinearity has been conducted by Hosseini and Zamanian (2013) [9]. The equations of motion are derived with the aid of the Hamilton principle and then transformed to the complex form. To analyze the free vibration, the method of multiple scales is directly applied to the partial differential equation of motion. It has been shown in that study that both forward and backward nonlinear natural frequencies are being excited. Furthermore, another study has stated that the forward and backward whirling modes are involved in investigating the free nonlinear vibrations of nonlinear slender rotating shaft [10]. It is found that the values of nonlinear forward and backward frequencies are lower in the second mode compared to first mode. 3D finite element-based model order reduction method has also been used to show the split of the sole resonant peak in case of gravity and unbalance condition due to anisotropy of bearings [11]. In addition, a cluster of additional resonant peaks appeared as both shaft and bearing are anisotropic. In rather complete modal analysis of a high speed rotor, Jalali et al. (2014) [12] used both 3D and 1D finite element method (FEM) to model the shaft. The natural frequencies and mode shapes of the rotor at rest under free–free boundary conditions are obtained, and the results are then compared to those obtained from the modal test. The Campbell diagram and the critical speeds are calculated using both FE models to evaluate the rotating system dynamics. The imbalance response of the system to the center of mass imbalance at the turbine is also calculated to investigate the dynamic behavior more practically and to verify the critical speeds obtained from the FEM.

The above studies show the progress in research on the characteristics of whirling frequencies of a shaft. However, except for Ref. [12], no other study has focused on high speed rotating shaft. At high rotating speed, a shaft may also carry high torque but the effect of this torque on whirling frequency of the shaft has been hardly considered. In their work to investigate the importance of considering the torsional effect of the shaft, Seshendra and Rao (2012) [13] have studied the whirling characteristics of a geared rotor system for combined torsional-lateral vibrations. The study demonstrated that the lateral motions interact significantly with the torsional degrees of freedom when the resonance frequencies are similar. As such, this study is aimed to investigate the torsional effect on the whirling frequency of high speed rotor system that consists of a shaft, discs and bearings. The main focus is on the effect of several disc parameters on the whirling behavior of the shaft. FE formulation based on the Nelson's finite element model [14] that considers the torsional effect is developed. The Bolotin's method [15] is applied to the derived Mathieu-Hill type of equation to get the quadratic eigenvalues problem equation that gives the forward and backward frequencies of the shaft for the Campbell's diagram to be plotted.

2. Material and Methods

Referring to Figure 3, the rotor system used in this study consists of a shaft, one or more discs and two bearings. Bearings are represented as the supports in Figure 3. d is the diameter of the shaft whereas t is the thickness of the disk. L is the length of the shaft while *l* is the distance of the disk from the left bearing. Furthermore, Table 1 shows the material properties and dimensions of the shaft and bearing components of the rotor system. Table 2, on the other hand, tabulates the material properties and also dimensions of the discs attached to the shaft.

This study employs FEM to estimate the whirling natural frequency of the rotating shaft system. The finite element formulation developed here is based on Rao (2011) [16] where Timoshenko beam theory and Nelson's FEM approach [14] that considers the torsional effect are applied. FEM model that applies 5 degrees of freedom (DOF) per node including the torsional DOF is called 5DOF model in comparison to 4DOF model that does not consider the torsional DOF. Derivation of the formulation is very similar to that for parametric instability problem of rotating shaft.

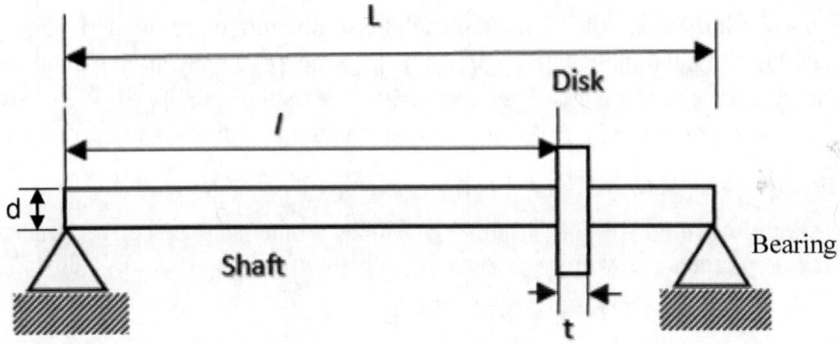

Figure 3: The rotor system being investigated

Table 1: Dimensions and material properties of the shaft and bearing of the rotor system

Shaft	
Young's modulus, E	207 GPa
Modulus of rigidity, G	79.6 GPa
Poisson's ratio, v	0.303
Density, ρ	7833 kg/m^3
Radius, r	round tube with r_{inner}= 0 m, r_{outer} =0.0508 m
Length, L	1.27 m
Shear factor, κ	0.9
Bearing	
Direct stiffness coefficient, K_{by},K_{bz}	7x10^7 N/m
Direct damping coefficient, C_{by},C_{bz}	0

Table 2: Dimensions and material properties of the disk correspond to the studies on the effect of the location of disc (Study 1), the mass of the disc (Study 2) and the number of discs attached (Study 3)

Disk in Study 1	
Young Modulus, E	207 GPa
Density, ρ	7833 kg/m^3
Thickness, t	0.03 m
Diameter, d	0.3 m
Disk in Study 2	
Young Modulus, E	207 GPa
Density, ρ	7833 kg/m^3
Diameter, d_1 and d_2	0.3 m
Thickness, nominal weight, t_1	0.03 m
Thickness, heavy weight, t_2	0.07 m
Diameter, d_3	0.16 m
Thickness, nominal weight, t_3	0.1055 m
Disk in Study 3	
Young Modulus, E	207 GPa
Density, ρ	7833 kg/m^3
Diameter, d_1, d_2	0.4 m
Thickness, t_1, t_2	0.005 m
Diameter, d_3	0.4 m
Thickness, t_3	0.006 m

Both types of whirling and the dynamic instability problem require the derivation of Mathiew-Hill equation such as in Equation 1, where $[M]$, $[G]$, $[K]$ and $[K_g]$ are the elemental mass, gyroscopic, stiffness and geometric stiffness matrices, respectively, Ω is the spin speed, $P(t)$ is the periodic axial force, α is the static load factor, β is the dynamic load factor and ϕ is the excitation frequency.

$$[M]\{\ddot{q}\} + \Omega[G]\{\dot{q}\} + ([K] - [K_g](\alpha P^* + \beta P^* \cos\phi T\,(t))\{q\} = 0 \quad (1)$$

Applying Bolotin's method [16], an infinite eigenvalues problem can be derived such as in Equation 2, which actually represents the whirling frequency problem to be solved in this study.

$$(\phi^2[M_E] + \phi[G_E] + [K_E])q = 0 \quad (2)$$

3. Results and Discussion

The effect of attaching disk to shaft on the whirling frequency of the shaft is investigated. In reference to Table 2, the effect of disk considered in this study constitutes of the effect of disk location with respect to shaft (Study 1), disk mass (Study 2) and disk number (Study 3) on the whirling frequency of the shaft.

Four cases are being compared here: Case 1 - shaft without any disk (no disk), Case 2 – shaft with a disk located at the middle of the shaft (disk @ 50%), Case 3 – shaft with a disk positioned near to the end of the shaft (disk @ 75%) and Case 4 – shaft with a disk positioned at the other end of the shaft with bearing inboard that results in overhung rotor (overhung). Disks with similar properties are used throughout the analysis as given in Table 2. The results from Case 1 are used as the benchmark for any differences that may occur in dynamic behaviour of the shaft when the disk is added. Rigid bearings are used in this study. Table 3 and Table 4 show the first and second modes of natural frequencies of static shaft for both 4DOF and 5DOF models, respectively. For the first mode, as tabulated in Table 3, the natural frequency for Case 2 is somewhat lower compared to that in Case 3. It should also be noted that, for the case of disk @ 75%, the disk is very near to the end of shaft where the radius of gyration of the shaft is smaller compared to that of the shaft with disk @ 50%. The reduction of the radius of gyration has decreased the moment of inertia of the disk. As a result, the natural frequency of the shaft is increased.

Table 3: First mode natural frequency of static shaft system for different disk locations

Disk location from end	**Natural frequency, Hz 4DOF**	**Natural frequency, Hz 5DOF**	**Differences (%)**
No disk	91.15	91.21	0.1
50%	79.94	79.98	0.1
75 %	83.34	83.38	0.0
Overhung	49.45	49.51	0.1

The opposite trend occurs to the second mode of frequency as the location of the disk changes. The natural frequency of disk in Case 2 of the shaft is higher than that in Case 3 such as shown in Table 4. For the case of overhung condition (Case 4), the natural frequencies are at the lowest for both first and second modes of frequency. This is because, in the overhung condition, the radius of gyration of the shaft is at its highest value. Thus the effect of moment of inertia of the disk that reduces the natural frequency is at its highest. On the other hand, in comparison between the 4DOF and 5DOF models, there is no significant effect of torsional motion on natural frequency of static shaft as the difference between the two models is only 0.3% at the highest. This occurs for both the first and second modes.

Table 4: Second mode natural frequency of static shaft system for different disk locations

Disk location from end	**Natural frequency, Hz 4DOF**	**Natural frequency, Hz 5DOF**	**Differences (%)**
No disk	203.62	204.16	0.3
50%	198.63	199.14	0.3
75 %	184.09	184.49	0.2
Overhung	157.17	157.42	0.2

Figure 4 and Figure 5 show the plots of first and second modes of vibration frequency against the rotational speed of the shaft, respectively, for the four cases of disk locations based on the 4DOF and 5DOF models. In general, the FF and BF natural frequencies are decreased for first and second modes as the disk is added to the shaft system. The effects of disk location on whirling speed for the first and second modes are shown in Table 5 and Table 6, respectively. It can be observed that the disk with overhung condition shows the smallest value of the critical speed, especially for the 5DOF model at 2865 RPM and 3084 RPM for BF and FF, respectively. As the disk is moved towards the end of the shaft, the whirling speed increases for both modes. This shows that the shaft with disk located near to its end can operate at a higher speed compared that with disk located far from the end. The effect of torsional motion considered in the 5DOF model is very significant. Both Figure 4 and Figure 5 show a big difference between the FF and BF natural frequencies of the 4DOF and 5DOF models. Similarly, critical speed differs for the 4DOF and 5DOF models, where the difference ranges from 2% to 8% as can be observed in Table 5 and Table 6. This shows that torsional motion needs to be considered in order to estimate accurately the whirling speed of rotating shaft with different locations of disk.

In the meantime, the effect of disk mass on the whirling frequency is also investigated here. A rotor with rigid bearings and centred single disk is used in this study. The comparison is between nominal disk, heavy disk and nominal thick disk to be attached to the shaft. While the masses of nominal disk and heavy disk differ, the mass of nominal disk and nominal thick disk is the same but the two differ in radius and thickness. Table 7 and Table 8 show the first and second modes of natural frequency of the shaft system, respectively.

IOP Conf. Series: Materials Science and Engineering **270** (2017) 012031 doi:10.1088/1757-899X/270/1/012031

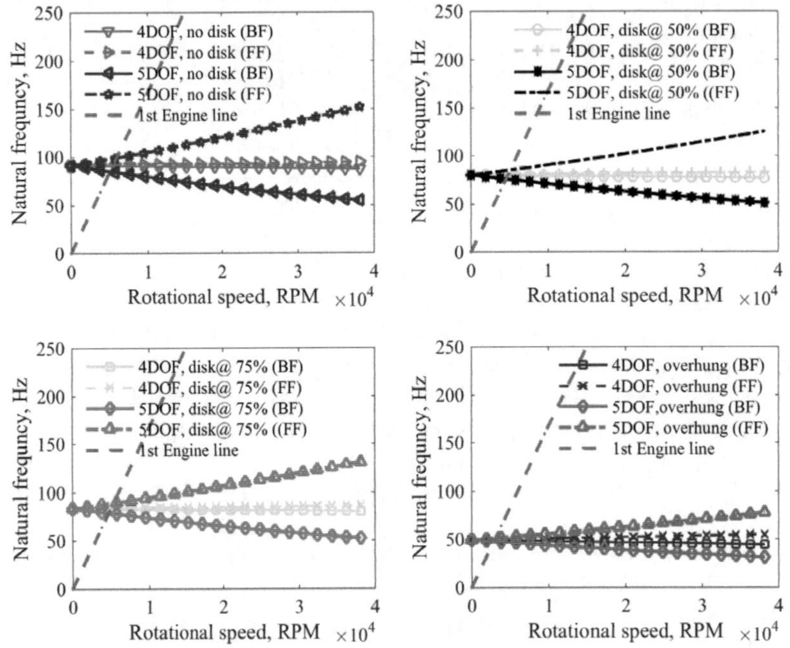

Figure 4: Effect of disk location on the whirling speed of shaft in the first mode

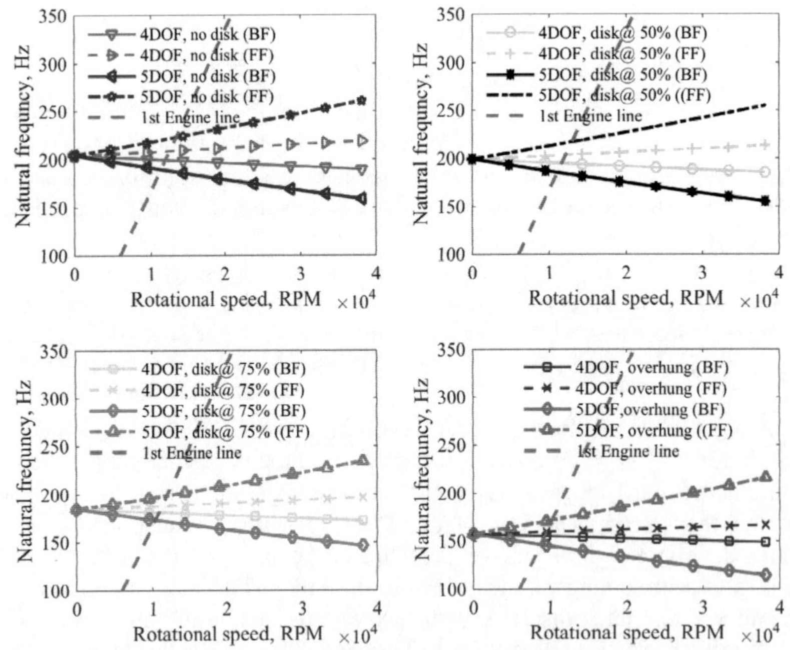

Figure 5: Effect of disk location on the whirling speed of shaft in the second mode

Table 5: Comparison of the first mode of whirling frequency for different disk locations

Disk location from end	Whirling direction	Whirling speed, RPM 4DOF	Whirling speed, RPM 5DOF	Differences, %
No disk	BF	5443	5099	-6.3
	FF	5500	5949	8.1
50 %	BF	4775	4545	-4.8
	FF	4822	5109	5.9
75 %	BF	4975	4708	-5.4
	FF	5023	5348	6.5
Overhung	BF	2960	2865	-3.2
	FF	2998	3084	2.9

Table 6: Comparison of the second mode of whirling frequency for different disk locations

Disk location from end	Whirling direction	Whirling speed, RPM 4DOF	Whirling speed, RPM 5DOF	Differences, %
No disk	BF	11940	11340	-5.0
	FF	12540	13380	6.7
50 %	BF	11690	11130	-4.8
	FF	12190	13020	6.8
75 %	BF	10840	10380	-4.2
	FF	11280	11960	6.0
Overhung	BF	9320	8785	-5.7
	FF	9597	10340	7.7

In Table 7, the first mode natural frequency of heavy disk is lower compared to that of the nominal disk in static condition. Table 7 also highlights that, since nominal disk and nominal thick disk have a similar mass, the natural frequencies for both conditions are similar. Similar results are obtained for the second mode of frequency of the shaft when the comparison is made between the nominal and the heavy disks such as shown in Table 8. However, in contrast to the case of the first mode, the second mode of natural frequency for nominal thick disk is higher than that of the nominal disk. For the first and second modes of the nominal and heavy disks, the mass of the disk might increase but the mass moment of inertia is unchanged because of their similar radius of gyration. Thus the natural frequency decreases as the mass increases. Similar condition occurs for the nominal and nominal thick disks that have the same mass in the first mode. With similar mass and unchanged mass moment of inertia, the natural frequencies are similar for both conditions. For the second mode, the polar moment of inertia changes as nominal thick disk has smaller diameter than nominal disk. Although both conditions have similar mass, the smaller disk reduces the distance of second moment of mass from the spinning axis. Hence, this condition increases the natural frequency of the nominal thick disk as compared to that of the nominal disk.

Table 7: First mode natural frequency of shaft system with different disk mass

Disk mass	Natural frequency, Hz 4DOF	Natural frequency, Hz 5DOF	Differences (%)
Nominal	102.21	102.24	0.0
Heavy	88.08	88.09	0.0
Nominal thick	102.22	102.24	0.0

Table 8: Second mode natural frequency of shaft system with different disk mass

Disk mass	Natural frequency, Hz 4DOF	Natural frequency, Hz 5DOF	Differences (%)
Nominal	228.18	229.03	0.4
Heavy	220.32	221.12	0.4
Nominal thick	231.42	232.33	0.4

Figure 6 and Figure 7 depict the Campbell diagrams for the first and second whirling frequency, respectively, for three different disk masses modelled using the 4DOF and 5DOF models. Referring to Figure 6, the first vibration frequency mode corresponds to the 5DOF model changes greatly with the increase of speed as compared to the first vibration frequency of the 4DOF model. However, for the second mode, both 4DOF and 5DOF models show similar changes of FF and BF as the rotating speed increases. This occurs for each disk mass. Table 9 and Table 10 show the effect of disk mass towards critical speed., in which the nominal and nominal thick disk give slightly similar critical speed for the first and second modes. Due to more added mass in heavy disk, the whirling speed is lower compared to that of the nominal disk. The differences between the 4DOF and 5DOF models are significant for the first mode that gives the highest difference of 5.3% in the FF of the nominal thick disk. The lowest difference of 3.4% is for BF of heavy disk. Similar frequencies are obtained for nominal and nominal thick disks since both have similar mass with different geometry. However, as the mode of frequency increases to second mode, significance of torsional motion is decreasing with the highest difference of 1.6% is for FF of nominal thick disk. These results show that torsional motion needs to be considered in studying the effect of attaching disk with different mass, at least for the lowest mode of whirling frequency estimation. As the degree of mode increases, the torsional motion becomes less significant.

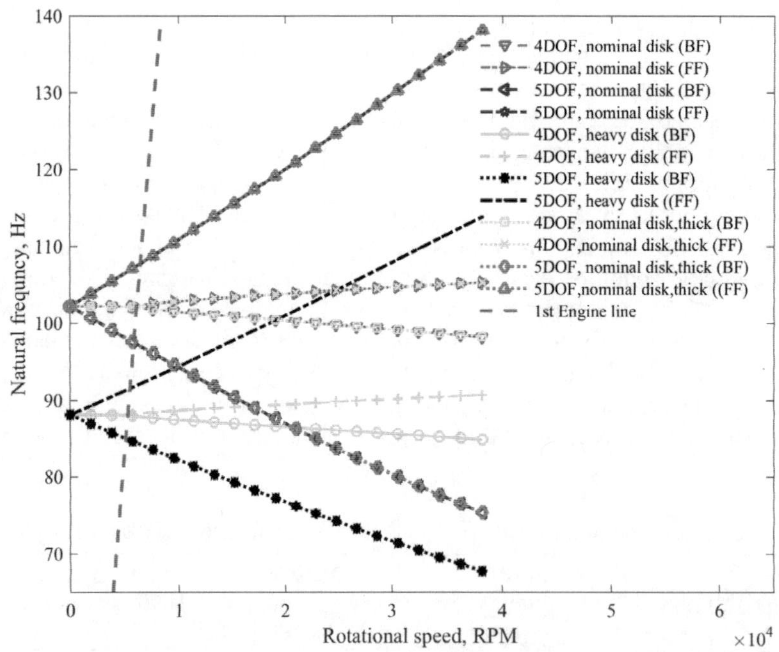

Figure 6: Effect of disk mass on the first mode of whirling frequency of shaft

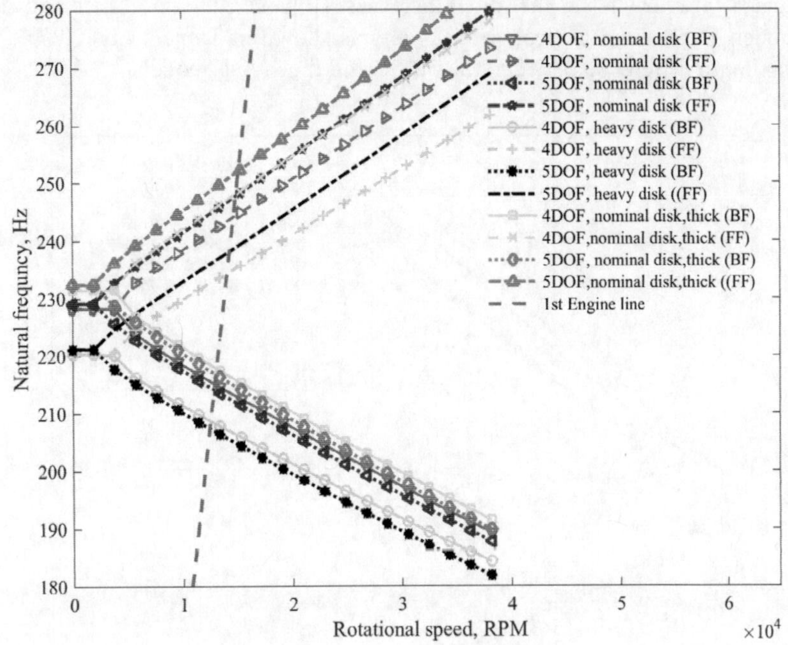

Figure 7: Effect of disk mass on the second mode of whirling frequency of shaft

Table 9: Comparison of whirling speed for different disk mass at first mode

Disk mass	Whirling direction	Whirling speed, RPM 4DOF	Whirling speed, RPM 5DOF	Differences, %
Nominal	BF	6140	5854	-4.66
	FF	6147	6465	5.17
Heavy	BF	5290	5109	-3.42
	FF	5296	5491	3.68
Nominal, thick	BF	6143	5857	-4.66
	FF	6145	6470	5.29

Table 10: The comparison of whirling speed for different disk mass at 2nd mode

Disk mass	Whirling direction	Whirling speed, RPM 4DOF	Whirling speed, RPM 5DOF	Differences, %
Nominal	BF	12920	12840	-0.62
	FF	14660	14860	1.36
Heavy	BF	12530	12460	-0.56
	FF	14070	14290	1.56
Nominal, thick	BF	13090	13010	-0.61
	FF	14910	15150	1.61

Last but not least, the effect of the number of disk attached to shaft on its free vibration is studied. Three disks with similar mass are added one after another into the shaft system. The disk is located at the same distance from each other. Figure 8 and Figure 9 show the plots of the first natural frequency mode against the rotational speed of the shafts having 1, 2 and 3 disks, respectively, using the 4DOF and 5DOF models. The plots show a big difference between the 4DOF and 5DOF models with regards to FF and BF frequencies at the same value of speed. The 5DOF model shows much wider difference between the FF and BF frequencies as the speed increases. However, the difference between the 4DOF and the 5DOF models in terms of critical speed is less as shown in Table 11 and Table 12. Table 11 shows that the maximum difference between the 5DOF and the 4DOF models is 14%.

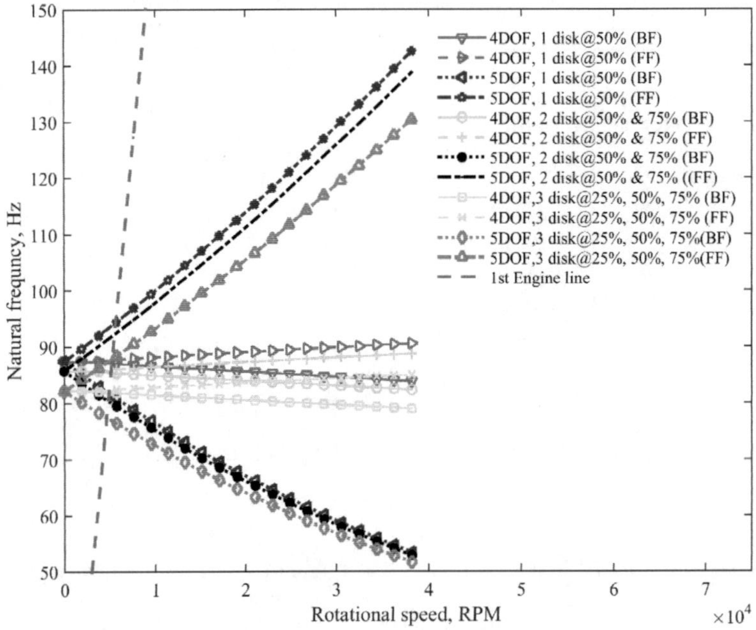

Figure 8: Effect of the number of disks on the first mode whirling frequency of shaft

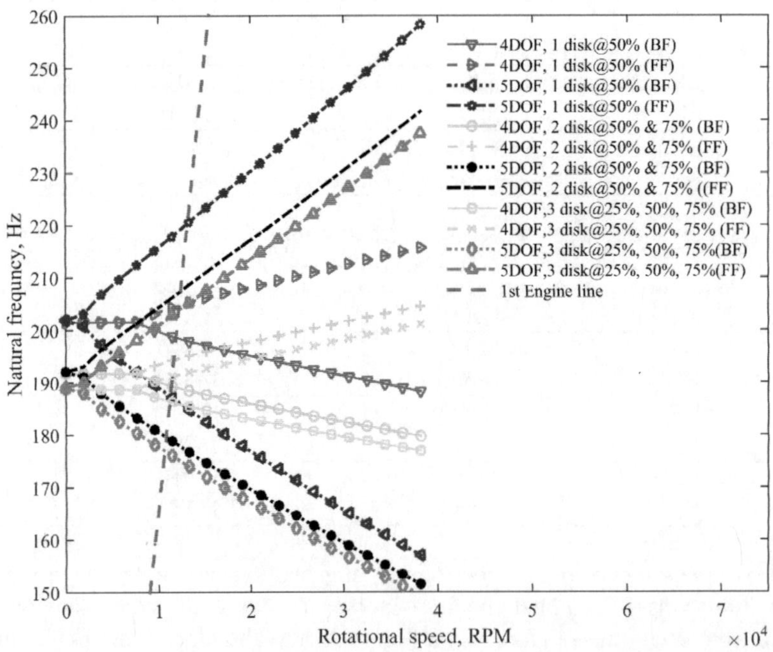

Figure 9: Effect of the number of disks on the second mode whirling frequency of shaft

Table 11: First mode of whirling speed for shaft having different number of disks

Disk quantity	Whirling direction	Whirling speed, RPM 4DOF	Whirling speed, RPM 5DOF	Differences, %
1	BF	5223	4908	-6.03
	FF	5262	5663	7.62
2	BF	5128	4832	-5.77
	FF	5175	5539	7.03
3	BF	4918	5651	14.90
	FF	4947	5281	6.75

Table 12: Second mode of whirling speed for shaft having different number of disks

Disk quantity	Whirling direction	Whirling speed, RPM 4DOF	Whirling speed, RPM 5DOF	Differences, %
1	BF	11910	11240	-5.63
	FF	12290	13230	7.65
2	BF	11350	10780	-5.02
	FF	11670	12440	6.60
3	BF	11180	10630	-4.92
	FF	11480	12230	6.53

4. Conclusion

A study has been conducted on the effect of disk parameters on the whirling frequency of high speed shaft while considering the torsional effect of the shaft. Finite element formulation in the form of the Mathieu-Hill type of equation is developed. The Bolotin's method is applied to obtain the quadratic eigenvalues equation that gives the Campbell's diagram. Some of the findings from this study include the effect of the location and mass of the disk on the natural frequency of shaft in static condition. In addition, it is also found that changing the geometry of the disk while maintaining its mass gives no effect to the natural frequency. The natural frequency of the shaft is lower for disk located at the centre of the shaft. Furthermore, the effect of torsional motion considered in the 5DOF model is significant as reflected by the wide difference between the FF and BF of the shaft. The difference in critical speed corresponding to 4DOF and 5DOF models highlights the importance to consider the torsional motion in studying the whirling frequency of the shaft with added disks.

Acknowledgement

The authors acknowledge the support in this study from Malaysia-Japan Institute of Technology and Universiti Teknologi Malaysia through the provision of the GUP Tier 2 Grant – PY/2017/00768.

References

[1] Pei Y C 2009 *European Journal of Mechanics and Solids* **28** 891–6
[2] Gerada D, Mebarki A, Brown N L, Gerada C, Cavagnino A and Boglietti A *IEEE Trans. Ind. Electron* **61** 2946–59
[3] Silber S, Sloupensky J, Dirnberger P, Moravec M, Amrhein A and Reisinger M *IEEE Trans. Ind. Electron* **61** 2990–7
[4] Crescimbini F, Lidozzi A, Calzo L and Solero G 2014 *IEEE Trans. Ind. Electron* **61** 2998–3011
[5] Tenconi A, Vaschetto S and Vigliani A 2014 *IEEE Trans. Ind. Electron* **61** 3022–9
[6] Ambekar A G 2006 *Mechanical Vibration and Noise Engineering* PHI Learning Pvt. Ltd.
[7] Lien-Wen C. and Der-Ming K 1991 *Comput. Struct.* **40** 741–7
[8] Karunendiran S and Zu J W 1999 *J. Vib. Acoust.* **121** 256–8
[9] Hosseini S A A and Zamanian M 2013 *Scientia Iranica* **20** 131-40
[10] Shahgholi M, Khadem S E and Bab S 2014 *Mech. Mach. Theory* **82** 128–40

[11] Wang S, Wang Y, Zi Y and He Z 2015 *J. Sound Vib.* **359** 116–35
[12] Jalali M H, Ghayour M, Ziaei-Rad S and Shahriari B 2014 *Measurement* **53** 1–9
[13] Seshendra K V S and Rao S B 2012 *Int. J. Mech. Appl.* **2** 10–13
[14] Nelson H D 1980 *J. Mech. Des.* **102** 793
[15] Bolotin V 1964 *The Dynamic Stability of Elastic Systems* Holden−Day
[16] Rao J S 2011 *Finite Element Methods for Rotor Dynamics* in *History of Rotating Machinery Dynamics* Springer Netherlands

IOP Conf. Series: Materials Science and Engineering **270** (2017) 012032 doi:10.1088/1757-899X/270/1/012032

Volumetric calculation using low cost unmanned aerial vehicle (UAV) approach

A A Ab Rahman[1,*], K N Abdul Maulud[1,2], F A Mohd[2], O Jaafar[2] and K N Tahar[3]

[1]Earth Observation Centre, Institute of Climate Change (IPI), Universiti Kebangsaan Malaysia, 43600 Bangi, Selangor, Malaysia
[2]Department of Civil and Structural Engineering, Faculty of Engineering and Built Environment, Universiti Kebangsaan Malaysia, 43600 Bangi, Selangor, Malaysia
[3]Department of Surveying Science and Geomatic, Faculty of Architecture, Planning and Surveying, Universiti Teknologi MARA, Malaysia

*p90513@siswa.ukm.edu.my

Abstract. Unmanned Aerial Vehicles (UAV) technology has evolved dramatically in the 21st century. It is used by both military and general public for recreational purposes and mapping work. Operating cost for UAV is much cheaper compared to that of normal aircraft and it does not require a large work space. The UAV systems have similar functions with the LIDAR and satellite images technologies. These systems require a huge cost, labour and time consumption to produce elevation and dimension data. Measurement of difficult objects such as water tank can also be done by using UAV. The purpose of this paper is to show the capability of UAV to compute the volume of water tank based on a different number of images and control points. The results were compared with the actual volume of the tank to validate the measurement. In this study, the image acquisition was done using Phantom 3 Professional, which is a low cost UAV. The analysis in this study is based on different volume computations using two and four control points with variety set of UAV images. The results show that more images will provide a better quality measurement. With 95 images and four GCP, the error percentage to the actual volume is about 5%. Four controls are enough to get good results but more images are needed, estimated about 115 until 220 images. All in all, it can be concluded that the low cost UAV has a potential to be used for volume of water and dimension measurement.

1. Introduction

Nowadays, all development work has to perform surveying process, which consists of reconnaissance and topographical tasks. The problem with conventional surveying method is related to the calculation of volume for an object or area. In short, volume is defined as the amount of three dimensional space by some closed boundary. For example, the space that a substance such as solid, liquid, gas or plasma or shape occupies. In the other hand, volume of a container is generally understood to be the capacity of the container. The current method requires a team of people, which usually consists of the surveyor, his/her assistances and a draft person to process the data, to manually perform the surveying task. This task is time consuming, requires huge cost and laborious.

It is believed that unmanned aerial vehicle (UAV) can be alternatively used for obtaining the aerial images. This can be a much better alternative to the current method and makes the surveying process becomes more efficient, less time consuming and also cost effective for small budget projects. In fact, the use of all resources to generate the volume at the interested area is expected to be reduced using UAV. In this study, the task conducted with the UAV to calculate the volume is similar to that of the land survey method. It is anticipated that by using UAV, the surveying process can be completed with less time, cost and manpower. The common conventional method to calculate volume in surveying is tacheometry method. A study that has been conducted with this method showed that the user needs to determine which area is significant and obtain the height of the surface at that survey area [1]. All data is recorded using equipment such as total station and mini prism. After the site survey is completed, all collected data is processed using a software. Commonly, land can be represented as an irregularly shaped polygon. Different methods such as graphical method, coordinate method and planimeter can be used to determine the area of irregular polygons. Among these three methods, the most popular method is the coordinate method, which is typically applied to compute catchment area, drainage area, volume and others [2]. The volume can then be generated by a software after calculating the area and contour. The volume calculation using total station equipment requires the height on the surface or on the polygon area [3].

Nowadays, the development of the UAV has increased its potential use in surveying work. UAV is capable of performing complex tasks in limited environment, small survey area with low cost budget [4]. It has a great potential in many applications such as remote sensing, disaster effect, search and rescue, surveillance, atmospheric survey and others [5]. For this study, the aim is to highlight the UAV capability in photogrammetry to become a suitable alternative way to generate volume for surveying process, which is done by taking images at the site area.

2. Literature Review

Close range photogrammetry has close connections with other measurement techniques and also with fundamental sciences such as mathematics, physics, information sciences or biology. It has substantial relations with aspects of graphics and photographic science [6]. Its product is the sequences of images during the image acquisition. For instance, these include computer graphics, computer vision, digital image processing, computer aided design (CAD), geographic information systems (GIS) and also cartography [7]. Traditionally, close range techniques also have strong associations with the surveying methods, particularly in areas of adjustment methods and engineering surveying. There is increasing application of photogrammetry in industrial metrology and quality control [8, 9].

The data obtained from engineering design surveys are used to calculate earthwork quantities. The standard method for calculating earthwork quantities is the average end area method [10]. The area of a triangle is the surface within three sides. One of the objectives of land surveys is to determine the area of a parcel of land. There are a number of methods used for calculating areas, including the use of double meridian distances and coordinates. This is the straight forming method to calculate the volume of objects and now, digital close-range photogrammetry is an alternative method to volume calculation [11]. Time and accuracy are two important factors that reflect on the perfection of volume calculation for road project, mining enterprise, geological works and building applications [12]. In Geodetic, current surveying methods have been insufficient for volume calculation of the objects where it needs to accomplish the task in a short time or risk area. Alternatively, multiple overlapping images taken from different perspectives can produce measurements to create accurate three dimensional models of objects [13]. Information on the position of the camera is not necessary because the geometry of the object is established directly from the images [14]. With this in mind, the photogrammetry techniques will allow the user to convert images of an object into a three dimensional model. A study has shown that a minimum of two pictures of an object is needed to accomplish this using a digital camera with known characteristics (i.e. lens focal length, imager size and a number of pixels) [15]. If the user can indicate the same three object points in the two images and known dimensions, other three dimension points in the images can be determined [16]. The output of this method will then be the volume unit.

IOP Conf. Series: Materials Science and Engineering **270** (2017) 012032 doi:10.1088/1757-899X/270/1/012032

3. Material and Method

This study is splited into four principle phases: Phase 1 is about preliminary study, Phase 2 is about data collection, Phase 3 is about data processing and finally, Phase 4 is about data analysis. Figure 1 depicts the stream outline of the methodology applied in this study. Flight planning is planned after the site reconnaissance stage. In this study, the site area is a water tank and the route for image acquisition is around the water tank. Images were taken at the site, which is located nearby Kolej Mawar, UiTM Shah Alam using a UAV drone. The flight altitude was set at 20m from the ground and the orientation was using the point of interest (POI) to navigate the route of the drone. The image acquisitions were taken automatically.

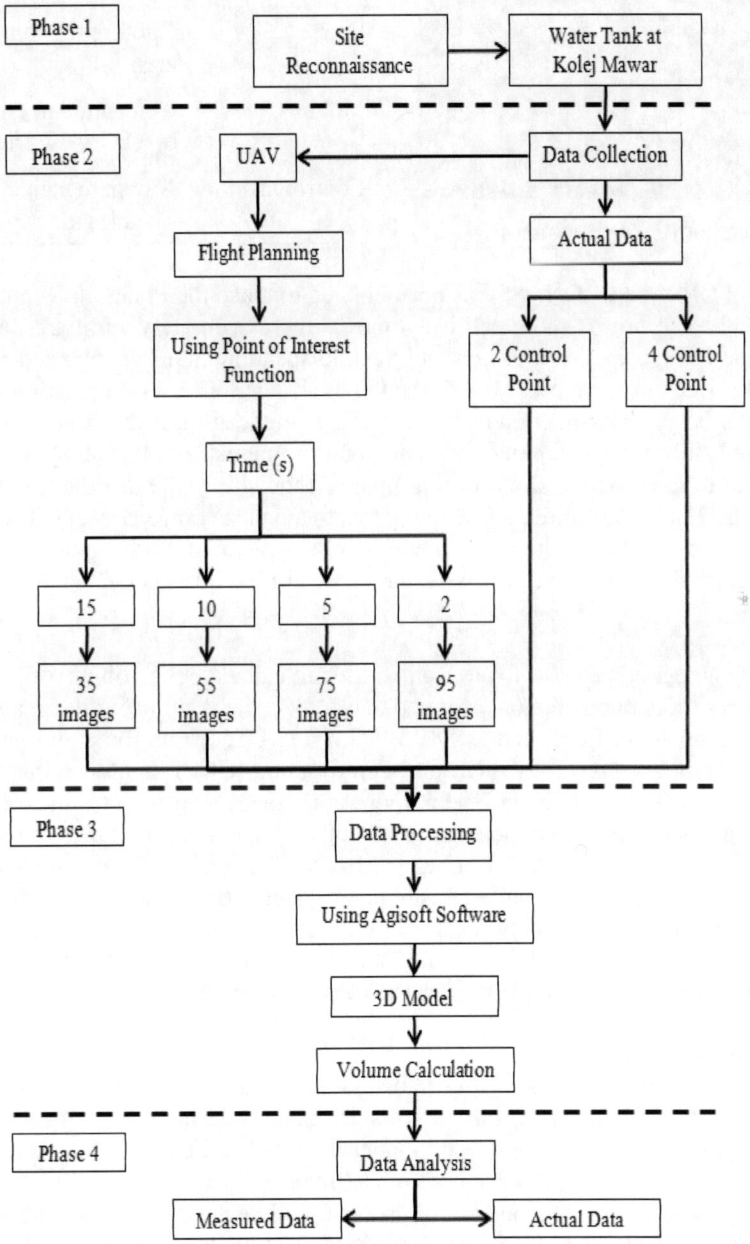

Figure 1: Methodology flow chart

3.1 Equipment

Phantom 3 Professional UAV, as shown in Figure 2, used for the image acquisition must be supported with a laptop and a controller. This is to ensure that any problems can be detected immediately if there is any damage or system failure during the flight mission. Checking of all hardware before starting the work is important to make sure the image acquisition process is successful. The component of UAV for this research is tabulated in Table 1.

Figure 2: Phantom 3 Professional [17]

Table 1: UAV components and function

Components	**Function**
Drone	Capture the images of the site area
Mobile	Connected to the drone as a display view from the drone.
Controller	Navigate manually the flight planning during the drone fly.

Combination of GPS and GLONASS has been utilized to make the Phantom 3 completely aware of its location to the user. It hovers more precisely, moves more accurately and locks onto the satellites faster. Through the DJI Go application, user can track its location on a live map and record its takeoff point such that the user can bring it back with the tap of a finger. The vision positioning system in the Phantom 3 literally helps to fly it. Moreover, for DJi Go application, it consists of several functions: follow me, course lock, waypoints, home lock and point of interest. For this study, the function "point of interest" is used to navigate the flight during images acquisition. Meanwhile, the flight is planned around the water tank in vertical direction with regards to the close range concept. The specification of the UAV camera is 35mm, hence the tank features can be detected by the captured images. After the data acquisition stage, the images are processed into three dimensional model using Agisoft Software.

3.2 Software

There are a lot of photogrammetric software can be used to calculate the volume of the object. In fact, the capability of these software is not only limited to calculate the volume but it can also produce three dimensional modelling application such as DEM, DSM and DTM from the aerial photo [10]. In this study, in order to obtain good results, the Ground Control Point (GCP) should be observed on the earth surface and recorded WGS84 projection. The purpose of these GCPs is to get x and y coordinates to tie with the images. In this research approach, the minimum and maximum numbers of GCPs are two and four, respectively. These coordinates were obtained from the traversing technique. The software used is a stand-alone software package that performs photogrammetric processing of digital images and generates three dimensional spatial data for use in GIS applications, cultural heritage documentation, and visual effects production, as well as for the indirect measurement of objects with various scales. Meanwhile, Agisoft software is used to perform the volume computation.

4. Results and Discussions

The developed three dimensional model appeared smoothly at the bottom of the water tank. Based on the result, the identified problem is the missing of some images during the image acquisition process. The result of the developed three dimensional model is shown in Figure 3. The volume of the water tank was calculated after the three dimensional modelling stage was successfully done. The first step to generate the volume is to select the feature or area that will be generated. The calculation of volume based on 35 images with two control points gives 4837.84 m^3. Compared with the actual volume of 4027.659 m^3, it seems that 35 images were not enough to get an accurate estimation. Therefore, this step is repeated with a different number of images. Next, four control points are used with different numbers of UAV images captured to get the volume estimation.

Figure 3. Constructed three dimensional model of the water tank

From the results shown in Figure 4 and Table 2, the number of control points and UAV images will affect the results of the processed images in measuring the water tank's volume. Based on the results, it shows that increasing the number of control points gives a big improvement in the volume estimation. In similar manner, increasing the number of UAV captured images gives the same impact. However, the estimated volume is still far from the actual value of 4027.659m^3.

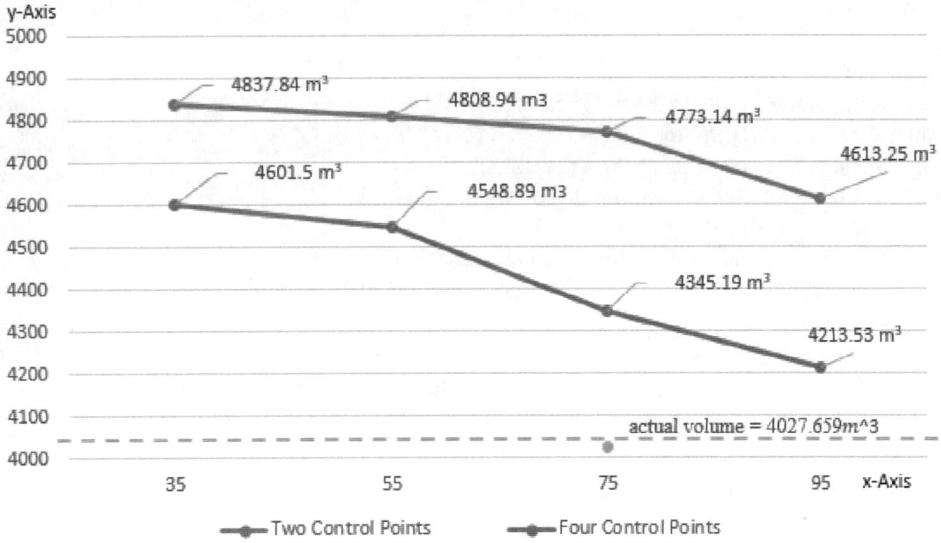

Figure 4: Relationship between number of UAV captured images and estimated volume

In Table 2, the volume difference refers to the difference between the estimated and actual volume. From the tabulated data, it implies that having 95 UAV captured images with four control points has the lowest difference value (i.e. 185.87m^3 or 5% of error). In comparison, with the same number of images but with two control points, the resultant difference between the estimation and actual volume is 810.18m^3. Furthermore, the more images captured, better estimation results can be obtained.

Table 2: Error percentage comparison

Number of UAV Images	Volume difference (m^3)		Error (%)	
	Two GCPs	Four GCPs	Two GCPs	Four GCPs
35	810.18	573.84	20	14
55	781.28	521.23	19	13
75	745.48	321.53	18	8
95	585.59	185.87	14	5

5. Conclusion

This paper demonstrates the use of a low cost UAV in image acquisition process at the water tank site located near Kolej Mawar, UiTM Shah Alam. Volume of the water tank has been calculated using the Agisoft Software. All in all, the results show that having more captured images will give better volume estimation. In addition, the higher number of GCP used will also tremendously improve the estimation results. With 95 images and four GCPs, the estimation error of the water tank's volume is reduced to only 5%. It is projected that 115 to 220 images are possibly needed to get to the actual volume. From this study, it has been shown that low cost UAV has the potential to calculate the volume of water and dimension measurement.

Acknowledgement

The authors acknowledge the funding for this study from Trans Disciplinary Research Grant Scheme (TRGS/1/2015/UKM/02/5/1) and Top Down Research Grant (TD-2016-012). In addition, the authors also want to acknowledge the contribution by Earth Observation Centre, Institute of Climate Change, UKM in this study.

References

[1] Remondino F and Fraser C 2006 *ISPRS* **47** 266-71
[2] Tahar K N 2012 *IJRRAS* **46** 780-5
[3] Gulgen F and Gokgoz T 2009 *ISPRS* **4** 92-5
[4] Perez M, Aguera F and Carvajal F 2013 *ISPRS* **56** 311-5
[5] Takagi M 2008 *ISPRS* **75** 25-30
[6] Tahar K N and Ahmad A 2011 *PERTANIKA* **28** 387-396
[7] ESRI 2011 *http://gis.stackexchange.com/questions/5701/what-is-the-difference-between-dem-dsm-and-dtm*
[8] Tahar K N and Ahmad A 2013 *World Academy of Science, Engineering & Technology* **7** 48-52
[9] Akay A E, Karas I R, Sessions J, Yuksel A and Bozali N 2008 *ISPRS* **5** 78-81
[10] Blaha M, Eisenbeiss H, Grimm D and Limpach P 2011 *IAPRSS* **7** 131-6
[11] T Bektas 2006 *Omega* **34** 209-19
[12] Tahar K N and Ahmad A 2012 *Internal Journal of Physical Science* **7** 1300-6
[13] Ruzgiene B, Berteska T, Gecyte S and Jakubauskiene E 2014 *ELSEVIER* **5** 619-27
[14] Kraus T and Reinartz P 2007 *ISPRS* **65** 78-81
[15] Takagi M 2009 *ISPRS* **2** 56-60
[16] Papasaika H, Poli D and Baltsavias E 2008 *IAPRSS* **21** 811-7
[17] Wu C 2011 *http://www.cs.washington.edu/homes/ccwu/vsfm/*

IOP Conf. Series: Materials Science and Engineering **270** (2017) 012033 doi:10.1088/1757-899X/270/1/012033

Study of air traffic over KLFIR

I F Nusyirwan* **and J Mohd Rohani**

Faculty of Mechanical Engineering, Universiti Teknologi Malaysia, Malaysia.

*istaz@utm.my

Abstract. This paper shares the overview of the work currently being conducted with the Department of Civil Aviation Malaysia related to the air traffic. The aim is to study air traffic performance over KL and KK FIR, and the area of interest in this paper is the Kuala Lumpur Flight Information Region (KLFIR). The air traffic performance parameters includes general air traffic movement such as level allocation, number of movements, sector load analysis and also more specific parameters such as airborne delays, effects of weather to the air movements as well as ground delays. To achieve this, a huge effort has been undertaken that includes live data collection algorithm development and real time statistical analysis algorithm development. The main outcome from this multi-disciplinary work is the long-term analysis on the air traffic performance in Malaysia, which will put the country at par in the aviation community, namely the International Civil Aviation Organization (ICAO).

1. Introduction

The aviation industry is huge. There is a daily movement of more than 2000 flights over the Malaysian airspace. Such big traffic can create significant challenge in air traffic management. The solution is to have short-term, mid-term and long-term strategies [1]. Short-term strategy involves understanding the current available technologies, limitations and technical skills. It is important to identify the strengths, weaknesses and potential opportunities, which will help to prepare for mid-term strategy that involves development of the required human capital and technology. Having the right human capital and also technology enables the independent strive for long-term strategy in improving air traffic management procedures, processes and capabilities.

Among common issues in air traffic management include bad weather, airspace closure, aerodrome closure and inefficient schedules. In general, this can be seen as a complex and multi-faceted problem. The International Civil Aviation Organization (ICAO) is currently spearheading several initiatives to improve efficiency of the air traffic such as Air Traffic Flow Management and Collaborative Decision Making (ATFM/CDM) [4]. For ATFM/CDM initiatives to succeed, all relevant data is required before it can be processed and analyzed to obtain the right information. With the right information, planning of air traffic movements can be made more efficient and seamless between airports, and subsequently delays can be minimized and improved. With this in mind, the aim of this paper is to share a general overview of the work done on the air traffic data in Kuala Lumpur FIR that is illustrated in Figure 1. A brief explanation on the methodology is also provided and some selected outputs are shared.

2. Methodology

Several tools have been developed in this study, which will be utilized for data collection, data storage and data analysis. The methodology that is followed in this study is as shown in Figure 2.

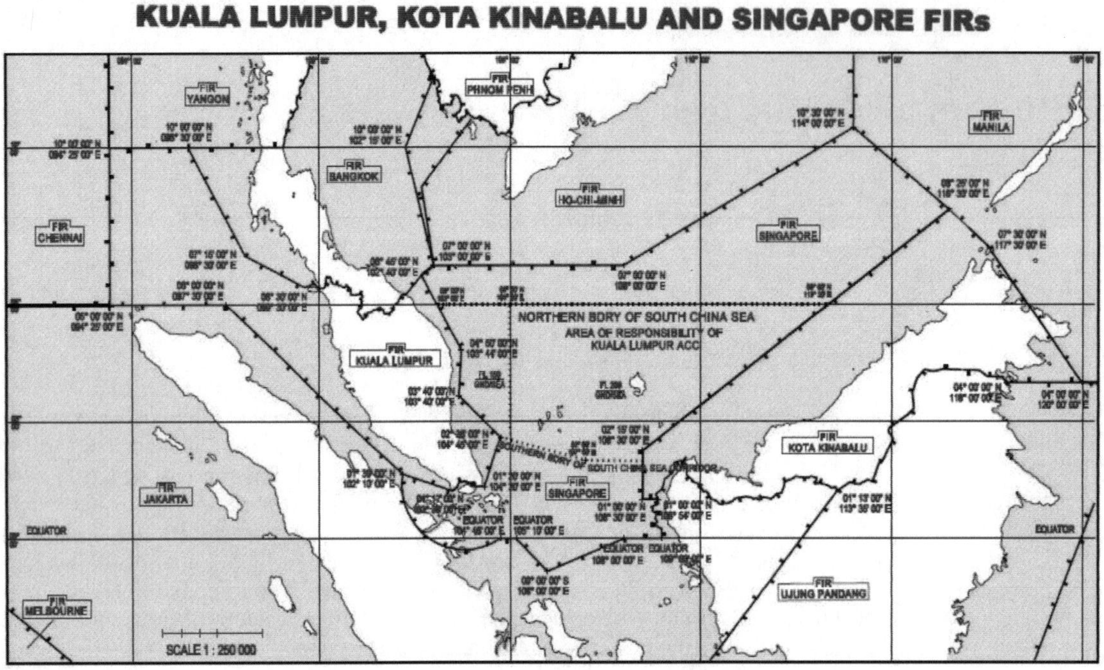

Figure 1: Kuala Lumpur FIR

Figure 2: Flowchart of the data analysis process

The developed data collection tools are a set of computer programs that read and decode raw data from various sensors in real time. To decode these raw format data, the protocol used has to be known. For raw radar data, the protocol is the EuroControl ASTERIX [2]. Some of the data are in the form of specialized text-based AFTN format such as the flight plans, NOTAMs and METARs. The decoded data are then stored into a dedicated database. In this project, MySQL is used as the database server.

On the other hand, the data analysis tools are consisted of computer programs that perform various statistical analyses. These tools are developed based on evolving requirements that are tailored to the given inputs by the Air Traffic Control (ATC) and airlines. An example of data analysis is the number of aircraft being delayed during their approach to the Kuala Lumpur International Airport (KLIA). To obtain this information, specific algorithm has to be designed. The algorithm will perform the required data collection that would give enough data to perform the analysis.

3. Overview of Malaysian Airspace Air Traffic Performance

The number of air traffic movements over KLFIR is depicted in Figure 3. The project started its data collection initiatives in year 2012. Since then, an increase in the number of air traffic movements over KLFIR is observed. This increase is contributed to several factors, including addition of new aircraft into the fleet of the major airlines within the region (not only Malaysia-based airlines but also airlines

IOP Conf. Series: Materials Science and Engineering **270** (2017) 012033 doi:10.1088/1757-899X/270/1/012033

from neighboring countries) and increase of flight destinations due to increase of passengers' demands. New low-cost airlines also contribute to the increase of traffic in the region.

On daily basis, the average number of instantaneous aircraft counts (IAC) observed within KLFIR is as shown in Figure 4. IAC is defined as the number of aircraft in a given time instant. Since the time is recorded in Universal Time Coordinated (UTC), 0000 UTC is 0800 local Malaysia time. From 6am, the number of detected aircraft started to increase until it reached the peak period around 2-4pm. The AIC then started to decrease after 9pm and reached its lowest point in between 2-3am in the morning. This data is mainly used to determine the suitable workload for the air traffic controllers. An example is given in Figure 5, which shows the IAC for Sector 3. Sector 3 is an area in KLFIR that is located in the eastern part of Johor. A group of air traffic controllers is responsible to manage all aircraft entering Sector 3. In order to ensure that the workload of the controllers is within permissible limit, the IAC in Sector 3 is monitored in real time. As indicated in Figure 5, the peak time (i.e. the IAC is more than 11 aircraft) was between 12pm and 11pm. At this period of time, the workload of the air traffic controller is at the highest point and special attention is required to ensure that the controllers are able to work efficiently.

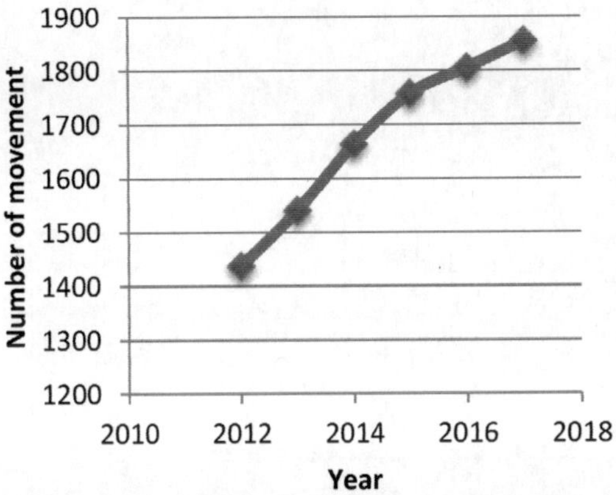

Figure 3: Average increase of movement in KLFIR

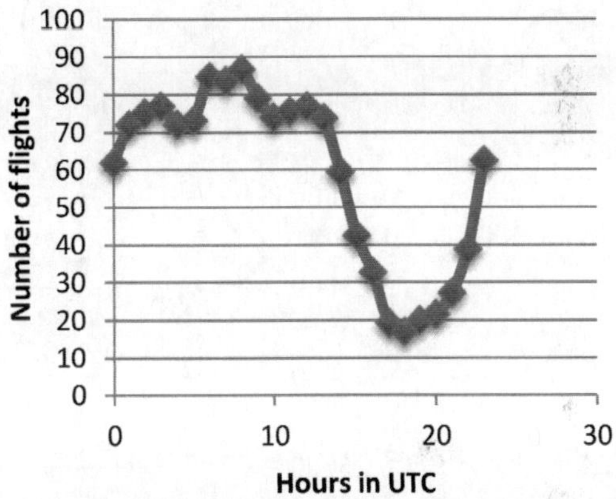

Figure 4: Daily movements in KLFIR (data in 2017)

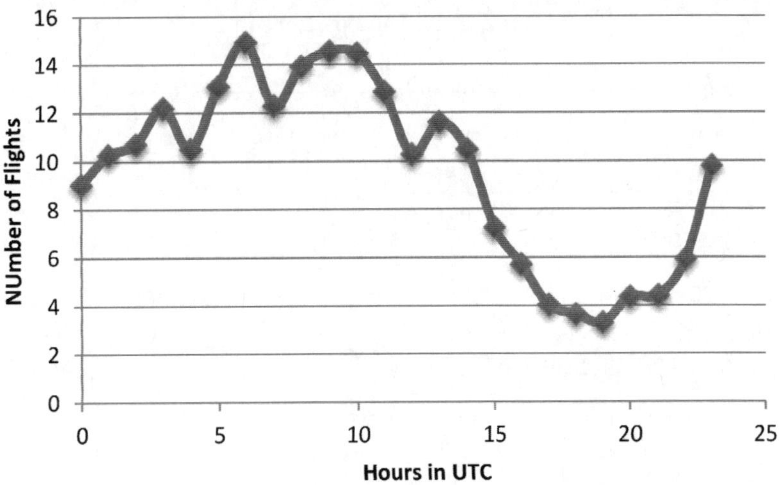

Figure 5: IAC recorded in Sector 3 KLFIR

It is observed in Figure 6 and Figure 7 that there is an increase in the number of departures between the years 2014 and 2017. The introduction of the Kuala Lumpur International Airport 2 (KLIA2) has increased the capacity of KLIA to accommodate more aircraft. The rationale behind the move can be seen in Figure 8. Figure 8 shows the number of flights experiencing airborne delays between 10 to 30 minutes. Prior to KLIA2, there were only two runways in operation: one for departure and the other one for arrival. This configuration has created a demand-capacity problem since demands for arrivals and departures were more than the actual capacity of the airport.

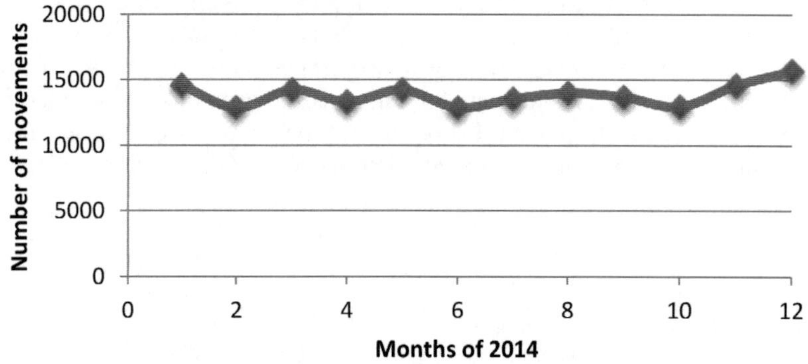

Figure 6: KLIA departures in 2014

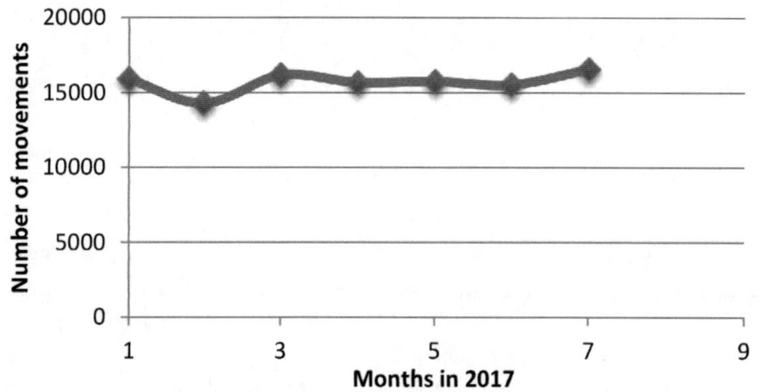

Figure 7: KLIA departures in 2017

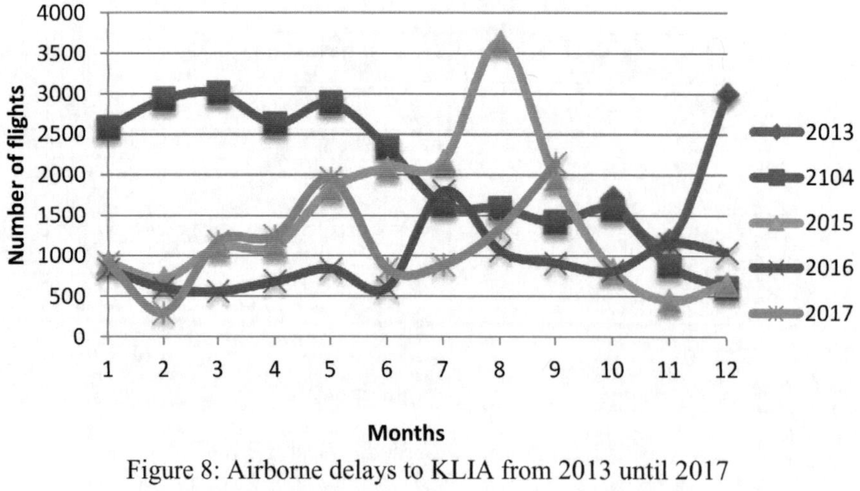

Figure 8: Airborne delays to KLIA from 2013 until 2017

IOP Conf. Series: Materials Science and Engineering **270** (2017) 012033 doi:10.1088/1757-899X/270/1/012033

On 1st May 2014, KLIA2 was officially in operation. This in turn has changed the operation of the runways around the airport. Previously, KLIA used runway 32L/14R and 32R/14L for departures and arrivals, respectively. Depending on the wind conditions, one runway is used exclusively for departure and the other one is for arrival. After KLIA2 is operational, runway 33/15 is used for both departures and arrivals. Despite the mixed mode operation, the overall capacity of the airport has improved. Low-cost airlines used runway 33/15 most of the time, and occasionally the other runways if the conditions permit. Figure 8 shows a significant drop of flight delays that are over 10 minutes. The transition was on the 1st May 2014, which coincides with the introduction of KLIA2. An improvement from delays involving 2000 or more flights per month to less than 2000 flights per month is an achievement.

In Figure 8, it can be deduced that there is a spike in the number of delays in the months of June until August. Similar trends are observed in years 2015, 2016 and 2017. The spike was most probably caused by the increase in number of flights over KLIA due to summer schedule. Airborne delays in 2015 were observed to be extraordinary. The obvious reason was the haze season. As Time magazine reported, the haze problem in 2015 was the worst in history [3]. Many aircraft were put on a holding pattern at longer period of time or being diverted elsewhere. This is a good example where the data collection and analysis system was able to correlate airborne delays with major event in KLFIR.

4. Conclusion

The development of data collection and analysis system is a difficult task. It takes a group of experts from multiple disciplines to work together to develop such a complex system at a large scale, which include features such as live data feed, raw data processing and real-time analysis with minimal data loss. It took around six years of development to get from the conceptual idea to a working model. The understanding of aviation technology is paramount for this project to succeed. The data is stored in a large database for further processing. Algorithms have been developed to find optimal methods so that the processing time is kept to the minimum. The use of load balancing is important for this project. Each task is equally distributed to keep the computing load as low as possible. The output from each analysis is stored in another database for fast retrieval. A good knowledge is database programming is essential. An optimal database design would allow for fast storage and retrieval of data. The project has managed to collect a lot of information related to air traffic movement, some of which are shared in this paper.

Acknowledgement

The project is a joint research work with Department of Civil Aviation (DCA) Malaysia. The project team wishes to thank DCA Malaysia for the cooperation and ideas that make this project a reality. The project is still far from its objective and more work still need to be done.

References

[1] International Civil Aviation Organisation (ICAO) 2017 *Global Air Navigation Plan*
[2] EuroControl 2012 *EUROCONTROL Specification for Surveillance Data Exchange - Part 1 All Purpose Structured EUROCONTROL Surveillance Information Exchange (ASTERIX)*
[3] Jenkins N 2015 *http://time.com/4060786/haze-singapore-indonesia-malaysia-pollution/*
[4] International Civil Aviation Organisation (ICAO) 2012 *Manual on Collaborative Decision Making (CDM)*

IOP Conf. Series: Materials Science and Engineering **270** (2017) 012034 doi:10.1088/1757-899X/270/1/012034

A review on feasibility study of ultrasonic assisted machining on aircraft component manufacturing

M S A Hafiz[1], M H A Kawaz[1], W N F Mohamad[2], M S Kasim[1,*], R Izamshah[1], J B Saedon[3] and S B Mohamed[2]

[1]Faculty of Manufacturing Engineering, Universiti Teknikal Malaysia Melaka, Hang Tuah Jaya, 76100 Durian Tunggal, Melaka, Malaysia.
[2]Department of Manufacturing Technology, Faculty of Innovative Design and Technology, Universiti Sultan Zainal Abidin, Terengganu, Malaysia.
[3]Faculty of Mechanical Engineering, Universiti Teknologi MARA, Shah Alam, Selangor, Malaysia.

*shahir@utem.edu.my

Abstract. Inconel 718 has been widely used in aerospace because of its excellent mechanical properties such as good corrosion resistance, strong creep resistance and high fatigue strength. However, these excellent properties also lead to heavy tool damage and high cutting force in the milling process. There is no reported investigation on ultrasonic assisted machining (UAM) of Inconel 718 parts. In this paper, UAM is proposed as the potential technique to reduce tool damage and the cutting force of Inconel 718 parts. This review paper provides an overview of UAM to investigate the relationship between the tool wear and the cutting force with ultrasonic vibration compared to without ultrasonic vibration assisted. Throughout the study, the UAM scopes are related to the tool life of coated carbide insert, the force generated during the cutting process and also the final surface finish of the workpiece by using various parameters during the machining activity.

1. Introduction

Superalloy has been highly used in extreme applications due to its good mechanical properties. 80% of the superalloy usage such as Inconel 718 and Titanium is in the aerospace industry. Gas turbine engine (shown in Figure 1) uses nickel alloys for approximately 45-50% of its material [1-2]. Due to superior properties such as good corrosion resistance, strong creep resistance and high fatigue strength, nickel-based alloys is preferred to be used for the rotating and static components in the hottest section of aero engines such are the blades and disks in high-pressure turbine and compressor [3]. Inconel 718 (53.3% Ni, 18.5% Fe, 4.98% Nb, 3.03% Mo) [4] is considered as one of the metals that is very difficult to cut. Throughout the years, machinability of Inconel 718 still is the challenge for many researchers to fulfill higher manufacturing efficiency nowadays. Due to increased demands, this leads to a great innovation of advanced machining process. Cutting force is the crucial variable applied as the tool wear indicator among the indirect on-line tool wear monitoring methods. Measurement and study of the relationship and correlation between the means of tool wear and the cutting force is important for the development of the effective process or tool condition strategy [5-7].

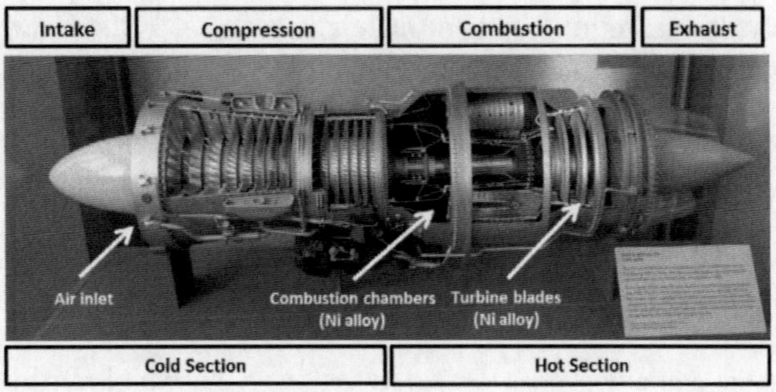

Figure 1: Cross section of Pratt & Whitney JT3 turbojet engine [8]

Chatter is defined as regenerative vibration when the width of the chip is too large in comparison to the dynamic structure in a particular spindle speed [9]. Regenerative vibration is generated as a result of interaction between the cutting force and structural dynamics of the machine [10], and is considered as a dominant type of chatter in milling operation. If the mechanism is properly applied, the advanced technology such as Ultrasonic Assisted Machining (UAM) can provide positive effects on vibration suspension [11]. A recent study implies that ultrasonic vibration on the milling tool holder decreases the measured cutting force by 62%, which is small enough that the tool does not measurably deflect, enabling it to mill the surface square. This cutting force reduction leads to various potential benefits. In addition to potential improvement in feed rate, tool life and accuracy, less cutting force also means reduced power consumption [12].

As Inconel 718 is hard to cut, UAM will help the tool life where the cutting tools tend to get worn out faster during its conventional milling machining. UAM has shown a positive impact for the hard material surface roughness on finishing such as diamond turning of stainless steel [13] and ceramics [14]. In this paper, a review on the feasibility of UAM is done through previous experimental research and their results compared to conventional milling (CM) process on the Inconel 718 superalloy.

2. Machining of Inconel 718

Inconel 718 is a nickel-based superalloy that exhibits desirable properties over wide temperature range and it is widely used in industry. However, its excellent mechanical properties such as good corrosion resistance, strong creep resistance and high fatigue strength [15] are compromised by shear strength, strong work hardening and high-temperature tensile strength that shorten the tool life and cause severe work-surface damage during CM [16]. To improve machinability of superalloys, ultrasonic vibration assisted technique has been employed in the milling [17], drilling [18], grinding [19] and turning [20] processes. Although this technique has been implemented, it still has not been carried out for Inconel 718 in milling process yet.

2.1. Traditional milling process

Traditional milling process is divided into two types: climb and conventional milling. Conventional milling has a significant benefit when machining hard materials such as cast iron or hot rolled steel. In contrast, performing a climb mill on hot-rolled steel can result in chipped cutting edges because of the hardness of that outer layer and the more aggressive way the tool engages the rough surface, causing more deflection and potentially heavier chips. Since the tool deflection with a conventional mill tends to be parallel to the tool, it engages the rough surface at a more forgiving rate. Comparison between both milling types on Inconel 718 has been done [21, 22]. Through the research, it has been found that the tool flank wear propagation in the up milling (conventional milling) operation is more rapid than that in the down milling operations (climb milling) as shown in Figure 2. On the other hand, the chip morphology is different in both operations, where the up-milling operation produces a segmented chip with typical saw-tooth shape and the down-milling operation produces a discontinuous serrated chip. In general, tool wear is increased with increasing cutting speed, feed rate and depth of cut (DOC).

The results of conventional milling on hard materials have led to the UAM technique, which could be the solution to some of the problems encountered by CM. During the CM of Inconel 718, it is found that the tool wear is increased as the DOC, feed rate and cutting speed are increased [21]. The results have shown that the tool flank wear rapidly occurred, thus affecting the tool performance, tool life and workpiece surface. Significant notch and pitting are the predominant failure modes that are located near the line of DOC. Since Inconel 718 is a super-heat-resisting alloy, the tool life is shorter during the interrupted than continuous cutting process since the built-up edge (BUE) is removed. This causes the coating film to flake on the cutting tool, particularly during the machining of high tensile strength [23, 24].

The cutting tool is the equipment that is used to remove a material layer by layer. At a certain time, through the contact between the cutting tool and the workpiece, flaking and chipping (tool wear) will occur due to the interrupting cutting force (e.g. major failure mode during end milling of Inconel 718 due to interrupted cutting force as illustrated in Figure 3). During the dry cutting of Inconel 718, the flank faces and rake from welding and adhesion of workpiece are dominant wear modes [25] while the predominant failure modes of cemented carbide tool insert is found on flank wear and microchipping that affects the tool's life and performance [26].

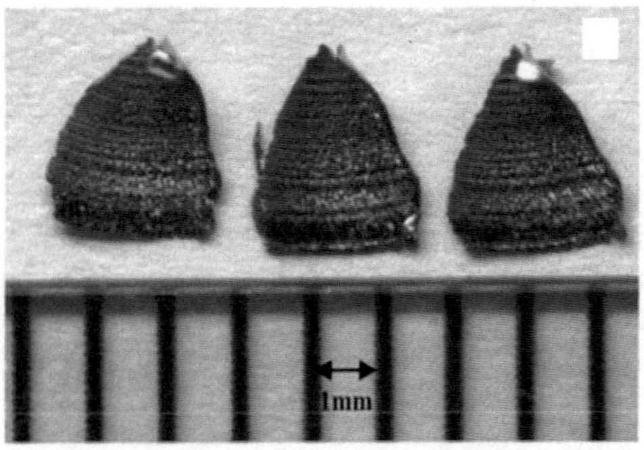

Figure 2: Discontinuous serrated chip during end milling [21]

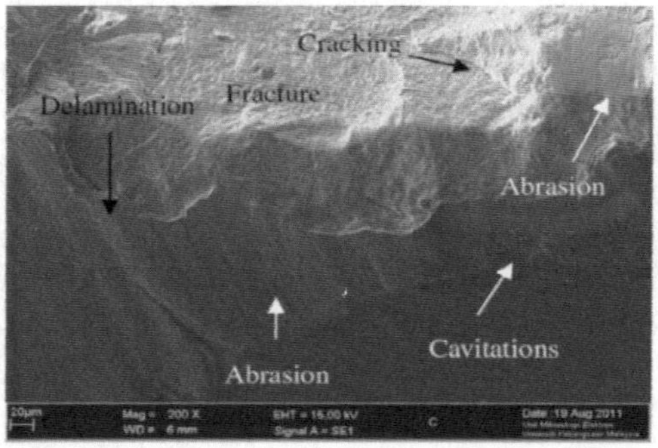

Figure 3: Major failure mode during end milling of Inconel 718 due to interrupted cutting force [24]

Figure 4 shows the chipping that occurred on the rake face due to interrupted nature of the cutting process. During the cutting of Inconel 718 using band-saw, the dominant wear mode is the flank wear (shown in Figure 4(a)). It has also been found that adhesion of the workpiece material onto the worn flanks promotes the adhesive-type wear flank wear to be developed at the cutting edge due to abrasive action between the flank face and machined workpiece (as shown in Figure 4(b)).

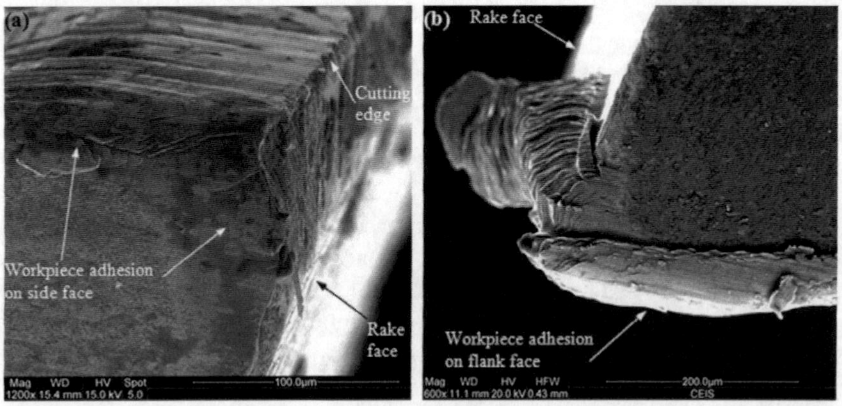

Figure 4: Adhesion of the workpiece on the: (a) side face of tooth corner and (b) flank face [27]

IOP Conf. Series: Materials Science and Engineering **270** (2017) 012034 doi:10.1088/1757-899X/270/1/012034

Since Inconel 718 is difficult to cut, a lot of heat is generated while machining it at high speeds. During the investigation of Inconel 718 machining in high-speed, it is found that at high cutting speed, heavy notching of major tool failure occurred because the tool got worn-out faster [28]. Wear patterns observed at these cutting speeds are uniform along the nose radius with some traces of peeling coating. In the same investigation, very high amount of heat in the cutting zone is generated while burns marks are visible on the cutting tool. The heat is diverted into the cutting tool because of the poor thermal conductivity of Inconel 718 and this could be one of the major reasons that leads to tool wear.

The illustration for the effect of tool geometry on residual stresses in machining IN-718 is shown in Figure 5. The cutting edge geometry also produces a ploughing effect due to the honed radius (cutting edge radius) compared to the chamfered tools that are said to increase the compressive nature of the residual stresses [29]. Coelho et al. (2004) [30] have also studied the effects of the tool chamfer edge geometry on residual stress in turning process of Inconel 718.

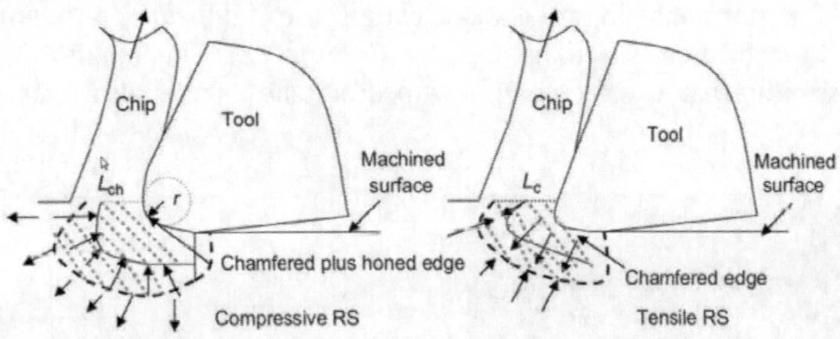

Figure 5: Effect of tool geometry on residual stresses in machining of Inconel 718 [30]

It is a general fact that the forces are strong indicators of the cutting performance, which are related to tool wear, surface finish and specific cutting energy (e.g. the variation in cutting length at different depths of cut is shown in Figure 6). The wear in cutting edge and the gradual degradation are indicated by the force components that increase steadily through the increment of cut length. The cutting tool is relatively sharp at the beginning, hence produces lower cutting force [31]. The increased wear of the edge leads to a blunt edge with increased contact area between the edge and workpiece, thus produces higher cutting forces [27]. The forces then served as a good indicator of the degradation of the cutting edge throughout its life. Similar results have been observed during the machining of steels with HSS band saws [32, 33].

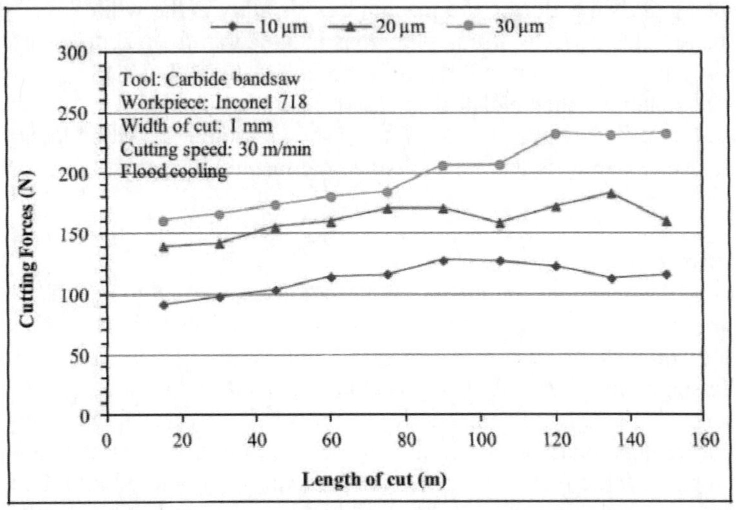

Figure 6: The variation in cutting length at different depths of cut [31]

IOP Conf. Series: Materials Science and Engineering **270** (2017) 012034 doi:10.1088/1757-899X/270/1/012034

Combination of adhesive and abrasive wear with degrees of plastic deformation is the mechanism of flank wear. The reduction in the machining efficiency and increases in force components are led by the increment of the edge cutting radius and the BUE that formed during machining. Even with the removal of BUE, the force components will still decrease periodically as the carbide tip geometry has been significantly altered due to the loss of edge material through chipping or welding with BUE. The chipping and fracture occur due to the formation of micro-cracks. The general trend of specific cutting energies with the length of cut is shown in Figure 6 and it continues to increase by the degradation and wear of the cutting edge [33].

In a machining operation, the critical objective is to have a good surface finish. Surface finish is an important criterion in machining process and selection of proper machining parameters is important to obtain good surface finish. The presented image in Figure 7 shows a damaged surface of Inconel 718 at V = 150m/min, f = 0.15 mm/rev, DOC = 0.50mm and dry condition. In Inconel 718 case, the lower surface roughness is rarely achieved through CM due to interrupted cutting force phenomenon that is affecting the machine performance like tool wear, cutting force and surface roughness (e.g. variation in surface roughness for different speeds is depicted in Figure 8). This condition is because, at high speed, there is a possibility that a surface flaw is wiped-out and there is also a presence of heat in the cutting tool [28].

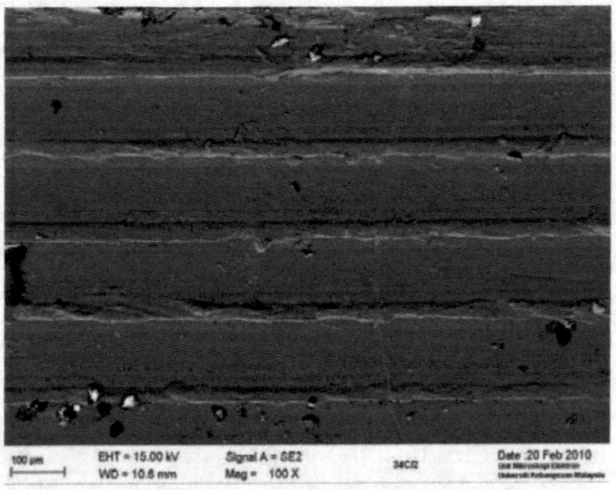

Figure 7: Surface damages of Inconel 718 during machining [34]

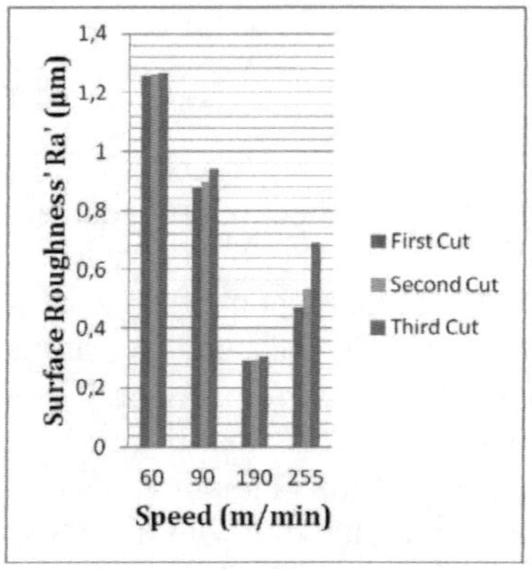

Figure 8: Variation in surface roughness for different speeds [28]

One of the problematic issues during the machining this alloy is the white layer. The requirements of Inconel 718 to be fulfilled in aerospace applications include that the machine surface roughness has to be less than 0.8 µm, no recast layer and the strain needs less than 10 µm [35]. However, turning and drilling machining on materials such nickel-iron based superalloys have led to the formation of a white layer on the surface [36, 37]. Kasim et al. (2003) have shown there is no white layer during machining of Inconel materials using ball nose cutting tool since the machining temperature was less than 650°C [38].

2.2. *Ultrasonic Assisted Application*

Azlan et al. (2016) [39] carried out the ultrasonic vibration assisted experiments to obtain the effects of machining parameters on surface roughness of hardened steel material for slot milling operation. The experiments are conducted to compare surface roughness magnitudes obtained with and without the presence of ultrasonic vibration. UAM is a machining operation that uses vibration tool oscillating at ultrasonic frequencies (18-20 kHz) and is aided by abrasive slurry that flows freely between the tool and the workpiece to remove material from the latter. The vibration tool is oscillating with amplitude 10 to 40 mm and imposes static pressure on the abrasive grains, hammered into the workpiece surface to cause chipping of fine particles and formed the required tool shape [39]. The advantage of UAM is

IOP Conf. Series: Materials Science and Engineering **270** (2017) 012034 doi:10.1088/1757-899X/270/1/012034

the great reduction in cutting force [40], in which a force of approximately 10% lower than CM has been reported [41]. By applying the ultrasonic vibration, it will increase the heat removal rate and thus increasing the tool life and decreasing the tool temperature [42]. It has also been found that the use of UAM is significantly advantageous to the productivity, which include about 30% reduction of the machining cost and 5-10% of manufacturing time [43]. Figure 9 shows the setup of UAM on the tool holder in the experiments.

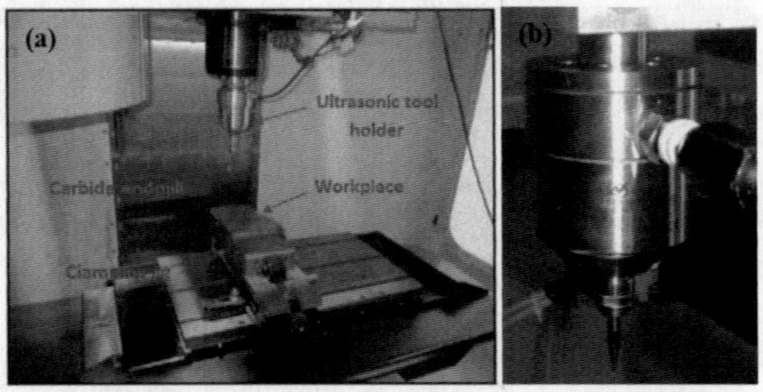

Figure 9: UAM setup on tool holder: (a) by Azlan [39] and (b) by Lian [44]

Ultrasonic machining refers to specialized ultrasonic machine tools that oscillate an abrasive tool. The Acoustech system is an application that makes use of the controlled ultrasonic vibration on any existing machining centre using any standard drill or end mill. The vibration comes from a device that is essentially a tool holder with an ultrasonic actuator built-in. The same device also incorporates a compensating vibration-cancellation mechanism that prevents the ultrasonic actuation from affecting the machine tool itself. The vibration is "ultrasonic" because its frequency is above what the human ear can hear. Audible noise occurs at 15 to 18 kilohertz. Specific effect of this vibration is expansion and contraction of the tool through controlled amplitude [12]. Tool wear still occur during application of ultrasonic vibration due to the interrupted cutting process that occurs during machining between the cutting tool and the workpiece. However, the tool wear is found to be reduced by approximately one-fifth when the UVC process is used to cut the workpiece under all cutting conditions [45].

The experimental results shown in Figure 10 are for both conventional turning (CT) and ultrasonic vibration cutting (UVC) on low alloy steel (DF2) workpiece using cubic boron nitride (CBN) inserts as the cutting tool [45]. The results are for feed rate of 0.1 mm/rev (in Figure 10(a)) and for feed rate of 0.2 mm/rev (in Figure 10(b)), where a, b, c is the cutting speeds of 50, 70, 90 m/min, respectively. It is observed that UAM can help prolong the tool life since the tool wear is lower for longer period of machining [41]. Only micro cracks in the diamond grits of the cutting tool are found after 10 m of the machining length of CFRP with UAM while in CM, grain pull-out and wear flats in the diamond grits are observed.

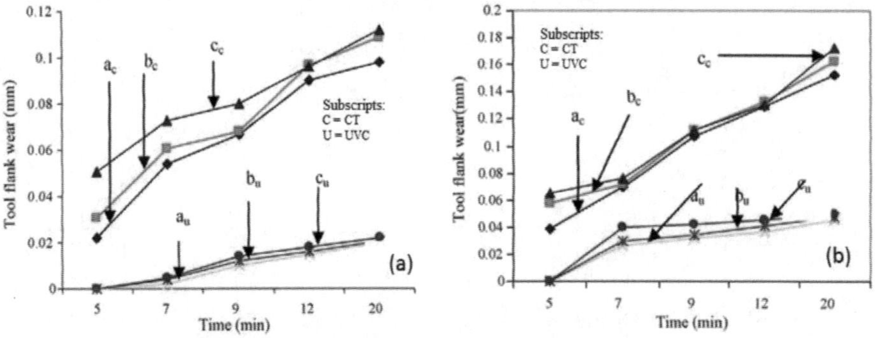

Figure 10: Tool wears condition at three cutting speeds for both CT and UVC processes: (a) feed rate at 0.1 mm/rev and (b) feed rate at 0.2 mm/rev [45]

Force is the energy required by the machining process to perform the cutting process. Since Inconel is a material that *is* hard to cut, high amount of energy is needed to perform the cutting process. The feed force (Fx) recorded by UAM has been shown to be up to 20% of force reduction as compared to CM on carbon fiber reinforced plastic (CFRP) using the nickel layer electroplated cutting tool with average of 420μm diamond grit [41]. The results shown in Figure 11 are the comparison of machining forces in y-direction (Fy) and feed force in x-direction (Fx) between CM and UAM. The machining force increases as the machining length is increased for both methods. During this time, the cutting tool started flaking and chipping due to interrupted cutting force phenomenon. However, the result indicates that UAM has lower machining forces through the 10m of machining length. This took place due to a larger plastically-deformed matrix transferred to the tool in CM than in UAM, causing higher obstruction in machining.

Ma et al. (2005) [43] analyzed the improvements gained with ultrasonic on a turning process using a CT. The thrust force is applied in the bending direction of the spindle, which is perpendicular to the long axis of the workpiece. Small forces from the vibration-assisted machining (VAM), with their application at ultrasonic frequencies that are well above the natural frequency of the lathe structure, is found to greatly reduce the tool tip vibration in the radial (depth of cut) direction.

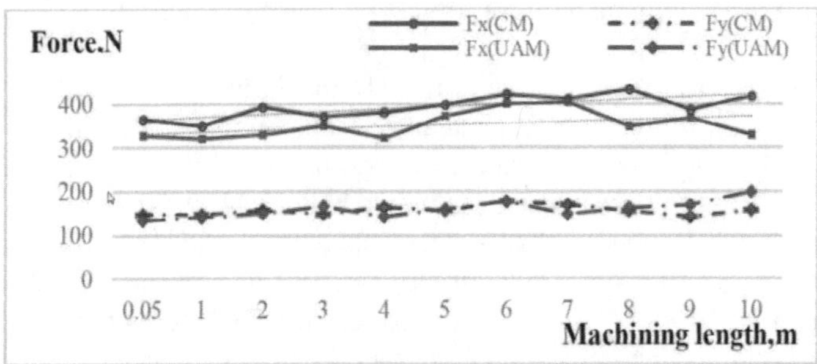

Figure 11: Maximum machining force in CM and UAM [41]

A low surface finish is one of the challenging aspects to be achieved in primary machining process without undergoing any secondary or finishing process. Nevertheless, it is still possible to eliminate the finishing process when optimum machining technique is implemented. An experimental research of ultrasonic vibration assisted micro-milling on Al6061 at different amplitudes has been conducted and the results are depicted in Figure 12. This study has proven that the application of UAM improved the Ra value than that by ordinary micro-milling [44]. Machining marks can be detected on machine surface at 0μm amplitude. Figure 13 depicts overall result of the study.

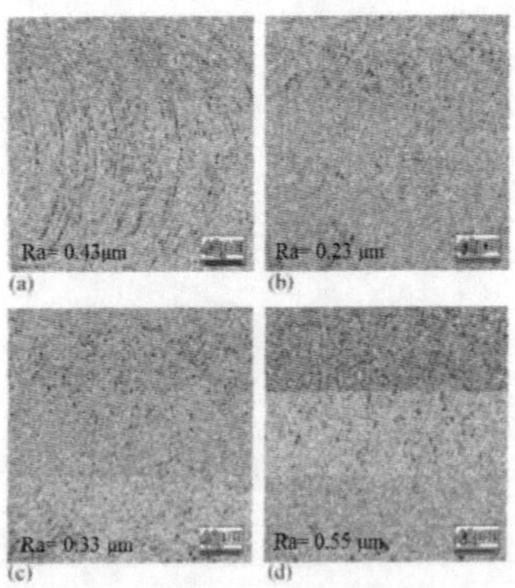

Figure 12: The surface finish of micro-milling ultrasonic vibration at 38 kHz: (a) 0μm (b) 11μm (c) 15μm and (d) 19μm amplitude [44]

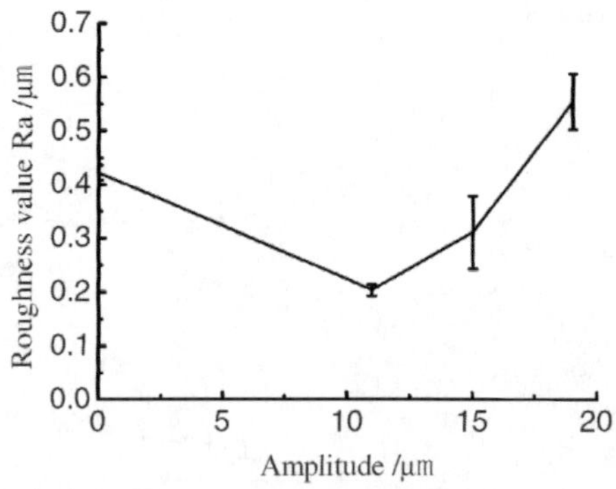

Figure 13: Effect of UAM amplitude on surface roughness [44]

IOP Conf. Series: Materials Science and Engineering **270** (2017) 012034 doi:10.1088/1757-899X/270/1/012034

From the amplitude parameter range, the minimum surface finish can be achieved at a specific amplitude. Too high or too low amplitude can cause an increase in the surface finish. The best surface roughness of 0.23μm is obtained when the cutting amplitude at 11μm. Further investigation has found that hammer effect induced by ultrasonic milling causes poor results. With high-frequency vibration, the cutter teeth will hammer the workpiece constantly. This result is in line with findings from another study that demonstrated as the feed increases in UAM process, the chip curvature also decreases [46]. The type of chips has an influence on the workpiece surface roughness and the findings from a study on the ultrasonic-assisted dry grinding (UADG) have led to significant improvements on the Rz and Ra parameter [47]. It is assumed that the improvement in these parameters has been due to the fact that the grit in UADG has higher chance to cut the peak of the surface due to the feed ultrasonic oscillation and increasing the possibility of interaction of the grit and workpiece surface in each contact length.

3. Experimental Technique and Input Parameter

Based on previous findings, machining of superalloys like Inconel 718 using CM remains a challenge until today. Many have tried to find the empirical relationship between the input and output parameters through experimental results since analytical solutions are often difficult to formulate and FEM-based solutions might take time. The combinations of cutting speed (V), feed rate (fz), depth of cut (DOC) as tabulated in Table 1 and Table 2 have been used as input parameters to analyse and predict the white layer thickness [48], surface roughness [49], residual stresses [50-53] and the yield strength of plastic strain on workpiece material [29]. Results from these studies can be applied with analytical predictions [50, 52, 54].

Since information on implementation of UAM technique is available through experimental results, analytical solutions and FEM-based solutions, the input parameter on cutting speed (V), spindle speed (n), feed rate (fz), depth of cut (DOC) and vibration frequency (f) can readily be established and also applied during the machining process. The outcomes such as tool life, cutting force and the surface roughness can be compared to those from CM process on the Inconel 718. Table 3 tabulates previous research and findings through assisted by ultrasonic vibration.

Table 1: Material, process, and parameter information for Inconel group of workpiece

Reference	Workpiece Material	Process and Cutting Tool	Parameters	Remarks
Derrien and Vigneau [59]	IN-718	Dry milling, Uncoated carbide	V = 16, 200 m/min fz = 0.04 mm/tooth DOC = 0.5 mm	Increasing V from 16-200m/min will increase the sub-surface tensile residual stress from 400-1400MPa.
Guerville and Vigneau [60]	IN-718	Point milling, Uncoated carbide	V = 18, 200 m/min fz = 0.08 mm/tooth DOC = 0.2 mm	Increasing V from18-200 m/min will increase the surface tensile residual stress from 250-750 MPa and the peak compressive residual stress from 200-450 MPa.
Arunachalam et al. [61]	IN-718 (36HRC)	Facing, CBN and mixed ceramic	V = 150-450 m/min fz = 0.15 mm/rev DOC = 0.05-0.5 mm	Increasing V will increase surface roughness and change residual stresses from compressive to tensile stress (-100 to 500 MPa)

Table 2: Material, process and parameter information for titanium group of workpiece

Reference	Workpiece Material	Process and Cutting Tool	Parameters	Remarks
Nurul-Amin et al. [55]	Ti-64	End milling, Uncoated carbide and PCD	V = 40-160, 120-250 m/min fz = 0.1 mm/tooth DOC = 1 mm	For both uncoated carbide and PCD tools, increasing V from 40-160 and from 120-250 m/min, respectively, will increase the surface roughness.
Sun and Guo [56]	Ti-64	End milling, Coated carbide (TiAlN)	V = 50-110 m/min fz = 0.06-0.14 mm/tooth DOC = 1.5 mm	Increasing fz, tensile residual stresses at the surface increased. Surface roughness increased when the DOC increased due to increased overlap between cutting paths.
Ginting and Nouari [57]	Ti-6242S	Milling, Uncoated and Coated carbide (TiN, TiC,TiCN)	V = 100-125 m/min fz = 0.15-0.2 mm/tooth DOC = 2-2.5 mm	Increasing V will increase the surface roughness with uncoated tools, but decrease lightly with coated tools. Increasing fz will increase the surface roughness under all cutting conditions.
Sridhar et al. [50]	Ti-IMI 834	Milling, Coated carbide (TiN)	V = 11.56 m/min fz = 0.056, 0.1 mm/tooth DOC = 0.25, 2 mm	At low V (11m/min), increasing fz will reduce sub-surface compressive residual stresses. At high V (56m/min), increasing fz will increase sub-surface compressive residual stresses. The compressive magnitude of residual stresses decreases when DOC is increased.
Mantle and Aspinwall [58]	γ-TiAl	Milling, Uncoated carbide	V = 70-120 m/min fz = 0.06-0.12 mm/rev DOC = 0.2-0.5 mm	Increasing V will decrease the amount of compressive residual stresses observed.

The general finding of these studies is that, depending on the experiments, the output parameters are found to depend heavily or slightly on the input parameters. However, through the reviews on the previous research of ultrasonic vibration assisted technique that has been employed in the milling [17], drilling [18], grinding [19] and turning [20] processes, the subsequent impact has been more positive than negative. This proves that the UAM is feasible to be explored in further study and research.

Table 3: Material, process and parameter information for machining assisted by ultrasonic vibration

Reference	Workpiece Material	Process and Cutting Tool	Parameters	Remarks
Tabatabaei et al. [62]	Special conical shape for workpiece	Turning, Triangular TNMG	n = 1200-1400 rpm fz = 0.3 mm/tooth DOC = 5.85- 9.8 mm *f* = 228.8 Hz, 20 kHz	Ultrasonic vibration can improve the stability for some cutting conditions while degrading the stability in some other conditions
Yasmine et al. [63]	Soda glass	Milling, Diamond	n = 2000-10000 rpm fz = 10-100 mm/min DOC = 0.02-0.1 mm *f* = 24.6, 25.4 kHz,	Ultrasonic vibration maximizes the axial cutting force and minimizes the moment. However, lower cutting forces are obtained in some experiments using CM
Nor Farah Huda et al. [64]	Carbon Fibre Reinforced Plastic (CFRP)	Milling, Polycrystalline Diamond (PCD) with three cutting flute	V = 500 mm/min fz = 0.8 mm/tooth DOC = 2 mm *f* = 39 kHz Amplitude = 5 μm	Application of UAM resulted in reduced forces (up to 20 %) and temperatures (up to 15 %). However, surface roughness increases (up to 5 %) and higher tool wear (106 μm) compared to CM (80 μm) after 10m machining length.
Abootorabi Zarchi et al. [46]	X20Cr13 Stainless Steel	Milling, 4-flute TiN coated end mill made of high-speed steel (HSS)	n = 500 rpm fz = 0.05 mm/tooth DoC = 5.0 mm *f* = 16-40 kHz Amplitude = 2-30 μm	As feed increases in UAM process, the chip curvature decreases. The chips have an influence on workpiece Ra.
Lian et al. [44]	Al6061	Micro-milling, Micro-milling tool (Unidentified type)	n = 150000 rpm fz = 0.5 mm/s DoC = 5.0 mm *f* = 38 kHz Amplitude = 11, 15, 19 μm	The surface processed by assisted ultrasonic vibration is more uniform than that processed by micro-milling without vibration, the latter has clear machining marks are left on the surface
Taghi and Bahman [47]	42CrMo4 (85 HRB)	Grinding, Diamond single point dresser	V = 60 m/s fz = 500, 1000, 1500, 2000 mm/min DoC = 0.010- 0.030 mm *f* = 23 kHz Amplitude = 10 μm	Assisted ultrasonic vibration reduces the grinding forces, frictional effect and plastic deformation zone. It leads to significant improvements on the Rz and Ra parameter.
Ning et al. [65]	Inconel 718	Laser engineered net shaping (LENS)	V = 508, 635 m/s hatch space = 0.3 mm f = 0, 41 kHz	Ultrasonic vibration on microstructures and microhardness gives lower porosity, smaller grain size, and uniform laves phases of the as-deposited Inconel 718 parts.

4. Conclusion

The performance and quality of a product are influenced by surface integrity from final manufacturing processes. Surface integrity such as residual stress, microstructure and surface roughness possibly can be improved through UAM recent technique. This review paper provides an overview of machining-induced surface integrity in Inconel 718 superalloys. The following are specific conclusions reached in this review.

- UAM on the CNC milling process will improve tool life, achieve lower forces (thrust force) and improve the surface quality of a workpiece.
- In terms of the cutting forces, the maximum cutting forces observed in UAM are lower (10 %) compared to CM on CFRP workpiece where tool vibration aided the material removal process in UAM.
- Reduction of cutting forces is attributed to vibration amplitude influencing the gap between the cutting tool and the workpiece material resulting in improved chip-breaking conditions.
- By applying ultrasonic vibrations on milling process makes formed chips thinner and smaller, and leads to flatter machined surface where the effect of UAM on AISI 1020 steel improves the surface roughness by up to 12.9%.
- UAM implementation can be significantly advantageous to the productivity due to reduction of about 30% of machining cost and 5-10% of manufacturing time.

The difficulties involved in CM that are addressed in UAM, improvements in machining process and the reasons for these improvements have been presented in this paper. UAM is found to be effective in advanced machining techniques in improving the process stability and quality of the products. UAM can be applied to almost all precision machining techniques. Nonetheless, the proper mechanics of the machining process to support the improvements in machining have to be developed and also studied in depth.

Acknowledgement
The authors wish to thank Universiti Teknikal Malaysia Melaka for the facility and monetary support in this study through the research grant number PJP/2016/FKP/HI6/S01485.

References

[1] Arunachalam R and Mannan M A 2000 *Mach. Sci. Technol.* **4** 127-68
[2] Krain H, Sharman A R and Ridgway K 2007 *The Journal of Materials Processing Technology* **189** 153-61
[3] Locq D and Caron P 2011 *The Onera Journal Aerospace Lab* **3** 1-9
[4] Jawaid A, Koksal S and Sharif S 2001 *Journal of Materials Processing Technology* **116** 2-9
[5] Fong A M, Chen X Q and Li H Z 2002 *Overview of Material Processing Automation* in *Advanced Automation Techniques in Adaptive Material Processing* World Scientific Publishing
[6] Li H Z, Chen X Q and Zeng H 2004 *International Conference on Presicion Engineering*
[7] Prickett P W and Johns C 1999 *Int. J. Tool Manufacturing* **39** 105-22
[8] Archarya S 2011 *https://commons.wikimedia.org/wiki/File:Pratt_%26_Whitney_JT3.jpg*
[9] Tlusty J 2000 *Manufacturing Process and Equipment* Prentice-Hall
[10] Merrit H E 1965 *ASME Journal of Engineering for Industry* **87** 447-54
[11] Tabatabaei S, Behbahani S and Mirian S M 2013 *Journal of Materials Processing Technology* **213** 418-25
[12] Zelinski P 2014 *https://www.mmsonline.com*
[13] Moriwaki T and Shamoto E 1991 *CIRP Annals - Manufacturing Technology* **40** 559-62
[14] Kumabe J, Fuchizawa K, Soutome T and Nishimoto Y 1989 *Precision Engineering* **11** 71-7
[15] Thakur D, Ramamoorthy B and Vijayaraghavan L 2009 *Materials and Manufacturing Processes* **24** 497-503
[16] Zhou J, Bushlya V, Avdovic P and Stahl J E 2012 *The International Journal of Advanced*

Manufacturing Technology **58** 141-51
[17] Hsu C, Huang C and Wu C 2007 *The International Journal of Advanced Manufacturing Technology* **34** 857-66
[18] Baghlani V, Mehbudi P, Akbari J and Sohrabi M 2013 *Procedia CIRP* **6** 572-7
[19] Bhaduri D, Soo S L, Novovic D, Aspinwall D K, Harden P, Waterhouse C, Bohr S, Mathieson A C and Lucas M 2013 *Procedia CIRP* **6** 616-21
[20] Wang Q, YongboWub W Y, Guc J, Lud D, Jie Y and Nomurab M 2016 *18th CIRP Conference on Electro Physical and Chemical Machining*
[21] Hadi M A, Ghani J A, CheHaron C H and Kasim M S 2013 *Procedia Engineering* **68** 647-53
[22] Li H Z, Zeng H and Chen X Q 2006 *Journal of Materials Processing Technology* **180** 296-304
[23] Yamamoto K, Kuroda M, Omokawa H and Itakura K 2001 *Mitsubishi Heavy Industries Ltd. Technical Review* **38** 12-6
[24] Kasim M S, CheHaron C H, Ghani J A, Md Ali M A, Sulaiman M A and Aziz M S A 2013 *Malaysian Technical Universities Conference on Engineering & Technology*
[25] Devillez A, Le Coz G, Dominiak S and Dudzinski D 2007 *Wear* **262** 931-42
[26] Thakur D G, Ramamoorthy B and Vijayaraghavan L 2009 *International Journal Advance Manufacturing Technology* **45** 421-9
[27] Khan F N, Haider J and Persson M 2017 *Proc IMechE Part B: J Engineering Manufacture*
[28] D'Addona D M, Raykar S J and Narke M M 2017 *Procedia CIRP* **62** 269-74
[29] Pawade R S, Joshi S S and Brahmankar P K 2008 *International Journal of Machine Tools and Manufacture* **48** 15-28
[30] Coelho R T, Silva L R, Braghini Jr. A and Bezerra A A 2004 *Journal of Materials Processing Technology* **148** 147-53
[31] Wang J, Huang C Z and Song W G 2003 *JMater Process Tech* **142** 338-46
[32] Sarwar M, Persson M and Hellbergh H 2005 *Wear* **259** 1144-50
[33] Sarwar M, Persson M and Hellbergh H 2005 *Wear* **263** 1438–41
[34] Yazid M Z A, Ibrahim G A, Said A Y M, CheHaron C H and Ghani J A 2011 *Procedia Engineering* **19** 396-401
[35] Kappmeyer G, Hubig C, Hardy M, Witty M and Busch M 2012 *5th CIRP Conference on High Performance Cutting*
[36] Hetherington C, Bushlya V, Zhou J M and Ståhl J 2012 *http://www.emc2012.org.uk*
[37] Herbert C, Axinte D, Hardy M and Brown P D 2012 *Machining Science and Technology* **16** 40-52
[38] Kasim M S, CheHaron C H, Ghani J A, Sulaiman M A and Yazid M Z 2013 *Wear* **302** 1171-9
[39] Azlan R, Izamshah R, Hadzley M, Kasim M S, Arfauz M and Akmal M 2016 *Mechanical Engineering Research Day*
[40] Wang L J and Zhao J 1987 *International Journal of Machine Tools and Manufacture* **27** 181-90
[41] Nor Farah Huda A H, Helen A and Stuart B 2016 *Procedia CIRP* **46** 488-91
[42] Kumabe J and Masuko M 1958 *Transactions on JSME* **24** 109-14
[43] Ma C, Shamoto E, Moriwaki T, Zhang Y and Wang L 2005 *International Journal of Machine Tools and Manufacture* **45** 1295-300
[44] Lian H, Guo Z, Huang Z, Tang Y and Song J 2013 *Procedia CIRP* **6** 561-4
[45] Chandra N, Rahman M and Andrew S S K 2007 *Journal of Materials Processing Technology* **192-193** 159-65
[46] Abootorabi Zarchi M M, Razfar M R and Abdullah A 2013 *ASME Manufacturing Science and Engineering Conference*
[47] Taghi T and Bahman A 2008 *6th International Conference on Manufacturing Research*
[48] Shi J, Wang J Y and Liu C R 2006 *Journal of Engineering Manufacture* **220** 119-28
[49] Axinte D A and Dewes R C 2002 *Journal of Materials Processing Technology* **127** 325-35
[50] Sridhar B R, Devananda G, Ramachandra K and Bhat R 2003 *Journal of Materials Processing Technology* **139** 628-34

[51] Jacobus K, DeVor R and Kapoor S G 2000 *Journal of Manufacturing Science and Engineering* **122** 20-31
[52] Capello E 2005 *Journal of Materials Processing Technology* **160** 221-8
[53] El-Axir M H 2002 *International Journal of Machine Tools and Manufacture* **42** 1055-63
[54] Pawade R S, Joshi S S, Brahmankar P K and Rahman M 2007 *Journal of Materials Processing Technology* **192–193** 159
[55] Nurul-Amin A K, Ismail A F and Nor Khairusshima M K 2007 *Journal of Materials Processing Technology* **192-193** 147–58
[56] Sun J and Guo Y B 2009 *Journal of Materials Processing Technology* **209** 4036-42
[57] Ginting A and Nouari M 2009 *International Journal of Machine Tools and Manufacture* **49** 325-32
[58] Mantle A L and Aspinwall D K 2001 *Journal of Materials Processing Technology* **118** 143-50
[59] Derrien S and Vigneau J 2004 *International Journal of Machine Tools and Manufacture* **44** 439-56
[60] Guerville L and Vigneau J 2004 *International Journal of Machine Tools and Manufacture* **44** 439-56
[61] Arunachalam R M, Mannan M A and Spowage A C *2004 International Journal of Machine Tools and Manufacture* **44** 879-87
[62] Tabatabaei S, Behbahani S and Mirian S M 2013 *Journal of Materials Processing Technology* **213** 418-25
[63] Yasmine E T, Mohab H and Hassan E 2017 *16th CIRP Conference on Modelling of Machining Operations*
[64] Nor Farah Huda A H, Helen A and Stuart B 2017 *Advances in Material & Processing Technologies Conference*
[65] Ning F, Hu Y, Liu Z, Cong W, Li Y and Wang X 2017 *SME North American Manufacturing Research Conference*

IOP Conf. Series: Materials Science and Engineering **270** (2017) 012035 doi:10.1088/1757-899X/270/1/012035

Detection of fastener loosening in simple lap joint based on ultrasonic wavefield imaging

M I Gooda Sahib[1], S J Leong[1], C C Chia[1,*] and F Mustapha[1,2]
[1]Department of Aerospace Engineering, Faculty of Engineering, Universiti Putra Malaysia, 43400 UPM Serdang, Selangor, Malaysia.
[2] Institute of Tropical Forestry and Forest Products (INTROP), Universiti Putra Malaysia, 43400 UPM Serdang, Selangor, Malaysia.

* chia_cc@upm.edu.my

Abstract. Joints in aero-mechanical structures are critical elements that ensure the structural integrity but they are prone to damages. Inspection of such joints that have no prior baseline data is really challenging but it can be possibly done using the Ultrasonic Propagation Imager (UPI). The feasibility of applying UPI for detection of loosened fastener is investigated in this study. A simple lap joint specimen made by connecting two pieces of 2.5mm thick SAE304 stainless steel plates using five M6 screws and nuts has been used in this study. All fasteners are tightened to 10Nm but one of them is completely loosened to simulate the damage. The wavefield data is processed into ultrasonic wavefield propagation video and a series of spectral amplitude images. The spectral images showed noticeable amplitude difference at the loosened fastener, hence confirmed the feasibility of using UPI for structural joints inspection. A simple contrast maximization method is also introduced to improve the result.

1. Introduction

Joints in aero-mechanical structures are critical elements that ensure the structural integrity. The joints unite and assemble parts to enable them to collectively function as a structural system. Performance and effects of the joints on structural stability and dynamic response are still a concern under active investigation [1]. The inherent material discontinuity at the joints make them a damage hotspot. In general, the joints often suffer from damages such as cracks, corrosion, loosening of fasteners, snap of improperly tightened fasteners, as well as other deviations from original structural condition [2]. The situation is worsened by environmental factors during their service life, which is expected to last for decades due to the high capital cost of respective aero-mechanical structures. Notorious examples of structural failure primarily due to joint damage include the collapse of I-35W Bridge over Mississippi River in the USA [3] and the failure of Nipigon River Bridge in Canada [4]. In fact, the field statistics have shown that 58% of bridge failures are due to fastener loosening and joint failure [5].

Inspection and monitoring of fastened joints is imperatively needed to ensure their good condition and safe continual use. The resonant frequency shift method [6-8] and the ultrasonic time-of-flight (ToF) method [9-11] have been exploited for this purpose. Both methods involve the injection of the acoustoelastic or ultrasonic waves into a fastener, followed by respective measurement of resonant frequency or ToF of the waves in the fastener. Fastener preload could then be calculated by exploiting the linear proportionality of the measurand to the stress in fastener. Major drawback of these methods is the need of baseline data for previous known states of each fastener, which is difficult to maintain or unavailable. On the other hand, the ultrasound velocity ratio method [12, 13] has circumvented this need and can measure fastener preload without a baseline data. Anyhow, comparison with a baseline is still necessary in order to infer the condition of the fastener. Furthermore, there are also piezoelectric impedance-based methods [14, 15] that generally rely on permanently bonded piezoelectric transducer on the fasteners or structural components being monitored for data acquisition. They exploit linear relationship of the fastener preload with the piezoelectric impedance frequency for result processing. Moreover, a novel optical measurement system known as the Fiber Optic Bolt Loosening Monitoring System has also been previously developed [16]. The system basically consists of one optical fiber strategically fixed to all fasteners under monitoring and an optical time-domain reflectometer (OTDR) to measure the intensity of light transmitted through the fiber. Any loosening of the fastener as small as one degree turn will cause additional bending to the fiber, hence partial loss of optical transmission. The signature of the loss can be associated with the fiber length position in OTDR and this can reveal the loosened fastener.

The methods discussed above have worked excellently under their respective working conditions but generally suffer from one or both of the following issues over a larger perspective. First of all, the methods require either manual attachment of transducers during inspection or permanently bonded transducers for continuous monitoring. Attaching the transducer as needed is a typical task for non-destructive inspection but the work is tedious, especially when the fasteners that require inspection are difficult to access. In the meantime, permanent bonding of transducer is common for structural health monitoring applications but such bonding often suffers from environmental factors and will degrade over time until it adversely affects the result. Secondly, many of the methods require a reference to the baseline data, which is also known as temporal referencing. The baseline is needed either for result processing or loosening inference. Regardless the purpose, the baseline data is difficult to maintain, if not available at all, and also has a low reliability due to susceptibility to the environmental effects such as temperature variation.

An alternative referencing strategy is therefore needed in case of unavailability of baseline data such as for the first inspection of an existing structure with suspected damage. One good strategy is the spatial referencing, which is the intrinsic nature of the ultrasonic propagation imager (UPI). Result processing and conclusion about existence of damage can be done by UPI with reference to data taken from different grid points of the same inspection. This has been applied successfully for detection and evaluation of damages in various structures and materials, including metallic pipelines [17], composite wings and wind turbine blades [18-20]. It should be noted that all of these inspected specimens did not have individual constituent parts that possess strong material discontinuity as in a fastened lap joint. One exception is the inspection of a composite aircraft wing with multiple rivets [21]. However, the focus of that application is the evaluation of skin-spar debond and not the condition of the fasteners.

In this study, the feasibility of applying UPI for detection of fastener loosening in a simple lap joint specimen is explored. The data is first processed into an ultrasonic wavefield propagation video to analyze if any change in the wavefield propagation pattern associated to fastener loosening can be identified. Spectral amplitude analysis is then performed by segregating ultrasonic energy into discrete frequency bins and visualizing their amplitude distribution at the fasteners. To improve the visibility of loosened fastener in the final result, a simple contrast maximization method is applied.

2. Materials and Wavefield Data Acquisition

A screw-nut fastened simple lap joint specimen is designed based on the typical engineering practice [22]. Two constituent plates of width 100.5mm and height 201mm are CNC machined from 2.5mm thick SAE304 stainless steel. The plates are then fastened together using five M6 socket head cap screws, and plain washer and nuts on the opposite side. The fasteners are tightened to 10Nm using a torque wrench, except for the second one from the left that is slightly tightened with bare hand only to simulate the loosening case of a fastener. A custom-made diode-pumped solid state Nd:YAG pulsed laser is used for the ultrasonic wave generation through rapid, localized, thermoelastic expansion–

contraction in specimen material. The laser has a wavelength, beam diameter at the laser exit port, divergence and pulse duration of 532nm, 4mm, 0.5m rad and 30ns, respectively. Its output energy is adjusted to approximately 1mJ per pulse in order to keep the wave generation process within non-destructive, linear elastodynamic regime of the specimen material.

Ultrasonic wavefield imaging is performed over 95mm by 100mm area, enclosing all five fasteners of the specimen, is performed. Individual waves are generated over a uniform grid of 0.5mm pitch in both horizontal and vertical directions. This results in a 190 by 200 pixel grid. The 2D grid scanning is realized using a galvanometric optical scanner at 1m standoff and a pulse repetition frequency of 10Hz. Ultrasonic wave generated at each grid point is then measured using a broadband piezoelectric sensor, fixed position at 100mm above the middle fastener. This sensor can be replaced by non-contact sensors such as laser Doppler vibrometer when a remote, on-site measurement is needed. The signals are then sampled using an oscilloscope without any filtering so that they contain all possible ultrasonic modes with as much spectral information as possible. Signal at each spatial point is recorded with 500 samples at a sampling rate of 5 MHz.

Complete ultrasonic wavefield is reconstructed by synchronizing all waves traveling from the laser excitation points and sensed by the fixed sensor in a time-reversal fashion. This is done such that the wavefield can be interpreted as if generated at, and traveled away from, the sensor position. Such a reconstruction is more intuitive and valid as long as the elastodynamic reciprocity holds. It produces high signal-to-noise ratio measurements with broadband frequency contents, facilitating the frequency analysis for evaluation of fastener preload. Visualizing changes of wavefield with time and analyzing variation of its frequency information are realized, respectively, by processing data using Ultrasonic Wavefield Propagation Imaging (UWPI) method [23] and Ultrasonic Spectral Imaging (USI) method [24]. The processes are summarized in a flow diagram given in Figure 1.

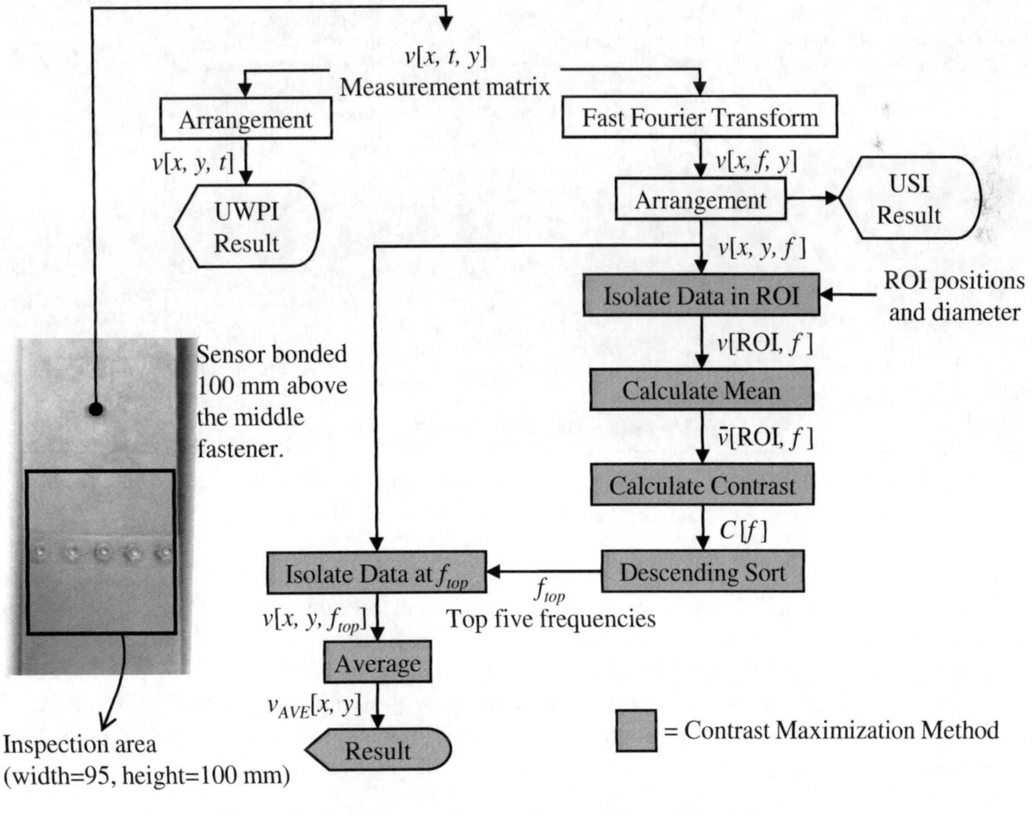

Figure 1: Specimen and result processing flow diagram

IOP Conf. Series: Materials Science and Engineering **270** (2017) 012035 doi:10.1088/1757-899X/270/1/012035

3. Results and Discussions

The result generated by UWPI method is a video of wavefield propagation over the inspection area. Snapshots captured from this video at selected instances of the wavefield propagation are shown in Figure 2. Snapshot at 20.0μs shows incident wavefield of S0 and A0 modes concentric at the sensing position. The continuation of the S0 mode is broken when the wavefield encounters the fasteners at 23.8μs near the center. Similar discontinuation occurs when the slower A0 mode meets the fasteners at 35.2μs. The fasteners at regular interval caused periodic reflection-transmission of incident wavefield, generating the complex wavefield interference pattern as shown in snapshots at 42.4μs and 47.0μs. On top of that, the strong side reflections of A0 mode further complicate the wavefield until it reaches an indistinguishable complex state as shown in snapshot of 90.2μs. Even though the second fastener has been completely loosened, it did not produce any noticeable difference in the wavefield pattern even for trained eyes. Another result processing method that can extract and present the relevant loosening information, perhaps in a different domain, is needed.

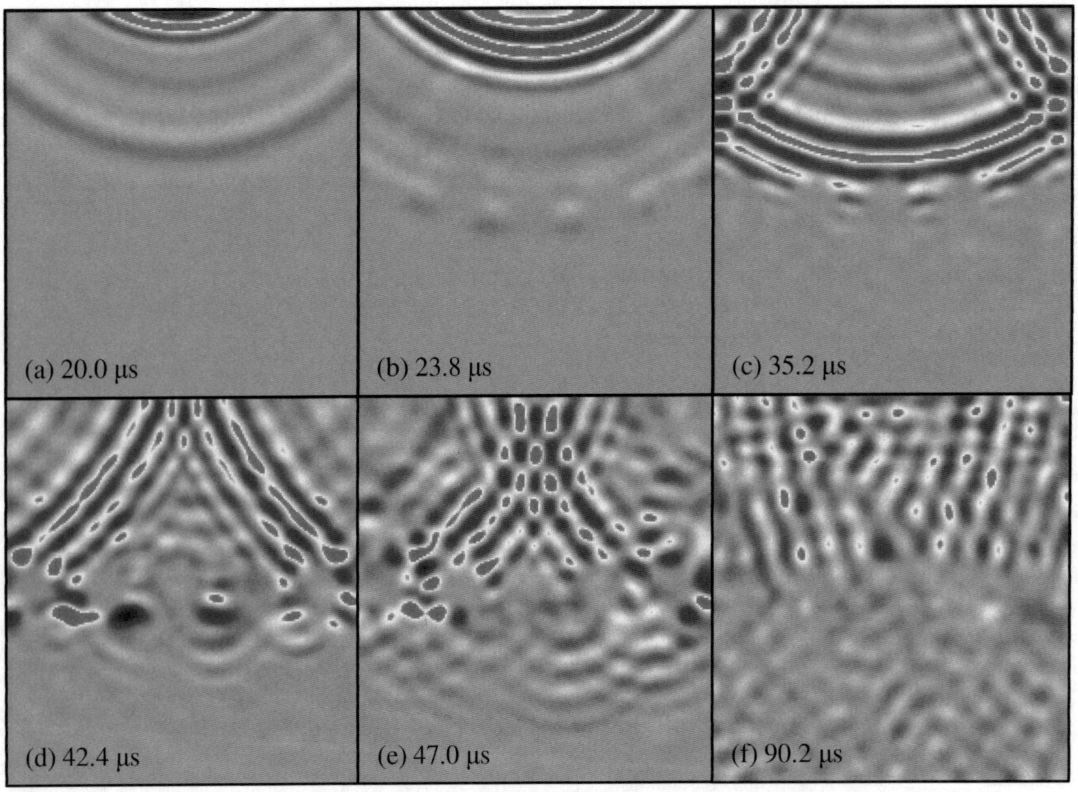

Figure 2: Snapshots of ultrasonic wavefield propagation video at (a) 20.0μs, (b) 23.8μs, (c) 35.2μs, (d) 42.4μs, (e) 47.0μs and (f) 90.2μs

Analysis in the frequency domain is performed by processing the same set of wavefield data using USI method. The result is a series of images showing the spectral amplitude response of the specimen at discrete frequencies of the periodic interval. Figure 3 depicts three of them at 260kHz, 290kHz and 360kHz, respectively. Five dotted circles are included in Figure 3(a) to indicate actual positions and head size of the fasteners. Figure 3 represents the images with noticeable amplitude difference (using rainbow color scale) between the properly tightened fasteners and the loosened one (second fastener from the left).

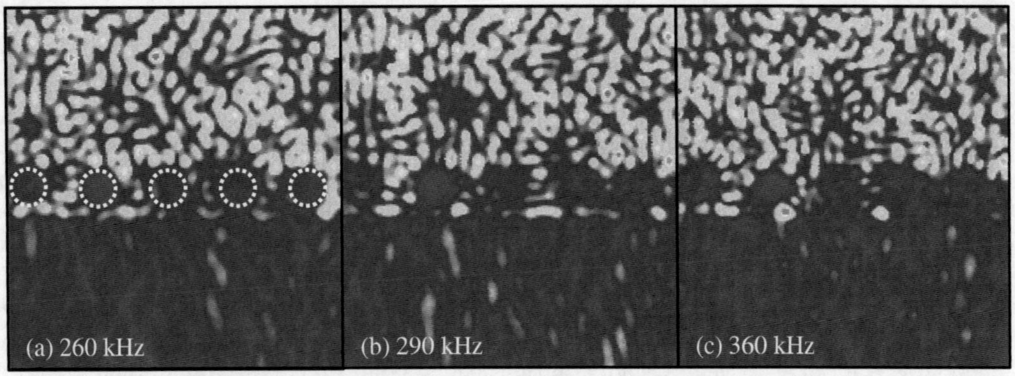

Figure 3: Selected images of spectral amplitude response at (a) 260kHz, (b) 290kHz and (c) 360 kHz. The dotted circles in (a) indicate positions and head size of the five fasteners.

For clearer results, contrast maximization based on the mean amplitude analysis is performed. First of all, five regions of interest (ROI) having the same relative position and size of the dotted circles in Figure 3(a) are established on all spectral amplitude images. The data within these ROI at respective frequency, f are isolated as v [ROI, f]. The mean amplitude of all data within the respective ROIs is calculated, giving five values of mean amplitude for each frequency as $\bar{v}$[ROI, f]. The values at some selected frequencies are plotted and shown in Figure 4, after self-normalization against the amplitude range of all values used in the plot. They exhibit a high amplitude for properly tightened fasteners but a low amplitude for the loosened fastener. This trend is consistent with the lower transmissibility of ultrasound across the fastener-plate interface at the loosened fastener. Specifically, the interface looks flat and smooth but contains many surface asperities at the microscopic scale. The transmissibility of ultrasound across the interface depends on the true contact area of the asperities, which in turn has direct proportionality with the stress at interface and hence the tightness of fastener [25]. Figure 4 also shows high amplitude consistency for the first and second fasteners but the values are quite disperse for the other fasteners. The inconsistency could be caused by inaccurate machining, although unlikely for CNC machining, subjected to further investigation.

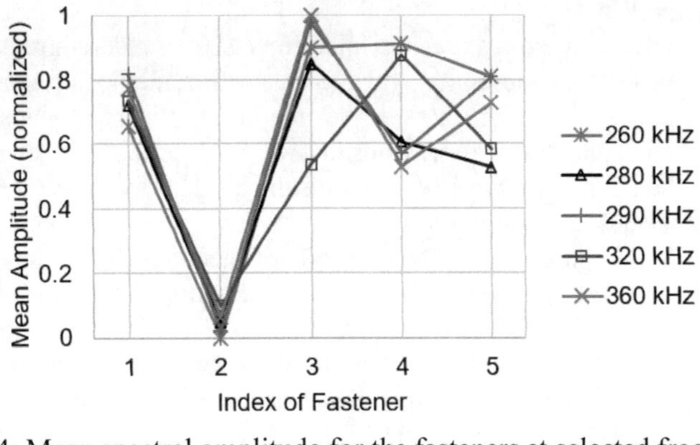

Figure 4: Mean spectral amplitude for the fasteners at selected frequencies

In order to improve the consistency, the contrast of the mean amplitude at respective frequency are calculated using Equation 1, where $C[f]$ represents the contrast at frequency, f, and $\bar{v}_{ROI=i}$ represents mean amplitude at ROI number i.

$$C[f] = \min|\bar{v}_{\mathrm{ROI}=i} - \bar{v}_{\mathrm{ROI}=2}|,\ i = 1,3,4,5 \tag{1}$$

Images corresponding to the five frequencies that provide the highest contrast $C[f]$ are selected for a pixel-wise averaging. The result is an averaged image $v_{AVE}[x, y]$ as shown in Figure 5, in which more consistent amplitude distribution could be observed, making the specimen features such as fasteners, upper plate and plate boundary highly distinguishable. First of all, a significant lower amplitude could be observed (using rainbow-coloured scale with red as the highest amplitude) at the loosened fastener as compared to the other four intact fasteners. The loosening causes poorer contact at the fastener-plate interface, resulting in a higher loss of wave amplitude. The lower stress level at the loosened fastener and plate material at close proximity will change their natural frequency. Nonetheless, this effect is indistinguishable in this result. Perhaps, a new result processing method that can map the natural frequency or dominant frequency of the wavefield data should be developed for this purpose. Other observations include high spectral amplitude over the upper plate, which provides material continuity with minimal loss for the waves to reach the sensor bonded on it. Much lower amplitude is shown on the lower plate that is due primarily to the material discontinuity of the plate and the wave reflection by the fasteners. Apparently, 20.1mm overlapping of the plates does not provide a sufficient continuity for the waves to travel across the joint.

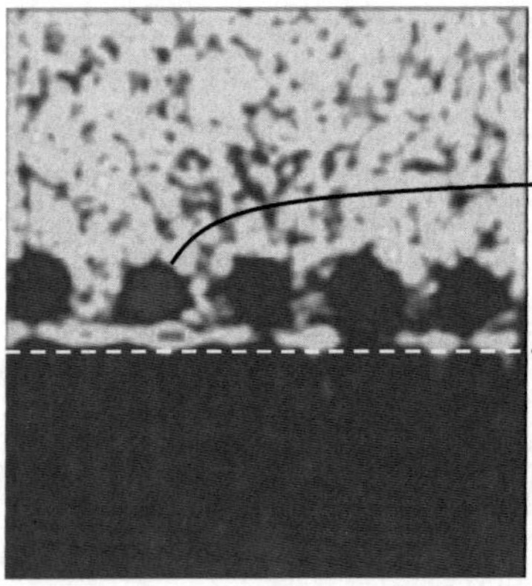

Figure 5: Averaged spectral amplitude image after contrast maximization

4. Summary and Future Works

The joints in aero-mechanical structures are critical elements that ensure structural integrity but they are prone to damages like fastener loosening. Various inspection methods have been proven effective and reported in the literatures. However, they are not readily applicable for inspection of an existing structure that has no baseline data taken at previous known good condition. UPI that does not need any baseline data could be the solution under such circumstances. The feasibility of applying the UPI for detection of fastener loosening in a simple lap joint specimen has been investigated in this study. All fasteners in the specimen are tightened to 10Nm but one of them has been slightly tightened with bare hand only to simulate the loosening case of a fastener. The wavefield data acquired from the specimen is reconstructed as an ultrasonic wavefield propagation video. The video has shown the propagation of wavefield across the fastened joint but it does not show any abnormal wavefield pattern that can be associated with the loosened fastener. The same set of data is then processed into a series of images that show the spectral amplitude response of the specimen at discrete frequencies of periodic interval. Some of the images at specific frequencies show noticeable amplitude difference between the properly tightened fasteners and the loosened one. This confirms feasibility of using UPI for structural joints inspection. Lastly, a contrast maximization method is introduced to improve the visibility of loosened fastener and the consistency in the final result.

There are many opportunities to expand or improve this work. For example, inspection of fastened specimen with various levels of loosening should be done in order to understand the sensitivity of the method. The incident wavefield can be also removed and the wavefield related to fasteners only can be highlighted, facilitating an easier interpretation of the wavefield changes due to the fastener loosening. Furthermore, the wavefield data can be processed into natural or dominant frequency map such that the changes of frequency due to different stress-strain level at the loosened fastener can be identified. Perhaps with such type of frequency analysis, direct measurement of axial stress in the fastener could be done in the future. In addition, investigation on how this method can be expanded for evaluation of joints and fasteners that have been damaged by corrosion is also interesting.

Acknowledgement
This study is supported by Universiti Putra Malaysia through research grant: GP-IPB/2013/9415402.

References
[1] Zaman I, Khalid A, Manshoor B, Araby S and Ghazali M I 2013 *IOP Conf Ser Mater Sci Eng* **50** 12018
[2] Hasiotis T, Badogiannis E and Tsouvalis N G 2007 *4th Int. Conf. NDT*
[3] The USA National Transportation Safety Board 2008 *Accident Report on the Collapse of I-35W Highway Bridge, Minneapolis, Minnesota, August 1, 2007*
[4] The Ontario Ministry of Transportation 2016 *Nipigon River Bridge – Bolt Testing Fact Sheet*
[5] Park J, Kim T, Lee K, Nguyen T C and Kim J 2015 *World Congr. Adv. Struct. Eng. Mech.*
[6] Conradi M S, Miller J G and Heyman J S 1974 *Rev Sci Instrum* **45** 358–60
[7] Heyman J S 1977 *Exp Mech* **17** 183–7
[8] Joshi S G and Pathare R G 1984 *Ultrasonics* **22** 261–9
[9] Hirao M, Ogi H and Yasui H 2001 *NDT E Int* **34** 179–83
[10] Jhang K Y, Quan H H, Ha J and Kim N Y 2006 *Ultrasonics* **44** 1339–42
[11] Chaki S and Bourse G 2009 *Exp Mech* **49** 673–81
[12] Johnson G, Holt A and Cunningham B 1986 *J Test Eval* **14** 253–9
[13] Chaki S, Corneloup G, Lillamand I and Walaszek H 2006 *J Press Vessel Technol* **129** 383–90
[14] Shao J, Wang T, Yin H, Yang D and Li Y 2016 *Appl Sci* **6**
[15] Lim H J, Kim M K, Sohn H and Park C Y 2011 *NDT E Int* **44** 740–50
[16] Shin H-J, Lee J-R and Park C-Y 2014 *J Intell Mater Syst Struct* **25** 647–53
[17] Lee J-R, Jeong H, Chia C C, Yoon D-J and Lee S-S 2010 *Nucl Eng Des* **240** 3513–20
[18] Chia C C, Lee J-R, Park C-Y and Jeong H-M 2012 *Opt Laser Technol* **44** 428–40
[19] Lee J-R, Shin H-J, Chia C C, Dhital D, Yoon D-J and Huh Y-H 2011 *Opt Lasers Eng* **49** 1361–71
[20] Jung-Ryul L, See Yenn C, Nitam S and Chan Yik P 2013 *Meas Sci Technol* **24** 85201
[21] Lee J R, Cho C M, Park C Y, Truong C T, Shin H J, Jeong H and Flynn E B 2015 *Aerosp Sci Technol* **45** 180–5
[22] Timings R L 2006 *Mechanical Engineer's Pocket Book* Newnes
[23] Chia C C, Lee J-R and Shin H-J 2009 *Meas Sci Technol* **20** 8
[24] Chia C C, Lee J R and Park C Y 2012 *Compos Part B Eng* **43** 2898–906
[25] Wang T, Song G, Wang Z and Li Y 2013 *Smart Mater Struct* **22** 87001

IOP Conf. Series: Materials Science and Engineering **270** (2017) 012036 doi:10.1088/1757-899X/270/1/012036

The root cause of ability and inability to assemble and install components using written manual with or without diagrams among non-native English speakers: Root cause analysis

S Ahmad Shukri[1,*], R Millar[2], G Gratton[3], M Garner[4] and H Mohd Noh[1]

[1]Universiti Kuala Lumpur-Malaysian Institute of Aviation Technology, Kampung Jenderam Hulu, Dengkil, Selangor, Malaysia.
[2]The University of Aberdeen, King's College, Aberdeen AB24 3FX, UK
[3]Brunel University, West London, Uxbridge, UB8 3PH, UK
[4]University of Roehampton, London, SW15 4JD, UK

*suhailah@unikl.edu.my

Abstract. Documentation errors and human errors are often claimed to be the contributory factors for aircraft maintenance mistakes. This paper highlights the preliminary results of the third phase of a four-phased research on communication media that are utilised in an aircraft maintenance organisation. The second phase has looked into the probability of success and failure in completing a task by 60 subjects while in this third phase, the same subjects have been interviewed immediately after completing the task by using Root Cause Analysis (RCA) method. It is discovered that the root cause of their inability to finish the task while using only written manual is the absence of diagrams. However, haste is identified to be the root cause for the incompletion of the task when both manual and diagram are given to the participants. It is observed that those who are able to complete the task is due to their reference to both manual and diagram, simultaneously.

1. Introduction

Communication plays the pivotal role in conveying the critical information in any aircraft maintenance organisation. Many previously conducted researches suggest that human mistakes and documentation errors have contributed to the aircraft delays, incidents and accidents [1-11]. Human errors are related to poor language proficiency, inexperience, lack of knowledge, misunderstood or unused information and time constraint. In the meantime, documentation errors can be attributed to incorrect, insufficient or unavailable information. Written media such as manuals, emails and software have a higher risk in conveying critical information in an aircraft maintenance organisation compared to verbal media [12]. Misunderstanding transpires from written information that has been transferred between departments could affect the aircraft's safety and efficiency. This is supported by the argument that the richness of information through written media is lacking, thus it creates ambiguity [13, 14]. The vagueness of the message in written media can have serious effects when maintenance personnel perform their task by referring to these task cards and maintenance manuals that contain unclear, incorrect and insufficient information. With this notion in mind, this study looks into the root cause of the ability and inability to assemble and install the components by referring to manual with or without diagram.

2. Methodology

As mentioned, the main purpose of this study is to investigate the root cause of the ability and inability of assembling and installing components with or without diagram among non-native English speakers through the approach of Root Cause Analysis (RCA). RCA is widely used in investigating incidents and accidents, not only in aviation industries but also in other critical safety management systems such as railway, healthcare, road transport, military strategy, constructions and many more [15]. There are six different methods of RCA and interview is one of the ways to obtain valuable insights from those involved in the incidents by reflection of their personal experiences during the event. In this study, the interviews are conducted immediately after the subjects have completed two experimental situational tasks using Lego mini aircraft. The Experiment Condition 1 (EC1) involves assembling and installing components by referring to a written manual only whereas Experiment Condition 2 (EC2) involves the similar type of tasks but by referring to both the written manual and diagrams. The overall research methodology flowchart is illustrated in Figure 1.

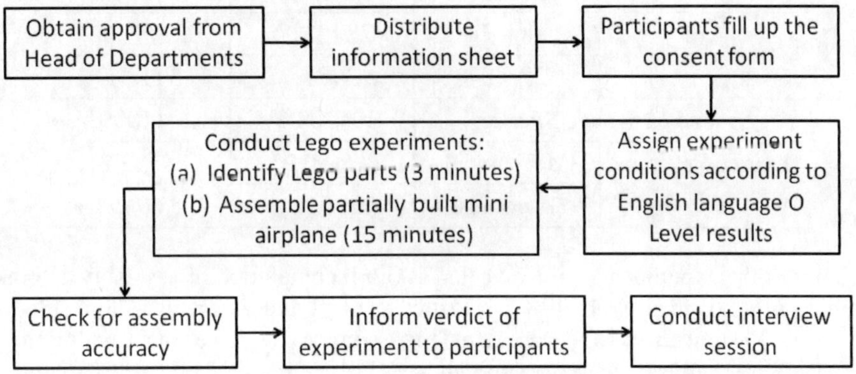

Figure 1: Research flowchart

This research utilised qualitative research methodology where 15 interview questions are designed and self-administered. These questions cover both contents of the written manual given to the subjects and their experiences in assembling and installing the Lego mini aircraft within 15 minutes. Based on the results, they can provide some insights on the root causes of aircraft maintenance errors, whether documentation errors or human mistakes. The recruitment of the subjects for this study involved 60 semester six Bachelor degree students, aged 18 and above at the Universiti Kuala Lumpur - Malaysian Institute of Aviation Technology (UniKL-MIAT). The students are chosen as they have been exposed to theoretical and practical knowledge of aircraft engineering technology in mechanical and avionics, and their learning at the technical university has exposed them in assembling and installing aircraft components in reference to aircraft maintenance manuals during their hands-on lessons in workshops and hangar. Since these experiments are taken to simulate the real maintenance activities: maintaining, repairing and overhauling aircraft, which involve referring to the written manuals and diagrams, their selection as the subjects in this study seems highly practical.

The interview answers are coded and classified into themes, which are later inserted into Statistical Package of Social Science (SPSS). The interview data is first regarded as continuous data, which later is converted into categorical data as the answers are categorised thematically, and the correlation coefficient test has been conducted. To obtain adequate estimate of the group average exposure for observation, the minimum sample size of 15 to 25 is acceptable.

3. Results and Discussion

The experiments are divided into two categories: EC1 and EC2. Several variables are of interest in this study and they are the subjects' O level/ GCSE English language results, time constraint, experiment condition, experience in assembly and install of components, and their feedback on the causes of their success and failure in the experiments. Relationship of these variables with the experiment outcomes is identified to find out whether or not they have significant association with the outcomes.

Total number of semester six students at the institution during the study was 159: 119 mechanical engineering students and 40 avionics students. All seven classes have been approached and invited to take part in the experiment. However, only 60 students agreed to volunteer while the others declined for reasons of the clash of class schedule and/or lack of interest. Table 1 tabulates the breakdown of the subjects for both EC1 and EC2 experiments with regards to their gender, age, language proficiency and program of the study, with total number of subjects, $n = 60$.

Table 1: Distribution of subjects in EC1 and EC2

Characteristics		**EC1**	**EC2**
Gender	Male	27 (90%)	27 (90%)
	Female	3 (10%)	3 (10%)
Age	18 – 24	28 (93%)	27 (90%)
	25 – 34	2 (7%)	3 (10%)
Second Language Speaker		20 (100%)	20 (100%)
Degree Program	Mechanical	27 (90%)	25 (83%)
	Avionics	3 (10%)	5 (17%)

The outcomes of the experiments have revealed 100% incompletion of assembly and installation of components by written manual while 40% by written manual and diagrams. The results presented in Figure 2 shows all 30 subjects in EC1 were unable to complete the experiment within the given time. This includes 16 (36%) subjects, who obtained above 85% (exceptional) in the English proficiency O level/GCSE result. On contrary, as for EC2 shown in Figure 3, it is unexpected to see seven (50%) out of 14 subjects who failed in EC2 are those who obtained an exceptional result of English Language proficiency. Only five (31%) of the subjects who have above 85% (exceptional) in English proficiency O level/GCSE result was able to successfully complete the task. However, it is discovered that there is no relationship between the students' O Level English language results with the success and failure of both experiments with $p = 0.982$. This shows that one's level of English language is not the root cause one's ability or inability to complete a task that involves assembly and installation of components.

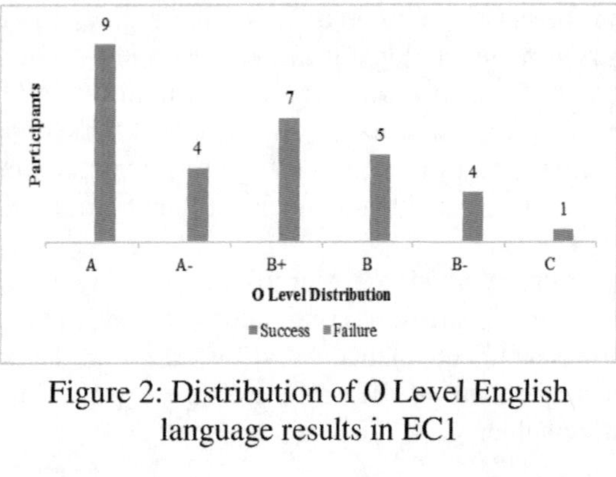

Figure 2: Distribution of O Level English language results in EC1

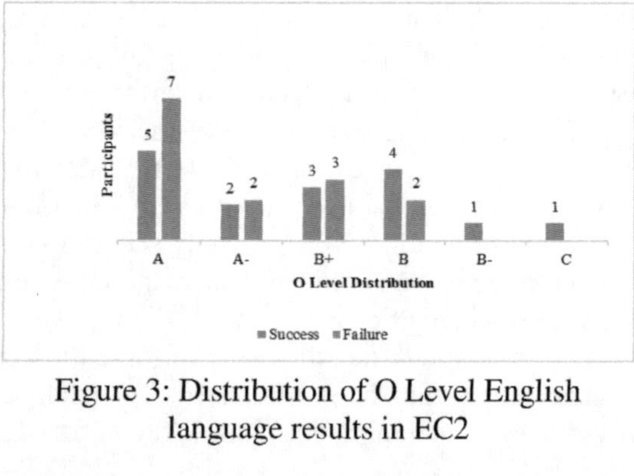

Figure 3: Distribution of O Level English language results in EC2

Figure 4 indicates time spent by those who were unable to complete the experiment. All 30 (100%) subjects in EC1 took the full 15 minutes for the tasks while, for subjects in EC2, five (36%) of them spent eight minutes, three (21%) took 15 minutes, three (21%) took 10 minutes, three (21%) spent 15 minutes and one (7%) took seven minutes. Figure 5, on the other hand, shows those who were able to complete, which are all in EC2 experiments. Only five (31%) spent a full 15 minutes, four (25%) spent eight minutes, three (21%) spent 10 minutes, another two (14%) subjects spent nine minutes and only one (7%) took seven minutes, respectively, in completing the tasks. The findings hinghlight that there is a significant relationship between the duration of time given in both experiments with the results of those experiments with $p = 0.002$.

Both the experiment condition and duration have significant associations with the outcomes of the experiment. Figure 5 has indicated that all 30 (100%) EC1 subjects spent 15 minutes in completing the task whereas the majority of EC2 subjects (30%) spent eight minutes assembling the Lego. For EC2, eight subjects spent 15 minutes, six subjects spent 10 minutes, four subjects spent nine minutes and three subjects spent seven minutes of the given time. The results also show that there is a significant relationship between the two variables with $p = 0.000$ as shown in Figure 6. Furthermore, in Figure 7, it is found that that 13 (43%) subjects who were unable to complete EC1 had previous experience in Lego building whereas 17 (57%) did not have any prior experience with Lego. As for those who were unsuccessful in EC2, it is discovered that 11 (79%) did have experience in Lego building while only three (21%) did not. However, there is no significant relationship between having experience in Lego building with the results of both experiments with $p = 0.386$.

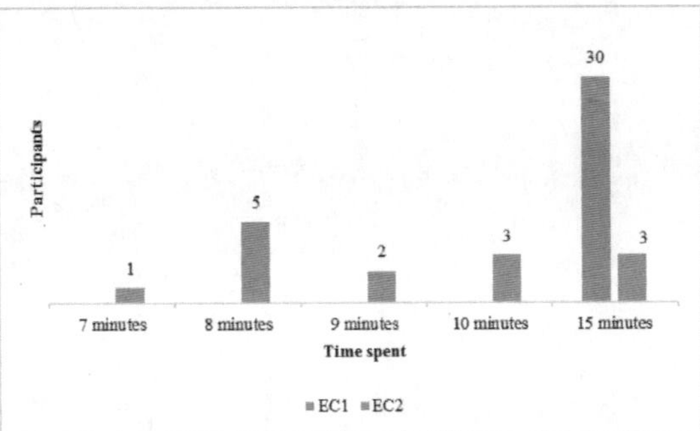

Figure 4: Time spent to complete EC1 and EC2 (unsuccessful outcome)

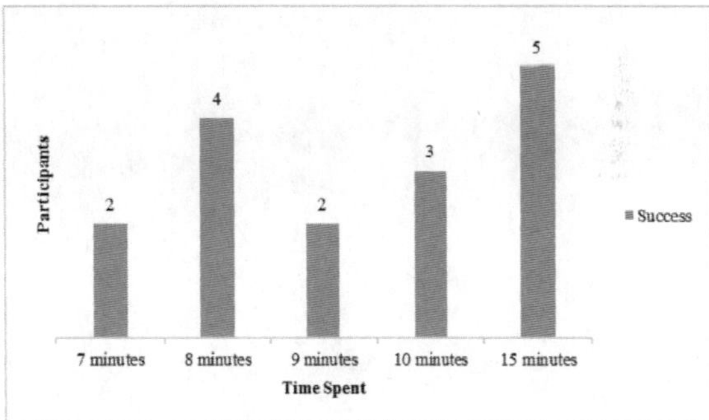

Figure 5: Time spent in EC2 (successful outcome)

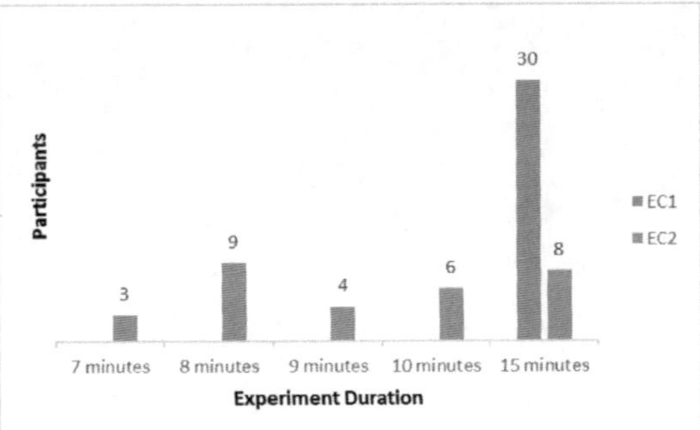

Figure 6: Experiment condition and experiment duration

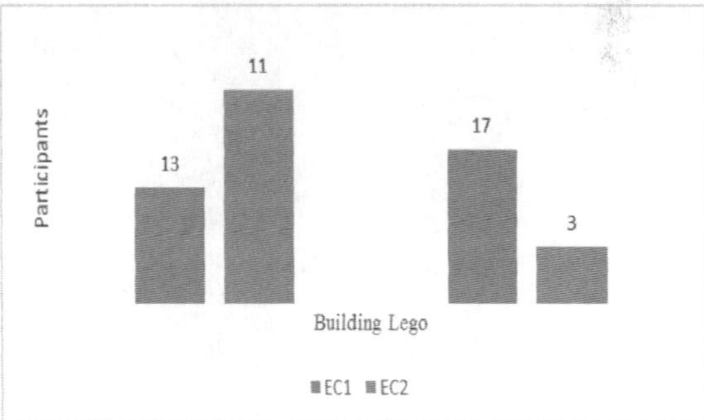

Figure 7: Experience in building Lego (unsuccessful outcome)

After the completion of each experiment, the subjects are interviewed to find out reasons for their success or failure in the given task as indicated in Figure 8. With EC1, it has been discovered that all 30 (100%) subjects were unable to complete the task given within fifteen minutes. 22 (73%) of them mentioned that their inability to understand the written manual contributed to their failure while 11 (37%) said that they were not familiar with the components of Lego, five (17%) claimed that there was no diagram provided to help them and another two (7%) thought that the time given was not enough.

The results concur with previous findings that "information not understandable" and "information not available" and "written documents" have contributed to maintenance errors [1-3].

On the other hand, in EC2, only 16 subjects managed to complete the experiment successfully in 15 minutes. The reasons for their success, as illustrated in Figure 9, are because five (31%) of the subjects referred to both written manual and diagram simultaneously while another six (38%) claimed that they referred to the diagram without reading the manual. Another five (31%) claimed that they referred the diagram more than they read the manual. Although the outcome is successful, the act of referring to diagram alone would be considered as "routine violation" in aircraft maintenance industry [11]. This is because it is clearly stated in the experiment instruction that the subjects have to refer to both manual and diagrams during the experiment. It is observed that the participants did not put in an extra effort in reading the written manual as they pushed it aside and focused on the diagram, aiming to complete the experiment before the fifteen minutes ended.

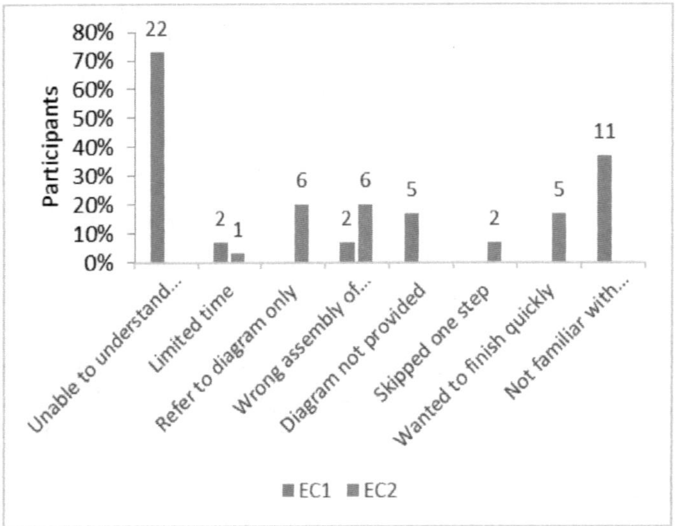

Figure 8: Reasons given for inability to complete EC1 and EC2

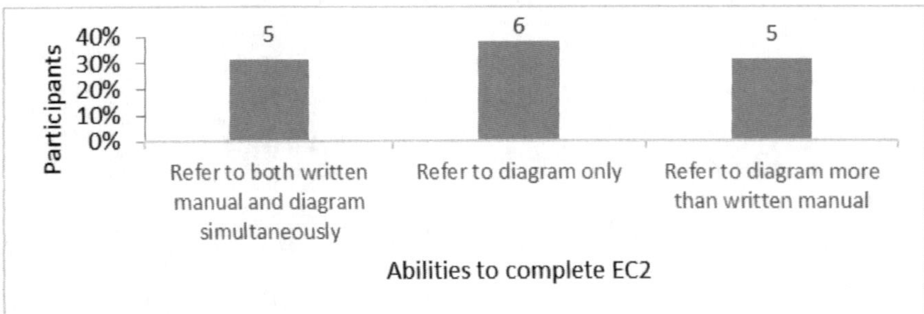

Figure 9: Abilities to complete EC2

4. Conclusion

The interview findings present a new discovery in searching for the root cause of success and failure among non-native English speakers in carrying out aircraft maintenance tasks, which include assembly and installation of components by referring to written manual with or without the presence of diagram. The first phase of this study suggests that written media namely software, E-mail and documentation pose a higher risk of misunderstanding critical information. This is due to incorrect, insufficient and unavailable information of which, if not detected or corrected, will result in maintenance errors and aircraft delay. The second and third phases of the study unveil detailed account on the potentially high risk of misunderstanding of written manual based on two conditions: documentation errors and human errors. Though the Lego written instructions that simulate Aircraft Maintenance Manual are validated and verified as complete, correct and clear manual, the results show that they are not comprehended by all 30 participants in EC1. The language ability of the subjects nor their inexperience did not have any significant relationship with the outcome in both experimental conditions. It is concluded that the root cause of their failure in EC1 is unavailability of information (i.e. the absence of diagram). This led to the misunderstanding of vocabulary, sentence structure, unfamiliarity of components, wrong selection of components and time constraints. They took more time to repeatedly read the instructions in order to understand the meaning of the critical information to complete the task. Meanwhile, the root cause of the failures in completing EC2 is observed as haste. The consequences of being hasty, wanting to finish the task quickly, led the participants to ignore reading the written instructions. Instead, they spent more time digesting information from only the diagrams. This creates a chain reaction where they skipped a step that led to wrong assembly of the components. From analysing the contributory factors of failure, the findings suggest that the success factor among non-native English speakers in completing a maintenance task is to have comprehensive critical information in the documents and to simultaneously refer to both manual and diagram while performing maintenance tasks. This is vital such that each step of the instructions is carefully read and the labeled diagram is accurately followed in order to select the right components and place them in the precise position with the condition that the personnel is not in a state of a hurry to complete the task. It is also important to always read the instructions again and inspecting the product to confirm the completion of work done.

Acknowledgements
The authors like to thank Universiti Kuala Lumpur and University of Aberdeen for the strong support in this study.

References

[1] Hobbs A 1999 *Maintenance Mistakes and System Solutions* Asia-Pacific Air Safety
[2] Lattanzio D, Patankar K and Kanki B 2008 *The International Journal of Aviation Psychology* **18** 17-29
[3] Drury C G, Ma J and Marin C 2005 *Language Error in Aviation Maintenance* Federal Aviation Authority
[4] Chen C C, Chen J and Lin P C 2009 *Journal of Air Transport Management* **15** 261-3
[5] Drury C G 2009 *Handbook of Aviation Human Factors* CRC Press
[6] Safety Regulation Group 2003 *Aviation Maintenance Human Factors (EASA/JAR145 Approved Organisations)* Civil Aviation Authority
[7] Zafiharimalala H, Robin D and Tricot A 2014 *The International Journal of Aviation Psychology* **24** 190-209
[8] Civil Aviation Safety Authority 2013 *HF Engineers Resource Guide*
[9] Reason J T and Hobbs A 2009 *Managing Maintenance Error: A Practical Guide* Ashgate
[10] Civil Aviation Authority Bostwana 2013 *Advisory Human Factors in Aircraft*
[11] Shepherd W T 2005 *Human Error In Aviation Maintenance*
[12] Shukri S A, Millar R, Gratton G and Garner M 2016 *IOP Conference Series: Materials Science and Engineering* **152** 012044
[13] Daft R L, Lengel R H and Trevino L K 1987 *JSTOR* **11** 355–66
[14] Daft R L and Lengel R H 1986 *JSTOR* **32** 554–71
[15] Salmon P M, Stanton N A, Rafferty L and Walker G H 2011 *Human Factors Methods and Accident Analysis: Practical Guidance and Case Study Applications* Ashgate

IOP Conf. Series: Materials Science and Engineering **270** (2017) 012037 doi:10.1088/1757-899X/270/1/012037

Implementation of learning outcome attainment measurement system in aviation engineering higher education

I Mohd Salleh* and M Mat Rani

Mechanical Section, Universiti Kuala Lumpur – Malaysian Institute of Aviation Technology, Jenderam Hulu, 43800 Dengkil, Selangor, Malaysia.

* imransalleh@unikl.edu.my

Abstract. This paper aims to discuss the effectiveness of the Learning Outcome Attainment Measurement System in assisting Outcome Based Education (OBE) for Aviation Engineering Higher Education in Malaysia. Direct assessments are discussed to show the implementation processes that become a key role in the successful outcome measurement system. A case study presented in this paper involves investigation on the implementation of the system in Aircraft Structure course for Bachelor in Aircraft Engineering Technology program in UniKL-MIAT. The data has been collected for five semesters, starting from July 2014 until July 2016. The study instruments used include the report generated in Learning Outcomes Measurements System (LOAMS) that contains information on the course learning outcomes (CLO) individual and course average performance reports. The report derived from LOAMS is analyzed and the data analysis has revealed that there is a positive significant correlation between the individual performance and the average performance reports. The results for analysis of variance has further revealed that there is a significant difference in OBE grade score among the report. Independent samples F-test results, on the other hand, indicate that the variances of the two populations are unequal.

1. Introduction

The University Kuala Lumpur-Malaysian Institute of Aviation Technology (else better known UniKL-MIAT) has been gradually growing to become the leading aviation training institution that specialises in aircraft maintenance technology in Malaysia. This is perfectly in line with the increasing needs of national aviation development programmes as laid down in the Malaysian government blueprint. With world class infrastructures, state-of-art facilities, highly trained and multi-skilled teaching personnel, and comprehensive training that complies with the technical standard and industry requirements, the graduates from UniKL-MIAT are expected to be able to meet the demand of current and future aircraft maintenance technology.

UniKL-MIAT is the pioneer aviation institution in Malaysia and is the first Maintenance Training Organisation (MTO) to be approved by Department of Civil Aviation (DCA) Malaysia to offer aircraft maintenance technology programs, therefore it is an Approved Training Organization (ATO) Part 147 in Malaysia. With all these, UniKL-MIAT is responsible to educate and train students, and to prepare them with enough experience to adapt with the real working environment condition to ensure aircraft airworthiness. Besides that, UniKL-MIAT has to embark on the expansion of its programs in order to be accredited under the Malaysian Qualification Agency (MQA) and ensure that the establishment of its programs is recognized and accredited worldwide. This is important in order for it to produce high quality and marketability graduates. UniKL-MIAT has to comply and meet the requirements imposed by the Malaysian Qualifications Framework (MQF) to obtain the program qualifications and produces higher quality graduates in relation to its education system. However, to monitor the students' learning assessment, MQA has implemented Outcome Based Education (OBE) system to assess the student's Continuous Quality Improvement (CQI). In short, OBE can be considered as an assessment driven and the assessment is used to determine whether or not a qualification, condition, criterion, skill has been achieved. The OBE approach is practically used to determine whether the learning has been successful (i.e. whether the students know what they have learnt or not) and to achieve this, the learners must be assessed using the assessment criteria of the intended outcome.

2. Outcome Based Education (OBE) Implementation at UniKL-MIAT

Ministry of Higher Education Malaysia (MOHE) has advocated OBE as the basis for higher education in Malaysia. Numerous researches have been done to study the needs of emerging and movement from instructional objectives to learning outcomes implementation in higher private institutions in Malaysia. It has been realized that many graduates are jobless as they have nothing outstanding to offer to the job market though they possess degrees and diplomas from recognized institutions. The curriculum review has long recognized the value of analyzing the subject matter to be learned in terms of the intended learning outcomes. In the 1960s, the concept of instructional objectives have attracted attention in the education world [1, 2, 3]. Outcome based does not mean a curriculum-based with outcomes sprinkled on top. It is a transformational way of doing business in education [4]. In brief, OBE is an approach to education in which decisions about the curriculum are driven by the learning outcomes that students should be able to display at the end of the course. These decisions, among others, include curriculum content, educational strategies, student selection and also assessment. In other words, OBE is basically education based on producing particular educational outcomes that focus on what students can actually do after they are taught and expect all learners or students to successfully achieve the certain particular (sometimes minimum) level of knowledge and abilities.

Education that is outcome-based is a learner centred, results oriented system founded on the belief that all individuals can learn. In an effort to create a higher institution provider with the highest quality graduates, the Malaysian government has started to implement MQF at the end of year 2007. With its role as the reference place of higher learning, MQF emphasizes education based on learning outcomes and decides to adopt OBE in higher private institutions with the guaranteed quality control. MQF is an instrument to develop and classify qualifications based on a set of criteria that are approved nationally and benchmarked against international best practices. The criteria clarify the earned academic levels, learning outcomes of study areas and credit system based on the student's academic load as guided in Malaysian Qualification Act (2007). These criteria have been accepted and used for all qualifications awarded by recognized higher education providers in Malaysia. MQF provides educational pathways that systematically link the qualifications, which will enable the individual to progress through credit transfers and accreditation of prior experiential learning in the context of lifelong learning.

UniKL-MIAT has also gone through several changes in the curriculum and syllabus of its programs to obtain recognition either from MQA and DCA Malaysia to ensure its graduates are marketable and capable to perform the job tasks through hands-on training methods. Based on the collected responses from a conducted survey among UniKL-MIAT alumni, it is shown that UniKL-MIAT has successfully produced graduates who are supporting the local and international aviation industry, and only a few of them are filling vacancies in non-aviation sectors. Up until now, the curriculum has been continually evolved to ensure that the graduates are well-equipped with the knowledge and experience to enter the industry [5]. This approach is known as Continuous Quality Improvement (CQI), which can be seen as a real benefit of OBE implementation. Furthermore, the quality of a programme is ultimately assessed by the ability of the learner to carry out their expected roles and responsibilities. This requires the programme to have a clear statement of the learning outcomes to be achieved by the learner as guided by Code of Practice for Programme Accreditation (COPPA) [6]. The learning outcomes cumulatively should reflect on eight (8) domains of learning outcomes that are significant for Malaysia according to

IOP Conf. Series: Materials Science and Engineering **270** (2017) 012037 doi:10.1088/1757-899X/270/1/012037

MQF, which are knowledge; practical skills; communication; social skills and responsibilities; values, attitudes and professionalism; information management and lifelong learning skills; leadership and team skills; problem solving and scientific skills; and managerial and entrepreneurial skills. Learning outcome is necessary in any education institution since it is written with specific intention and needs to be addressed by the program. The idea of learning outcome has some common features with the move to instructional objectives that became fashionable in the 1960s. However, five important differences between learning outcomes and instructional objectives can be recognized as follow:

- Learning outcomes, when appropriately set out, are intuitive and user friendly. They can be used easily in curriculum planning, in teaching and learning and in assessment.
- Learning outcomes are broad statements and typically designed around a framework of 8–12 higher order outcomes.
- The outcomes recognize the authentic interaction and also integration in clinical practice of knowledge, skills and attitudes and the artificiality of separating them.
- Learning outcomes represent what is achieved and assessed at the end of a course of study and not only the aspirations or what is intended to be achieved.
- A design-down approach encourages ownership of the outcomes by teachers and students.

In the meantime, the principle of OBE can be defined as follows [7]:

- Clarity of focus about outcomes: Always have the outcomes as the focus and let the students know what they are aiming for.
- Designing backwards: Design curriculum backward by using the major outcomes as the focus and linking all planning, teaching and assessment decisions directly to these outcomes.
- Consistent, high expectations of success: Set the expectation that OBE is for ALL learners and expect students to succeed by providing them encouragement to engage deeply with the issues they are learning and to achieve the high challenging standard set [4].
- Expanded opportunity: Develop curriculum to give scope to every learner to learn in his/her own pace and cater for individual needs and differences, for example, expansion of available time and resources so that all students succeed in reaching the exit outcomes.

Assessment of students is a crucial aspect of the quality assurance since it drives student learning and is one of the measures to show achievement of learning outcomes stipulated for the programme that is the basis in awarding qualifications. Hence, methods of student assessment have to be clear, consistent, effective, reliable, in line with current practices and must clearly support the achievement of the learning outcomes. The methods of assessment will depend on the specific requirements of each module. However, as a general guide, the following must be considered:

- The usage of summative and formative assessments.
- Knowledge and understanding (the cognitive domain) should be accessed through written, oral or other suitable means, but practical skills should be assessed by practical evaluation such as laboratory, workshop, computer-based simulation and project work.
- For modules requiring significant practical skills, a pass in practical evaluation is compulsory. A pass indicates that the examiner, using an appropriate assessment tool, is satisfied that the candidate has met the learning outcomes of the particular module.

Lecturers in UniKL-MIAT have been encouraged to apply a variety of methods and tools that are appropriate for measuring learning outcomes since the method of studies comprises of theoretical and practical parts with more hands-on training provided for the students to prepare them with adequate knowledge and skills in-line with the technology advancement. CQI of the graduates can be accessed through their assessment and also exit survey. A comprehensive overview of the history and evolution of the CQI methodology is provided in Ref. [8], which is generally a process of constantly introducing small incremental changes in order to improve quality and/or efficiency. One of the methods that can

IOP Conf. Series: Materials Science and Engineering **270** (2017) 012037 doi:10.1088/1757-899X/270/1/012037

be adapted for this purpose is Plan-Do-Check-Act (PDCA) Cycle, which is a four-phase, fact-based approach. The four phases are defined as follow [9]:

- Plan: define purpose, goals and objectives, and collect data.
- Do: identify needs, propose change, implement.
- Check: monitor, evaluate and analyze change, and compare old and new data.
- Act: adjust strategies for improvement; Refine and reinstitute.

Sometimes called a team involvement tool, the PDCA Cycle requires a commitment and "continuous conversations with as many stakeholders as possible. . . (it) is a constant process" [10]. Stakeholders here include professionals and academics [11, 12, 13]. CQI requires functional teams and utilizes the team approach to critically assess the process and to devise solutions to problems and new products. This PDCA cycle or an alternative methodology is employed in process analysis, problem solving, and implementation of solutions. The ability to trust and completely interact with others (interdependence) is a desirable characteristic to serve as an individual basis for CQI.

UniKL-MIAT teaching and learning process flow is shown in Figure 1. It is an approach to quality management that builds on UniKL-MIAT vision to be the premier entrepreneurial technical university and its mission to produce enterprising global technopreneurs. OBE has already been introduced and implemented since 2011. One of the measures of its effectiveness is to assess the Program Educational Objective (PEO), Program Learning Outcome (PLO) and intended Course Learning Outcomes (CLO) through the systematic direct assessment using the Learning Outcome Attainment System (LOAMS) and indirect assessment using Graduate Exit Survey (GES). This assessment provides proper strategic plan to ensure the best action may be taken by the management team.

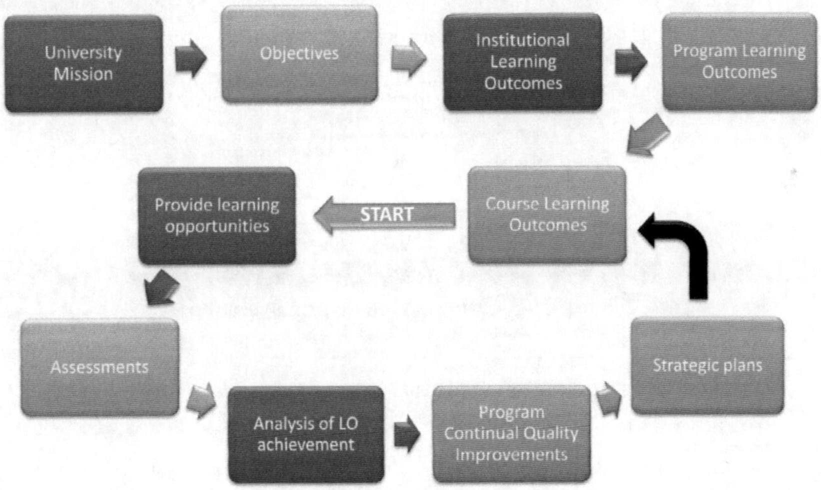

Figure 1: UniKL-MIAT teaching and learning process flow

Currently, course developments in UniKL MIAT are done based upon some steps as the following list;

1. Identify the PLOs and PEOs
2. Set the CLOs
3. Rationalize inclusion of course within the programme.
4. Map CLOs to PLOs and PEOs.
5. Identify transferable skills.
6. Identify mode of delivery and its Learning and Teaching Assessment (LTA).
7. Identify types of assessments.
8. Create Student Learning Time (SLT) for content topics and outline for the course.
9. Allocate the appropriate SLT for each topic and calculate the total SLT.
10. Calculate the credit hours for the course.

IOP Conf. Series: Materials Science and Engineering **270** (2017) 012037 doi:10.1088/1757-899X/270/1/012037

LOAMS is an automated system to measure attainment of learning outcomes for all undergraduate programmes in UniKL and it is controlled by the Centre for Instructional Technology and Curriculum Development (CITC). This is one of the university's efforts to systematically implement the concept and philosophy of OBE with respect to curriculum continual improvements. In conjunction to this, Curriculum and Delivery Handbook has been developed by the CITC unit to guide all teaching staff in preparing their course syllabus, which is accessible through e-citie system. The CITC has encouraged teaching staffs to update their e-citie system and has also performed mentoring sessions through briefing and regular meeting with teaching staff and management group. Teaching staff should update all the assessment in the e-citie system and this is continuously monitored by CITC. Figure 2 shows the E-Citie Service Setting System whereas Figure 3 shows the setting up of LOAMS in the system. Prior to the utilization of LOAMS, all teaching staff must understand the concept of OBE that includes alignment of assessments principles, methods and practices to the learning outcomes and programme content delivery, and ensure that assessments must be consistent with the levels defined in the 8 MQF learning domains. With LOAMS, the result of each subject may be generated by each lecturer after the submission all assessment marks including quizzes, assignments, practical mark assessments, midterm examination and final exam. In the system, OBR001 is the corresponding CLO result for each type of assessment while OBR002 shows the course analysis for each CLO and its attainment percentage.

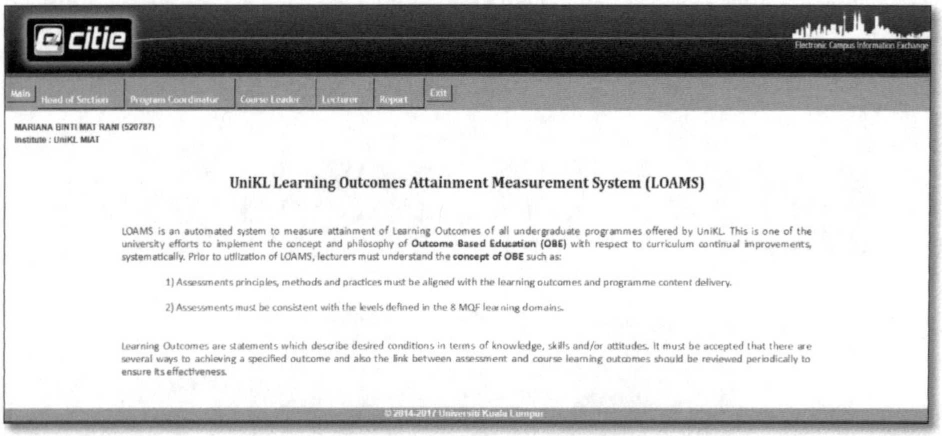

Figure 2: E-Citie service setting system

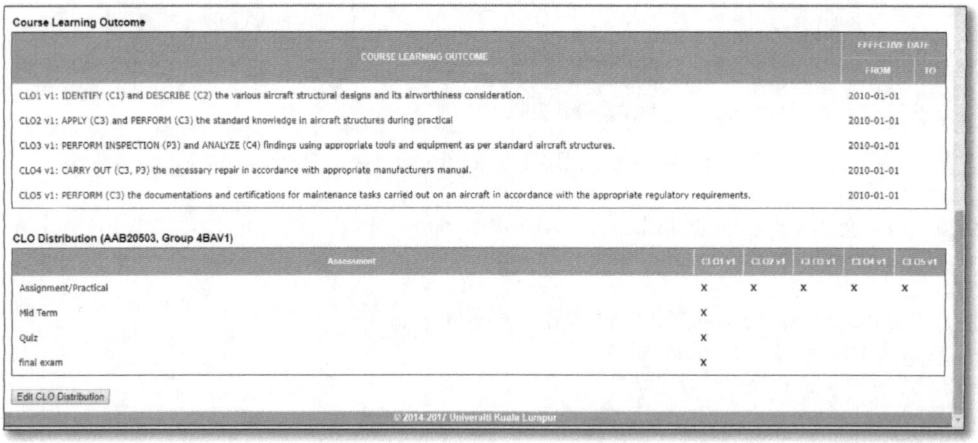

Figure 3: E-Citie LOAMS set up

3. Implementation Case Study: Course AAB20503

To better highlight the practicality of LOAMS implementation, an example case of Course AAB20503 is presented. This Aircraft Structure course has been conducted for session between January 2015 and January 2016, and its contents include the sub-modules of 7.8, 7.14, 7.15, 7.18, 11.2 and 11.3 as per Department of Civil Aviation Malaysia Airworthiness Notices (AN)1101 Issue 1 dated on 1st January 2011. This Airworthiness Notice is issued in pursuant to Section 24 of the Civil Aviation Act 1969 for DCAM Part 66 Aircraft Maintenance License. This case study involved more than 99 students from the same cohort. The OBE report generated from LOAMS, including information regarding the course learning outcome of individual performance and average gap, has been rated by the lecturer. Reports on each of its course learning outcomes have been analysed and the instrument has been standardized for all OBE implementation in discipline core subjects at UniKL-MIAT.

As a standard practice, each lecturer needs to key in the CLO mapping with current PEO and PLO before the first week of the academic semester. The lecturer should perform the lectures as per course portfolio and the teaching portfolio prepared by the lecturer will be approved by their respective Head of Section. Any change on the subject content should be endorsed by Quality Assurance Unit, with the approval of their expert work group and Head of Section, through the submission of the curriculum amendment form. Students are assessed by some assessment methods including assignment, quizzes, practical worksheet, midterm examination and final examination. Final examination is conducted at the end of each semester and it is controlled by the approved exam unit. All final examination question banks are controlled to ensure the quality of the students' results before they are made eligible to sit for their licensing examination in order for them to become license aircraft engineer. The results from OBE are applied to identify the level of students' knowledge to determine whether they are qualified and competent to sit for the licensing examination. By doing so, the failure rate of students in the exam will be minimized and this provides a good result for the aviation training organization industry. Other than that, UniKL-MIAT has also been practicing blended learning using its e-learning platform, which can be accessed by the students once they are enrolled under the subject. Students may download and retrieve important information necessary for the subject including training notes, students' handout, presentation slides, practical worksheet, assignment, quiz and other supporting materials prepared by their respective lecturer. Some of the assessments may be accessed online and students may use the forum column to discuss some issues related with the subject.

The assessment of the students will be done in the ninth week of the technical academic calendar and the final marks should be submitted a week after the final exam date. The OBE performance is calculated based on individual performance, average rating score and by the headcount to evaluate the course learning outcome and lecturer should ensure that all CLOs are attained. If the result shows that any CLO is not attained, a moderation form should be prepared by the subject matter expert who acts as the subject leader and it needs to be agreed by their respective group before any amendment is done to the subject. The last part to complete the OBE and CQI cycle is the GES that is often filled by the students at the end of the semester.

For the sample course AAB20503, the descriptive analysis including frequency, mean and standard deviation is to indicate and conclude the data collected from LOAMS. Pearson correlation analysis, analysis of variance (ANOVA) and independent samples t-test have been carried out to evaluate the significance of the variables under study. As shown in Table 1, a correlation coefficient of +1 indicates a perfect positive correlation between individual performance and average performance parameters. As individual performance increases, average performance also increases, and vice versa. On the other hand, the results of ANOVA is shown in Table 2 that highlight the significant difference between the individual performance and average performance parameters. Nonetheless, the ANOVA results do not indicate where the difference lies, hence t-test should be conducted to test each pair of means since the outcomes of ANOVA should contribute the expected values of the errors are zero, the variances of all errors are equal to each other, the errors are independent and they are normally distributed. Descriptive statistics of the data is tabulated in Table 3 and they show that the mode, mean and standard deviation of individual performance is greater than average performance.

IOP Conf. Series: Materials Science and Engineering **270** (2017) 012037 doi:10.1088/1757-899X/270/1/012037

Table 1: Correlation between individual performance and average performance

	Individual Performance	Average Performance
Individual Performance	1	
Average Performance	0.082250554	1

Table 2: ANOVA results

SUMMARY						
Groups		**Count**	**Sum**	**Average**	**Variance**	
Individual Performance		100	8226.9	82.269	183.5582	
Average Performance		100	8348.9	83.489	3.0526	
ANOVA						
Source of Variation	**Sum of Squares**	**df**	**Mean Square**	**F**	***p*-value**	**F_{crit}**
Between Groups	74.395602	1	74.3956	0.797334	0.372976	3.888853
Within Groups	18474.46936	198	93.3054			
Total	18548.86496	199				

Table 3: Descriptive statistics on OBE individual performance and average performance

Individual Performance		**Average Performance**	
Mean	82.269	Mean	83.4888
Standard Error	1.354836406	Standard Error	0.174717855
Median	84.8	Median	83.66
Mode	85.1	Mode	82.62
Standard Deviation	13.54836406	Standard Deviation	1.747178551
Sample Variance	183.5581687	Sample Variance	3.052632889
Kurtosis	27.74619209	Kurtosis	0.946135038
Skewness	-5.255829247	Skewness	0.467918685
Range	90.3	Range	7.1
Minimum	0	Minimum	80.3
Maximum	90.3	Maximum	87.4
Sum	8226.9	Sum	8348.88
Count	100	Count	100
Confidence Level (95%)	2.688289294	Confidence Level (95%)	0.346678121

In addition, Table 4 shows the results of F-test two-sample for variances used to compare statistical models that have been fitted to the data set to identify the model that best fits the population from which the data are sampled. After that, the t-test is used to test the null hypothesis that the means of two populations are equal. Therefore, in this case, individual performance and average performance of 99 students are equal is known as the null hypothesis. The results of this t-test are tabulated in Table 5,

IOP Conf. Series: Materials Science and Engineering **270** (2017) 012037 doi:10.1088/1757-899X/270/1/012037

which show that variance of individual performance is higher than that of average performance. In this case, the null hypothesis is rejected implying that variances of the two populations are not equal to each other. Moreover, a two-tail *t*-test (inequality) is then conducted in which the null hypothesis is not rejected, indicating that observed difference between the sample means is not convincing enough to say that the average number of individual performance and average performance are significantly different.

Table 4: Result of *F*-test two-sample for variances

	Individual Performance	**Average Performance**
Mean	82.78245614	83.22666667
Variance	230.2768296	0.412390476
Observations	57	57
df	56	56
F	558.3951203	
P(F<=f) one-tail	4.23E-62	
F Critical one-tail	1.557933377	

Table 5: Result of *t*-test two-sample assuming unequal variances

	Individual Performance	**Average Performance**
Mean	82.78245614	83.22666667
Variance	230.2768296	0.412390476
Observations	57	57
Hypothesized Mean Difference	0	
df	56	
t Stat	-0.220806891	
P(T<=t) one-tail	0.413022945	
t Critical one-tail	1.672522304	
P(T<=t) two-tail	0.82604589	
t Critical two-tail	2.003240704	

4. Conclusion

It can be concluded that there is a perfect positive correlation between the individual performance and average performance of OBE as retrieved in LOAMS data. However, as shown in few other analyses done in this study, it appears that the individual performance is slightly related to average performance of OBE. However, this finding also suggests that some lower grades of individual performance will be hidden once the average performance is taken and the attainment on each CLO statement is not 100%. Independent samples *F*-test have revealed that the variances of the two populations are unequal and *t*-test results show that they are not significantly related. The *t*-test shows that the difference between the sample means is not plausible enough to conclude that the average number of individual performance and average performance is differing significantly. This use of LOAMS has shown a positive feedback from the lecturers and students. This systematic approach is considered as direct measurement method that provides a structured way to analyze the attainment outcomes for each course. Using this system, the lecturer could key in the relevant data and assess the CO attainment for their subject. The analysis

IOP Conf. Series: Materials Science and Engineering **270** (2017) 012037 doi:10.1088/1757-899X/270/1/012037

and result will be automatically produced. This system is continuously improved and helpful features are added.

References

[1] Bloom B S, Englehart M D, Furst E J, Hill W H and Kratwohl D R 1956 *A Taxonomy of Educational Objectives Handbook I: Cognitive Domain* David McKay Company Inc.

[2] Kratwohl D R, Bloom B S and Masia B B 1964 *Taxonomy of Educational Objectives Handbook II: Affective Domain* David McKay Company Inc.

[3] Popham W J 1969 *Objectives and Instruction* in *Instructional Objectives* Rand McNally

[4] Spady W 1994 *Outcomes Based Education: Critical Issues and Answers* American Association of School Administration

[5] Mohd Aris K D and Baidzawi I J 2005 *Proceedings of the 2005 Regional Conference on Engineering Education*

[6] *http://www.mqa.gov.my/portal2012/garispanduan/coppa/COPPA 2nd Edition Dec 2008.pdf*

[7] Acharya C 2003 *Outcome-based Education (OBE): A New Paradigm for Learning* Centre for Development of Teaching and Learning

[8] Bhulyan N and Baghel A 2005 *Management Decision* **43** 761-71

[9] Brown J F and Marshall B L 2008 *Nursing Education Perspective* **29** 205-11

[10] National Council of Teachers of English and Council of Writing Program Administrators 2008 *NCTE-WPA White Paper on Writing Assessment in Colleges and Universities in Assessment of Writing*

[11] Joshi A W 2009 *Journal of Marketing* **73** 133-50

[12] Birnbaum R 1988 *How Colleges Work: The Cybernetics of Academic Organization and Leadership* Jossey-Bass

[13] Finkelstein M J 1984 *The American Academic Profession: A Synthesis of Social Scientific Inquiry since WorldWar II* Ohio State University Press

IOP Conf. Series: Materials Science and Engineering **270** (2017) 012038 doi:10.1088/1757-899X/270/1/012038

Analysis of face milling performance on Inconel 718 using FEM and historical data of RSM

C H Chan[1], M S Kasim[1,*], R Izamshah[1], H A Bakar[1], S A Sundi[1], K A Zakaria[2], C H C Haron[3], J A Ghani[3] and M S A Hafiz[1]

[1]Precision Machining Group, Advance Manufacturing Center, Faculty of Manufacturing Engineering, Universiti Teknikal Malaysia Melaka, Malaysia.
[2]High Performance Structure Research Group, Center for Advanced Research Energy, Faculty of Mechanical Engineering, Universiti Teknikal Malaysia Melaka, Malaysia.
[3]Advance Manufacturing Research Group, Faculty of Engineering and Built Environment, Universiti Kebangsaan Malaysia, Malaysia.

*shahir@utem.edu.my

Abstract. This study involves conducting finite element (FE) analysis on face milling of Inconel 718 and investigating the effect of cutting parameters to resultant force and tool temperature. The cutting parameters considered include cutting speed in the range of 20 to 40 m/min, feed rate in the range of 0.10 to 0.20 mm/tooth and depth of cut in the range of 0.1 to 0.2 mm. The experimental approach is carried out using Analysis of Variance (ANOVA) and Response Surface Methodology (RSM). According to the results, the significant factors that affect the resultant force are feed rate and depth of cut. On the other hand, cutting speed has no significant impact on the resultant force. The interaction between feed rate and depth of cut has the most influence on the resultant force. Meanwhile, all cutting parameters considered significantly affect the tool temperature. Through the RSM method, the obtained optimal setting of the parameters is 23.13 m/min, 0.10 mm/tooth and 0.10 mm for cutting speed, feed rate and depth of cut, respectively. The simulated results are then validated by an experiment. It is found that the percentage of relative error between the FE simulation and the actual experiment for resultant force and tool temperature is 10% and 31%, respectively. Considering all possible errors, the overall trend of the simulated results is reasonably in a good agreement with the experimental results.

1. Introduction

Inconel 718 is a nickel-based alloy that exhibits good mechanical and chemical properties, and possesses special features such as corrosion resistance, high temperature resistance, creep resistance and also high strength to weight ratio [1, 2]. It can be applied in a temperature range of -217°C to 700°C [3]. Due to its remarkable characteristics, this material has been selected as one the main components in the aerospace industry. More than 50% of the structural material in various modern aircraft jet engines and space shuttle main engines (SSMEs) are made of Inconel 718, especially in combustion section that undergoes extreme temperatures and pressures [4, 5]. Nevertheless, the drawback of this material is that it is hard to machine. The machinability rating of Inconel and titanium is 0.09-0.3 as compared to 0.4, 1.2 and 1.9 of stainless steel 304, 6061 and 7075 aluminium alloy, respectively [6]. Abrasiveness, gummy property, poor thermal conductivity and work hardening are the main issues reported in the body of literatures. Premature failure of the cutting tool occurs due to hardening surface layer, attributable to stress at tool-workpiece interface, thermal softening, adhesion, diffusion, notching and thermal cracking [7,8]. Other problems reported are the undesirable alteration of machined surfaces to increase the hardened layer as a grain structure and the workpiece easily deforms in response to the machining induced strain loading on the sub-surface [9]. In addition, the cutting tool is prone to chipping, abrasion, adhesion and attrition on cutting edge, flank face and rake surface [10]. Inconel 718 is widely used as a critical component of turbine engine components and the material is required to meet the following criteria: surface finish less than 0.8μm, material strain less than 0.01mm, no formation of white layer, amorphour or recast layer, no redeposit material and no non-parent material [11].

Face milling is among the machining strategies to form a flat surface from the component. It has been indicated that the type of coating is more significant as compared to cutting speed [8]. The CBN tool has shown severe wear pattern and higher wear value compared to ceramic and carbide tools. The application of coated cutting tool of TiAlN-TiN has been shown to over perform the TiAlN and double the tool life time [12]. Several studies associated with the cutting force have reported that it increases as the feed rate and depth of cut increases. Bhatt et al. (2006) measured the cutting force in an investigation of Inconel 718 under dry cutting conditions [13]. The researchers observed that the cutting force increases as the feed rate and the axial depth of cut increase, which contributes to the increase of the resultant force. This is due to the increase in volume of cut per tooth. Higher forces are required during the shearing process of the work material [14]. A similar result is found by another study that highlights the increase of chip load per tooth as the feed rate increases is the reason for the increase of the resultant force in the machining process [15]. The temperature also increases due to friction and adiabatic processes. However, the depth of cut is found to be more significant on heat being generated rather than feed rate and cutting speed [16].

There are numerous works claiming the success of employing finite element method (FEM) to model the machining process. FEM can eliminate shop floor trials that are often very costly and labour intensive [17]. However, the big issue involved in modelling machining process with FEM is long computational analysis time. Depending on the complexity of the problem, the computational analysis time can be days or even weeks. This is due to the nature of FEM calculation that analyses all over the workpiece at every feed step and angular increment of the cutter. This long computational analysis time limits its application for industry practice that needs to manufacture parts in a few days. Therefore, there exists an opportunity to improve the modelling efficiency. This paper proposes a new modelling methodology by taking the advantages of FEM and combining them with statistical analysis that can increase the analysis time for it to become practical and feasible for industry use. In this study, the finite element package AdvantEdge is combined with statistical Response Surface Methodology (RSM) historical data to simulate face milling of Inconel 718. The simulation aims to investigate the magnitude of the cutting forces and the effect of the tool temperature for different machining parameters.

2. Methodology

In this study, the face milling of Inconel 718 is modelled as orthogonal cutting using FEM AdvantEdge software. The boundary conditions are defined by assigning the cutting process, cutting parameter and the process kinematics. During this study, face milling process is modelled and analysed to evaluate on the magnitude of cutting force and cutting temperature. The Lagrangian method is adopted in the machining simulation, in which each individual node of the mesh follows the corresponding material particle during the motion. The cutting tool and workpiece are modelled with six-nodes triangular elements. Considering performance and accuracy of simulation, the minimum element edge length for chip bulk and cutting edge are set to 36μm and 28μm, respectively. Due to large deformation on the machining simulation, adaptive re-meshing method is employed to avoid any extreme element distortions. The re-meshing takes place whenever elements from the cutting edge area are changed from their initial shape. The radius of refined region is set 0.137mm. The mesh refinement factor, mesh coursing factor and chip refinement factor of adaptive re-meshing are all set to 1, 5 and 1, respectively. The result is set with 30 output frames.

Inconel 718 is used as the workpiece material. The length, width and height of the workpiece used are 10mm, 15mm and 2mm, respectively. The diameter of the face cutter tool is set at 50mm. Round inserts of

IOP Conf. Series: Materials Science and Engineering **270** (2017) 012038 doi:10.1088/1757-899X/270/1/012038

10mm in diameter are used. The cutting parameters such as cutting speed, feed rate and depth of cut are set for the machining process. Table 1 shows the cutting conditions applied in this study while Figure 1 depicts the geometry of the indexable inserts.

Table 1: Cutting conditions

Cutting Condition	**Details**
Cutting type	Face milling
Workpiece material	Inconel 718
Type of insert	PVD coated with TiAlN
Coolant condition	Dry with chilled air

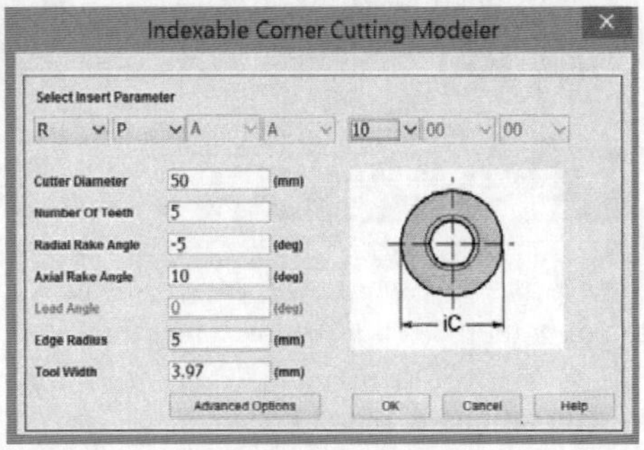

Figure 1: Geometry of the indexable inserts

A total of 18 runs are simulated with a combination of various parameter ranges for the cutting speed (Vc), feed rate (fz) and depth of cut (DOC) as shown in Table 2. An angle of rotation of 360° is applied, which represents one turn of the face cutter. The initial default temperature is set at 24°C. The simulation results are collected to determine the best combination of parameters that produce the lowest resultant force and temperature of the cutting tool. Next, the simulated results obtained are validated with the actual experimental results.

Table 2: Design of experiment matrix for simulation

Level	**Cutting speed (m/min)**	**Feed rate (mm/tooth)**	**Depth of Cut (mm)**
1	20	0.10	0.10
2	300	0.15	0.2
3	40	0.20	

In the FE simulations, the force values are obtained by comparing the before and after tool-workpiece engagement. The force values of feed force (F_x), cutting force (F_y) and thrust force (F_z) can be calculated by using Equation 1 to Equation 3, respectively. The total of these forces is combined into the resultant force (F_r), which is obtained by Equation 4 [18]. Equation 5 is used to measure the tool temperature, T.

$$F_X = F_{Xmax} - F_{Xmin} \tag{1}$$

$$F_Y = F_{Ymax} - F_{Ymin} \tag{2}$$

$$F_Z = F_{Zmax} - F_{Zmin} \tag{3}$$

$$Fr = \sqrt{Fx^2 + Fy^2 + Fz^2} \tag{4}$$

$$T = T_{max} - T_{min} \tag{5}$$

The data is analysed by analysis of variance (ANOVA) to evaluate the significance of the model and the parameters. The significant terms can be identified when their p-value is less than 0.05. A historical data of RSM is used to find the optimum response, which yields the lowest resultant force and cutting temperature. Validation experiments are conducted to investigate the simulation results with the actual machining process. The experiment is set up using CNC milling machine shown in Figure 2. The Kistler dynamometer is used to measure the feed force (F_x), cutting force (F_y) and thrust force (F_z). A thermal imager Fluke Ti400 is employed to measure the temperature of the cutting tool during machining. The emissivity is set at 0.19 for Inconel 718. Image of the temperature is analysed by InsideIR 4.0 software.

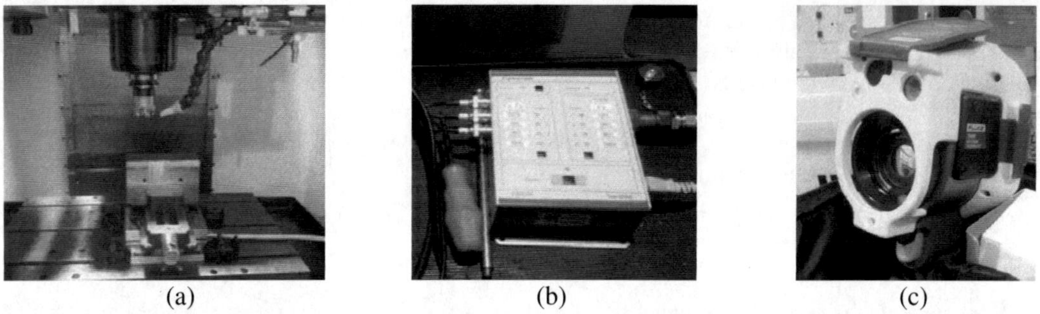

Figure 2: Experiment set-up during validation: (a) Inconel 718 on the CNC milling, (b) Kistler dynamometer set for force measurement, (c) Fluke thermal imager to capture cutting temperature

3. Result and Discussion

Based on the results of FE simulation, Figure 3 to Figure 5 show the profile forces exerted on the cutting tool in different axes. The graph profile manifests the force exerted onto the rake face of the cutting tool during the circumferential tool motion. Because the number of inserts is five, one rotation generates five waveforms. The waveform increases when the cutter removes maximum material and equally decreases when minimum.

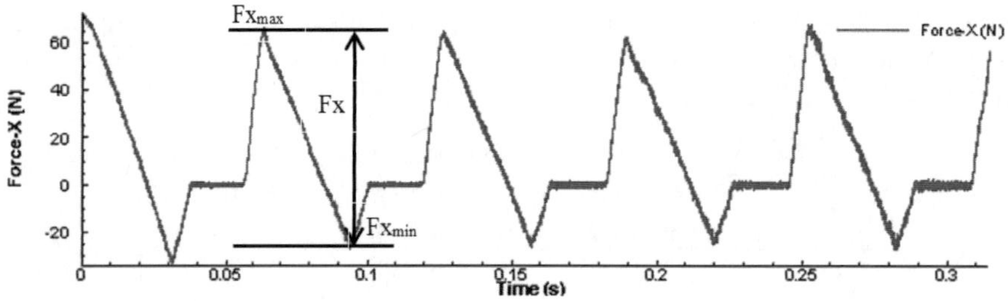

Figure 3: Graph of feed force (F_x)

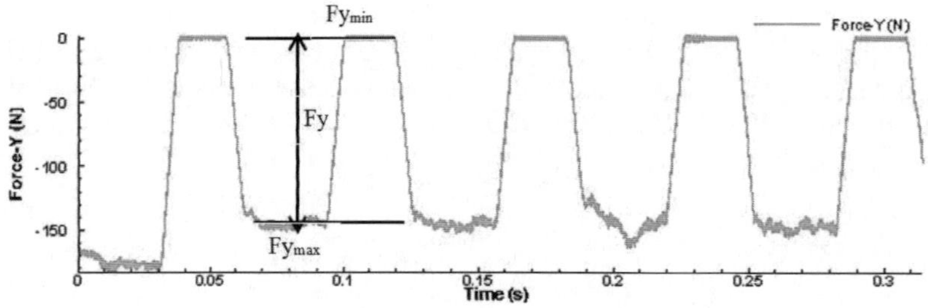

Figure 4: Graph of cutting force (F_y)

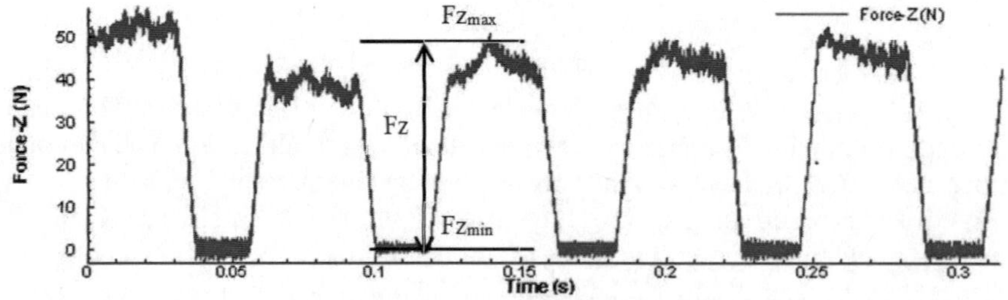

Figure 5: Graph of thrust force (F_z)

IOP Conf. Series: Materials Science and Engineering **270** (2017) 012038 doi:10.1088/1757-899X/270/1/012038

Figure 6 depicts the simulation results of the cutting temperature. It shows that the heat is localized at the tool tip due to friction between the chip and the cutting tool. The simulation also shows that most of the temperature is dissipated through the chip formation and the cutting temperature during the face mill is relatively low. Figure 7 depicts the time history of the tool temperature during one cycle of tool rotation. Table 3 tabulates the collected data after all simulations are done. Each run corresponds to a different combination of cutting parameters to evaluate their effect on the resultant force and tool temperature.

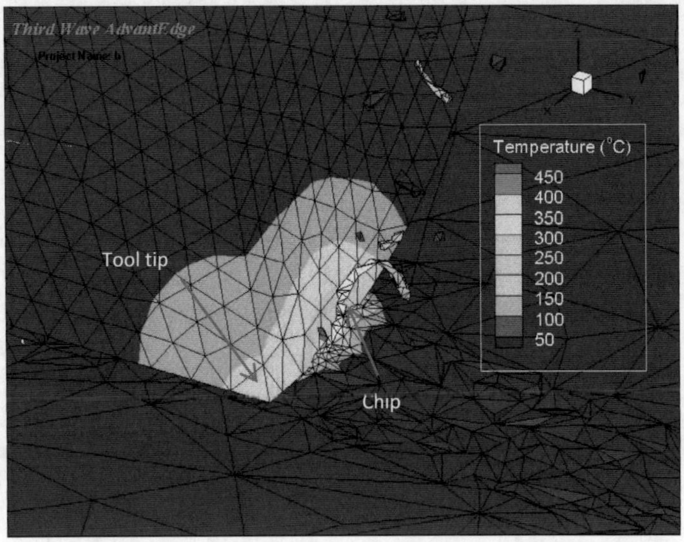

Figure 6: 3D temperature generated model by simulation

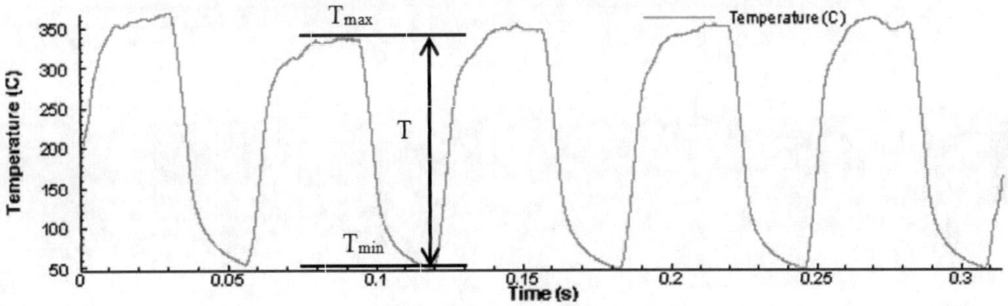

Figure 7: Graph of tool temperature

The resultant force for each of the simulation runs is calculated using Equation 4 and they are plotted in Figure 8. The graph indicates that the increasing feed rate will increase the resultant force and the depth of cut. This trend continues at various cutting speeds for each of the three runs. On the other hand, to evaluate the effect on the tool temperature, each simulation run has a different combination of cutting parameters. A graph for the tool temperature is plotted as shown in Figure 9. The graph indicates that the increasing feed rate will increase the tool temperature and the depth of cut. This trend continues at various cutting speeds for each of the three runs.

Table 3: Simulation results under different cutting conditions

No.	Cutting speed, Vc (m/min)	Feed rate, fz (mm/tooth)	Depth of cut, ap (mm)	Feed Force, Fx (N)	Cutting Force, Fy (N)	Thrust Force, Fz (N)	Resultant force, Fr (N)	Tool temperature (°C)
1	20.00	0.10	0.10	41.00	116.43	34.59	128.19	200.35
2	20.00	0.15	0.10	65.94	179.50	62.52	201.19	257.34
3	20.00	0.20	0.10	92.91	240.19	125.31	286.40	309.48
4	30.00	0.10	0.10	39.28	111.78	36.12	123.86	253.22
5	30.00	0.15	0.10	68.33	168.09	59.18	190.85	323.35
6	30.00	0.20	0.10	82.19	228.58	109.48	266.44	377.89
7	40.00	0.10	0.10	38.85	111.10	30.49	121.58	285.98
8	40.00	0.15	0.10	73.91	168.63	61.68	194.17	379.72
9	40.00	0.20	0.10	96.58	253.59	116.74	295.40	440.86
10	20.00	0.10	0.20	65.94	150.11	47.67	170.74	278.90
11	20.00	0.15	0.20	92.56	262.08	80.48	289.36	345.27
12	20.00	0.20	0.20	124.58	322.93	130.45	369.89	396.85
13	30.00	0.10	0.20	65.63	150.76	45.13	170.51	337.39
14	30.00	0.15	0.20	90.24	250.96	84.52	279.76	411.40
15	30.00	0.20	0.20	166.00	365.73	178.29	439.43	541.86
16	40.00	0.10	0.20	67.35	162.41	48.88	182.49	402.59
17	40.00	0.15	0.20	91.94	247.09	83.17	276.45	467.49
18	40.00	0.20	0.20	129.11	358.08	133.91	403.51	551.50

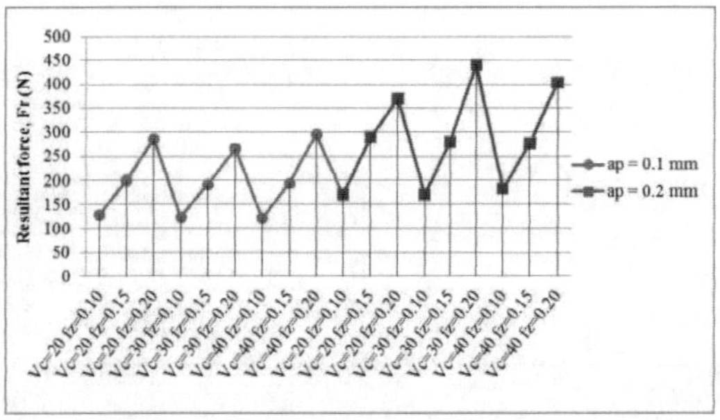

Figure 8: Graph of resultant force for 18 runs of simulation

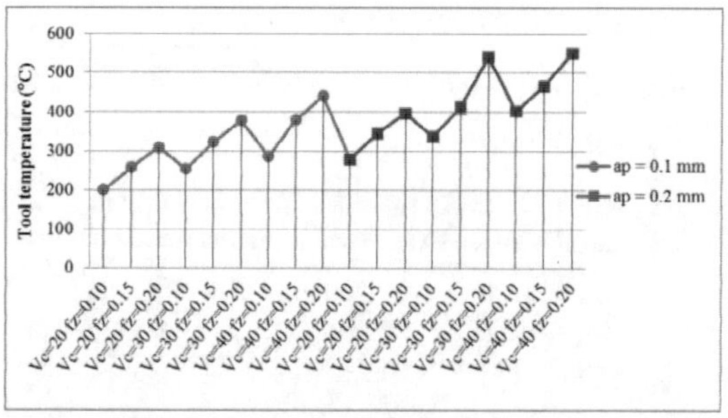

Figure 9: Graph of tool temperature for 18 runs of simulation

In order to indicate the influence of the cutting parameters on the resultant force and tool temperature, ANOVA method is used. Two-factor interaction (2FI) model is found to be significant for the resultant force. The *p*-value of <0.0001 and *F*-value of 91.38 imply that the model is significant as shown in Table 4. The value of Prob > F, which is less than 0.05, indicates that the model terms are significant. The ANOVA results show that the feed rate (B), depth of cut (C) and interaction between feed rate and depth of cut (BC) are significant model terms. In addition, the values greater than 0.10 indicate that the model terms are insignificant. The results show that the cutting speed (A), interaction between cutting speed and feed rate (AB), and depth of cut (AC) are insignificant model terms.

Table 4: ANOVA table for resultant force

Source	Sum of Squares	DF	Mean Square	F Value	Prob > F	
Model	1.503E+005	6	25054.20	91.38	<0.0001	significant
A	64.54	1	64.54	0.24	0.6371	
B	1.128E+005	1	1.128E+005	411.59	<0.0001	
C	33287.16	1	33287.16	121.41	<0.0001	
AB	175.59	1	175.59	0.64	0.4405	
AC	114.64	1	114.64	0.42	0.5311	
BC	3833.47	1	3833.47	13.98	0.0033	
Residual	3015.98	9	274.18			
Cor Total	1.533E+005	17				

Based on an analysis of the variance decomposition, the R^2 of the model is 0.98, demonstrating that the variability of the response data is around its mean. The prediction of the resultant force (F_r) is denoted by Equation 6.

$$F_r = 47.85 - 2.1V_C + 586F_z - 397.78ap + 9.37V_CF_Z + 6.18V_Cap + 7149.33F_Zap \qquad (6)$$

Figure 10 shows the effect of each cutting parameter on the resultant force. The interaction between the cutting speed and feed rate (AB) is insignificant as shown in Figure 10(a). As the feed rate increases with the increasing cutting speed, the resultant force only slightly increases. Figure 10(b) shows the same condition of the interaction between cutting speed and depth of cut (AC). The effect of the depth of cut on the resultant force is also almost constant when interacting with the cutting speed due to the insignificant effect of the cutting speed on the resultant force. The interaction between the feed rate and depth of cut is significant on the resultant force as shown in Figure 10(c).

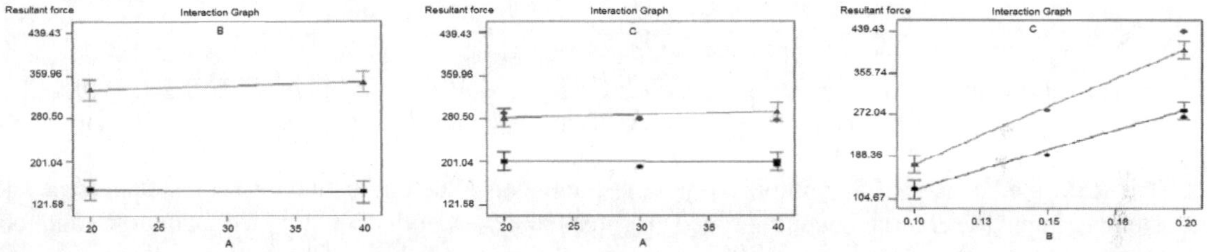

(a) (b) (c)

Figure 10: Interaction analysis response to resultant force between: (a) cutting speed and feed rate (AB), (b) cutting speed and depth of cut (AC), (c) feed rate and depth of cut (BC)

Meanwhile, linear model is suggested for the tool temperature in the software. It implies that the linear model is significant with p-value of <0.0001 and F-value of 133.72. The value of Prob > F, which is less than 0.05, indicates that the model terms are significant. The ANOVA results in Table 5 show that the cutting speed (A), feed rate (B) and depth of cut (C) are all significant model terms.

Table 5: ANOVA table for tool temperature

Source	Sum of Squares	DF	Mean Square	F Value	Prob > F	
Model	1.528E+005	3	50923.12	133.72	<0.0001	Significant
A	45627.17	1	45627.17	119.81	<0.0001	
B	61634.77	1	61634.77	161.84	<0.0001	
C	45507.42	1	45507.42	119.49	<0.0001	
Residual	5351.32	14	382.24			
Cor Total	1.581E+005	17				

Based on the ANOVA, the R^2 of the model is 0.96, showing that the model fits the data. The prediction of the tool temperature (T) is given by Equation 7.

$$T = -186.31 + 6.17V_C + 1433.35F_z + 1005.62ap \quad (7)$$

The graph of the tool temperature against the cutting speed is plotted in Figure 11. The temperature increases as the cutting speed increases since the heat generated per unit time increases with the cutting speed. This is in line with findings by other researchers [14]. The reduction in force is due to an increase in the cutting speed, which can be attributed to the tool mean face temperature. As cutting speed increases, the temperature increases, making the material softer. The shear plane temperature and the temperature rise are mainly due to friction. Meanwhile, the plot of the tool temperature against the feed rate is shown in Figure 12. It can be observed that the temperature increases as the feed rate increases, and the graph linearly increases. One of the reasons for this is the tool-chip contact length and heat distributed off by the chip increases at the mean time. Besides this, the chip deformation coefficient decreases and the material removal rate increases, which leads to a higher temperature. The feed rate is among the dominant factors to generate a high temperature. Last but not least, the graph of the tool temperature against the depth of cut is depicted in Figure 13. With the increase of the cutting depth, both the cutting force and the cutting heat generated also increase. This is because area of cut per tooth increases as the depth of cut increases. The greater the contact area and the friction between the workpiece and the cutting tool, the higher the temperature will be.

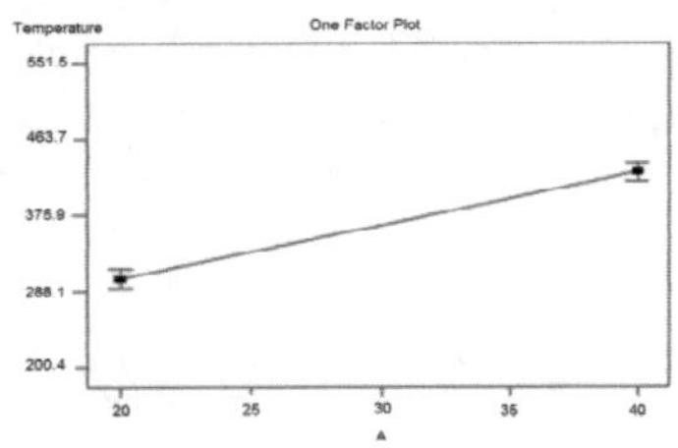

Figure 11: Graph of tool temperature versus cutting speed

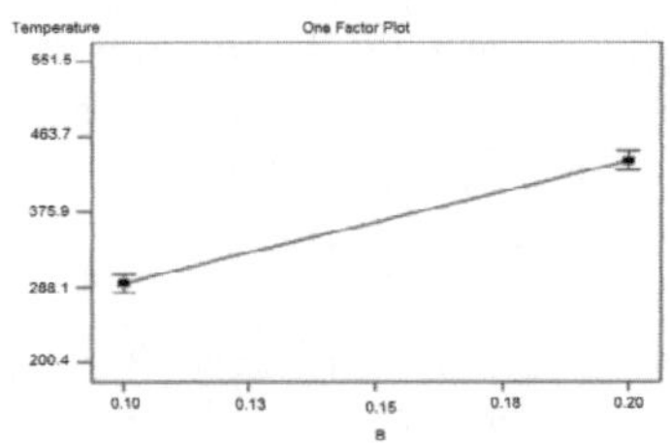

Figure 12: Graph of tool temperature versus feed rate

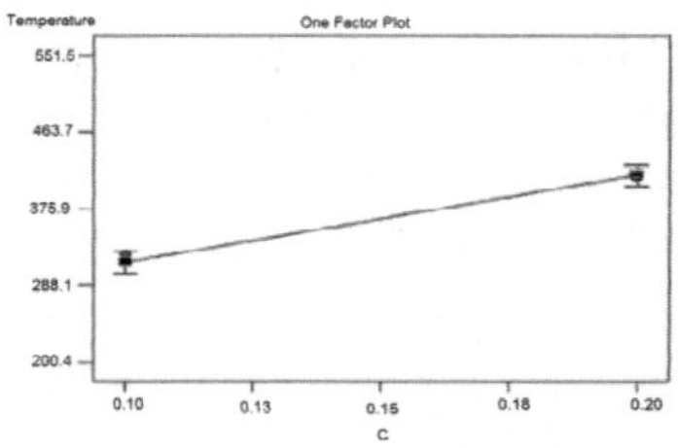

Figure 13: Graph of tool temperature versus depth of cut

The RSM method is used for optimization of the parameters to determine the lowest resultant force and tool temperature. The criteria for cutting speed (V_C), feed rate (f_Z), depth of cut (ap), resultant force and tool temperature are set according to the goals to be achieved shown in Table 6. The lowest resultant force and tool temperature to be obtained during machining is the main goal of this study.

Table 6: Criteria for obtaining the optimum response

Constraints		**Lower Limit**	**Upper Limit**
Name	**Goal**		
Cutting speed, V_C	is in range	20.0	40.0
Feed rate, f_Z	is in range	0.10	0.20
Depth of cut, ap	is in range	0.10	0.20
Resultant force	minimize	121.58	439.43
Temperature	minimize	200.35	551.50

The minimum value of cutting force and temperature can be obtained by the following combination: cutting speed of 23.13 m/min, feed rate of 0.10 mm/tooth and depth of cut of 0.10 mm, which has the highest desirability of 0.994 as shown in Figure 14. The ramp function shows the optimization parameters are set within parameter range. Predicted value of the resultant force is 125.67N and the tool temperature is 200.34°C.

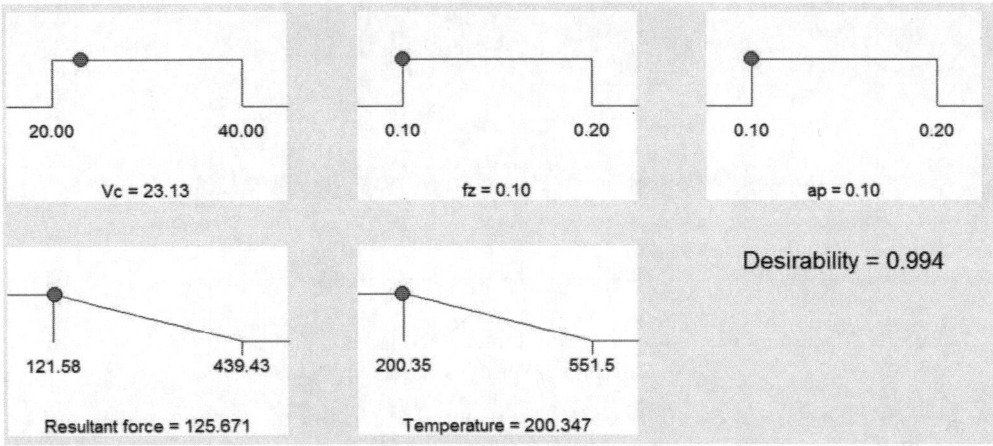

Figure 14: Ramp function graph

Figure 15 shows the result of all the forces recorded by the dynamometer during one path of 100mm tool travel length. Due to the orientation dynamometer during the actual cutting process, the cutting force, feed force and thrust force were labelled as *Fx, Fy* and *Fz*, respectively. The details of the forces can be clearly seen when the graph is zoomed in into one cycle. Figure 16 shows the force within one rotation of the cutting tool. There are five waveforms that indicate the number of inserts used for the cutting tool. It took 0.4 seconds to complete one cycle. The graph shows that the cutting force (*Fx*) dominates the feed force (*Fy*) and thrust force (*Fz*). Based on this observation, the located cutter rotation angle affects the magnitude of the forces. This is evidenced by the direction of the *Fx* axis being perpendicular to the feed direction. In contrast, *Fy* is minimum due to the force axis being parallel to the feed direction. The thrust force is also minimum due to minimum of rake angle to push the material downwards. It is also reported that the first waveform has higher force values compared to subsequent waveforms. This is because the insert initiates to cut the fresh uncut work piece. Higher cutting forces are required during the beginning of the machining because the surface area and volume of material removed are greater. The subsequent force is slightly lower due to the reduction in the amount of material to be cut.

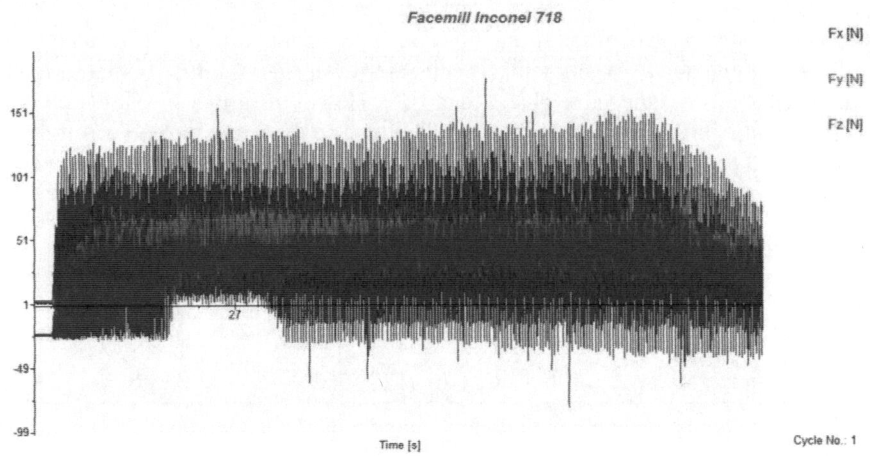

Figure 15: Full machining process with cutting forces

IOP Conf. Series: Materials Science and Engineering **270** (2017) 012038 doi:10.1088/1757-899X/270/1/012038

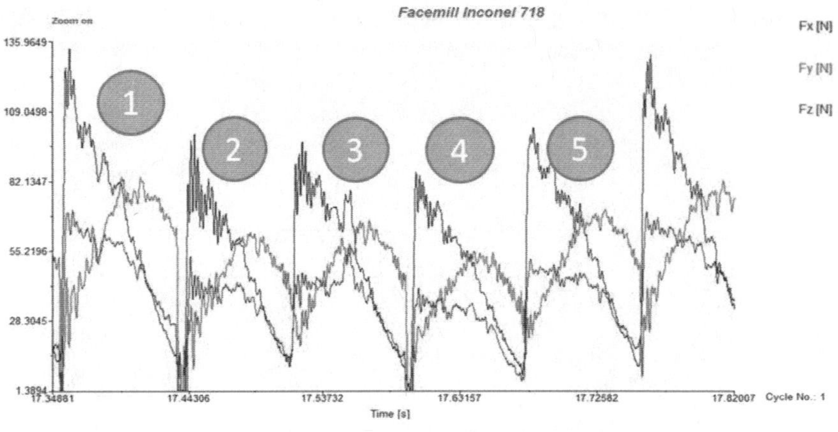

Figure 16: Machining process with cutting forces, feed force and thrust force of one cycle from the Kistler dynamometer

After the experiments, post recording process is required to identify the maximum temperature. The temperature analysis is done with the help of the thermal imager software. Figure 17 shows the spectrum image of the cutting tool and workpiece. The measured temperature of the cutting tool is highlighted with a small rectangle box in the tool area of Figure 17(b).

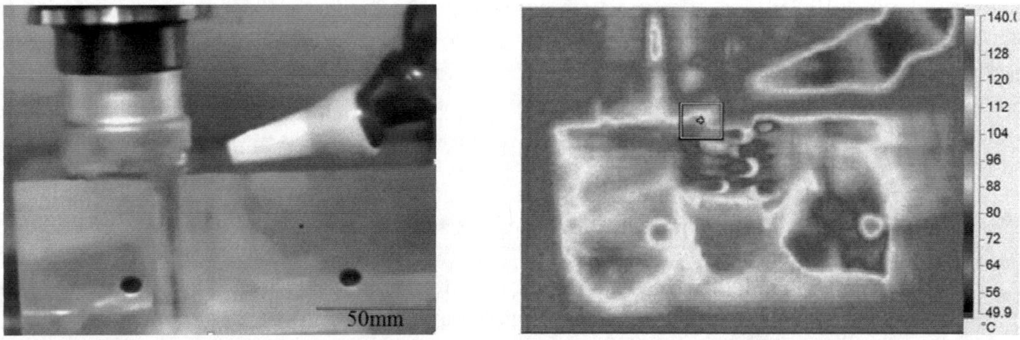

(a) (b)

Figure 17: Infrared image of the sample: (a) actual picture, (b) temperature spectrum

Table 7 shows the comparison between simulation and experimental runs for the resultant force and tool temperature. Generally, the actual experimental results are lower than the FE simulations. The error for cutting force is acceptable since the value is less than 10%. However, further investigation is required for the tool temperature. The influence of high error is likely due to the small traceable cutting region to be captured by the thermal camera. Inaccurate results could be due to the material property being based on a software database that requires appropriate customization via subroutines.

Table 7: Measurement between simulation and experimental run

Parameter	**Simulation**	**Experiment**	**Error, R_e (%)**
Resultant force	121.32 N	109.22 N	9.97
Tool temperature	186.53°C	128.55°C	31.08

4. Conclusion

In this study, AdvantEdge software is used as a simulation tool for face milling of Inconel 718. The effect of cutting speed, feed rate and depth of cut on resultant force and tool temperature are evaluated. Based on 18 simulation runs, the resultant force recorded ranges from 121N to 439N while the tool temperature ranges from 200°C to 551°C. The most dominating factor is found to be the feed rate, followed by depth of cut, while cutting speed has no significant difference on the resultant force. On the other hand, the cutting speed, feed rate and depth of cut are the dominating factors that affect the tool temperature during machining. According to the validation, the resultant force, which is composed of cutting force, feed force and thrust force, is accurately modelled with a percentage of relative error of less than 10% between FEM simulation and actual experiments. However, the temperature model error is 31%.

Acknowledgement

The authors acknowledge the support from both Universiti Teknikal Malaysia Melaka and Universiti Kebangsaan Malaysia for enabling this work to be successfully carried out and funding through research grant no: PJP/2016/FKP/HI6/S01485.

References

[1] Costes J P, Guillet Y, Poulachon G and Dessoly M 2007 *Int. J. Machine Tools and Manufacture* **47** 1081–7
[2] Pawade R S, Joshi S S and Brahmankar P K 2008 *Int. J. Machine Tools and Manufacture* **48** 15-28
[3] Alauddin M, Maziq M and Baradi M 1998 *J. Mat. Proc. Tech.* **77** 153–9
[4] Arunachalam R M, Mannan M A and Spowage A C 2004 *Int. J. Machine Tools and Manufacture* **44** 1481-91
[5] Kalluri S and Halford G R 1995 *Fatigue Behavior and Deformation Mechanisms in Inconel 718 Superalloy Investigated* NASA Glenn's Research & Technology
[6] Quaker Chemical 2013 *https://www.quakerchem.com*
[7] Tanaka H, Sugihara T and Enomoto T 2016 *Proc. CIRP* **46** 545–8
[8] Anthony X M, Manohar M, Jeyapandiarajan P and Patil M M 2017 *Proc. Eng.* **174** 1000–8
[9] Ulutan D and Ozel T 2011 *Int. J. of Machine Tools & Manufacture* **51** 250-80
[10] Kasim M S , Haron C H C , Ghani J A, Sulaiman M A, Ibrahim G A and Yazid M Z A 2013 *Sains Malaysiana* **42** 1721-6
[11] Kappmeyer G, Hubig C, Hardy M, Witty M and Busch M 2012 *Proc. CIRP* **1** 28-43
[12] Motorcu A R, Kus A, Arslan R, Tekin Y and Ezentas R 2013 *Tech. Gazette* **5** 765-74
[13] Bhatt A 2006 *Effect of Cutting Tool Coating on the Machinability of Inconel 718* Master Thesis, McGill University
[14] Yang Z, Zhang D and Huang X 2010 *Key Eng. Mat.* **458** 149–54
[15] Singaravel B and Selvaraj T 2016 *Management and Production Engineering Review* **7** 71–6
[16] Kasim M S, Che Haron C H and Ghani J A 2014 *J. Sci. Ind. Res* **73** 62-5
[17] Izamshah R, Mo J and Ding S 2011 *Proc. Inst. Mech Eng Part B, J. Eng. Manuf.* **226** 592-605
[18] Alauddin M, El Baradie M A and Hashmi M S J 1996 *J. Mat. Proc. Tech.* **58** 100-8

IOP Conf. Series: Materials Science and Engineering **270** (2017) 012039 doi:10.1088/1757-899X/270/1/012039

A review of Kuala Lumpur International Airport (KLIA) as a competitive South-East Asia hub

A M Bardai*, A Z Er, M K Johari and A A Mohd Noor

Universiti Kuala Lumpur – Malaysian Institute Aviation Technology, Dengkil, Selangor, Malaysia

* ahdmaulan@unikl.edu.my

Abstract. This paper is aimed to determine the strengths and weaknesses of Kuala Lumpur International Airport (KUL) against its competitors in Southeast Asia countries. Due to the geographical and market relevance issues, several airport hubs around KUL are chosen for comparison: Changi International Airport (SIN), Suvarnabhumi International Airport (BKK) and Soekarno-Hatta International Airport (CGK). Strengths, weaknesses, opportunities and threats (SWOT) analysis is used to evaluate the comparative situation between the airports in relation to changing industry and market environment. The study discovers some competitive edges for KUL. Nonetheless, the airport still has to be improved to face future challenges and it is under imminent threat of new aircraft technology.

1. Introduction

The airport or aerodrome has been defined by the International Civil Aviation Organization (ICAO) as an area intended to be used either wholly or in part for the arrival, departure and the surface movement of aircraft. Airport can be considered as an operation system comprising of infrastructure, equipment, personnel, operating and management procedures that collectively provide services to its customer. The airport standard reflects on its own management efficiency, whether it is adaptable to the current changes or maintains its own management system. Management decisions on future strategic direction will influence the infrastructure facilities and capacities, and the future of the airport [1].

The main objective of this study is to establish the strengths and weaknesses of the Kuala Lumpur International Airport (KUL) against its competitors within the Southeast Asia region. Few competitor airport hubs have been chosen for comparison, which include the Changi International Airport (SIN), Suvarnabhumi International Airport (BKK) and the Soekarno-Hatta International Airport (CGK). The factors that contribute to the strengths and weaknesses are also identified.

2. Methodology

To conduct this study, the strengths, weaknesses, opportunities and threats (SWOT) analysis is applied as suggested by Ref. [2], which adopts both the environment focus school of thought [3] and resources based view [4]. The idea is to identify each airport's competitive factors based on the identified criteria for a good airport hub. In evaluating these factors, the industry environment and resources are assessed as an opportunity or a threat that the airport management needs to mitigate to enhance its strength in the weakness areas or address the threats.

There are different methods to measure the service quality. It is rather insufficient to only review the economic performance of an airport to evaluate its overall performance. An airport hub acts as a centre point that connects anyone to anywhere. Airports Commissions London states that airport hubs enable a virtuous circle of connectivity by aggregating the demand for passengers and freight, drawing on origin/destination and transfer traffic, and a comprehensive mix of short-haul and long-haul flights [5]. Hubs continue to play a key role in aviation because they serve the economic interests of airlines and meet the connectivity needs of passengers and freight. An effective hub airport with the sufficient spare capacity will maximise passenger choices and also promote airline competition, enabling new entrants, routes and frequencies [5]. According to Airport Council International (ACI), the best airport can be examined according to the passenger numbers and its excellence and achievement in facilities, customer services, operations, retail, environmental awareness and community relations [6]. However, in this study, the review is done through the selective sub-areas as tabulated in Table 1.

Table 1: Selective criteria of best airport by ACI Europe

Criteria	Description
Facilities	Accessibility to all users; assistance to and accessibility for persons with reduced mobility; public transport access; cleaning facilities; parking facilities; check-in processes, including self-service check-in; information desks and VIP areas.
Customer Service (Passengers)	Food and beverage; banking facilities; lounges/meeting infrastructure; internet access; toilets; luggage trolleys and baggage reclaim; WIFI Internet Access; variety of additional services, for example, children's nurseries/post office/beauty facilities; car park/park & ride facilities; staffing levels; monitoring of customer satisfaction and customer service improvements and specific initiatives undertaken to enhance the passenger experience.
Customer Service (Airlines)	Developing long term partnerships; attracting new airline partners; marketing support and incentives offered

Meanwhile, a successful airport hub has to possess a certain quality level to ensure its operational efficiency and attractiveness towards the airlines and passengers. Airports Commissions from London has outlined some characteristics that determine the effectiveness of airport hub operation and they are sufficient capacity, optimized facilities and also capability to support relatively unrestricted 24 hours operations, without dire impacts on local communities [5]. The capacity of an airport hub is crucial since this allows the launching of new routes and flight frequencies, which subsequently enables the hub carriers to organise their flights in connecting flights that minimise transfer times for passengers. Without sufficient capacity, the airport hub cannot provide such slots. One of the factors that airlines are concerned with is the runway capacity of the airport that is measured to provide an estimate on how many aircraft movements can be performed on the runway system during specific unit of time (especially in one hour) [7]. The runway configurations also affect the types of aircraft that could be served in the airport and the required separation between aircraft needs to be considered for safety reasons. Therefore, to achieve a successful hub, the airport operators need to be good in planning its operation 10 to 20 years ahead in the future. Furthermore, facilities are among the services provided by the airport and they have become one of the competitive factors among rival airports. Optimizing the facilities includes minimizing the costs, turnaround times for airlines and the processing times for freight and passengers. This should help the airport hub to maximize its operation and provide better quality of experiences that passengers expect. Last but not least, it is essential for a hub to be effective in supporting passenger and freight operations, specifically early morning long-haul arrivals and night-time freighter services. Mostly applied in the European countries, the standard eight hours night period applied to most noise sources, also known as curfew (including road and rail traffic), is between 2300-0700 hours. Specifically for this paper, selective scopes of facilities and operational categories of an airport hub are chosen for comparison and they include connectivity, facilities, traffic flows, locations and durations, free Wi-Fi, banking, baggage, duty free outlet, transportation and medical centre.

3. Results and Discussion

Connectivity says a lot about an airport. Table 2 shows the connections of each considered airport to each various regions' busiest airport hubs in a year. The data are collected from the airport's departure schedule, followed with the airlines and aircraft model that the airlines used.

Table 2: Flight frequencies and capacity of each airport in a year [8-11]

Airport	Europe (London)	Middle East (Dubai)	United States (Los Angeles)	Australia (Sydney)
KUL	2,704 flights 1,099,280 pax	2,548 flights 1,179,828 pax	-	1,872 flights 642,096 pax
CGK	52 flights 16,536 pax	832 flights 317,304 pax	-	780 flights 205,452 pax
BKK	1,456 flights 531,076 pax	3,640 flights 1,264,900 pax	-	936 flights 321,516 pax
SIN	2,288 flights 859,976 pax	2,184 flights 892,528 pax	728 flights 272,636 pax	2,964 flights 983,216 pax

The direct route to Los Angeles is advantageous to SIN since it is the only hub to currently provide this connection. The next highest flight frequency provided from SIN is to Sydney. This shows SIN has high connectivity in two regions compared to other airports with only one. The highest frequency flights to Europe (London) are from KUL whereas the most frequent route to Dubai is offered from BKK. In contrast, CGK's flights to each region are not as frequent as others and one of the reason for this situation is that CGK has more domestic flights than international flights. Operators should be aware that getting a connection to the region's busiest airport hub is not easy. There is a reason for passengers heading for the busiest airport, whether it is for business, further study or VFR. Having a connection to the busiest airport is a confirmation in the industry that the airport is growing and has the capability to cater the increasing demands. KUL should attract more international routes to cover its weakness of not flying to Los Angeles even though it already has the highest frequency for London. This can be done by increasing the frequencies for existing routes or making marketing segmentation for future demands.

Furthermore, Figure 1 illustrates the comparison of population, tourists flow and passenger handled among KUL, SIN, CGK and BKK airports. The passengers handled in KUL and SIN have been higher than their own population and tourists flow, respectively. This proves that the airport users are getting more mobility than ever. Passengers now can access any destination easily through the airport hub's connectivity and can reach any airport by direct or indirect connectivity of a hub. As discussed before, one of the keys to improve an airport's performance is to enhance its connectivity. The more choices the airport can provide, the higher the willingness of passengers to choose the airport. In other words, the level of convenience of an airport affects its performance. The level of convenience that an airport hub brings to passenger does not only limited to the provision of various routes but also includes the facilities and services in the hub. Since local population and tourists have been shown to be in contrast with the passenger flow, the airport operator should target the purpose of every passenger. Different types of passengers have different demands and the operator has to identify which categories of the passengers occupy the bigger ratio of the total passenger flow such that strategies could be planned. The airport revenue does not only depending on aeronautical but also non-aeronautical. Therefore, the segmentation of target market (passengers) is vital to generate more revenue. As passengers increase their dwelling time in airport, the higher the possibility for them to spent money in airport. So, various outlets that can fulfil requirements of different types of passengers (business traveller, leisure, further studies, etc.) should be provided in airport. For the other two airport hubs, CGK and BKK, the number of tourists and passenger handled are lower than their own population. Based on the results, it is more suitable for CGK and BKK to build a hub in its own population to cater the high number of its own population to increase their competitiveness. KUL and BKK have more advantages to focus in tourism aspect as the tourist flows are higher than SIN and CGK. The right marketing strategy must be planned to attract passengers such that the airport could compete with others. Referring to the forecast from Airbus and Boeing, Asia Pacific has the highest potential market growth and based on this projection, KUL should consider some focused routes to China or India.

On the other hand, cargo movements for each airport did not show any big changes. Based on the data provided, all airports have maintained their rank. Some cargo movements of airports like BKK and KUL have fluctuated a little but that did not affect their overall performance. Cargo movements for most airports have suffered a slight down slope in 2016 as indicated in Figure 2 but SIN is the only airport to increase its cargo movements from 1,847,423 tons to 1,867,784.

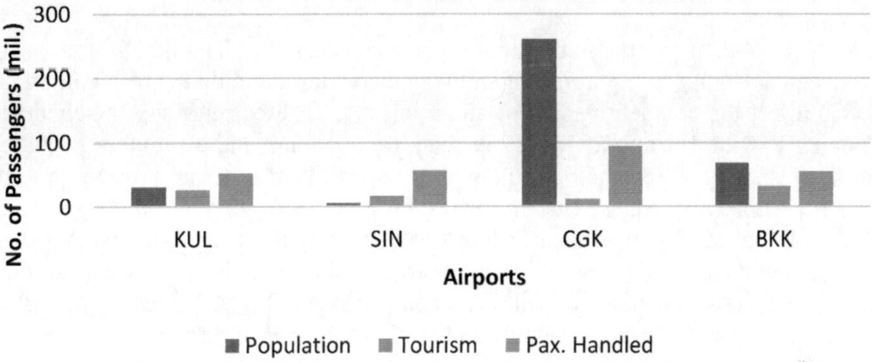

Figure 1: Comparison of population, tourists flow and passenger movements for 2016 [8-11]

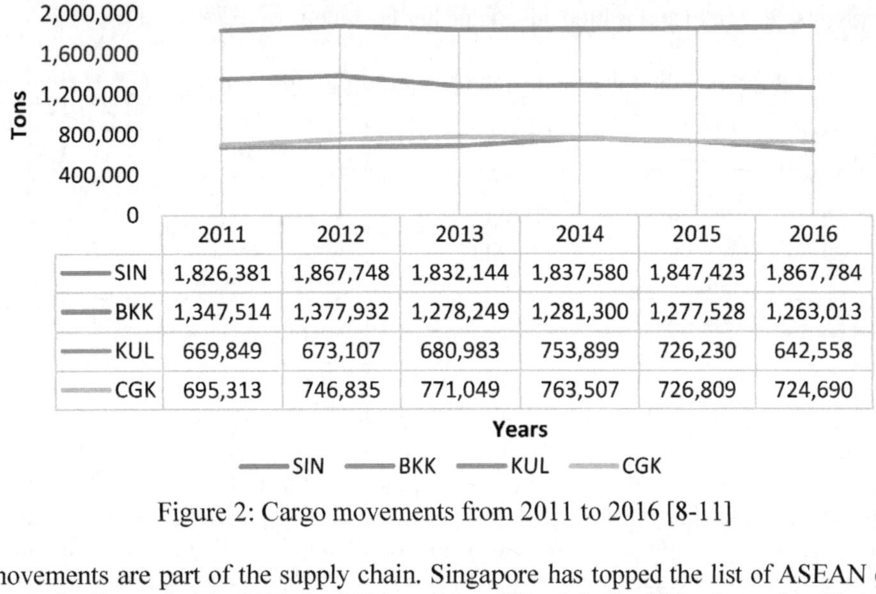

	2011	2012	2013	2014	2015	2016
SIN	1,826,381	1,867,748	1,832,144	1,837,580	1,847,423	1,867,784
BKK	1,347,514	1,377,932	1,278,249	1,281,300	1,277,528	1,263,013
KUL	669,849	673,107	680,983	753,899	726,230	642,558
CGK	695,313	746,835	771,049	763,507	726,809	724,690

Figure 2: Cargo movements from 2011 to 2016 [8-11]

Cargo movements are part of the supply chain. Singapore has topped the list of ASEAN countries on Logistics Performance Index (LPI) for 2016, ranking fifth globally. This shows that Singapore has major logistic firms operating their regional and global functions there, and their consistent efforts to drive the operational excellence through public-private collaborations [12]. Although this includes the other modes of transportation, air cargo performance is the best proof for the firms' logistic industry. According to latest biannual report published by World Bank Group, Singapore, Malaysia, Thailand and Indonesia have all outperformed the East Asia and Pacific regional average. Indonesia, Vietnam and Philippines are ranked among the top 10 of lower middle-income countries [13]. Table 3 tabulates the global LPI ranking for Malaysia, Thailand, Singapore and Indonesia in 2016. It should be noted that this international score uses six key dimensions to benchmark the performance: infrastructure, customs, international shipments, logistics competence, tracking and tracing, and timeliness.

Table 3: Global LPI ranking and score [12]

Country	LPI Rank	LPI Score
Malaysia	32	3.43
Singapore	5	4.14
Thailand	45	3.26
Indonesia	63	2.98

In addition, Table 4 tabulates the flight duration between each hub. The flight duration shown is for direct and connecting flights. The purpose of showing these flight duration is to highlight how aircraft technology will affect the competitiveness of an airport hub. The current longest scheduled flight by flying the Boeing 777-200LR aircraft with 259 seats for 17 hours and 30 minutes, 9,032nm is from Doha hub to Auckland, New Zealand. It is the new service offered by Qatar Airways on 5th February 2017 [14]. As for Airbus aircraft, the current longest range is by A340-500 with capacity of 293 seats and 9,000nm range that can fly more than 18 hours [15]. Boeing is currently innovating B777x, which is expected to be introduced in the market by 2019 [16]. As the technology in aviation industry further develops, it also indirectly endangers the airport's competitiveness. The longer an aircraft can fly, the lower the competitiveness of an airport will be, especially for transit airports. For instance, consider the case study of Colombo Airport in Ref. [1]. The airport is in a geographically advantageous position to become a hub in South Asia since its geographical location makes it suitable to cater lots of airlines that need technical stop or transit. As the aircraft performance improves by time, the duration of long-haul flights become longer and the possibility of serving long-haul direct flights is also getting higher. Subsequently, its geographical location advantage has become invalid.

Table 4: Flight duration from each airport to different region [8-11]

AIRPORT	LONDON (LHR)	DUBAI (DXB)	SYDNEY (SYD)	LOS ANGELES (LAX)
KUL	13hrs 30mins – 17hrs 50mins	7hrs – 10hrs 20mins	8hrs 15mins – 11hrs 45m	17hrs 40mins – 23hrs 50mins
CGK	16hrs 15mins – 20hrs 50mins	8hrs – 10hrs 40mins	6hrs 50mins – 10hrs 55mins	18hrs 35mins – 24hrs
BKK	12hrs 25mins – 6hrs 30mins	6hrs 40mins – 9hrs 30mins	9hrs – 12hrs 20mins	16hrs 55mins – 22hrs
SIN	13hrs 35mins – 17hrs 45mins	7hrs 20mins – 10hrs 30mins	7hrs 45mins – 11hrs 10mins	18hrs – 23hrs 30mins

For KUL, it is located in the centre of these four airports and thus has the advantage of the flight duration to become a hub between Europe and Australia. The flight from KUL to London needs 13 hours and 30 minutes while the flight from KUL to Sydney requires 8 hours and 15 minutes. The total flight duration of both destinations is within the current aircraft range, which makes KUL a more suitable stop for operating airlines. However, for BKK and CGK, total flight duration for both regions exceeds the aircraft range. In other words, KUL is indeed strategically placed in between these two regions. With new aircraft technology that enables airlines to have only one stop during the flight, many passengers prefer direct flights than connecting flights since the former will save more time and provide a hustle-free long-haul flight experience. An aircraft that could operate longer distance and time will benefit the airlines by saving their expenses in airport charges and resources. However, without a stop in between the entire flight, transit airports will lose their competitiveness as airlines will stop flying into them. In nutshell, airport hubs might face a crisis in the market for the transit or technical stops. A proper strategy must be planned to prevent this situation from happening to support the hub's revenue and performance.

In the meantime, Figure 3 shows the passenger flow for each airport from 2011 to 2016, with CGK is ranked in first place by handling 95,100,000 passengers in 2016. All airports have shown increments of the passenger flow but the gap between them and CGK is maintained. Airports other than CGK have similar number of passenger flows for 2016, which is about 50 million passengers. This six-year trend shows some changes in the passenger flow for each airport. BKK was second among the airports at the beginning before it suffered a slight drop in 2014 and finally, it is placed behind SIN in 2016. The performance of SIN has gradually increased within the past six years. As for KUL, the increment is very little and this indicates that its performance is slightly weaker than the others.

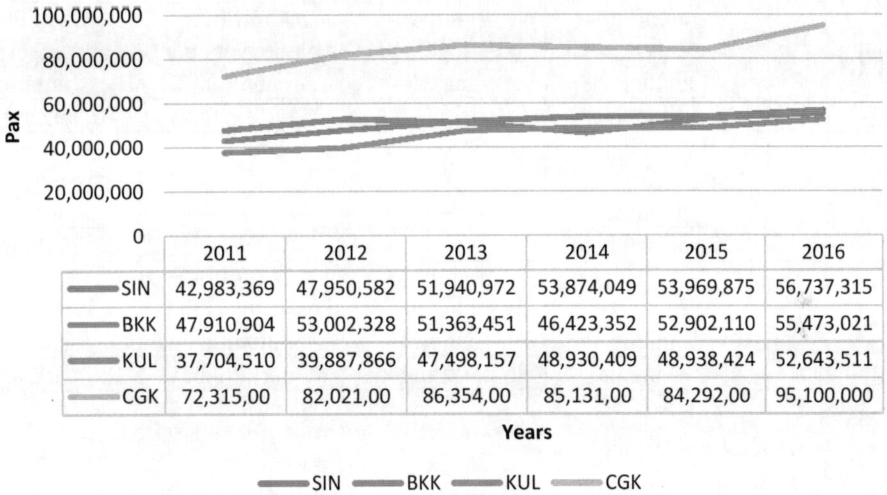

	2011	2012	2013	2014	2015	2016
SIN	42,983,369	47,950,582	51,940,972	53,874,049	53,969,875	56,737,315
BKK	47,910,904	53,002,328	51,363,451	46,423,352	52,902,110	55,473,021
KUL	37,704,510	39,887,866	47,498,157	48,930,409	48,938,424	52,643,511
CGK	72,315,00	82,021,00	86,354,00	85,131,00	84,292,00	95,100,000

Figure 3: Passenger movements from 2011 to 2016 [8-11]

Table 5 lists the capacity of passengers and aircraft in each airport. The airport that owns the largest capacity is SIN, having 67 million passenger capacity per year and 123 parking bays. In second place is BKK, with capacity of 45 million passengers per year and 120 parking bays, followed by KUL and CGK. Comparing the capacity and the actual number of passengers handled for the airports, both CGK and KUL are currently handling doubled the amount of its passenger capacity. This situation is further illustrated by Figure 4. One of the airport hub criteria is to have adequate capacity and able to expand or increase it if necessary. A sufficient capacity helps in maximizing passengers' choice and promoting airline competition, enabling new entrants, routes and frequencies [17]. Among the four airport hubs, the only airport that is over capacity is SIN, which reaches 85% in 2016. For the other three airports, CGK has the highest over utilization rate with 432%. This is followed by KUL, with over utilization rate of 227% and BKK with 124%. Over utilization is an issue that disturbs the whole flying progress. The consequences could be dreadful for both the airport and its users. For instance, London Heathrow lost a lot of traffic to Dubai due to its inability to serve the growing connections [18]. Facing the same problems, Jeddah airport has been demanded by its authority to avoid delays and crowded problems

IOP Conf. Series: Materials Science and Engineering **270** (2017) 012039 doi:10.1088/1757-899X/270/1/012039

[19]. Insufficient capacity of an airport will decrease its capability to reach high service quality and it will slowly lose one of the major criteria as the best airport hub. Flight delays, crowded airport and the long queues are consequential issues that could affect airport performance. Flight delays and crowded airport frustrate both the airlines and passengers easily. Capacity in an airport affects the queuing time and terminal seating, which are the two major contributing factors to rate an airport in SKYTRAX. SIN has been rated as a five star airport, BKK and CGK both have three star rating while KUL has not been rated. As for the queuing time and seating, BKK, CGK and KUL are rated with three stars. This shows that capacity in airport plays an important role in airport performance. One of the criteria for a good airport is to be flexible enough to expand its terminal to cater for future demands. KUL should forecast the demands and have suitable terminal expansion plan to avoid overutilization of its terminal.

Table 5: Terminal capacity for the considered airports [8-11]

Airport	Terminal 1	Terminal 2	Terminal 3	Total Capacity
KUL	5 million passengers per annum	(Satellite) 20 million passengers per annum	-	25 million passengers per annum
	43 parking bays	41 parking bays	-	84 parking bays
CGK	9 million passengers per annum	9 million passengers per annum	4 million passengers per annum	22 million passengers per annum
	21 parking bays	21 parking bays	28 parking bays	70 parking bays
BKK	45 million passengers per annum	-	-	45 million passengers per annum
	120 parking bays	-	-	120 parking bays
SIN	21 million passengers per annum	23 million passengers per annum	22 million passengers per annum	67 million passengers per annum
	45 parking bays	36 parking bays	42 parking bays	123 parking bays

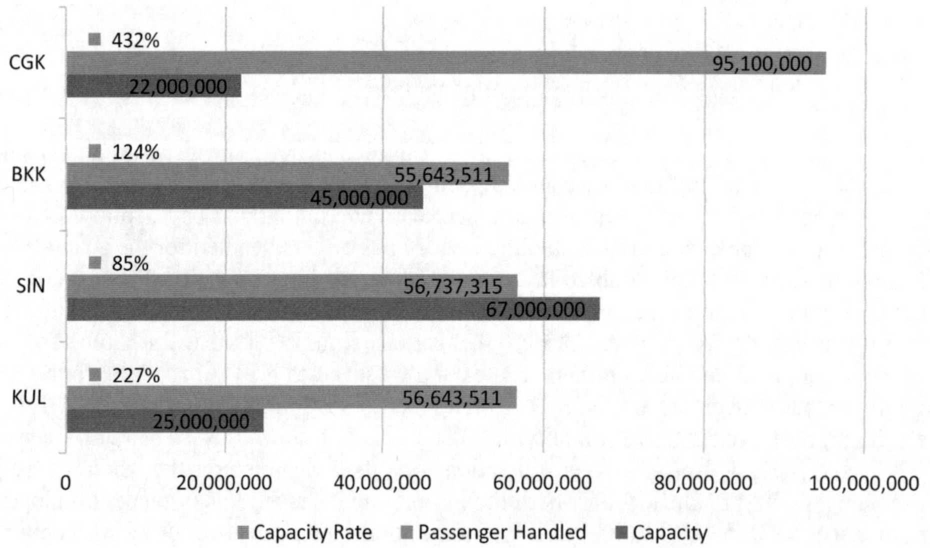

Figure 4: Terminal capacity and passenger handled for each airport [8-11]

Figure 5 represents all aircraft movements for the airports in six years. In 2016, CGK is ranked in first place with 763,395 aircraft movements, followed by KUL with 356,614, SIN with 351,059 and lastly BKK with 333,263. Meanwhile, the capacity for runway per hour is shown in Table 6 and it can be seen that the highest utilization among these airports is 89% for BKK. However, SIN is yet to open its third runway to cater for growing demands. The lowest rate of runway utilization is KUL, which is only 65% that could be categorized as average utilization. Comparing aircraft movements with cargo and passenger flows, the data shows a contrast especially for KUL. The aircraft movements are higher than BKK and SIN, but the passengers and cargo movements for KUL are the lowest among these four airports. This proves that the aircraft using KUL are not as fully loaded as others or the aircraft model used are smaller.

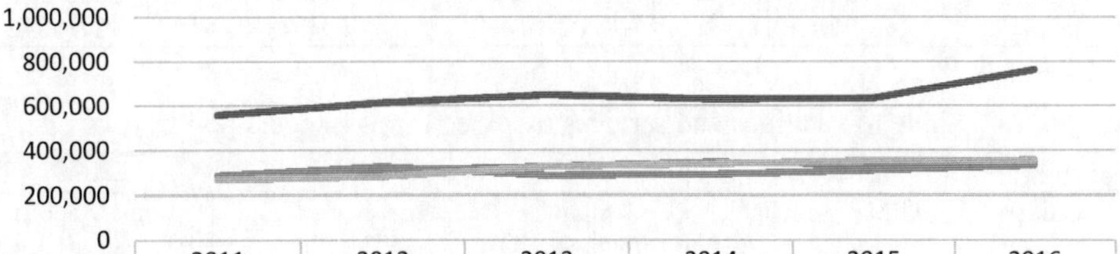

	2011	2012	2013	2014	2015	2016
CGK	559,183	612,850	648,561	630,584	632,418	763,395
SIN	271,286	310,763	327,819	347,458	339,191	351,059
BKK	288,540	326,970	288,004	292,932	310,870	333,263
KUL	269,509	283,352	326,678	340,821	354,519	356,614

CGK SIN BKK KUL

Figure 5: Aircraft movements from 2011 to 2016 [8-11]

Table 6: Runway capacity and actual handled aircraft per hour [8-10]

Airport	KUL	CGK	BKK	SIN
Runway Capacity (per hour)	120 (78 in actual) 65%	86 (72 in actual) 84%	76 (68 in actual) 89%	N/A (42 in actual) -

All airports have been documented to have free Wi-Fi, duty-free shopping outlets, medical centers, and necessary facilities for special-needs visitors. Other facilities, however, differ in terms of not just the availability, but also in terms of the names of companies / brands offered. The basic facilities and services provided for airlines and passengers are listed in Table 7 and 8, respectively. It can be seen that all airports provide similar services and facilities to airlines and passengers. When every airport is doing the same thing, how can they compete among themselves? Therefore, SIN has started to focus into its passenger's experience. Due to high capacity of SIN, the airport provides a total of 18 lounges that caters for different categories of airlines passengers. In addition to that, various types of gardens, entertainment deck, movie theatres and others are built in the airport to maximize the passengers time spent in the terminal. A modernized airport focused on how a passenger or airport user feels when they are in the airport. The operator will try their best to make them feel comfortable and enjoyable while spending time in an airport. Airport operator should know the way to help passengers to kill their time while waiting to board their flights, and also for non-passenger and arriving passengers. Since airport experience is one of the major challenges in the aviation industry, it is suggested KUL could provide facilities or services that can make passengers to prefer staying a longer time in the terminal.

Table 7: Facilities and services prepared for airlines [8-11]

Facility	KUL	SIN	BKK	CGK
Lounges	12	18	5	11
Curfew	None	None	None	None
Aircraft Maintenance	KLAS Engineering Services Sdn. Bhd. Sepang Aircraft Engineering Sdn. Bhd.	SIA Engineering Company Singapore JAMCO Services Pte Ltd	THAI Technical	GMF AeroAsia FL Technics
Runway Type	2 Parallel 32L 14R 32R 14L	2 Parallel 02L 20R 02C 20C	2 Parallel 01R 19L 01L 19R	2 Parallel 07L 25R 07R 25L

Table 8: Facilities and services prepared for passengers [8-11]

Facility	KUL	SIN	BKK	CGK
Banking (money exchange, ATM)	CIMB, Maybank, Hong Leong, Bank Muamalat, Al Rajhi Bank, Public Bank	OCBC, Maybank, UOB, HSBC, DBS, POSB	KasiKorn Bank, The Siam Commercial Bank	Bank Mandiri, PT. Bank Rakyat, Bank Bca, Bank Bukopin, CIMB, Bank Danamon
Baggage Services	Super Strap, Baggage Storage	Porter Service, Baggage Storage	Bellugg, Wrapping services	Porter Service
Transportation	Bus, Taxi, Car Rental, ERL, Car Park	Bus, Taxi, Car Rental, Train, Car Park	AOT Limousine, Bus, Public Van, Airport Rail Link, Car Rental, Car Park, Shuttle Service	Shuttle Service, Bus, Car rental
Passengers Experience	Jungle Boardwalk	Gardens, Swimming Pool, Entertainment Deck, Movie Theatre, The Slides, The Social Tree	None	None

4. Conclusion

This paper reviews the selective facilities and operational aspects of an airport hub focusing on KUL amongst its competitors. Throughout the paper, it highlights that the existing competitiveness of KUL is inadequate to face future challenges, especially in the designed capacity and passenger's experience. Better strategies are needed to protect and further enhance this competitiveness.

References

[1] Jayalath J D and Bandara J S 2001 *Journal of Air Transportation World Wide* **6** 117-28
[2] Andrews K R 1971 *The Concept of Corporate Strategy* Dow Jones
[3] Porter M E 1980 *Competitive Strategy: Techniques for Analyzing Industry and Competitors* The Free Press
[4] Penrose E L 1959 *The Theory of the Growth of the Firm* Wiley
[5] Airport Commission 2013 *Airport Operational Model: Transport for London*
[6] Airport Council International Europe 2016 *12th ACI Europe 2016 Awards*

[7] Neufville R D and Odoni A R 2013 *Airports System: Planning, Design and Management* McGraw-Hill
[8] Malaysia Airport Holding Berhad 2016 *Annual Report 2015*
[9] PT Angkasa Pura II 2016 *Annual Report 2015*
[10] Airport of Thailand Plc 2016 *Annual Report 2015*
[11] Changi Airport Group 2016 *Annual Report*
[12] The World Bank 2016 *http://www.worldbank.org/en/news/press-release/2016/09/15/singapore-tops-list-of-asean-countries-on-2016-logistics-performance-index-ranked-fifth-globally.*
[13] Burnson B 2017 *http://www.logisticsmgmt.com/article/asean_logistics_building_collectively*
[14] Mutzabaugh B 2017 *https://www.usatoday.com/story/travel/flights/todayinthesky/2017/02/05/17-hour-flight-now-longest-world/97536010/*
[15] Airbus *http://www.airbus.com*
[16] Boeing *http://www.boeing.com*
[17] Airports Commissions 2013 *Airport Operational Models*
[18] Newcombe T 2015 *https://buyingbusinesstravel.com/news/1323629-dubai-warns-heathrow-over-capacity-issues*
[19] Arab News 2016 *http://www.arabnews.com/node/985961/saudi-arabia*